Developments in Mathematics

Volume 44

More information about this series at http://www.springer.com/series/5834

Saburou Saitoh • Yoshihiro Sawano

Theory of Reproducing Kernels and Applications

Springer

Saburou Saitoh
Professor Emeritus
Gunma University
Kiryu, Gunma, Japan

Yoshihiro Sawano
Department of Mathematics
and Information Science
Tokyo Metropolitan University
Hachioji, Tokyo, Japan

ISSN 1389-2177 ISSN 2197-795X (electronic)
Developments in Mathematics
ISBN 978-981-10-0529-9 ISBN 978-981-10-0530-5 (eBook)
DOI 10.1007/978-981-10-0530-5

Library of Congress Control Number: 2016951457

Printed on acid-free paper

This Springer imprint is published by Springer Nature
The registered company is Springer Science+Business Media Singapore Pte Ltd.

Preface

The theory of reproducing kernels started with two papers of 1921 [449] and 1922 [45] which dealt with typical reproducing kernels of Szegö and Bergman, and since then the theory has been developed into a large and deep theory in complex analysis by many mathematicians. However, precisely, reproducing kernels appeared previously during the first decade of the twentieth century by S. Zaremba [497] in his work on boundary value problems for harmonic and biharmonic functions. But he did not develop any further theory for the reproducing property. Furthermore, in fact, we know many concrete reproducing kernels for spaces of polynomials and trigonometric functions from much older days, as we will see in this book. Meanwhile, the general theory of reproducing kernels was established in a complete form by N. Aronszajn [28] in 1950. Furthermore, L. Schwartz [428], who is a Fields medalist and founded distribution theory, developed the general theory remarkably in 1964 with a paper of over 140 pages.

The general theory is certainly beautiful. It seems, however, that for a long time we have overlooked the importance of the general theory of reproducing kernels. We were not able to find an essential reason why the theory is important. Indeed, it was an abstract theory, and from the theory, we were not able to derive any definite results and any essential developments in mathematics. The theory by Schwartz is great; however, its importance remained unnoticed for a long time: It is still ignored.

When we consider linear mappings in the framework of Hilbert spaces, we will encounter in a natural way the concept of reproducing kernels; then the general theory is not restricted to Bergman and Szegö kernels, but the general theory is as important as the concept of Hilbert spaces. It is a fundamental concept and important mathematics. The general theory of reproducing kernels is based on elementary theorems on Hilbert spaces. The theory of Hilbert spaces is the minimum core of functional analysis. However, when the general theory is combined with linear mappings on Hilbert spaces, it will have many relations in various fields, and its fruitful applications will spread over to differential equations, integral equations, generalizations of the Pythagorean theorem, inverse problems, sampling theory, nonlinear transforms in connection with linear mappings, various operators among

Hilbert spaces, and many other broad fields. Furthermore, when we apply the general theory of reproducing kernels to the Tikhonov regularization, it produces approximate solutions for equations on Hilbert spaces which contain bounded linear operators. Looking from the point of view of computer users at numerical solutions, we will see that they are fundamental and have practical applications.

Concrete reproducing kernels like Bergman and Szegö kernels will produce many wide and broad results in complex analysis. They developed some deep theory and lead to profound results in complex analysis containing several complex variables. Meanwhile, the formal general theory by Aronszajn also has favorable connections with various fields like learning theory, support vector machines, stochastic theory, and operator theory on Hilbert spaces.

In this book, we will concentrate on the general theory of reproducing kernels developed by Aronszajn while keeping in mind the theory combined with linear mappings and applications of the general theory to the Tikhonov regularization. We will present many concrete applications from the point of view of numerical solutions for computer use. These topics will be general and fundamental for many mathematical scientists beyond mathematicians as in calculus and linear algebra in an undergraduate course.

One of our strong motivations for writing this book was provided by the historical success of numerical and real inversion formulas of the Laplace transform, which is a famous ill-posed and difficult problem, and, in fact, we will give their mathematical theory and formulas, as clear evidence of the definite power of the theory of reproducing kernels by combining the Tikhonov regularization. For the algorithm based on the theory, Hiroshi Fujiwara made the software and we can use it through his helpful guide.

The web [159] is an open source to his inverse Laplace transform.

For these topics, we will need background materials like integration theory, fundamental Hilbert space theory, the Fourier transform, and the Laplace transform.

In Chap. 1, we will give many concrete reproducing kernels first, and in Chap. 2, we develop the general theory of reproducing kernels with general and broad applications by combining it with linear mappings.

In Chap. 3, we will apply the general and global theory of reproducing kernels to the Tikhonov regularization in a lucid manner. We stand on the point of view of numerical solutions of bounded linear operator equations on Hilbert spaces for computer use in a definite and self-contained way.

Chapter 4 is intended as an introduction to what Hiroshi Fujiwara did. In particular, Fujiwara solved linear simultaneous equations with 6,000 unknowns by means of discretization of a Fredholm integral equation of the second kind. This integral equation of the second kind was derived by the Tikhonov regularization and the reproducing kernel method in the above real inversion formula. At this moment, theoretically we will use all the data of the output—in fact, 6,000 pieces of data. Fujiwara gave solutions in **600 digits** precision with the data of **10 GB** for solutions. This fact had a great impact on the authors. Computer power and its algorithms will improve year by year. Meanwhile, we can practically obtain a finite amount of observation data, and so we expect to obtain solutions in terms of a finite number of data for various forward and inverse problems. Thanks to the power of computers, we will be able to realize more direct and simple algorithms, and so we have included results based on a finite amount of observation data. This method will give a new discretization principle.

Chapter 5 deals with the applications to ordinary differential equations such as fundamental equations $y'' + \alpha y' + \beta y = 0$, where α and β can be general functions. Sometimes, we consider the case when the boundary condition comes into play.

As one main substance of new results, in Chap. 6, we present many concrete results for various fundamental partial differential equations. Here we take up the Poisson equation, the Laplace equation, the heat equation, and the wave equation.

Similarly, in Chap. 7, we deal with integral equations. We will consider typical singular integral equations, convolution equations, convolution integral equations, and integral equations with the mixed Toeplitz and Hankel kernel.

In Chap. 8, we refer to specially hot topics and important materials on reproducing kernels, namely, norm inequalities, convolution inequalities, inversion of an arbitrary matrix, representation of inverse mappings, identification of nonlinear systems, sampling theory, statistical learning theory, and membership problems. This will yield a new method of how to catch analyticity and smoothing properties of functions by computers. Furthermore, we will see basic relationships among eigenfunctions, initial value problems for linear partial differential equations, and reproducing kernels, and we will refer to a new type of general sampling theory with numerical experiments. In the last two subsections, we added new fundamental results on generalized reproducing kernels, generalized delta functions, generalized reproducing kernel Hilbert spaces, and general integral transform theory. In particular, any separable Hilbert space consisting of functions may be viewed as generalized reproducing kernel Hilbert spaces, and the general integral transform theory may be extended to a general framework.

Finally, an appendix is provided. In Sect. A.1, we introduce the theory of Akira Yamada discussing equality problems in nonlinear norm inequalities in reproducing kernel Hilbert spaces; indeed, we may be surprised at his general theory of reproducing kernels. In Sect. A.2, we introduce Yamadafs unified and generalized inequalities for Opialfs inequalities. Similar but different generalizations were independently published by Nguyen Du Vi Nhan, Dinh Thanh Duc, and Vu Kim Tuan, in the same year. In Sect. A.3, we introduce concrete integral representations of implicit functions. We rely upon the implicit function theory guaranteeing the existence of implicit functions. The fundamental result was obtained as a great development of a general abstract theory of reproducing kernels.

Kiryu, Japan — Saburou Saitoh
Hachioji, Japan — Yoshihiro Sawano
November 2015

Acknowledgments

The authors thank the following authors of the textbooks:

1. Alain Berlinet and Christine Thomas-Agnan: *Reproducing Kernel Hilbert Spaces in Probability and Statistics*
2. Baver Okutmustur and Aurelian Gheondea: *Reproducing Kernel Hilbert Spaces: The Basics, Bergman Spaces, and Interpolation Problems* on reproducing kernels that were very instructive for our book

Professor H.G.W. Begehr encouraged the publication of this book.

The three referees gave valuable comments and suggestions to the first draft of this book.

The following mathematicians kindly sent their papers or their text files or kind suggestions for our book publication:

Luis Daniel Abreu, Kaname Amano, Joseph A. Ball, P. L. Butzer, L.P. Castro, Minggen Cui, Hiroshi Fujiwara, Antonio G. Garcia, J. R. Higgins, Hiromichi Itou, M. T. Garayev, Kenji Fukumizu, Tsutomu Matsuura, Yan Mo, J. Morais, Nguyen Du Vi Nhan, Masaharu Nishio, Takeo Ohsawa, Hidemitsu Ogawa, Tao Qian, A. G. Ramm, M. M. Rodrigues, Michio Seto, Fethi Soltani, N. S. Stylianopoulos, Mariko Takagi, Akira Yamada, Masato Yamada, Hiroyuki Yamagishi, Nguyen Minh Tuan, Vu Kim Tuan, Masahiro Yukawa, and Kohtaro Watanabe.

This work of the first author was supported in part by Portuguese funds through the CIDMA (Center for Research and Development in Mathematics and Applications) and the Portuguese Foundation for Science and Technology (FCT), within project PEst-OE/MAT/UI4106/2014.

The first author was also supported in part by the Grant-in-Aid for the Scientific Research (C)(2)(No. 21540111, 24540113) from the Japan Society for the Promotion of Science, and the second author was supported by the Grant-in-Aid for Young Scientists (B) (No.21740104, 24740085) from the Japan Society for the Promotion of Science.

Contents

1 Definitions and Examples of Reproducing Kernel Hilbert Spaces 1
1.1 What Is an RKHS? 1
1.1.1 Definition 1
1.1.2 Orientation of Chap. 1 2
1.2 Paley Wiener Reproducing Kernels 3
1.2.1 Paley Wiener Space 3
1.2.2 A Characterization Using the Fourier Transform 6
1.3 RKHS of Sobolev Type 8
1.3.1 Weak-Derivatives and Sobolev Spaces 8
1.3.2 1-Dimensional Case 10
1.3.3 In Connection with 1-Dimensional Wave Equations 18
1.3.4 Higher Regularity RKHS 21
1.3.5 Higher-Dimensional Case 22
1.4 RKHS and Complex Analysis on $\mathbb{C}$ 24
1.4.1 RKHS on $\Delta(1)$ 25
1.4.2 RKHS on $\mathbb{C}$ 29
1.4.3 RKHS on $\mathbb{C} \setminus \overline{\Delta(1)}$ 31
1.4.4 RKHS on a Small Neighborhood of the Origin 34
1.4.5 Bergman Kernel on $\Delta(1)$ 35
1.4.6 Bergman Selberg Reproducing Kernel 37
1.4.7 Pullback of Bergman Spaces 39
1.5 RKHS in the Spaces of Polynomials 41
1.5.1 General Properties of Orthonormal Systems 41
1.5.2 Examples of Orthonormal Systems 46
1.5.3 Polynomial Reproducing Kernel Hilbert Spaces 49
1.5.4 RKHS Constructed by Meixner-Type Polynomials 51
1.5.5 RKHS for Trigonometric Polynomials 52
1.6 Graphs and Reproducing Kernels 53
1.6.1 RKHS $\mathscr{H}_G$ 54
1.6.2 Gram Matrices 58

1.7 Green's Functions and Reproducing Kernels 60
1.7.1 Basic Concepts of Dirac's Delta Function, Green's Functions and Reproducing Kernels 60
1.7.2 Relationship Between Dirac's Delta Function, Green's Functions and Reproducing Kernels 61
1.7.3 The Simplest Example 62

2 Fundamental Properties of RKHS 65
2.1 Basic Properties of RKHS 65
2.1.1 Elementary Properties of RKHS 65
2.1.2 Positive Definite Quadratic form Functions 67
2.1.3 Smoothness of Functions and Kernels of RKHS 73
2.2 Operations of One Reproducing Kernel Hilbert Spaces 78
2.2.1 Restriction of a Reproducing Kernel 78
2.2.2 Pullback of a Reproducing Kernel by Any Mapping 81
2.2.3 Balloon of Reproducing Kernel Hilbert Spaces 82
2.2.4 Squeezing of RKHS 86
2.2.5 Transforms of RKHS by Operators 89
2.2.6 Pullback of Reproducing Kernels by a Hilbert Space-Valued Function 92
2.3 Operations of More Than One Reproducing Kernel Hilbert Space 93
2.3.1 Sum of Reproducing Kernels 93
2.3.2 Sum of Arbitrary Abstract Hilbert Spaces 100
2.3.3 Pasting of Reproducing Kernel Hilbert Spaces 103
2.3.4 Tensor Product of Hilbert Spaces 105
2.3.5 Increasing Sequences of RKHS 109
2.3.6 Wedge Product of RKHS 113
2.4 Construction of Reproducing Kernel Hilbert Spaces 115
2.4.1 Use Linear Mappings and CONS 115
2.4.2 Use the Fourier Integral 119
2.4.3 Mercer Expansion: A General Approach 120
2.4.4 Realization by Discrete Points 125
2.4.5 Aveiro Discretization Method 131
2.5 RKHS and Linear Mappings 133
2.5.1 Identification of the Images of Linear Transforms in Terms of the Adjoint 134
2.5.2 Identification of the Image of a Linear Mapping 134
2.5.3 Image Identification by Reproducing Kernels 141
2.5.4 Integro-Differential Equations 147
2.5.5 Inversion Mapping from Many Types of Information Data 148
2.5.6 Random Fields Estimations 150
2.5.7 Support Vector Machines and Probability Theory for Data Analysis 151

2.5.8 Inverse Formulas Using CONS 153
2.5.9 A General Fundamental Theorem on Inversions 154
2.5.10 Nonharmonic Transforms 158
2.5.11 Determinations of Linear Systems 159

3 Moore Penrose Generalized Inverses and Tikhonov Regularization ... 161
3.1 The Best Approximations and the Moore Penrose Generalized Inverses 161
3.1.1 Frames 161
3.1.2 A General Example of the Moore Penrose Inverses 165
3.1.3 Best Approximation Problems 166
3.1.4 Applications to Best Approximation Problems 169
3.1.5 Applications to Operator Equations with a Parameter 174
3.2 Spectral Analysis and Tikhonov Regularization 179
3.2.1 Spectral Analysis 179
3.2.2 Tikhonov Regularization and Reproducing Kernels 184
3.2.3 Representations of the Solutions of the Tikhonov Functional Equation 186
3.2.4 Applications of Tikhonov Regularization: Approximations by Sobolev Spaces 190
3.3 General Fractional Functions 193
3.3.1 What Is a Fractional Function? 193
3.3.2 An Approach by Using RKHSs 194

4 Real Inversion Formulas of the Laplace Transform 197
4.1 Real Inversion Formulas of the Laplace Transform 197
4.1.1 Problem and Orientation 197
4.1.2 Known Real Inversion Formulas of the Laplace Transform 198
4.1.3 Compactness of the Modified Laplace Transform 200
4.1.4 Examples of the Weights 205
4.1.5 Real Inversion of the Laplace Transform Through Singular Values Decomposition 208
4.1.6 Real Inversion of the Laplace Transform Using Tikhonov Regularization 209
4.1.7 Optimal Real Inversion Formulas Using a Finite Number of Data 212

5 Applications to Ordinary Differential Equations 217
5.1 Use Tikhonov Regularization 217
5.1.1 Ordinary Linear Differential Equations with Constant Coefficients 217
5.1.2 Variable Coefficients Case 220
5.1.3 Finite Interval Cases 221
5.1.4 Exact Algorithm 222
5.1.5 One Point Boundary Condition Case 223

5.2 Discrete Ordinary Linear Differential Equations 224
5.2.1 Use a Finite Number of Data 224
5.2.2 Discrete Inverse Source Problems 229

6 Applications to Partial Differential Equations 231
6.1 Poisson's Equation 231
6.1.1 Construction of the Solutions 231
6.1.2 Source Inversions in the Poisson Equation 238
6.1.3 Constructing Solutions Satisfying Boundary Conditions ... 243
6.2 Laplace's Equation 245
6.2.1 Numerical Dirichlet Problems 245
6.2.2 Algorithm by the Iteration Method 247
6.2.3 The Cauchy Problem 249
6.2.4 Use a Finite Number of Data for the Cauchy Problem...... 250
6.2.5 Error Estimates.. 252
6.3 Heat Equation.. 253
6.3.1 Applications to the Inversion to 1-Dimensional Heat Conduction 253
6.3.2 Use Sobolev Spaces 255
6.3.3 Use of Paley Wiener Spaces (Sinc Methods)............... 257
6.3.4 Source Inversions of Heat Conduction Using a Finite Number of Data.. 262
6.3.5 Natural Outputs and Global Inputs of Linear Systems...... 269
6.4 Wave Equation... 273
6.4.1 Inverse Problems in the 1-Dimensional Wave Equation 273
6.4.2 Inverse Problems Using a Finite Number of Data........... 282
6.5 General Inhomogeneous PDEs on Whole Spaces 284
6.5.1 Approximate Solutions by Tikhonov Regularization 284
6.5.2 Convergence Properties of the Approximate Solutions 287
6.5.3 General Discrete Partial Differential Equations and Inverse Formulas 289
6.5.4 PDEs and Inverse Problems: A General Approach 293

7 Applications to Integral Equations 295
7.1 Singular Integral Equations 295
7.1.1 Mathematical Formulation and an Approach via the Paley Wiener Spaces..................................... 295
7.1.2 The Complex Constant Coefficients Case on the Whole Line 297
7.1.3 Construction of Approximate Solutions by Using a CONS Expansion.................................. 303
7.1.4 General Singular Integral Equations......................... 306
7.1.5 Discrete Singular Integral Equations 308
7.1.6 Inverse Source Problems by a Finite Number of Data 311

7.2 Convolution Integral Equations 313
7.2.1 Example of Convolution Integral Equations 313
7.2.2 An Approach Using RKHSs 313
7.2.3 Integral Equations with Mixed Toeplitz Hankel Kernel 315

8 Special Topics on Reproducing Kernels 321
8.1 Norm Inequalities 321
8.1.1 The Bergman Norm and the Szegö Norm: An Overview ... 321
8.1.2 Examples of Norm Inequalities via RHKSs 323
8.1.3 General Nonlinear Transforms of Reproducing Kernels 325
8.1.4 Convolution Norm Inequalities 330
8.2 Inequalities for Gram Matrices 334
8.3 Inversions 337
8.3.1 Inversion for Any Matrix by Tikhonov Regularization 337
8.3.2 Representation of Inverse Functions 340
8.3.3 Identifications of Nonlinear Systems 346
8.4 Sampling Theory 347
8.4.1 The Information: Loss Error in Sampling Theory 350
8.4.2 A General Kramer-Type Lemma in Sampling Theory 354
8.5 Error and Convergence Rate Estimates in Statistical Learning Theory 357
8.6 Membership Problems for RKHSs 363
8.6.1 How to Obtain Smoothing Properties and Analyticity of Functions Using Computers? 363
8.6.2 Band Preserving, Phase Retrieval and Related Problems ... 365
8.7 Eigenfunctions, Initial Value Problems, and Reproducing Kernels .. 366
8.7.1 Formulation of the Problem 366
8.7.2 The Derivative Operator and the Corresponding Exponential Function Case 370
8.7.3 The Case of the Operator $L(D) = D^2$ 372
8.7.4 How to Obtain Eigenfunctions 373
8.8 Generalized Reproducing Kernels and Generalized Delta Functions 378
8.8.1 What Is a Reproducing Kernel? 378
8.8.2 A Generalized Reproducing Kernel 379
8.8.3 Generalized Reproducing Kernels 380
8.9 General Integral Transforms by the Concept of Generalized Reproducing Kernels 381
8.9.1 Formulation of a Fundamental Problem 381
8.9.2 Completion Property 383
8.9.3 Convergence of $f_t(x) = \langle F, h_x \rangle_{L^2(I_t, dm)}$ for $F \in L^2(I, dm)$ ... 384
8.9.4 Inversion of Integral Transforms 384
8.9.5 Discrete Versions 385

Appendices 387
A.1 Equality Problems for Norm Inequalities 387
A.1.1 Introduction of the Results by Akira Yamada 387
A.1.2 Equality Conditions for the Norm Inequalities 395
A.1.3 The Case of Polynomial Ring 399
A.1.4 Algebra of Meromorphic Functions 400
A.1.5 Applications 403
A.2 Generalizations of Opial's Inequality 406
A.2.1 Main Inequality 407
A.2.2 Equality Condition 409
A.2.3 Applications 411
A.3 Explicit Integral Representations of Implicit Functions 413
A.3.1 2-Dimensional Case 413
A.3.2 Representations of Implicit Functions 415
A.3.3 The 2-Dimensional Case: $n = 1, k = 1$ 419
A.3.4 The 3-Dimensional Cases: $n = 1, k = 2$ or $n = 2, k = 1$ .. 421
A.4 Overview on the Theory of Reproducing Kernels 422

References 425

Index 447

List of Notation

1. The symbol $\delta_{\cdot_1 \cdot_2}$ denotes the Kronecker delta.
2. Let $K : E \times E \to \mathbb{C}$ be a function on the cross product $E \times E$, where E is a set. Then for $p \in E$, we write $K_p \equiv K(\cdot, p)$.
3. The pointwise product of the functions f and g defined on a set E is denoted by $f \cdot g$, which means $f \cdot g(x) \equiv f(x)g(x)$ for $x \in E$.
4. The tensor product of f and g defined on sets E and F, respectively, is denoted by $f \otimes g$, which means $f \otimes g(x, y) \equiv f(x)g(y)$ for $x \in E$ and $y \in F$.
5. Throughout the whole paper, we denote by C a *positive constant* which is independent of the main parameters, but it may vary from line to line, while $C(\alpha, \beta, \ldots)$ denotes a positive constant depending on the parameters $\alpha, \beta, \ldots$. The symbol $A \lesssim B$ means that $A \le CB$. If $A \lesssim B$ and $B \lesssim A$, then we write $A \sim B$.
6. If E is a subset of $\mathbb{R}^n$, we denote by χ_E the characteristic function of E. For all $a, b \in \mathbb{R}$, let $a \vee b \equiv \max\{a, b\}$ and $a \wedge b \equiv \min\{a, b\}$.
7. For $r > 0$, define $\Delta(r) \equiv \{z \in \mathbb{C} : |z| < r\}$ and $\partial\Delta(r) \equiv \{z \in \mathbb{C} : |z| = r\}$.
8. For an open set $D \subset \mathbb{C}$, the set $\mathcal{O}(D)$ denotes the set of all holomorphic functions on D.
9. Let $\mathbb{K} = \mathbb{R}, \ \mathbb{C}$.
10. Let X be a topological space. Define

$$\mathcal{B}(X) \equiv \{F : X \to \mathbb{C} : F \text{ is bounded continuous}\}. \tag{0.1}$$

11. We use the standard multi-index notation: If we are given points

$$z = (z_1, z_2, \ldots, z_n) \in \mathbb{C}^n \text{ and } \alpha = (\alpha_1, \alpha_2, \ldots, \alpha_n) \in \{0, 1, 2, \ldots\}^n,$$

then $|\alpha| \equiv \alpha_1 + \cdots + \alpha_n$ and $z^\alpha \equiv \Pi_{i=1}^n z_i^{\alpha_i}$.

12. The space AC$[a, b]$ denotes the space of real-valued absolutely continuous functions on $[a, b]$.
13. Let X and Y be topological spaces such that $X \subset Y$ as a set. If $O \cap X$ is an open set in X for any open set $O \subset Y$, then we write $X \hookrightarrow Y$.
14. Let V be a linear space and E a subset of V. *The span* Span(E) denotes the smallest linear subspace of E containing V.

Chapter 1
Definitions and Examples of Reproducing Kernel Hilbert Spaces

In Chap. 1 we introduce many examples of reproducing kernel Hilbert spaces (RKHSs for short). The reader may skip Chap. 1 and later go back to it if needed. We will consider RKHSs of Sobolev type, of Paley Wiener type, on the complex plane $\mathbb{C}$, and of polynomial and graph types. We will use the examples for later considerations.

1.1 What Is an RKHS?

Let us begin with the definition of the main topic of this book.

The prerequisite is the fundamental knowledge of Hilbert spaces, say, [495, Chapters 1–4].

1.1.1 Definition

Reproducing kernels will represent many important properties of their reproducing kernel Hilbert spaces, and so concrete forms of reproducing kernels themselves are important and valuable.

Definition 1.1. Let E be an arbitrary abstract (non-empty) set. Denote by $\mathscr{F}(E)$ the set of all complex-valued functions on E. *A reproducing kernel Hilbert space* (RKHS for short) on the set E is a Hilbert space $\mathscr{H} \subset \mathscr{F}(E)$ with a function $K : E \times E \to \mathscr{H}$, which is called *the reproducing kernel*, enjoying *the reproducing property*

$$K_p \equiv K(\cdot, p) \in \mathscr{H}, \tag{1.1}$$

S. Saitoh, Y. Sawano, *Theory of Reproducing Kernels and Applications*,
Developments in Mathematics 44, DOI 10.1007/978-981-10-0530-5_1

for all $p \in E$ and

$$f(p) = \langle f, K_p \rangle_{\mathscr{H}} \tag{1.2}$$

holds for all $p \in E$ and all $f \in \mathscr{H}$. Denote by $(H_K =)H_K(E)$ the Hilbert space $\mathscr{H}$ whose corresponding reproducing kernel function is K.

We will see that the correspondence of the reproducing kernel K and the reproducing kernel Hilbert space $H_K(E)$ is one-to-one (see Theorem 2.2 for the existence and Proposition 2.2 for the uniqueness).

A simple example follows.

Example 1.1. Let $E = \{x_1, x_2, \ldots, x_N\}$ be a set of N distinct points. Suppose that $A \equiv \{a_{ij}\}_{i,j=1}^n$ is strictly positive definite $N \times N$-Hermitian as matrix. Write $B = \{b_{ij}\}_{i,j=1}^n \equiv A^{-1}$ as its inverse. Then with the inner product

$$\langle f, g \rangle_{H_A(E)} \equiv \sum_{i,j=1}^{N} f(x_i) b_{ji} \overline{g(x_j)} = \overline{(g(x_1), g(x_2), \ldots, g(x_N))} B \begin{pmatrix} f(x_1) \\ f(x_2) \\ \vdots \\ f(x_N) \end{pmatrix},$$

$\mathscr{F}(E)$, the set of all complex-valued functions on E, is a reproducing kernel Hilbert (complex Euclidean) space whose kernel satisfies $K(x_i, x_j) = a_{ij}$.

Indeed, (1.1) follows trivially. Let us check that (1.2) holds as well. By the definition, we have

$$\langle f, K(\cdot, x_j) \rangle_{H_{K_A}\{x_1, x_2, \ldots, x_N\}} = \sum_{i,k=1}^{n} f(x_i) b_{ki} \overline{K(x_k, x_j)} = \sum_{i,k=1}^{n} f(x_i) b_{ki} a_{jk} = f(x_j),$$

proving that $H_A(E)$ is the reproducing kernel Hilbert space $H_K(E)$ with norm-coincidence.

Remark 1.1. If $a > 0$ and $f \in H_K(E)$, then $H_K(E) = H_{aK}(E)$ and $\|a \cdot f\|_{H_{aK}(E)} = a \cdot \|f\|_{H_K(E)}$.

1.1.2 *Orientation of Chap. 1*

We have many concrete reproducing kernels that are found in the solutions of many partial differential equations, in Sobolev spaces consisting of real-valued functions, and in complex analysis of one variable and several variables. (When we restrict in complex analysis reproducing kernels to the real spaces, their restrictions are also reproducing kernels, so that we have many problems in analytic continuations.) Meanwhile, the theory of reproducing kernels may be considered being the Bergman kernel and the Szegö kernel. In being these we have many references and profound results on these typical reproducing kernels as well as several technical books on them. However, interest in their results will be restricted to those in complex analysis or some specialists. So in Chap. 1, we will present typical reproducing kernels and their elementary properties of general interest.

Definition 1.2. A complex-valued function $k : E \times E \to \mathbb{C}$ is called a *positive definite quadratic form function* on the set E, or *positive definite function*, when it enjoys the property that for an arbitrary function $X : E \to \mathbb{C}$ and for any finite subset F of E,

$$\sum_{p,q \in F} \overline{X(p)} X(q) k(p,q) \geq 0. \tag{1.3}$$

We can see immediately that any reproducing kernel K satisfying (1.1) and (1.2) is a positive definite function. The important theorem is the converse; that is, for any positive definite function, there exists a uniquely determined Hilbert space $\mathscr{H}$ satisfying (1.1) and (1.2). This theorem is a core result in the general theory of reproducing kernels, whose proof will be discussed in Chap. 2. We have the following expression of $\mathscr{H}$:

Proposition 1.1. *Let K be a function satisfying* (1.1) *and* (1.2). *Then $\{K_q\}_{q \in E}$ spans a dense space of $\mathscr{H}$.*

The proof is simple; indeed, the only function perpendicular to $\{K_q\}_{q \in E}$ is zero by (1.2), proving Proposition 1.1. Thus, to specify $H_K(E)$, we need only to check (1.1) and (1.2).

Of course, in our main part Chap. 2 we will refer to detailed properties of these facts.

1.2 Paley Wiener Reproducing Kernels

1.2.1 Paley Wiener Space

Let $h > 0$ and $x \in \mathbb{R}$. We define

$$H_x(t) \equiv \frac{1}{\sqrt{2\pi}} \exp(ixt), \ |t| \leq \frac{\pi}{h}. \tag{1.4}$$

In Sect. 1.2 we consider the integral transform generated by $\{H_x\}_{x \in \mathbb{R}}$. To this end, we first recall that $\mathrm{sinc}(x)$ is an analytic function given by

$$\mathrm{sinc}(x) \equiv \begin{cases} x^{-1} \sin x, & x \in \mathbb{R} \setminus \{0\} \\ 1, & x = 0. \end{cases} \tag{1.5}$$

We define

$$K_h(x,x') \equiv \langle H_{x'}, H_x\rangle_{L^2(-\frac{\pi}{h},\frac{\pi}{h})} = \int_{-\frac{\pi}{h}}^{\frac{\pi}{h}} \exp\left(i(x-x')t\right)\frac{dt}{2\pi} = \frac{1}{h}\mathrm{sinc}\frac{\pi}{h}(x-x') \tag{1.6}$$

for $x, x' \in \mathbb{R}$ and

$$Lg(x) \equiv \frac{1}{\sqrt{2\pi}}\int_{-\frac{\pi}{h}}^{\frac{\pi}{h}} g(t)\exp(-ixt)\,dt \quad \left(x \in \mathbb{R},\ g \in L^2\left(-\frac{\pi}{h},\frac{\pi}{h}\right)\right). \tag{1.7}$$

Let us characterize the Paley Wiener space in terms of the Fourier transform. We use the standard notation $z = x + iy$ at a point $z \in \mathbb{C}$.

Theorem 1.1. *Let $h > 0$. Then $H_{K_h}(\mathbb{R})$ is a closed subspace of $L^2(\mathbb{R})$ and we have*

$$\begin{aligned}
&H_{K_h}(\mathbb{R})\\
&= \bigcup_{F\in\mathscr{O}(\mathbb{C})}\bigcap_{\kappa\in(\pi,\infty)}\left\{f \in L^2(\mathbb{R}) : f = F|\mathbb{R} \text{ and } \sup_{x,y\in\mathbb{R}}|F(x+iy)|\exp\left(-\frac{\kappa|y|}{h}\right) < \infty\right\}\\
&= \left\{f \in L^2(\mathbb{R}) : \mathrm{supp}(\mathscr{F}f) \subset \left[-\frac{\pi}{h},\frac{\pi}{h}\right]\right\},
\end{aligned} \tag{1.8}$$

where $\mathscr{O}(\mathbb{C})$ denotes the set of entire functions.

Proof. We can easily verify that the last term coincides with $H_{K_h}(\mathbb{R})$ by checking (1.1) and (1.2) as a result $H_{K_h}(\mathbb{R})$ is a closed subspace of $L^2(\mathbb{R})$.

Assume that $f \in L^2(\mathbb{R})$ whose frequency support is contained in $\left[-\frac{\pi}{h},\frac{\pi}{h}\right]$. Then we define

$$F(z) \equiv \frac{1}{2\pi i}\int_{-\frac{\pi}{h}}^{\frac{\pi}{h}} \mathscr{F}f(t)\exp(itz)\,dt. \tag{1.9}$$

From this expression, it is easy to see that for any $\varepsilon > 0$,

$$F \in \mathscr{O}(\mathbb{C}), \quad f = F|\mathbb{R}, \quad \sup_{x,y\in\mathbb{C}}|F(x+iy)|\exp\left(-\frac{(\pi+\varepsilon)|y|}{h}\right) < \infty. \tag{1.10}$$

This implies that $H_{K_h}(\mathbb{R})$ is contained in the middle term of (1.8).

Assume instead that f is a restriction of $F \in \mathscr{O}(\mathbb{C})$ to $\mathbb{R}$ satisfying

$$\sup_{x,y\in\mathbb{R}}|F(x+iy)|\exp\left(-\frac{(\pi+\varepsilon)|y|}{h}\right) < \infty \tag{1.11}$$

for all $\varepsilon > 0$. Let us choose a smooth function g supported outside of $\left[-\frac{\pi}{h},\frac{\pi}{h}\right]$ and consider

$$A \equiv \int_{\mathbb{R}} g(\xi)\mathscr{F}f(\xi)\,d\xi. \tag{1.12}$$

We will prove that $A = 0$, which immediately yields, together with the fact that $C_c^\infty(\mathbb{R})$ is dense in $L^2(\mathbb{R})$, that $\mathscr{F}f$ is supported on $\left[-\frac{\pi}{h}, \frac{\pi}{h}\right]$. For this purpose we shall decompose g into two functions supported on $\left(-\infty, -\frac{\pi}{h}\right)$ and $\left(\frac{\pi}{h}, \infty\right)$ respectively. By symmetry we may assume that g is supported on $\left(\frac{\pi}{h}, \infty\right)$. Then we have

$$A = \int_{\mathbb{R}} G(x)F(x)\,dx,\ G(x+iy) \equiv \frac{1}{\sqrt{2\pi}}\int_{\frac{\pi}{h}+\delta}^{\infty} g(\xi)\exp(i(x+iy)\cdot\xi)\,d\xi \tag{1.13}$$

for some $\delta > 0$. By the Cauchy theorem for the line integral, we have

$$A = \int_{\mathbb{R}} \mathscr{F}g(x+iy)F(x+iy)\,dx,\ y \geq 0 \tag{1.14}$$

in view of the estimate

$$|G(x+iy)| \lesssim \frac{1}{1+|x|^2}\exp\left(-\left(\frac{\pi}{h}+\delta\right)y\right). \tag{1.15}$$

Letting $y \to \infty$, we obtain

$$A = \lim_{y\to\infty}\int_{\mathbb{R}} G(x+iy)F(x+iy)\,dx = 0. \tag{1.16}$$

This is the desired result, which shows that the last term of (1.8) is contained in the middle term of (1.8).

The space $H_{K_h}(\mathbb{R})$ can be applied to sampling theory. Thanks to the Parseval formula, we have

$$\|g\|_{L^2(-\frac{\pi}{h},\frac{\pi}{h})} = \sqrt{h\sum_{j=-\infty}^{\infty}|Lg(jh)|^2} \quad \left(g \in L^2\left(-\frac{\pi}{h},\frac{\pi}{h}\right)\right). \tag{1.17}$$

Meanwhile by the Plancherel formula, we have

$$\|g\|_{L^2(-\frac{\pi}{h},\frac{\pi}{h})} = \|Lg\|_{L^2(\mathbb{R})} \quad \left(g \in L^2\left(-\frac{\pi}{h},\frac{\pi}{h}\right)\right). \tag{1.18}$$

Observe also that $f(x) = \langle f, (K_h)_x\rangle_{H_{K_h}(\mathbb{R})}$. With (1.17) and (1.18) we have

$$f(x) = \langle f, (K_h)_x\rangle_{L^2(\mathbb{R})} = \int_{\mathbb{R}} f(\xi)K_h(\xi,x)\,d\xi = h\sum_{j=-\infty}^{\infty} f(jh)K_h(jh,x)$$

for all $x \in \mathbb{R}$. This is a typical result in sampling theory.

With (1.7) let us investigate

$$g \in L^2\left(-\frac{\pi}{h}, \frac{\pi}{h}\right) \mapsto Lg \in H_{k_h}(\mathbb{R}). \tag{1.19}$$

Theorem 1.1 gives the well-known sampling theory statement that the complete information of $f(x)$ on the real line is represented by the discrete data $\{f(jh)\}_{j=-\infty}^{\infty}$. When we consider some finite number of data $\{f(jh)\}_{j\in J}$, its truncation error is also estimated in the framework of reproducing kernel Hilbert spaces. See [388, Section 4.2]. Its comprehensive generalizations were established by Higgins [206] and they are given in the Sect. 8.4.1. From the viewpoint of the theory of function spaces, this is also important; see [156, Lemma 2.1] and [157, Lemma 2.1] for the precise statement. Based on this expansion, many other expansions came about; see [462, (2.53)], [259, Theorem 1.3] and [422, Lemma 5.4].

1.2.2 A Characterization Using the Fourier Transform

We make use repeatedly of multidimensional Paley Wiener reproducing kernels. Recall that the space $H_{K_h}(\mathbb{R})$ is given by (1.6). So, we fix simple notation and state results.

Let $h > 0$. We will consider the integral transform for $g \in L^2((-\frac{\pi}{h}, \frac{\pi}{h})^n)$. We use the coordinates $z = (z_1, z_2, \ldots, z_n)$ and $t = (t_1, t_2, \ldots, t_n)$. We write

$$dt = dt_1 dt_2 \cdots dt_n, \; z \cdot t = z_1 t_1 + \ldots + z_n t_n. \tag{1.20}$$

Also, we set

$$\chi_h(t) \equiv \prod_{\nu=1}^{n} \chi_{(-\pi,\pi)}(ht_\nu) = \prod_{\nu=1}^{n} \chi_{(-\frac{\pi}{h},\frac{\pi}{h})}(t_\nu) = \chi_{(-\frac{\pi}{h},\frac{\pi}{h})^n}(t), \tag{1.21}$$

the characteristic function of $\left(-\frac{\pi}{h}, \frac{\pi}{h}\right)^n$. Then we consider the integral transform given by

$$f(z) \equiv \frac{1}{(2\pi)^n} \int_{\mathbb{R}^n} \chi_h(t) g(t) \exp(-iz \cdot t) dt \quad (z \in \mathbb{C}^n). \tag{1.22}$$

We write $u = (u_1, u_2, \ldots, u_n) \in \mathbb{C}^n$. In order to identify the image space, we form the reproducing kernel

$$K_h(z, \overline{u}) = \frac{1}{(2\pi)^n} \int_{\mathbb{R}^n} \chi_h(t) \exp(-iz \cdot t) \overline{\exp(-iu \cdot t)} dt = \prod_{\nu=1}^{n} \frac{1}{h} \mathrm{sinc}\left(\frac{\pi}{h}(z_\nu - \overline{u}_\nu)\right).$$

The image space of the mapping $f \mapsto F$ in (1.9) is the Paley Wiener space $W\left(\frac{\pi}{h}\right)$ $(\equiv W_h)$ consisting of all analytic functions of exponential type satisfying for each ν some constant C_ν, and as $z_\nu \to \infty$

$$|f(z_1, z_2, \ldots, z_\nu, z_{\nu+1}, \ldots, z_n)| \le C_\nu \exp\left(\frac{\pi |z_\nu|}{h}\right) \tag{1.23}$$

and

$$\int_{\mathbb{R}^n} |f(x)|^2 dx < \infty. \tag{1.24}$$

Denote by $\delta : \mathbb{Z} \times \mathbb{Z} \to \{0, 1\}$ Kronecker's delta. For a multi-index $j = (j_1, j_2, \ldots, j_n)$, we have

$$K_h(jh, j'h) = \prod_{\nu=1}^{n} \frac{1}{h} \delta(j_\nu, j'_\nu). \tag{1.25}$$

Since $\delta(j_\nu, j'_\nu)$ is the reproducing kernel for the Hilbert space $\ell^2(\mathbb{Z})$ for each ν, we have the isometric identities in (1.22):

$$\frac{1}{(2\pi)^n} \int_{-\frac{\pi}{h}}^{\frac{\pi}{h}} |g(t)|^2 dt = h^n \sum_{j=1}^{\infty} |f(jh)|^2 = \int_{\mathbb{R}} |f(x)|^2 dx. \tag{1.26}$$

That is, the reproducing kernel Hilbert space $H_{K_h}(\mathbb{R}^n)$ with kernel K_h is characterized as a space consisting of the Paley Wiener space $W_h(\mathbb{R}^n)$ and with the norms above in the sense of both the discrete and continuous versions. See (1.6) for the definition. Then the reproducing property of $K_h(z, \overline{u})$ states that

$$f(x) = \langle f, (K_h)_x \rangle_{H_{K_h}(\mathbb{R})} = h^n \sum_{j=1}^{\infty} f(jh) K_h(jh, x) = \int_{\mathbb{R}^n} f(\xi) K_h(\xi, x) d\xi, \tag{1.27}$$

in particular, on the real space x. This representation is the sampling theorem which represents all of the data $f(x)$ in terms of the discrete data $\{f(jh)\}_{j \in \mathbb{Z}^n}$.

By using 2×2 matrices, L. de Branges [60, Section 47] generalized to a large extent Paley Wiener spaces with reproducing kernels. We refer to J.R. Higgins [205, Chapters 4–6 and 10] and F. Stenger [446, Section 1.10] as well as [204] for the connection with sampling theory.

1.3 RKHS of Sobolev Type

In Theorems 1.2 and 1.3 we will consider RKHSs on compact intervals. In Theorem 1.4 we take up the one on the half line $(0,\infty)$. The spaces on the whole line $\mathbb{R}$ are dealt with as Theorems 1.7 and 1.11. In each case, the existence of $H_K(E)$ is a generality following from Theorem 2.2, however, together with the characterization we check that $H_K(E)$ does exist.

1.3.1 Weak-Derivatives and Sobolev Spaces

To prove Theorems 1.2, 1.3, 1.4, 1.5, 1.6 and 1.7, we collect some elementary properties of function spaces. First, the following property is fundamental, which justifies the meaning of $f(a)$, and so on in (1.35). Recall that $C_c^\infty(0,\infty)$ is the set of all smooth functions compactly supported on $(0,\infty)$. Using (1.29), we can prove the following lemma, whose proof is omitted.

Lemma 1.1. *Let I be an open interval and let $\{f_j\}_{j=1}^\infty \subset C_c^\infty(I)$ be a Cauchy sequence with respect to the Sobolev norm $\|\cdot\|_{W^{1,2}(I)}$, that is,*

$$\lim_{j,k\to\infty} \|f_j - f_k\|_{W^{1,2}(I)} = \lim_{j,k\to\infty} \sqrt{\|f_j - f_k\|_{L^2(I)}^2 + \|f_j' - f_k'\|_{L^2(I)}^2} = 0.$$

Then the following assertions hold:

1. *The limit*

$$f(x) \equiv \lim_{j\to\infty} f_j(x) \tag{1.28}$$

   exists for all $x \in I$.
2. *The limit $g(x) \equiv \lim_{j\to\infty} f_j'(x)$ exists for almost every $x \in I$, if we pass to a suitable subsequence.*
3. *The limit f does not depend on the choice of $\{f_j\}_{j=1}^\infty$ above and the limit g depends on the choice of $\{f_j\}_{j=1}^\infty$ only for sets of measure zero. More precisely, if $\{f_j\}_{j=1}^\infty \subset C_c^\infty(I)$ and $\{h_j\}_{j=1}^\infty \subset C_c^\infty(I)$ are Cauchy sequences such that*

$$\lim_{j\to\infty} \|f_j - h_j\|_{W^{1,2}(I)} = 0,$$

   then $\lim_{j\to\infty} f_j(x) = \lim_{j\to\infty} h_j(x)$ for all $x \in I$ and $\lim_{j\to\infty} f_j'(x) = \lim_{j\to\infty} h_j'(x)$ for almost all $x \in I$.
4. *The limit f given by (1.28) is continuous.*
5. *Suppose that $a \equiv \inf(I)$ is finite. Then the limit $f(a) \equiv \lim_{x\downarrow a} f(x)$ exists.*

The function spaces we take up here are the following:

Definition 1.3 (Sobolev spaces on an interval). Let I be a closed or open interval. Let us define $W^{1,2}(I)$, *the Sobolev space of differential order* 1 (based on $L^2(I)$), as the completion of $C_c^\infty(\mathrm{Int}(I))$ with respect to the norm

$$\|f\|_{W^{1,2}(I)} \equiv \sqrt{\|f'\|_{L^2(I)}^2 + \|f\|_{L^2(I)}^2}.$$

When $I = (a,b)$, reduce $W^{1,2}(I)$ to $W^{1,2}(a,b)$. When $I = [a,\infty)$, reduce $W^{1,2}(I)$ to $W^{1,2}[a,\infty)$.

In Definition 1.3, I can be finite or infinite. We can rephrase Definition 1.3 as follows: $W^{1,2}(I)$ denotes the set of all $L^2(I)$-functions f for which it is an L^2-limit of a $W^{1,2}(I)$-Cauchy sequence in $C_c^\infty(I)$, that is, one says f in Lemma 1.1 *1.* belongs to the class $W^{1,2}(I)$ and the limit g in Lemma 1.1 *2.* is denoted by f'.

The following lemma is useful in practical applications:

Lemma 1.2.

1. *Let* $f \in L^2(\mathbb{R})$ *and* $g \in L^1(\mathbb{R}) \cap L^2(\mathbb{R})$. *Assume that*

$$f(x) = \int_{-\infty}^{x} g(t)\, dt$$

for almost every $x \in \mathbb{R}$. *Then* $f \in W^{1,2}(\mathbb{R})$ *and* $f'(x) = g(x)$ *for almost every* $x \in \mathbb{R}$.

2. *Conversely,*

$$f(x) = \int_{-\infty}^{x} f'(t)\, dt$$

for all $x \in \mathbb{R}$ *whenever* $f \in W^{1,2}(\mathbb{R})$ *satisfies* $f' \in L^1(\mathbb{R})$.

3. *The space* $W^{1,2}(\mathbb{R})$ *is embedded into* $\mathrm{BC}(\mathbb{R})$, *the set of all bounded continuous functions.*

Proof.

1. Let us choose a $C_c^\infty(\mathbb{R})$-function ψ such that $\chi_{[-2,2]} \le \psi \le \chi_{[-1,1]}$. Also, choose a sequence $\{g_j\}_{j=1}^\infty$ of $C_c^\infty(\mathbb{R})$-functions such that $g_j \to g$ in the topology of $L^1(\mathbb{R}) \cap L^2(\mathbb{R})$. Define

$$f_j(x) \equiv \psi(j^{-1}x) \int_{-\infty}^{x} g(t)\, dt \quad (x \in \mathbb{R})$$

and

$$f_{j,l}(x) \equiv \psi(j^{-1}x) \int_{-\infty}^{x} g_l(t)\, dt \quad (x \in \mathbb{R}).$$

Then it is easy to see that $f_{j,l} \to f_j$ as $l \to \infty$ in the topology of $W^{1,2}(\mathbb{R})$. Also, it is easy to see that $\{f_j\}_{j=1}^{\infty}$ is a Cauchy sequence in $W^{1,2}(\mathbb{R})$. Since $f_j(x)$ converges to $f(x)$ for each $x \in \mathbb{R}$, it follows that $f \in W^{1,2}(\mathbb{R})$. Using sequences $\{f_j\}_{j=1}^{\infty}$ and $\{f_{j,l}\}_{j=1}^{\infty}$ for $l = 1, 2, \ldots$ we can deduce that $f' = g$ and that $f \in W^{1,2}(\mathbb{R})$.

2. We recall that we can choose a sequence $\{\varphi_j\}_{j=1}^{\infty}$ of $L^2(\mathbb{R})$-functions such that $\{\varphi_j'\}_{j=1}^{\infty}$ is convergent to f' in $L^1(\mathbb{R}) \cap L^2(\mathbb{R})$ and that $\{\varphi_j\}_{j=1}^{\infty}$ converges to f in $L^2(\mathbb{R})$. If we use this sequence, we can easily conclude that 2. holds.
3. This follows from the fact that the Fourier transform characterizes $W^{1,2}(\mathbb{R})$;

$$\|f\|_{W^{1,2}(\mathbb{R})} = \sqrt{\int_{\mathbb{R}} (1 + |\xi|^2)|\mathscr{F}f(\xi)|^2 \, d\xi} \gtrsim \int_{\mathbb{R}} |\mathscr{F}f(\xi)| \, d\xi \gtrsim \|f\|_{L^\infty} \tag{1.29}$$

for all $f \in W^{1,2}(\mathbb{R})$.

The next lemma concerns a quantitative estimate of the functions in $W^{1,2}(0, \infty)$.

Lemma 1.3. *Let $f \in W^{1,2}(0, \infty)$. Then*

1. *f is absolutely continuous;*
2. *for all $x \geq 1$, the estimate*

$$|f(x)| \leq |f(1)| + \|f\|_{W^{1,2}(0,\infty)} \sqrt{x-1} \tag{1.30}$$

holds.

Proof. The proof is routine. Indeed, if $f \in C_c^{\infty}(0, \infty)$, then (1.30) is clear from the Hölder inequality. The general case can be dealt with by passing to a limit in the above case.

1.3.2 1-Dimensional Case

Let $a < b$. First, we consider a positive definite quadratic form function

$$K_a(s,t) \equiv 1 - a + s \wedge t \tag{1.31}$$

on $s, t \in I \equiv [a, b]$. Note that $K_a(t,t) = 1 - a + t \geq 1$.

We have the following representation of RKHS given by the function K_a in (1.31): Denote by AC(I) the set of all absolutely continuous functions and fix a point $c \in I$, that is, $f \in$ AC(I) if and only if

$$f(x) = f(c) + \int_c^x g(t) \, dt \tag{1.32}$$

for some $g \in L^1_{\mathrm{loc}}(I)$.

Theorem 1.2 (RKHS on the closed interval (I)). *Define $K_a : I \times I \to \mathbb{R}$ by* (1.31), *where $I = [a, b]$. Then we have an expression of the reproducing kernel Hilbert space $H_{K_a}(I)$:*

$$H_{K_a}(I) = H_{K_a}[a, b] = \{f \in \mathrm{AC}(I) \,:\, f' \in L^2(I)\} \tag{1.33}$$

as a set and the norm is given by

$$\|f\|_{H_{K_a}(I)} \equiv \left(|f(a)|^2 + \int_a^b |f'(t)|^2 \, dt \right)^{\frac{1}{2}}. \tag{1.34}$$

Furthermore, as a set of functions, we have

$$H_{K_a}(I) = W^{1,2}(I). \tag{1.35}$$

Proof. Let H be the Hilbert space described on the right-hand side of (1.33). Equip H with the inner product given by (1.34). Then we have $(K_a)'_t \in L^2(I)$ for all t. As a result, (1.1) holds.

Let us check that (1.2) holds. Note that

$$\frac{d(K_a)_t}{ds}(s) = \chi_{[a,t]}(s) \text{ for almost every } s \in \mathbb{R} \tag{1.36}$$

by Lemma 1.1. Next, let us check the reproducing property. Let $f \in H$. Then we have

$$\langle f, (K_a)_t \rangle_H = f(a)K_a(a,t) + \int_a^t f'(u)\, du = f(a) + \int_a^t f'(u)\, du = f(t)$$

by the definition (1.31) and the norm (1.34).

Therefore, we obtain the desired reproducing property (1.2) and $H = H_{K_a}$ with inner product coincidence.

Equality (1.35) follows directly from (1.33).

On the closed interval $[a, b]$, we also have the following reproducing kernel Hilbert spaces:

Theorem 1.3 (RKHS of the first order on the closed interval II). *Let $-\infty < a < b < \infty$. Then $W^{1,2}[a, b]$ is realized by the reproducing kernel G given by*

$$G(x, y) \equiv \frac{1}{2} \exp(-|x - y|) \tag{1.37}$$

for $x, y \in [a, b]$. More precisely, G is positive definite and $H_G[a, b] = W^{1,2}[a, b]$ as a set of functions, and the norm is given by

$$\|f\|_{H_G[a,b]} \equiv \sqrt{\left(\int_a^b |f(x)|^2 + |f'(x)|^2\,dx\right) + |f(a)|^2 + |f(b)|^2} \tag{1.38}$$

for $f \in H_G[a,b]$.

Proof. The fact that G is positive definite is a direct consequence of the reproducing property; let us concentrate on proving $H_G[a,b] = W^{1,2}[a,b]$. Equip $W^{1,2}(0,\infty)$ with the norm given by (1.38). By Lemma 1.2, we have $W^{1,2}(0,\infty) \subset \mathscr{F}(0,\infty)$ in the sense that we can choose the continuous representative.

Let us fix $y_0 \in [a,b]$. Then a direct calculation shows

$$G'_{y_0} = \frac{1}{2}\exp(\cdot - y_0)\chi_{(-\infty,y_0)} - \frac{1}{2}\exp(y_0 - \cdot)\chi_{[y_0,\infty)}, \tag{1.39}$$

which proves (1.1). Let $f \in W^{1,2}[a,b]$. From (1.39), we have $G_{y_0} \in W^{1,2}[a,b]$ and

$$\begin{aligned}
&\langle f, G_{y_0}\rangle_{W^{1,2}[a,b]}\\
&= f(a)G_{y_0}(a) + f(b)G_{y_0}(b) + \langle f, G_{y_0}\rangle_{L^2[a,b]} + \langle f', G'_{y_0}\rangle_{L^2[a,b]}\\
&= \frac{1}{2}f(a)\exp(a-y_0) + \frac{1}{2}f(b)\exp(y_0-b) + \frac{1}{2}\int_a^b f(x)\exp(-|x-y_0|)\,dx\\
&\quad + \frac{1}{2}\int_a^{y_0} f'(x)\exp(y_0-x)\,dx - \frac{1}{2}\int_{y_0}^b f'(x)\exp(x-y_0)\,dx.
\end{aligned}$$

From the absolute continuity of the function f, we have

$$\begin{aligned}
&\langle f, G_{y_0}\rangle_{W^{1,2}[a,b]}\\
&= \frac{1}{2}f(a)\exp(a-y_0) + \frac{1}{2}f(b)\exp(y_0-b)\\
&\quad + \frac{1}{2}\int_a^{y_0} (f(x)+f'(x))\exp(x-y_0)\,dx + \frac{1}{2}\int_{y_0}^b (f(x)-f'(x))\exp(y_0-x)\,dx\\
&= \frac{f(a)\exp(a-y_0)+f(b)\exp(y_0-b)}{2}\\
&\quad + \frac{1}{2}\left[f(x)\exp(x-y_0)\right]_a^{y_0} + \frac{1}{2}\left[f(x)\exp(y_0-x)\right]_{y_0}^b = f(y_0),
\end{aligned}$$

proving (1.2).

Remark 1.2. One is convinced that a difference of the inner product between (1.34) and (1.38) yields an actual difference of the corresponding reproducing kernels.

On the half line, we have the following important reproducing kernels:

Theorem 1.4 (RKHS on the half line). *The Sobolev space* $W^{1,2}(0,\infty)$ *is realized by the reproducing kernel Hilbert space given by*

$$K(s,t) \equiv \int_0^\infty \frac{\cos(s\,u)\cos(t\,u)}{u^2+1}\,du = \frac{\pi}{4}\left(\exp(-|s-t|)+\exp(-s-t)\right) \tag{1.40}$$

for $s,t>0$. *More precisely, K is positive definite and we have a norm equivalence:*

$$\|f\|_{H_K(0,\infty)} = \sqrt{\frac{2}{\pi}\int_0^\infty (|f'(u)|^2+|f(u)|^2)\,du} \tag{1.41}$$

for $f \in H_K(0,\infty)$.

For the proof of Theorem 1.4, we recall Lemma 1.3.

Proof. Equip $H = W^{1,2}(0,\infty)$ with the norm given by (1.41) and concentrate on proving the reproducing property as before.

By Lemma 1.2, $W^{1,2}(0,\infty)$ is contained in the set $\mathrm{BC}(0,\infty)$ made up of all bounded continuous functions. Thus, $W^{1,2}(0,\infty) \subset \mathscr{F}(0,\infty)$. We also have, for all $t,s>0$,

$$\frac{dK_t}{ds}(t) = -\frac{\pi}{4}\exp(-s-t) + \frac{\pi}{4}\exp(s-t)\chi_{(-\infty,t)}(s) - \frac{\pi}{4}\exp(t-s)\chi_{(t,\infty)}(s). \tag{1.42}$$

again by Lemma 1.2. Thus, (1.1) follows.

From the definition of the inner product of H, we obtain

$$\begin{aligned}
&\left\langle f, \frac{dK_t}{ds}\right\rangle_H \\
&= \frac{1}{2}\int_0^t f'(s)(\exp(s-t)-\exp(-s-t)) + f(s)(\exp(s-t)+\exp(-s-t))\,ds \\
&\quad + \frac{1}{2}\int_t^\infty f'(s)(-\exp(t-s)-\exp(-s-t)) + f(s)(\exp(t-s)+\exp(-s-t))\,ds \\
&= -\frac{1}{2}\left[\exp(-s-t)f(s)\right]_0^\infty + \frac{1}{2}\left[\exp(s-t)f(s)\right]_0^t + \frac{1}{2}\left[-\exp(s-t)f(s)\right]_t^\infty.
\end{aligned}$$

By formula (1.30), we have $\left\langle f, \dfrac{dK_t}{ds}\right\rangle_H = f(t)$, proving the reproducing property (1.2).

If we specify the boundary condition, then the kernel undergoes a change as the following theorem shows:

Theorem 1.5 (RKHS on the half line with boundary condition). *Let*

$$K(s,t) \equiv \int_0^\infty \frac{\sin(s\,u)\sin(t\,u)}{u^2+1}\,du = \frac{\pi}{4}\left(\exp(-|s-t|)-\exp(-s-t)\right) \tag{1.43}$$

for $s, t > 0$. *Then* K *is positive definite and*

$$H_K(0, \infty) = \{f \in \mathrm{AC}(0, \infty) \,:\, f(0) = 0\} \tag{1.44}$$

as a set of functions. The norm is given by

$$\|f\|_{H_K(0,\infty)} = \sqrt{\frac{2}{\pi}\int_0^\infty (|f'(u)|^2 + |f(u)|^2)\, du} \tag{1.45}$$

for $f \in H_K(0, \infty)$.

Proof (of Theorem 1.5*).* Let us denote by H the Hilbert space given by the right-hand side of (1.44) whose norm is given by (1.45). As before, it is the reproducing property that counts.

It follows from (1.43) and Lemma 1.2 that we have, for all $t, s > 0$,

$$\frac{dK_t}{ds}(s) = \frac{\pi}{4}\exp(-s-t) + \frac{\pi}{4}\exp(s-t)\chi_{(-\infty,t)}(s) - \frac{\pi}{4}\exp(t-s)\chi_{[t,\infty)}(s). \tag{1.46}$$

This proves (1.1).

Keeping (1.46) in mind, we let $f \in W^{1,2}(0, \infty)$ satisfy $f(0+) = 0$ in the sense of Lemma 1.1. As a result, we obtain

$$\begin{aligned}
&2\left\langle f, \frac{dK_t}{ds}\right\rangle_H \\
&= \int_0^t f'(s)(\exp(s-t) + \exp(-s-t)) + f(s)(\exp(s-t) - \exp(-s-t))\, ds \\
&\quad + \int_t^\infty f'(s)(\exp(-s-t) - \exp(t-s)) + f(s)(\exp(t-s) - \exp(-s-t))\, ds \\
&= [\exp(-s-t)f(s)]_0^\infty + [\exp(s-t)f(s)]_0^t + [-\exp(s-t)f(s)]_t^\infty \\
&= 2f(t),
\end{aligned}$$

which proves (1.2).

We have the following variant, whose proof is analogous and omitted:

Theorem 1.6 (RKHS on the half line with boundary condition). *Let*

$$K(s, t) \equiv \min(s, t) \quad (s, t > 0). \tag{1.47}$$

Then K *is positive definite and*

$$H_K(0, \infty) = \left\{f \in W^{1,2}(0, \infty) \,:\, \lim_{\varepsilon \downarrow 0} f(\varepsilon) = 0\right\} \tag{1.48}$$

as a set of functions. The norm is given by

$$\|f\|_{H_K(0,\infty)} = \sqrt{\int_0^\infty |f'(u)|^2\,du} \tag{1.49}$$

for $f \in H_K(0,\infty)$.

As we saw in Remark 1.2, the kernel depends on the inner product even if the space $H_K(\mathbb{R})$ is invariant as a set of functions.

Theorem 1.7 (Parametrized RKHS on $\mathbb{R}$ of first order). *Let* $a, b > 0$. *Define*

$$G(s,t) \equiv \frac{1}{2ab}\exp\left(-\frac{b}{a}|s-t|\right) \quad (s,t \in \mathbb{R}). \tag{1.50}$$

Then G is positive definite, $H_G(\mathbb{R}) = W^{1,2}(\mathbb{R})$ *as a set of functions and the norm is given by*

$$\|f\|_{W^{1,2}(\mathbb{R})} = \sqrt{\int_{\mathbb{R}} (a^2|f'(x)|^2 + b^2|f(x)|^2)\,dx} \quad (f \in W^{1,2}(\mathbb{R}). \tag{1.51}$$

Proof. Observe that G is positive definite from Theorem 1.3. Fix $t \in \mathbb{R}$. Note that

$$G(s,t) = \frac{1}{2ab}\exp\left(-\frac{b}{a}(t-s)\right)\chi_{(-\infty,t)}(s) + \frac{1}{2ab}\exp\left(-\frac{b}{a}(s-t)\right)\chi_{(t,\infty)}(s)$$

$$\frac{dG_t}{ds}(s) = \frac{1}{2a^2}\exp\left(-\frac{b}{a}(t-s)\right)\chi_{(-\infty,t)}(s) + \frac{1}{2a^2}\exp\left(-\frac{b}{a}(s-t)\right)\chi_{(t,\infty)}(s)$$

for almost all s, which proves (1.1).

Let $f \in W^{1,2}(\mathbb{R})$. Then we have

$$\begin{aligned}
&\int_{\mathbb{R}} a^2 f'(s)\frac{dG_t}{ds}(s) + b^2 f(s)G(s,t)\,ds \\
&= \int_{-\infty}^t f'(s)\frac{dG_t}{ds}(s) + b^2 f(s)G(s,t)\,ds + \int_t^\infty a^2 f'(s)\frac{dG_t}{ds}(s) + b^2 f(s)G(s,t)\,ds \\
&= \frac{1}{2}\int_{-\infty}^t \left(\frac{b}{a}f(s) - f'(s)\right)\exp\left(\frac{b}{a}(s-t)\right)\,ds \\
&\quad + \frac{1}{2}\int_t^\infty \left(f'(s) + \frac{b}{a}f(s)\right)\exp\left(\frac{b}{a}(t-s)\right)\,ds \\
&= \frac{1}{2}\left[f(s)\exp\left(\frac{b}{a}(s-t)\right)\right]_{-\infty}^t - \frac{1}{2}\left[f(s)\exp\left(\frac{b}{a}(t-s)\right)\right]_t^\infty = f(t).
\end{aligned}$$

Here, for the last equality, we used Lemma 1.3. This is the desired reproducing property (1.2).

Example 1.2. The space $H_S(\mathbb{R})$ is made up of absolutely continuous functions F on $\mathbb{R}$ with the norm

$$\|F\|_{H_S(\mathbb{R})} \equiv \sqrt{\int_{\mathbb{R}} (F(x)^2 + F'(x)^2)dx}. \tag{1.52}$$

The Hilbert space $H_S(\mathbb{R})$ admits the reproducing kernel

$$K(x,y) \equiv \frac{1}{2\pi}\int_{\mathbb{R}} \frac{1}{1+\xi^2}\exp(i(x-y)\xi)d\xi = \frac{1}{2}e^{-|x-y|} \quad (x,y \in \mathbb{R}). \tag{1.53}$$

This example also gives us another first-order Sobolev Hilbert space $H_{S,e}(\mathbb{R})$ consisting of absolutely continuous functions F on $\mathbb{R}$, which is made up of even functions with respect to the origin having the norm.

We set

$$\|F\|_{H_{S,e}(\mathbb{R})} = \sqrt{\frac{2}{\pi}\int_0^\infty (F(x)^2 + F'(x)^2)dx}. \tag{1.54}$$

This Hilbert space admits the reproducing kernel

$$K_e(x,y) = \frac{\pi}{4}\left(e^{-|x-y|} + e^{-|x+y|}\right) = \int_0^\infty \frac{\cos x\xi \cos y\xi}{1+\xi^2}\, d\xi. \tag{1.55}$$

Here, we place ourselves once again in the setting of the half-open interval $I = [a,b)$ and we suppose that we are given a positive measurable function $\rho : I \to (0,\infty)$ such that

$$\rho|[a,x] \in L^1[a,x] \quad (x \in I) \tag{1.56}$$

for all $x \in I$. Denote by AC(I) the set of all absolutely continuous functions on an interval I.

Theorem 1.8. *Let $r \geq 1$ be a real number and let the function ρ satisfy* (1.56). *Let us set*

$$W(t) \equiv \int_a^t \rho(\xi)\,d\xi,\ K_\rho(s,t) \equiv \int_a^{s\wedge t} \rho(v)\,dv = W(s\wedge t) \quad (s,t \in I), \tag{1.57}$$

where $s \wedge t \equiv \min(s,t)$. Then $(K_\rho)^r$ is positive definite and the reproducing kernel Hilbert space $H_{(K_\rho)^r}(I)$ is given by

$$H_{(K_\rho)^r}(I) \equiv \left\{f \in \mathrm{AC}(I) : f(a) = 0, f' \in L^2(I, W^{1-r}\rho^{-1}\,dt)\right\}, \tag{1.58}$$

where the norm is given by

$$\|f\|_{H_{(K_\rho)^r}(I)} \equiv \sqrt{\int_I |f'(t)|^2 W(t)^{1-r} \rho(t)^{-1}\, dt} \tag{1.59}$$

for $f \in H_{(K_\rho)^r}(I)$.

Proof. The proof is routine; again the reproducing property counts. Let us define a Hilbert space H by

$$H \equiv \left\{f \in \mathrm{AC}(I) \, : \, f(a) = 0, f' \in L^2(I, W^{1-r}\rho^{-1}\, dt)\right\}, \tag{1.60}$$

where the norm is given by the right-hand side of (1.59).

Let $f \in H$. Then we have

$$|f(t)| = \left|\int_a^t f'(u)\, du\right| \leq \int_a^t |f'(u)|\, du \tag{1.61}$$

by the triangle inequality. By the Schwarz inequality we have

$$|f(t)| \leq \left(\frac{1}{r}\int_a^t |f'(u)|^2 W(u)^{1-r}\rho(u)^{-1}\, du\right)^{\frac{1}{2}} \left(r\int_a^t W(u)^{r-1}\rho(u)\, du\right)^{\frac{1}{2}}. \tag{1.62}$$

Recall that W is given by $W(u) \equiv \displaystyle\int_0^u \rho(t)\, dt$ for $u > 0$. In view of the fact that W is increasing, we obtain

$$|f(t)| \leq \left(\frac{1}{r}\int_I |f'(u)|^2 W(u)^{1-r}\rho(u)^{-1}\, du\right)^{\frac{1}{2}} r^{\frac{1}{2}} W(t)^{\frac{r}{2}}. \tag{1.63}$$

Therefore, the evaluation mapping $f \in H \mapsto f(t) \in \mathbb{C}$ is continuous.

It is easy to see that

$$((K_\rho)_t)^r = \int_a^{\min(\cdot,t)} rW(u \wedge t)^{r-1}\rho(u)\, du \tag{1.64}$$

because

$$\frac{d}{ds}(K_\rho(s,t)^r) = r\chi_{[a,t]}(s)\rho(s)W(s \wedge t)^{r-1} \in L^2[0,t]. \tag{1.65}$$

Finally we check the reproducing property. To do this, we take $f \in H$. Then we have

$$\langle f, (K_\rho)^r(\cdot,t)\rangle_H = \int_a^t f'(u)\rho(u)R(u)^{r-1}\frac{du}{R(u)^{r-1}\rho(u)} = \int_a^t f'(u)\, du = f(t),$$

which means the reproducing property.

The next example is not only a mere quest to generalize but it will be of importance when we consider the inverse of the Laplace transform. See Chap. 3.

Example 1.3. Let $w : [0,\infty) \to (0,\infty)$ be a positive measurable function and consider the Hilbert space $H_K(w)$, which is given by

$$H_K(w) = \left\{ f \in \mathrm{AC}[0,\infty) \,:\, f(0) = 0 \text{ and } \|f\|_{H_K(w)} = \sqrt{\int_0^\infty |f(t)'|^2 w(t)dt} < \infty \right\}.$$

We can realize the Hilbert space $H_K(w)$ using the reproducing kernel $K(s,t)$ given by

$$K(s,t) = \int_0^{\min(s,t)} w(\xi)^{-1}\, d\xi \quad (s,t \in [0,\infty)). \tag{1.66}$$

Indeed, for $t \in \mathbb{R}^+$, by the fundamental theorem of calculus we have

$$\frac{dK_t}{ds}(\xi) \equiv \chi_{(0,t)}(\xi) w(\xi)^{-1}. \tag{1.67}$$

Let $f \in H_K(w)$. Then from the definition of the inner product of $H_K(w)$, we have

$$\langle f, K_t \rangle_{H_K(w)} = \int_0^\infty \frac{dK_t}{ds}(\xi) f'(\xi) w(\xi)\, d\xi = \int_0^t f'(\xi)\, d\xi = f(t),$$

because $f(0) = 0$ by definition.

1.3.3 In Connection with 1-Dimensional Wave Equations

Next we consider reproducing kernel Hilbert spaces of Sobolev type in connection with the wave equation (6.167).

Let $K \equiv \{(x,t) \,:\, |x| \le t\} \subset \mathbb{R}^2$ be a closed cone and define

$$U(x,t) \equiv \chi_K(x,t) = \begin{cases} 1 & (|x| \le t), \\ 0 & (|x| > t), \end{cases} \quad (x \in \mathbb{R},\ t > 0) \tag{1.68}$$

and

$$K(x_1,x_2) \equiv \int_0^T U(x_2, T-\xi) U(x_1, T-\xi) d\xi \quad (x_1,x_2 \in [0,T]). \tag{1.69}$$

Then we have

$$K(x_1,x_2) = \min(T-x_1, T-x_2) \quad (x_1,x_2 \in [0,T]),$$

as the following direct calculation shows:

$$\begin{aligned}K(x_1, x_2) &= \int_0^T U(x_2, \xi)U(x_1, \xi)d\xi \\ &= \int_0^T \chi_{(x_2,\infty)}(\xi)\chi_{(x_1,\infty)}(\xi)\, d\xi \\ &= \int_0^T \chi_{(\max(x_1,x_2),\infty)}(\xi)\, d\xi \\ &= \min(T - x_1, T - x_2)\end{aligned}$$

on $(x_1, x_2) \in [0, T] \times [0, T]$.

Now we prove the following result for K given by (1.69):

Theorem 1.9 (RKHS of order 1 associated to the wave equation). *Let K be the kernel given by (1.69). Then the reproducing kernel Hilbert space $H_K[0, T]$ is composed of all functions $F \in \mathrm{AC}[0, T]$ such that*

$$F' \in L^2[0, T],\ F(T) = 0 \tag{1.70}$$

and the inner product is given by

$$\langle F_1, F_2\rangle_{H_K[0,T]} = \int_0^T F_1'(x)F_2'(x)dx. \tag{1.71}$$

Proof. As usual, denote by $\mathscr{H}$ the Hilbert space given by (1.70) and whose inner product is given by the right-hand side of (1.80). It is immediate that $\mathscr{H} \subset \mathscr{F}[0, T]$.

Fix $x \in (0, T)$. Since

$$\min(T - \cdot, T - x)' = -\chi_{(x,T)} \text{ almost everywhere,} \tag{1.72}$$

we have (1.1).

Let $f \in \mathscr{H}$. Then, since $f(T) = 0$, we have

$$\langle f, \min(T - \cdot, T - x)\rangle_{H_K[0,T]} = -\int_x^T f'(s)\, ds = f(x) - f(T) = f(x) \tag{1.73}$$

by Lemma 1.1. Thus, the reproducing property (1.2) is established.

Now given a function $f \in L^2[0, T]$, we define

$$F(x) \equiv \int_0^T f(\xi)U(x, T - \xi)d\xi \quad (x \in [0, T]). \tag{1.74}$$

Let $L_U : f \mapsto F$ denote the operator defined by the right-hand side of (1.74) on $L^2[0,T]$. Let us also denote by L_U^* the adjoint operator of L_U. Then we can realize the operator $L_U^* L_U$ on $L^2[0,T]$ by the integral transform

$$F(x) = \int_0^T f(\xi)K(x,\xi)d\xi. \tag{1.75}$$

We will consider the associated reproducing kernel $H_{\mathbb{K}}[0,T]$, on $[0,T]$, where

$$\mathbb{K}(x_1,x_2) \equiv \int_0^T K(x_2,\xi)K(x_1,\xi)d\xi \quad (x_1,x_2 \in [0,T]). \tag{1.76}$$

As for the function given by (1.76), we have the following result:

Theorem 1.10. *The reproducing kernel Hilbert space $H_{\mathbb{K}}[0,T]$ is composed of all functions F such that*

$$F, F' \in \mathrm{AC}[0,T], \tag{1.77}$$

$$F'' \in L^2[0,T], \tag{1.78}$$

$$F(T) = F'(0) = 0, \tag{1.79}$$

and the inner product is given by

$$\langle F_1, F_2 \rangle_{H_{\mathbb{K}}[0,T]} = \int_0^T F_1''(x)F_2''(x)dx \quad (F_1, F_2 \in H_{\mathbb{K}}[0,T]). \tag{1.80}$$

Proof. Denote by $\mathscr{H}$ as usual the Hilbert space of functions F satisfying (1.77), (1.78) and (1.79), and equip $\mathscr{H}$ with an inner product satisfying (1.80). Then we have

$$\begin{aligned}
\frac{d\mathbb{K}_{x_2}}{dx_1}(x_1) &= \int_0^T K(x_2,\xi)\frac{d}{dx_1}K(x_1,\xi)\, d\xi \\
&= -\int_0^T \chi_{(\xi,T)}(x_1)K(x_2,\xi)d\xi \\
&= \int_{x_1}^0 K(x_2,\xi)d\xi,
\end{aligned}$$

and hence, from (1.72),

$$\frac{d^2\mathbb{K}_{x_2}}{dx_1^2}(x_1) = -K(x_1,x_2) = -\min(T-x_1, T-x_2),$$

since $\mathbb{K}(\cdot, x_2)$ is a piecewise polynomial. Consequently, for $F \in \mathscr{H}$,

$$\begin{aligned}\langle F, \mathbb{K}_{x_2} \rangle_{\mathscr{H}} &= \int_0^T \min(T - x_1, T - x_2) F''(x_1)\, dx_1 \\ &= -\int_0^{x_2} (T - x_2) F''(x_1)\, dx_1 - \int_{x_2}^T (T - x_1) F''(x_1)\, dx_1.\end{aligned}$$

Since $F(T) = F'(0) = 0$, we have

$$\langle F, \mathbb{K}_{x_2} \rangle_{\mathscr{H}} = -(T - x_2) F'(x_2) - \left[(T - x_1) F'(x_1)\right]_{x_2}^T - \int_{x_2}^T F'(x_1)\, dx_1 = F(x_2),$$

proving the reproducing property.

1.3.4 Higher Regularity RKHS

The Sobolev space having higher regularity has the reproducing kernel.

Definition 1.4. The set $W^{2,2}(\mathbb{R})$ is defined to be *the completion* of $C_c^\infty(\mathbb{R})$ with respect to the norm

$$\|f\|_{W^{2,2}(\mathbb{R})} \equiv \sqrt{\|f''\|_{L^2(\mathbb{R})}^2 + 2\|f'\|_{L^2(\mathbb{R})}^2 + \|f\|_{L^2(\mathbb{R})}^2}. \tag{1.81}$$

We have an analogy to Lemma 1.1 for $W^{2,2}(\mathbb{R})$ to define f'' for $f \in W^{2,2}(\mathbb{R})$.

The second-order counterpart of Theorem 1.7 is applicable as well.

Theorem 1.11 (RKHS of second-order). *We can realize the second-order Sobolev space $W^{2,2}(\mathbb{R})$, equipped with the norm* (1.81)*, by the reproducing kernel given by*

$$G(s,t) \equiv \frac{1}{4} e^{-|s-t|}(1 + |s - t|) \quad (s, t \in \mathbb{R}).$$

As before, the reproducing property automatically shows that G is positive definite, whose proof is omitted.

Proof. By Lemma 1.1, we have $W^{2,2}(\mathbb{R}) \subset \mathscr{F}(\mathbb{R})$. By Lemma 1.2, we have

$$\begin{aligned}G_t(s) &= \frac{t - s + 1}{4} \exp(s - t) \chi_{(-\infty,t)}(s) + \frac{s - t + 1}{4} \exp(t - s) \chi_{[t,\infty)}(s) \\ \frac{dG_t}{ds}(s) &= \frac{t - s}{4} \exp(s - t) \chi_{(-\infty,t)}(s) - \frac{s - t}{4} \exp(t - s) \chi_{[t,\infty)}(s) \\ \frac{d^2 G_t}{ds^2}(s) &= \frac{t - s - 1}{4} \exp(s - t) \chi_{(-\infty,t)}(s) + \frac{s - t - 1}{4} \exp(t - s) \chi_{[t,\infty)}(s).\end{aligned} \tag{1.82}$$

This proves (1.1).

Let $f \in W^{2,2}(\mathbb{R})$. From (1.82) we calculate

$$\begin{aligned}
&\int_{-\infty}^{\infty} \left(f''(s)\frac{d^2G_t}{ds^2}(s) + 2f'(s)\frac{dG_t}{ds}(s) + f(s)G(s,t) \right) ds \\
&= \frac{1}{2}\int_t^{\infty} (-f''(s) + f(s)) \exp(t-s)\, ds + \frac{1}{2}\int_{-\infty}^{t} (-f''(s) + f(s)) \exp(s-t)\, ds \\
&\quad + \int_{-\infty}^{\infty} f(s)\frac{d^2G_t}{ds^2}(s) + 2f'(s)\frac{dG_t}{ds}(s) + f''(s)G(s,t)\, ds \\
&= \frac{1}{2}\int_t^{\infty} (-f''(s) + f'(s) - f'(s) + f(s)) \exp(t-s)\, ds \\
&\quad + \frac{1}{2}\int_{-\infty}^{t} (-f''(s) - f'(s) + f'(s) + f(s)) \exp(s-t)\, ds \\
&= \frac{1}{2}(f(t) + f'(t)) + \frac{1}{2}(f(t) - f'(t)) = f(t).
\end{aligned}$$

The proof is therefore complete.

1.3.5 Higher-Dimensional Case

In $\mathbb{R}$, we have seen in Lemmas 1.1 and 1.3 that we have full control of the value $f(x)$ when we are given Sobolev functions of order 1. However, if we are in $\mathbb{R}^n$, this is not the case: We no longer have any control of $f(x)$, which is inconvenient in considering reproducing kernel Hilbert spaces. What makes it possible to have control is to consider higher order Sobolev spaces. Let $m > \frac{n}{2}$ be an integer. Let $\mathbb{Z}_+^n$ be the set of all multiindexes. The *Sobolev space* $W^{m,2}(\mathbb{R}^n)$ *of order m* is given by the norm

$$\|F\|_{W^{m,2}(\mathbb{R}^n)} = \sqrt{\sum_{\nu=0}^{m} \sum_{\alpha \in \mathbb{Z}_+^n,\, |\alpha| \le \nu} \frac{\nu!}{\alpha!}\binom{m}{\nu} \int_{\mathbb{R}^n} \left| \frac{\partial^\nu F(x)}{\partial x^\nu} \right|^2 dx}. \tag{1.83}$$

A higher-dimensional higher order Sobolev-spaces variation of Theorem 1.5 is as follows (see also Definition 1.3):

Theorem 1.12 (RKHS of Sobolev type of order m**).** *Let* $m > \frac{n}{2}$*. Then we can realize the Sobolev space* $W^{m,2}(\mathbb{R}^n)$ *using the reproducing kernel K by*

$$K(x, y) \equiv \frac{1}{(2\pi)^n} \int_{\mathbb{R}} \frac{\exp(i(x-y)\cdot\xi)}{(1 + |\xi|^2)^m}\, d\xi \quad (x, y \in \mathbb{R}). \tag{1.84}$$

Proof. Observe that

$$\|F\|_{W^{m,2}(\mathbb{R}^n)} = \sqrt{\sum_{\nu=0}^{m} \sum_{\alpha \in \mathbb{Z}_+^n, |\alpha| \le \nu} \frac{\nu!}{\alpha!}\binom{m}{\nu} \int_{\mathbb{R}^n} \left| \mathscr{F}\left[\frac{\partial^\nu F(\cdot)}{\partial x^\nu}\right](\xi)\right|^2 d\xi}$$

$$= \sqrt{\sum_{\nu=0}^{m} \sum_{\alpha \in \mathbb{Z}_+^n, |\alpha| \le \nu} \frac{\nu!}{\alpha!}\binom{m}{\nu} \int_{\mathbb{R}^n} |\xi^\nu \mathscr{F}F(\xi)|^2 \, d\xi}$$

$$= \sqrt{\int_{\mathbb{R}^n} (1+|\xi|^2)^m \, |\mathscr{F}F(\xi)|^2 \, d\xi}$$

by binomial expansion. Since $\mathscr{F}K_x(\xi) = \dfrac{\exp(-ix\cdot\xi)}{(2\pi)^{\frac{n}{2}}(1+|\xi|^2)^m}$ for $\xi \in \mathbb{R}^n$, we see that $K_x \in W^{m,2}(\mathbb{R}^n)$.

By the Fourier inverse transform, we have

$$\begin{aligned}\langle F, K_x\rangle_{W^{m,2}(\mathbb{R}^n)} &= \int_{\mathbb{R}} (1+|\xi|^2)^m \mathscr{F}F(\xi) \overline{\left(\frac{\exp(-ix\cdot\xi)}{(2\pi)^{\frac{n}{2}}(1+|\xi|^2)^m}\right)} d\xi \\ &= \mathscr{F}^{-1}\mathscr{F}F(x) \\ &= F(x).\end{aligned}$$

This is the desired result.

Remark 1.3. We have the following analogy to Lemma 1.1; if $m > \frac{n}{2}$, then $W^{m,2}(\mathbb{R}^n)$ is embedded into $\mathrm{BC}(\mathbb{R}^n)$. Unlike Lemma 1.1, we use (1.29).

A generalization of Theorems 1.11 and 1.12, whose proof is slightly simpler, is as follows: Let $s > 0$. First we define

$$H^s(\mathbb{R}^n) \equiv \left\{ f \in L^2(\mathbb{R}^n) \; : \; \|f\|_{H^s(\mathbb{R}^n)} \equiv \left(\int_{\mathbb{R}^n} (1+|\xi|^2)^s |\mathscr{F}f(\xi)|^2 \, d\xi \right)^{\frac{1}{2}} \right\}.$$

The following theorem can be proved analogously to Theorems 1.11 and 1.12. The proof is left for interested readers.

Theorem 1.13 (Reproducing kernel Hilbert spaces of fractional order). *We can realize the reproducing kernel Hilbert space $H^s(\mathbb{R}^n)$ by the reproducing kernel K by*

$$K(x,y) \equiv \frac{1}{(2\pi)^n} \int_{\mathbb{R}^n} (1+|\xi|^2)^{-s} \exp(i(x-y)\cdot\xi) \, d\xi \quad (x, y \in \mathbb{R}^n), \tag{1.85}$$

whenever $s > \dfrac{n}{2}$ is a real number.

These reproducing kernels are introduced with other kernels in [388, Section 2.4], in particular, see [122, p.322 Ex. 7] for the relation between reproducing kernels and the Green functions. For further extensions, see G. Kimeldorf and G. Wahba [256, Section 2]: They solved explicitly the generalized Birkhoff interpolation and smoothing problems involving Tchebychev spline functions. In [256, Section 2], the authors considered the case where

$$Ly = \frac{d}{dt}\frac{1}{a_1}\left(\frac{d}{dt}\frac{1}{a_2}\cdots\left(\frac{d}{dt}\frac{y}{a_m}\right)\cdots\right).$$

For the relationship among the Bessel potentials and the fundamental solutions for the operators $(1-\triangle)^n$, see [30, Section 2].

See a series of papers for the Green's functions, Sobolev spaces, inequalities and ordinary differential equations; [245–247, 249, 251, 313, 350–352, 455–457, 478–480]. For example, Theorem 1.12 is generalized in [245, Theorem 2.1]. K. Watanabe, Y. Kametaka, A. Nagai, K. Takemura and H. Yamagishi obtained the best constant in [478, Theorem 1] by using the reproducing kernel Hilbert space. The reproducing kernel Hilbert space used there can be regarded as an extension of Theorem 1.3. [351, Theorem 1] considers a similar problem; look for the best constant C satisfying

$$\|u\|_{L^\infty(-1,1)} \le C\|u'''\|_{L^p(-1,1)}$$

for all $u \in C^1[-1,1]$ with $u''(\pm 1) = u(\pm 1) = 0$. In [479, Theorem 1.1] the authors considered the best constant for the embedding of the linear equation with constant coefficients

$$P\left(\frac{d}{dt}\right)u = f$$

subject to

$$u^{(i)}(0) = u^{(i)}(0)$$

for $i = 1, 2, \ldots, \deg(P) - 1$. In [248, p. 285], we can find many concrete examples of Green functions. The best constants related to the Schrödinger equations are obtained in [250, Theorems 1 and 2], where again the Green kernel constructed in [250, Section 3] plays a key role.

See [459] for some examples similar to the ones in this section together with their applications to statistics.

1.4 RKHS and Complex Analysis on $\mathbb{C}$

In complex analysis for one variable and several variables, the Bergman kernel and the Szegö kernel were developed in a broad and deep way in some general domains by S. Bergman, M. M. Schiffer, P. R. Garabedian, Z. Nehari, D. A. Hejhal, J. D. Fay and many others [46, 146, 202, 302, 322, 425]. For the interrelationship among the solutions, boundary values and reproducing kernels in partial differential equations, see [47, 171, Chapter 1] and [186, Section 3]. We refer to [40, Chapter V] for reproducing kernels of higher order operators and polyanalytic functions. We refer to [41, Chapter VI] and references therein for higher order systems and complex first order equations. The properties of the Hardy H_2-kernel and related kernels were stated in [388].

Many corresponding results in several complex variables were examined in a broad and deep way. See [18, 167, 168, 223] for example.

Furthermore, in Clifford analysis, Bergman kernels are considered; see [109, 110] for example.

Here, we will state elementary and fundamental results having general interest.

1.4.1 RKHS on $\Delta(1)$

Now we work on $\Delta(1) \equiv \{z \in \mathbb{C} : |z| < 1\}$, which has the measure $dxdy$. We require that readers be familiar to some fundamental complex analysis, see [17]. By definition, for an integrable function $F : \Delta(1) \to \mathbb{C}$,

$$\iint_{\Delta(1)} F(z)\, dx\, dy = \iint_{x^2+y^2\le 1} F(x+iy)\, dx\, dy.$$

If we change the weight in the norms, then the corresponding kernel undergoes a corresponding change as we shall see below.

Theorem 1.14 ([64]). *Let us set*

$$K(z,u) \equiv \frac{1+\overline{u}z}{(1-\overline{u}z)^2} \tag{1.86}$$

for $z, u \in \Delta(1)$. Then K is positive definite, and the corresponding reproducing kernel Hilbert space is

$$H_K(\Delta(1)) = \left\{ f \in \mathcal{O}(\Delta(1)) \,:\, \|f\|_{H_K(\Delta(1))} \equiv \sqrt{\frac{1}{2\pi}\iint_{\Delta(1)} |f(z)|^2 \frac{dx\,dy}{|z|}} < \infty \right\}.$$

Proof. Again positive definiteness of K follows automatically from the reproducing property of K. It is not so hard to see that $K_u \in H_K(\Delta(1))$ for all $u \in \Delta(1)$, so that (1.1) holds.

Let us show that K, by (1.86), has the reproducing property (1.2). To do this, we pick $f \in \mathcal{O}(\Delta(1))$ satisfying

$$\iint_{\Delta(1)} |f(z)|^2 \frac{dx\,dy}{|z|} < \infty. \tag{1.87}$$

By replacing f with $f(r\cdot)$, $r < 1$, we may assume that f is a holomorphic function defined outside of $\overline{\Delta(1)}$. Then

$$\begin{aligned}\langle f, K_u\rangle_{H_K(\Delta(1))} &= \frac{1}{2\pi}\iint_{\Delta(1)} f(z)\cdot\frac{1+u\bar{z}}{(1-u\bar{z})^2}\frac{dx\,dy}{|z|}\\ &= \sum_{j=0}^{\infty}\frac{1}{2\pi}\iint_{\Delta(1)} f(z)(j+1)(1+u\bar{z})(u\bar{z})^j\frac{dx\,dy}{|z|}\\ &= \sum_{j=0}^{\infty}\frac{1}{2\pi}\iint_{\Delta(1)} f(z)(2j+1)(u\bar{z})^j\frac{dx\,dy}{|z|}.\end{aligned}$$

Let us calculate each summand; fix $j \in \mathbb{Z}_+$. We expand f into the Taylor series to obtain

$$\begin{aligned}\frac{1}{2\pi}\iint_{\Delta(1)} f(z)(2j+1)(u\bar{z})^j\frac{dx\,dy}{|z|} &= \sum_{k=0}^{\infty}\frac{1}{2\pi\cdot k!}\iint_{\Delta(1)} f^{(k)}(0)z^k(2j+1)(u\bar{z})^j\frac{dx\,dy}{|z|}\\ &= \frac{1}{2\pi\cdot j!}\iint_{\Delta(1)} f^{(j)}(0)z^j(2j+1)(u\bar{z})^j\frac{dx\,dy}{|z|}\\ &= \frac{u^j}{2\pi\cdot j!}\iint_{\Delta(1)} f^{(j)}(0)|z|^{2j-1}(2j+1)\,dx\,dy\\ &= \frac{1}{j!}f^{(j)}(0)u^j.\end{aligned}$$

Therefore, we see that K enjoys the reproducing property.

These reproducing kernels and the reproducing kernel Hilbert spaces were naturally derived from the structures of the related reproducing kernels.

Here and below, log is defined on $\mathbb{C}\setminus(-\infty,0]$ so that $\log 1 = 0$. We use the coordinate

$$z = x + iy \tag{1.88}$$

with $x, y \in \mathbb{R}$. Let us denote by $\mathscr{H}(\Omega)$ the set of all harmonic functions in a domain $\Omega \subset \mathbb{C}$. We omit the proof of the positive definiteness of the kernel by a reason described before.

Theorem 1.15 ([29, 315]). *Let*

$$K(z,u) \equiv \frac{1}{\pi}\mathrm{Re}\left(\log\frac{1}{1-z\overline{u}}\right)$$

for $z, u \in \Delta(1)$. Then, $K : \Delta(1)\times\Delta(1)\to\mathbb{C}$ is positive definite and

$$H_K(\Delta(1)) = \left\{U \in \mathscr{H}(\Delta(1)) \, : \, \|u\|_{H_K(\Delta(1))} < \infty, \; U(0) = 0\right\}, \tag{1.89}$$

where the norm is

$$\|U\|_{H_K(\Delta(1))} \equiv \sqrt{\iint_{\Delta(1)} \left\{ \left|\frac{\partial U}{\partial x}(x,y)\right|^2 + \left|\frac{\partial U}{\partial y}(x,y)\right|^2 \right\} dx\,dy} \tag{1.90}$$

for $U \in H_K(\Delta(1))$ via the identification (1.88).

Proof. As usual we let $\mathscr{H}$ be the Hilbert space on the right-hand side of (1.89) with the inner product given by (1.90).

Let $u \in \Delta(1)$ be a fixed point. Since holomorphic functions are harmonic and

$$\partial_x K_u(z) = \frac{1}{\pi}\mathrm{Re}\left(\frac{-\overline{u}}{1 - z\overline{u}}\right), \; \partial_y K_u(z) = -\frac{1}{\pi}\mathrm{Im}\left(\frac{\overline{u}}{1 - z\overline{u}}\right), \tag{1.91}$$

we have (1.1), namely, $K_u \in \mathscr{H}$.

Now let us take $U \in \mathscr{H}$ to verify (1.2). We want to calculate

$$\mathrm{I} \equiv \frac{1}{\pi}\iint_{\Delta(1)} \left\{ \partial_x U(z)\overline{\left(\mathrm{Re}\frac{1}{1 - z\overline{u}}\right)} - \partial_y U(z)\overline{\left(\mathrm{Im}\frac{1}{1 - z\overline{u}}\right)}\right\} dx\,dy. \tag{1.92}$$

To do this, we take a harmonic conjugate V of U: V is a unique $\mathbb{R}$-valued function such that

$$f(z) \equiv U(z) + i\,V(z), \; z \in \mathscr{O}(\Delta(1)). \tag{1.93}$$

Inserting the Cauchy Riemann equation $\partial_y U(z) = \partial_x V(z)$ and using the Lebesgue convergence theorem, we obtain

$$\begin{aligned}
\mathrm{I} &= \frac{1}{\pi}\iint_{\Delta(1)} \left\{ \partial_x U(z)\mathrm{Re}\left(\frac{1}{1 - \overline{z}u}\right) - \partial_x V(z)\mathrm{Im}\left(\frac{1}{1 - \overline{z}u}\right)\right\} dx\,dy \\
&= \lim_{r \uparrow 1}\frac{1}{\pi}\iint_{\Delta(r)} \left\{ \partial_x U(z)\mathrm{Re}\left(\frac{1}{1 - \overline{z}u}\right) - \partial_x V(z)\mathrm{Im}\left(\frac{1}{1 - \overline{z}u}\right)\right\} dx\,dy \\
&= \lim_{r \uparrow 1}\frac{1}{\pi}\iint_{\Delta(r)} \mathrm{Re}\left(\frac{f'(z)}{1 - \overline{z}u}\right) dx\,dy.
\end{aligned}$$

Now we consider the Taylor expansion of $\dfrac{1}{1-\alpha}$ with $\alpha \in \Delta(1)$:

$$\mathrm{I} = \lim_{r \uparrow 1}\frac{1}{\pi}\sum_{j=0}^{\infty}\iint_{\Delta(r)} \mathrm{Re}\left(f'(z)\overline{z}^j u^j\right) dx\,dy = \lim_{r \uparrow 1}\frac{1}{\pi}\sum_{j=0}^{\infty}\mathrm{Re}\left(\iint_{\Delta(r)} f'(z)\overline{z}^j u^j \, dx\,dy\right).$$

We expand f into the Taylor series as well. Note that

$$\iint_{\Delta(r)} z^j \overline{z}^k \, dx\, dy = \pi r^{2j} \delta_{jk}, \ j, k \in \mathbb{Z}_+, \tag{1.94}$$

where δ_{jk} denotes the Dirac delta function. Using (1.94), we obtain

$$\mathrm{I} = \lim_{r\uparrow 1} \sum_{j=1}^{\infty} \mathrm{Re}\left(\frac{r^{2j}}{j!} f^{(j)}(0) u^j\right) = \lim_{r\uparrow 1} \mathrm{Re}(f(r^2 u)) = \mathrm{Re}(f(u)) = U(u).$$

This is the desired reproducing property (1.2).

In connection with the Poisson integral, we have the following result:

Theorem 1.16. *Let*

$$K(z, u) \equiv \frac{1}{2\pi} \mathrm{Re}\left(\frac{1 + \overline{u}\, z}{1 - \overline{u}\, z}\right) \tag{1.95}$$

for $z, u \in \Delta(1)$. *Then* K *is positive definite, and the corresponding RKHS is*

$$H_K(\Delta(1)) = \left\{U \in C^\infty(\Delta(1), \mathbb{R}) \ : \ U \text{ is a Poisson integral of } f \in L^2(\partial\Delta(1))\right\}.$$

Here, the norm is given by

$$\|U\|_{H_K(\Delta(1))} \equiv \left(\oint_{\partial\Delta(1)} |f(z)|^2 \, |dz|\right)^{\frac{1}{2}}, \tag{1.96}$$

when U *is a Poisson integral of* $f \in L^2(\partial\Delta(1))$.

Proof. Recall that the Poisson integral is the correspondence;

$$f \in L^2(\partial\Delta(1)) \mapsto u_f \in C^\infty(\Delta(1), \mathbb{R}), \tag{1.97}$$

where

$$u_f(z) \equiv \frac{1}{2\pi} \oint_{\partial\Delta(1)} f(z) \mathrm{Re}\left(\frac{1 + \overline{z}\, u}{1 - \overline{z}\, u}\right) |dz|. \tag{1.98}$$

Let us prove

$$K(u, w) = \frac{1}{(2\pi)^2} \oint_{\partial\Delta(1)} \mathrm{Re}\left(\frac{1 + \overline{z}\, u}{1 - \overline{z}\, u}\right) \mathrm{Re}\left(\frac{1 + \overline{w}\, z}{1 - \overline{w}\, z}\right) |dz| \tag{1.99}$$

to establish that K enjoys the reproducing property from (1.98).

To this end we write down the real part of the integrand above in full;

$$
\begin{aligned}
&R.H.S. \text{ of } (1.99)\\
&= \frac{1}{(2\pi)^2}\oint_{\partial\Delta(1)} \mathrm{Re}\left(\frac{1+\exp(-i\theta)\,u}{1-\exp(-i\theta)\,u}\right)\mathrm{Re}\left(\frac{1+\overline{w}\,\exp(i\theta)}{1-\overline{w}\exp(i\theta)}\right)d\theta\\
&= \frac{1}{(2\pi)^2}\oint_{\partial\Delta(1)} \frac{(1-|u|^2)(1-|w|^2)\,d\theta}{(1-\exp(-i\theta)u)(1-\exp(i\theta)\overline{u})(1-\exp(-i\theta)w)(1-\exp(i\theta)\overline{w})}\\
&= \frac{1}{(2\pi)^2}\oint_{\partial\Delta(1)} \frac{(1-|u|^2)(1-|w|^2)z}{(z-u)(z-w)(1-z\overline{u})(1-z\overline{w})}\frac{dz}{i}.
\end{aligned}
$$

Let us assume that $u \neq w$ for the time being. Then by the complex line integral, we have

$$
\begin{aligned}
R.H.S. \text{ of } (1.99) &= \frac{1}{2\pi}\left(\frac{(1-|w|^2)u}{(u-w)(1-u\overline{w})} - \frac{(1-|u|^2)w}{(u-w)(1-\overline{u}w)}\right)\\
&= \frac{(1-|w|^2)u(1-\overline{u}w)-(1-|u|^2)w(1-u\overline{w})}{2\pi(u-w)(1-u\overline{w})(1-w\overline{u})}\\
&= \frac{1-|w|^2|u|^2}{2\pi|1-\overline{w}u|^2} = K(u,w).
\end{aligned}
$$

Thus,

$$
R.H.S. \text{ of } (1.99) = K(u,w). \tag{1.100}
$$

Continuity argument allows us to prove (1.100) even when $u = w$. In view of (1.100) we have the desired result.

1.4.2 RKHS on $\mathbb{C}$

Let $t > 0$ be fixed. Let us set

$$
K = K(z,u;t) \equiv \frac{1}{2\sqrt{2\pi t}}\exp\left(-\frac{(z-\overline{u})^2}{8t}\right),\ z,u \in \mathbb{C}. \tag{1.101}
$$

We shall determine the RKHS admitting the function (1.101).

Theorem 1.17. *Let $t > 0$ be fixed. Writing $z = x + i\,y$ canonically, we have*

$$
H_K(\mathbb{C}) = \left\{ f:\mathbb{C}\to\mathbb{C} : \|f\|_{H_K(\mathbb{C})} \equiv \sqrt{\frac{1}{\sqrt{2\pi t}}\iint_{\mathbb{C}} |f(z)|^2 \exp\left(-\frac{y^2}{2t}\right)dx\,dy} < \infty \right\}.
$$

Proof. As usual let us set

$$\mathscr{H} \equiv \left\{ f : \mathbb{C} \to \mathbb{C} \,:\, \|f\|_{\mathscr{H}} \equiv \sqrt{\frac{1}{\sqrt{2\pi t}} \iint_{\mathbb{C}} |f(z)|^2 \exp\left(-\frac{y^2}{2t}\right) dx\,dy} < \infty \right\}.$$

Then it is easy to see

$$K(\cdot, u; t) = \frac{1}{2\sqrt{2\pi t}} \exp\left(-\frac{(\cdot - \overline{u})^2}{8t}\right) \in \mathscr{H}, \tag{1.102}$$

because we can write, with a polynomial $P_u(x, y)$ of degree 1,

$$\begin{aligned} |K(z, u; t)|^2 \exp\left(-\frac{y^2}{2t}\right) &= \frac{1}{8\pi t} \exp\left(-\frac{(x - \mathrm{Re}(u))^2 - (y + \mathrm{Im}(u))^2}{4t} - \frac{y^2}{2t}\right) \\ &= \frac{1}{8\pi t} \exp\left(-\frac{x^2 + y^2}{4t} + P_u(x, y)\right). \end{aligned}$$

Let us check that

$$\langle K(\cdot, u; t), K(\cdot, z; t)\rangle_{\mathscr{H}} = K(z, u; t). \tag{1.103}$$

From the definition of the norm we have

$$\langle K(\cdot, u; t), K(\cdot, z; t)\rangle_{\mathscr{H}} = \frac{1}{\sqrt{2\pi t}} \iint_{\mathbb{R}^2} K(x + iy, u; t)\overline{K(x + iy, z; t)} \exp\left(-\frac{y^2}{2t}\right) dx\,dy.$$

Let us use w to denote a complex variable. That is, we write $w = x + iy$. Let us write completely out the integrand:

$$\begin{aligned} &K(w, u; t)\overline{K(w, z; t)} \exp\left(-\frac{y^2}{2t}\right) \\ &= \frac{1}{8\pi t} \exp\left(\frac{(w - \overline{w})^2}{8t} - \frac{(w - \overline{u})^2 + (\overline{w} - z)^2}{8t}\right) \\ &= \frac{1}{8\pi t} \exp\left(\frac{(w - \overline{w})^2}{8t} - \frac{1}{4t}\left(\frac{w + \overline{w} - \overline{u} - z}{2}\right)^2 - \frac{1}{4t}\left(\frac{w - \overline{w} - \overline{u} + z}{2}\right)^2\right) \\ &= \frac{1}{8\pi t} \exp\left(\frac{(w - \overline{w} + z - \overline{u})^2}{16t} - \frac{(z - \overline{u})^2}{8t} - \frac{1}{16t}(w + \overline{w} - \overline{u} - z)^2\right). \end{aligned} \tag{1.104}$$

If we integrate (1.104) over $w \in \mathbf{C}$, we obtain

$$\iint_{\mathbb{C}} K(w, u; t)\overline{K(w, z; t)} \exp\left(-\frac{y^2}{2t}\right) dw = \frac{1}{4\sqrt{t}} \exp\left(-\frac{(z - \overline{u})^2}{8t}\right), \tag{1.105}$$

which yields (1.103).

This reproducing kernel Hilbert space was naturally and simply introduced from the property of the solution of the heat equation. For a short history and related topics, see [379, p. 5, Theorem 3.1] and [388, pp. 128–142].

1.4.3 RKHS on $\mathbb{C} \setminus \overline{\Delta(1)}$

Even for the exterior domain $\mathbb{C} \setminus \overline{\Delta(1)}$ we have a concrete result. As before the branch of $\log : \mathbb{C} \setminus (-\infty, 0]$ of (1.106) is taken so that $\log 1 = 0$.

Theorem 1.18. *Let*

$$K(z,u) \equiv -\frac{1}{\pi} \log\left(1 - \frac{1}{\overline{u}z}\right) \quad (u, z \in \mathbb{C} \setminus \overline{\Delta(1)}). \tag{1.106}$$

Then we have

$$H_K(\mathbb{C} \setminus \overline{\Delta(1)}) = \left\{ f \in \mathcal{O}(\mathbb{C} \setminus \overline{\Delta(1)}) \, : \, \|f\|_{H_K(\mathbb{C} \setminus \overline{\Delta(1)})} < \infty \right\}, \tag{1.107}$$

where the norm is

$$\|f\|_{H_K(\mathbb{C} \setminus \overline{\Delta(1)})} \equiv \sqrt{\iint_{\mathbb{C} \setminus \overline{\Delta(1)}} |f'(z)|^2 \, dx \, dy}$$

for $f \in H_K(\mathbb{C} \setminus \overline{\Delta(1)})$.

To prove Theorem 1.18, we collect some fundamental properties of the functions in $\mathcal{O}(\mathbb{C} \setminus \overline{\Delta(1)})$.

Lemma 1.4. *There is no function* $f \in \mathcal{O}(\mathbb{C} \setminus \overline{\Delta(1)})$ *such that*

$$f'(z) = \frac{1}{z} \tag{1.108}$$

for all $z \in \mathbb{C} \setminus \overline{\Delta(1)}$.

Proof. Assume that such a function f exists. Then integrate (1.108) on $\partial\Delta(2)$. Then we have

$$0 = \oint_{\partial\Delta(2)} f'(z) \, dz = \oint_{\partial\Delta(2)} \frac{dz}{z} = 2\pi i,$$

which is a contradiction.

The next lemma concerns the analytic extension of functions.

Lemma 1.5. *Assume that* $f \in \mathscr{O}(\mathbb{C} \setminus \overline{\Delta(1)})$ *satisfies*

$$\iint_{\mathbb{C}\setminus\overline{\Delta(1)}} |f'(z)|^2 \, dx \, dy < \infty. \tag{1.109}$$

Define $g(w) \equiv \frac{1}{w} f'\left(\frac{1}{w}\right)$ *for* $w \in \Delta(1) \setminus \{0\}$. *Then:*

1. *g satisfies*

$$g(w) \to 0 \tag{1.110}$$

as $w \to 0$,

2. *The limit*

$$f(\infty) \equiv \lim_{z\to\infty} f(z) \tag{1.111}$$

exists.

Proof. Let us set

$$w \equiv \frac{1}{z}, \; w \equiv \alpha + i\beta, \tag{1.112}$$

where (α, β) are real coordinates. Then $x = \frac{\alpha}{\alpha^2+\beta^2}$, $y = \frac{-\beta}{\alpha^2+\beta^2}$. Thus, we obtain

$$dx \, dy = \frac{d\alpha \, d\beta}{\alpha^2 + \beta^2}. \tag{1.113}$$

As a consequence, the assumption (1.109) can be rephrased as

$$\iint_{\Delta(1)\setminus\{0\}} |g(\alpha + i\beta)|^2 \, d\alpha \, d\beta < \infty. \tag{1.114}$$

This implies that the function $g \in \mathscr{O}(\Delta(1) \setminus \{0\})$ can be extended to $\mathscr{O}(\Delta(1))$, as is easily seen by the Laurent expansion. Let us expand $g(w)$ into the Taylor series: $g(w) = \sum_{j=0}^{\infty} a_j w^j$. Then under the coordinate (1.112) we have

$$\iint_{\Delta(1)} |g(w)|^2 \, d\alpha \, d\beta = \lim_{r\uparrow 1} \iint_{\Delta(r)} |g(w)|^2 \, d\alpha \, d\beta$$

by the monotone convergence theorem. Hence it follows that

$$\iint_{\Delta(1)} |g(w)|^2 \, d\alpha \, d\beta = \lim_{r\uparrow 1} \iint_{\Delta(r)} |a_j|^2 |w|^{2j} \, d\alpha \, d\beta = 2\pi \sum_{j=0}^{\infty} \frac{|a_j|^2}{2j+2} < \infty.$$

As a result, we have

$$\limsup_{j\to\infty} \sqrt[j]{|a_j|} \le 1. \tag{1.115}$$

A direct consequence of (1.115) is that the mapping

$$z \in \mathbb{C} \setminus \Delta(1) \mapsto \sum_{j=1}^{\infty} a_j\, z^{-j-1} \in \mathbb{C} \tag{1.116}$$

belongs to $\mathscr{O}(\mathbb{C} \setminus \Delta(1))$.

Let us set

$$G(z) \equiv f(z) + \sum_{j=1}^{\infty} \frac{1}{j} a_j\, z^{-j} \quad (z \in \mathbb{C} \setminus \Delta(1)). \tag{1.117}$$

Then $G \in \mathscr{O}(\mathbb{C} \setminus \Delta(1))$ and

$$G'(z) = \sum_{j=0}^{\infty} a_j\, z^{-j-1} - \sum_{j=1}^{\infty} a_j\, z^{-j-1} = \frac{a_0}{z}. \tag{1.118}$$

By Lemma 1.4 we see that $a_0 = 0$ and that (1.119) follows. As a consequence, we obtain

$$f'(z) = \sum_{j=1}^{\infty} a_j z^{-j-1} \text{ and } f(z) = C_0 - \sum_{j=1}^{\infty} \frac{a_j}{j} z^{-j} \tag{1.119}$$

for some $C_0 \in \mathbb{C}$. In view of (1.115) and (1.119) we conclude that $f(\infty)$ exists.

Now we prove Theorem 1.18.

Proof (of Theorem 1.18*).* We keep to the notation in Lemma 1.5. The proof being routine, we content ourselves with proving

$$f(z) = \iint_{\mathbb{C}\setminus\overline{\Delta(1)}} f'(z) \overline{\partial_z \left[\frac{-1}{\pi} \log\left(1 - \frac{1}{\overline{u}z}\right)\right]}\, dx\, dy, \tag{1.120}$$

for all f on the right-hand side of (1.107). We will calculate the right-hand side. First, we change variables: if we set $w = z^{-1}$, then we have

$$\iint_{\mathbb{C}\setminus\overline{\Delta(1)}} f'(z) \overline{\partial_z \left[\frac{-1}{\pi} \log\left(1 - \frac{1}{\overline{u}z}\right)\right]}\, dx\, dy = \lim_{r\uparrow 1} \iint_{\Delta(r)\setminus\{0\}} f'\left(\frac{1}{w}\right) \frac{d\alpha\, d\beta}{\pi w \overline{w} \left(1 - \frac{\overline{w}}{u}\right)}.$$

Let us set

$$g(w) \equiv \sum_{j=1}^{\infty} a_j\, w^j \equiv \frac{1}{w} f'\left(\frac{1}{w}\right) \tag{1.121}$$

with (1.111) in mind. Then by the boundedness of g and (1.119) we have

$$\begin{aligned}
&\lim_{r\uparrow 1} \iint_{\Delta(r)\setminus\{0\}} \frac{1}{\pi |w|^2} f'\left(\frac{1}{w}\right)\cdot\left(1-\frac{\overline{w}}{u}\right)^{-1} d\alpha\, d\beta \\
&= \lim_{r\uparrow 1} \frac{1}{\pi} \sum_{k=1}^{\infty} \iint_{\Delta(r)\setminus\{0\}} f'\left(\frac{1}{w}\right) \frac{\overline{w}^{k-1}}{w\, u^k}\, d\alpha\, d\beta \\
&= \lim_{r\uparrow 1} \frac{1}{\pi} \sum_{j,k=1}^{\infty} \iint_{\Delta(r)\setminus\{0\}} a_j\, w^j \frac{\overline{w}^{k-1}}{u^k}\, d\alpha\, d\beta \\
&= \sum_{j=1}^{\infty} \frac{a_j}{j\, u^{j+1}} = f(u).
\end{aligned}$$

This is the desired reproducing property (1.2).

1.4.4 RKHS on a Small Neighborhood of the Origin

Below we denote by $\mathcal{O}(\{0\})$ the set of all analytic functions defined on a neighborhood of the origin. Although R is given by (1.122) below, we are not really interested in the precise value of R.

Proposition 1.2. *Let* $\{C_j\}_{j=0}^{\infty}$ *be a positive sequence such that* $\limsup\limits_{j\to\infty} \sqrt[j]{C_j} < \infty$. *Set*

$$R \equiv \left(\limsup_{j\to\infty} \sqrt[j]{C_j}\right)^{-1} > 0. \tag{1.122}$$

Define a kernel K *by*

$$K(z,u) \equiv \sum_{j=0}^{\infty} C_j\, z^j\, \overline{u}^j \quad (z, u \in \Delta(\sqrt{R})). \tag{1.123}$$

Then we have

$$H_K(\Delta(\sqrt{R})) = \left\{ f \in \mathscr{O}(\Delta(\sqrt{R})) \, : \, \sqrt{\sum_{j=0}^{\infty} \frac{|f^{(j)}(0)|^2}{(j!)^2 C_j}} < \infty \right\} \tag{1.124}$$

and the norm is

$$\|f\|_{H_K(\Delta(\sqrt{R}))} = \sqrt{\sum_{j=0}^{\infty} \frac{|f^{(j)}(0)|^2}{(j!)^2 C_j}}$$

for $f \in H_K(\Delta(\sqrt{R}))$.

Proof. Let us denote by H the Hilbert space equipped with the norm $\|\cdot\|_H$ given by

$$\|f\|_H = \sqrt{\sum_{j=0}^{\infty} \frac{|f^{(j)}(0)|^2}{(j!)^2 C_j}}.$$

We trivially have $K_u \in H$ for all $u \in \Delta(\sqrt{R})$, so that (1.1) is true.

The reproducing property (1.2) follows immediately since a direct calculation shows

$$\langle f, K_u \rangle_H = \sum_{j=0}^{\infty} \frac{f^{(j)}(0)\overline{K_u^{(j)}(0)}}{(j!)^2 C_j} = \sum_{j=0}^{\infty} \frac{f^{(j)}(0)u^j}{j!} = f(u)$$

for $f \in H$ and $u \in \Delta(\sqrt{R})$.

1.4.5 Bergman Kernel on $\Delta(1)$

Here we introduce the Bergman kernel on the unit disk.

Theorem 1.19. *Let*

$$K(z, w) \equiv \frac{1}{(1 - z\overline{w})^{2+\alpha}} \tag{1.125}$$

for $z, w \in \Delta(1)$, *where* $0 \le \alpha < \infty$. *Then we have*

$$H_K(\Delta(1)) = \left\{ f \in \mathscr{O}(\Delta(1)) \, : \, \sqrt{\iint_{\Delta(1)} |f(w)|^2 (1 - |w|^2)^{\alpha} \, du \, dv} < \infty \right\},$$

and the norm is

$$\|f\|_{H_K(\Delta(1))} = \sqrt{\frac{\alpha+1}{\pi}\iint_{\Delta(1)} |f(w)|^2(1-|w|^2)^\alpha \, du\, dv} \tag{1.126}$$

for $f \in H_K(\Delta(1))$.

Proof. Denote by $\mathscr{H}$ the Hilbert space on the right-hand side and equip $\mathscr{H}$ with the norm given by the right-hand of (1.126). It is easy to see that $K_z \in \mathscr{H}$. Then, via the real coordinate $w = u + iv$, $u, v \in \mathbb{R}$, we have

$$\begin{aligned}
\langle f, K_z\rangle_{\mathscr{H}} &= \frac{\alpha+1}{\pi}\iint_{\Delta(1)} \frac{f(w)(1-|w|^2)^\alpha}{(1-z\overline{w})^{2+\alpha}}\, du\, dv \\
&= \sum_{j=0}^{\infty}\iint_{\Delta(1)} f(w)(1-|w|^2)^\alpha \frac{1}{\pi j!} z^j\overline{w^j}\prod_{l=1}^{j+1}(l+\alpha)\, du\, dv \\
&= \sum_{j=0}^{\infty}\iint_{\Delta(1)} \frac{f^{(j)}(0)(1-|w|^2)^\alpha}{\pi (j!)^2}\prod_{l=1}^{j+1}(l+\alpha) z^j |w|^{2j}\, du\, dv \\
&= \sum_{j=0}^{\infty} 2\left(\int_0^1 r^{2j+1}(1-r^2)^\alpha\, dr\right)\frac{1}{(j!)^2}\prod_{l=1}^{j+1}(l+\alpha) f^{(j)}(0) z^j.
\end{aligned}$$

A change of variables yields

$$2\int_0^1 r^{2j+1}(1-r^2)^\alpha\, dr = \frac{\Gamma(j+1)\Gamma(\alpha+1)}{\Gamma(j+\alpha+2)} = \frac{j!}{(j+\alpha+1)\cdots(\alpha+1)}. \tag{1.127}$$

Inserting (1.127) into the above formula, we see that K has the reproducing property

$$f(z) = \sum_{j=0}^{\infty}\frac{f^{(j)}(0)}{j!} z^j = \langle f, K_z\rangle_{\mathscr{H}}. \tag{1.128}$$

For more detailed properties in connection with a complete orthonormal system, see, for example, [201].

The Bergman spaces were examined also in operator theory in depth and in depth with the Berezin transform, see [26, 33, 448] for example. In [26], the authors dealt with the weighted Bergman space; the space in Theorem 1.19 is a typical example in [26]. The authors considered the relation between weighted Bergman spaces and the Hankel operator in [26]. We refer to [43] and [44, Section 4] for the original works, where we can find weighted Bergman spaces dealt with in Theorem 1.19.

1.4.6 Bergman Selberg Reproducing Kernel

Let $E \equiv \{z \in \mathbb{C} : \mathrm{Re}(z) > 0\}$ be the right-half space. The following is the explicit representation of Bergman Selberg reproducing kernels:

Theorem 1.20. *Define*

$$K_q(z,u) \equiv \frac{\Gamma(2q)}{(z+\overline{u})^{2q}} \quad (z,u \in E) \tag{1.129}$$

for a parameter $q > \dfrac{1}{2}$. *Then*

$$H_{K_q(E)} = \left\{ f \in \mathscr{O}(E) : \|f\|_{H_{K_q}(E)} < \infty \right\}, \tag{1.130}$$

and the norm is

$$\|f\|_{H_{K_q}(E)} = \sqrt{\frac{1}{\pi\Gamma(2q-1)} \iint_E |f(z)|^2 [2\mathrm{Re}(z)]^{2q-2}\,dx\,dy} \tag{1.131}$$

for $f \in H_{K_q}(E)$.

Proof. Let $u \in E$. Clearly, the function $(K_q)_u$ is holomorphic. If we substitute the definition (1.129) and change variables twice, we have

$$\begin{aligned}
\iint_E |(K_q)_u(z)|^2 [2\mathrm{Re}(z)]^{2q-2}\,dx\,dy &= \iint_{x>0,y\in\mathbb{R}} \frac{\Gamma(2q)^2}{|z+\overline{u}|^{4q}} [2\mathrm{Re}(z)]^{2q-2}\,dx\,dy \\
&= \iint_{x>0,y\in\mathbb{R}} \frac{\Gamma(2q)^2}{|z+\mathrm{Re}(u)|^{4q}} [2\mathrm{Re}(z)]^{2q-2}\,dx\,dy \\
&= \iint_{x>0,y\in\mathbb{R}} \frac{\Gamma(2q)^2 (2x)^{2q-2}}{|(x+\mathrm{Re}(u))^2+y^2|^{2q}}\,dx\,dy \\
&= \int_{-\infty}^{\infty} \frac{dy}{(1+y^2)^{2q}} \\
&\quad \times \int_0^\infty \frac{\Gamma(2q)^2 (2x)^{2q-2}}{(x+\mathrm{Re}(u))^{4q-1}}\,dx < \infty,
\end{aligned}$$

where for the last inequality, we used $q > 1/2$.

Let $R > 0$. Fix $z \in \mathbb{C}$. Given a function f defined on E, we define $F(Z) \equiv f(RZ+R)$. Then, writing $W = U + iV$ with $U, V \in \mathbb{R}$ and using Theorem 1.19, we have

$$
\begin{aligned}
f(z) &= F\left(\frac{z-R}{R}\right)\\
&= \frac{2q-1}{\pi}\iint_{\Delta(1)} F(W)(1-|W|^2)^{2q-2}\left(1-\left(\frac{z-R}{R}\right)\overline{W}\right)^{-2q} dU\,dV\\
&= \frac{2q-1}{\pi}\iint_{\Delta(1)} f(R\,W+R)(1-|W|^2)^{2q-2}\left(1-\left(\frac{z-R}{R}\right)\overline{W}\right)^{-2q} dU\,dV.
\end{aligned}
$$

If we change variables $w = u + iv = RW + R$, then we obtain

$$
\begin{aligned}
f(z) &= \frac{2q-1}{\pi}\iint_{|w-R|<R} \frac{\left(1-\left|\frac{w-R}{R}\right|^2\right)^{2q-2} f(w)}{R^2\left(1-\left(\frac{z-R}{R}\right)\overline{\left(\frac{w-R}{R}\right)}\right)^{2q}} du\,dv\\
&= \frac{2q-1}{\pi}\iint_{|w-R|<R} f(w)\left(\frac{2\mathrm{Re}(w)}{R}-\frac{|w|^2}{R^2}\right)^{2q-2}\left(\frac{z+\overline{w}}{R}-\frac{z\overline{w}}{R^2}\right)^{-2q}\frac{du\,dv}{R^2}.
\end{aligned}
$$

Let $q \geq 1$. Note that

$$
\left(\frac{2\mathrm{Re}(w)}{R}-\frac{|w|^2}{R^2}\right)^{2q-2} \leq \left(\frac{2\mathrm{Re}(w)}{R}\right)^{2q-2},
$$

$$
\left|\frac{z+\overline{w}}{R}-\frac{z\overline{w}}{R^2}\right| \geq C_z|w+1|
$$

as long as $|w-R| < R$ and R is large enough, say $R \geq R_z$ with R_z dependent only on z. Therefore, we are in the position to use the Lebesgue convergence theorem. If we pass to the limit as $R \to \infty$, then we obtain

$$
f(z) = \frac{2q-1}{\pi}\iint_E \frac{f(w)\,(2\mathrm{Re}(w))^{2q-2}}{(z+\overline{w})^{4q}} du\,dv = \langle f, (K_q)_z\rangle_{H_{K_q}}.
$$

This is the desired reproducing property (1.2) for $q \geq 1$. If $q < 1$, we use a similar argument using the decomposition of $|w-R| < R$ according to $2\mathrm{Re}(w) \geq R|w|$ or not.

We can consider also the case of $q = 1/2$ as the Szegö space and also even for $q > 0$. See, for example, [388].

The Bergman kernel is important not only in analysis but also in other fields of mathematics such as algebraic geometry [343]. Denote by $A^2(\Omega)$

$$
A^2(\Omega) \equiv L^2(\Omega) \cap \mathscr{O}(\Omega) \tag{1.132}
$$

the *Bergman space* when we have an open set $\Omega \subset \mathbb{C}$. The mean value property shows that $A^2(\Omega)$ is a closed subspace of $L^2(\Omega)$.

Theorem 1.21. *The Bergman space $A^2(\Delta(1))$ is realized as the reproducing kernel*

$$K_{\Delta(1)}(z,u) \equiv \frac{1}{\pi(1-z\overline{u})^2} \tag{1.133}$$

for $z, u \in \Delta(1)$, namely, $H_{K_{\Delta(1)}}(\Delta(1)) = A^2(\Delta(1))$ with norm-coincidence.

Proof. Let $f \in A^2(\Delta(1))$. Then we have

$$\langle f, K_z\rangle_{A^2(\Delta(1))} = \iint_{\Delta(1)} f(w)\overline{K_{\Delta(1)}(w,z)}\,dx\,dy = \lim_{r\uparrow 1}\iint_{\Delta(r)} f(w)\overline{K_{\Delta(1)}(w,z)}\,dx\,dy$$

by the Lebesgue convergence theorem. By Taylor expansion, we have

$$\begin{aligned}
\langle f, K_z\rangle_{A^2(\Delta(1))} &= \frac{1}{\pi}\lim_{r\uparrow 1}\iint_{\Delta(r)} \frac{f(w)}{(1-\overline{w}z)^2}\,dx\,dy\\
&= \frac{1}{\pi}\lim_{r\uparrow 1}\sum_{j=0}^{\infty}(j+1)\iint_{\Delta(r)} f(w)\overline{w^j}z^j\,dx\,dy\\
&= \frac{1}{\pi}\lim_{r\uparrow 1}\sum_{j=0}^{\infty}(j+1)\cdot 2\pi\cdot\frac{z^j}{2j+2}f(0)r^{2j+2}\\
&= \lim_{r\uparrow 1} r^2 f(r^2 z) = f(z).
\end{aligned}$$

Thus we have the desired result.

1.4.7 Pullback of Bergman Spaces

Given a domain Ω, we define K_Ω to be the reproducing kernel of $A^2(\Omega)$.

Theorem 1.22. *Suppose that φ is a biholomorphic mapping from a domain Ω_1 to a domain Ω_2. Define*

$$\varphi^* K_{\Omega_2}(z,w) \equiv K_{\Omega_2}(\varphi(z),\varphi(w)),\ z, w \in \Omega_1. \tag{1.134}$$

Then

$$H_{\varphi^* K_{\Omega_2}}(\Omega_1) = \left\{f \in \mathcal{O}(\Omega_1)\ :\ \|f\|_{H_{\varphi^* K_{\Omega_2}}(\Omega_1)} < \infty\right\}, \tag{1.135}$$

where the norm is

$$\|f\|_{H_{\varphi^* K_{\Omega_2}}(\Omega_1)} \equiv \sqrt{\iint_{\Omega_1} |f(z)|^2 |\varphi'(z)|^2 \, dx \, dy}$$

for $f \in H_{\varphi^* K_{\Omega_2}}(\Omega_1)$.

Proof. Let us fix $w \in \Omega_1$. A change of variables yields that

$$\begin{aligned}\|\varphi^* K_{\Omega_2}(\cdot, w)\|_{H_{\varphi^* K_{\Omega_2}}(\Omega_1)} &= \sqrt{\iint_{\Omega_1} |K_{\Omega_2}(\varphi(z), \varphi(w))|^2 |\varphi'(z)|^2 \, dx \, dy} \\ &= \sqrt{\iint_{\Omega_2} |K_{\Omega_2}(z, \varphi(w))|^2 \, dx \, dy} < \infty,\end{aligned}$$

proving (1.2) holds.

Another change of variables yields that

$$\begin{aligned}\langle f, \varphi^* K_{\Omega_2}(\cdot, w)\rangle_{H_{\varphi^* K_{\Omega_2}}(\Omega_1)} &= \iint_{\Omega_1} f(z)\overline{K_{\Omega_2}(\varphi(z), \varphi(w))} |\varphi'(z)|^2 \, dx \, dy \\ &= \iint_{\Omega_2} f(\varphi^{-1}(z))\overline{K_{\Omega_2}(z, \varphi(w))} \, dx \, dy \\ &= f(\varphi^{-1}(\varphi(w))) \\ &= f(w)\end{aligned}$$

for all $f \in \mathcal{O}(\Omega_1)$ with $\|f\|_{H_{\varphi^* K_{\Omega_2}}(\Omega_1)} < \infty$.

The Bergman kernel can be defined on a very general domain containing Riemann surfaces, because the norm is defined in a conformal invariant way. For example, we assume that G is a bounded Jordan domain in the complex plane $\mathbb{C}$ with piecewise analytic boundary. Then, S. Bergman gave a famous formula in his thesis in 1923 representing the Riemann mapping function that maps conformally onto a disk in terms of the Bergman reproducing kernel on the domain G [45]. The Bergman kernel may be represented by some complete orthonormal system on the domain G and the system can be constructed by the Gram Schmidt orthogonalization process on the polynomials. So, the reproducing kernel may be considered computationally and therefore, by these procedures the Riemann mapping function may be constructed using computers. Since the Riemann mapping function is fundamental in conformal mapping theory, we can see their long history on these topics. This delicate and deep theory involves convergence estimates based on the geometric property of the boundary of the domain G. See the classical book [46, Chapter XI Section 8] and for recent results see [279] and references therein. Meanwhile, for practical constructions of various canonical conformal mappings on multiply-connected domains, see [20]. We will be able to see the great development of the practical construction of canonical conformal mappings which may be considered as one dream of S. Bergman.

de Branges and Rovnyak [59, Problem 53] is a generalization based on the Blaschke product. We refer to [10, 12–14, 383] for complex interpolation and reproducing kernel Hilbert spaces. Annaby and Zayed [21, Theorem 3.1], [22, p. 174], [100, Corollary 2.4], [172, 173], and [498] are other examples of complex interpolation. We refer to [175, Section 7] for general facts on the interpolation.

1.5 RKHS in the Spaces of Polynomials

In Sect. 1.5, we will introduce concrete reproducing kernels and their quite deep results regarding the spaces of polynomials.

1.5.1 *General Properties of Orthonormal Systems*

General Results

Consider $L_w^2(I)$. Let $u_n(x)$ $(n \in \mathbb{Z}_+)$ be an orthonormal system obtained by Gram Schmidt orthogonalization for polynomials of x of degree n, as in Sect. 1.5.2. The coefficients of x^n and x^{n-1} in $u_n(x)$ are denoted by $k_n > 0,\ k_n'$ and, for

$$\alpha_n \equiv \langle xu_{n-1}, u_n\rangle_{L_w^2(I)} = \frac{k_{n-1}}{k_n} > 0 \tag{1.136}$$

$$\beta_n \equiv \langle xu_{n-1}, u_{n-1}\rangle_{L_w^2(I)} = \frac{k_n k_{n-1}' - k_n' k_{n-1}}{k_n k_{n-1}}, \tag{1.137}$$

we define

$$a_n \equiv \frac{1}{\alpha_n} = \frac{k_n}{k_{n-1}} > 0 \tag{1.138}$$

$$b_n \equiv -\frac{\beta_n}{\alpha_n} = \frac{k_{n-1}k_n' - k_{n-1}' k_n}{(k_{n-1})^2} \tag{1.139}$$

$$c_n \equiv \frac{a_n}{a_{n-1}} = \frac{\alpha_{n-1}}{\alpha_n} = \frac{k_n k_{n-2}}{(k_{n-1})^2} > 0. \tag{1.140}$$

The right inequalities in (1.136) and (1.137) are by no means trivial. But we have the following general results including right equalities in (1.136) and (1.137).

Theorem 1.23. *We have right equalities in (1.136) and (1.137). Furthermore any orthonormal system* $\{u_n\}_{n=0}^{\infty}$ *satisfies the following recurrence relation:*

$$u_n(x) = (a_n x + b_n)u_{n-1}(x) - c_n u_{n-2}(x) \quad (n = 2, 3, 4, \ldots). \tag{1.141}$$

Proof. We admit right equalities in (1.136) and (1.137) for the time being. We set

$$f(x) \equiv u_n(x) - (a_n x + b_n)u_{n-1}(x) + c_n u_{n-2}(x) \quad (x \in I) \tag{1.142}$$

and we need to show that $f(x) \equiv 0$.

For this purpose we consider proving that

$$\langle f, u_j \rangle_{L^2_w(I)} = 0 \tag{1.143}$$

for $j = 0, 1, 2, \ldots, n$. By the definition, for $j = 0, 1, 2, \ldots, n-3$, (1.143) are evident. So, for $j = n, n-1, n-2$, we will establish (1.143). From (1.136) and (1.138)

$$\langle f, u_n \rangle_{L^2_w(I)} = 1 - a_n \langle x u_{n-1}, u_n \rangle_{L^2_w(I)} = 1 - a_n \alpha_n = 0. \tag{1.144}$$

From (1.137) through (1.139)

$$\langle f, u_{n-1} \rangle_{L^2_w(I)} = -a_n \langle x u_{n-1}, u_{n-1} \rangle_{L^2_w(I)} - b_n = -a_n \beta_n - b_n = 0. \tag{1.145}$$

Similarly, we deduce

$$\langle f, u_{n-2} \rangle_{L^2_w(I)} = c_n - a_n \langle x u_{n-1}, u_{n-2} \rangle_{L^2_w(I)} = c_n - a_n \alpha_{n-1} = 0$$

from equality $\langle x u_{n-1}, u_{n-2} \rangle_{L^2_w(I)} = \langle x u_{n-2}, u_{n-1} \rangle_{L^2_w(I)}$, (1.136), (1.138) and (1.140). We thus obtain the desired result $f(x) \equiv 0$.

Let $x \in I$. We will derive (1.136) with α_n and k_n. We set

$$u_n(x) \equiv k_n x^n + v_n(x), \tag{1.146}$$

where v_n is a polynomial with degree $n-1$. Then

$$\langle k_n x^n, u_n \rangle_{L^2_w(I)} = \langle u_n - v_n, u_n \rangle_{L^2_w(I)} = 1, \tag{1.147}$$

and so

$$\langle x^n, u_n \rangle_{L^2_w(I)} = \frac{1}{k_n}. \tag{1.148}$$

Therefore, we obtain

$$\alpha_n = \langle x u_{n-1}, u_n \rangle_{L^2_w(I)} = \langle x(k_{n-1} x^{n-1} + v_{n-1}), u_n \rangle_{L^2_w(I)} = k_{n-1} \langle x^n, u_n \rangle_{L^2_w(I)} = \frac{k_{n-1}}{k_n}.$$

Next we prove (1.137) with β_n and $k_n,\ k_n'$. We set

$$u_n(x) \equiv k_n x^n + k_n' x^{n-1} + w_n(x), \tag{1.149}$$

where w_n is a polynomial with degree $n-2$. Then, from (1.148)

$$\langle k_n x^n, u_{n-1}\rangle_{L^2_w(I)} = \langle u_n - k_n' x^{n-1} - w_n, u_{n-1}\rangle_{L^2_w(I)} = -k_n' \langle x^{n-1}, u_{n-1}\rangle_{L^2_w(I)} = \frac{-k_n'}{k_{n-1}}.$$

So it follows that

$$\langle x^n, u_{n-1}\rangle_{L^2_w(I)} = \frac{-k_n'}{k_n k_{n-1}}.$$

Therefore, using (1.148) we obtain

$$\beta_n = \langle x u_{n-1}, u_{n-1}\rangle_{L^2_w(I)} = \langle x(k_{n-1}x^{n-1} + k_{n-1}' x^{n-2} + w_{n-1}), u_{n-1}\rangle_{L^2_w(I)}. \tag{1.150}$$

We calculate

$$\beta_n = k_{n-1}\langle x^n, u_{n-1}\rangle_{L^2_w(I)} + k_{n-1}'\langle x^{n-1}, u_{n-1}\rangle_{L^2_w(I)} = \frac{k_{n-1}'}{k_{n-1}} - \frac{k_n'}{k_n} = \frac{k_n k_{n-1}' - k_n' k_{n-1}}{k_n k_{n-1}}.$$

The proof is therefore complete.

From the recurrent formula (1.141), we can obtain the Christoffel Darboux formula that shows the explicit reproducing kernel on the finite dimensional space of the polynomials spanned by $\{u_j\}_{j=0}^n$:

Theorem 1.24 (The Christoffel Darboux formula). *For the orthonormal system $\{u_n\}_{n=0}^\infty$ and for*

$$K(x,y) \equiv \sum_{j=0}^n u_j(x)u_j(y) \quad (x, y \in I), \tag{1.151}$$

we have

$$K(x,y) = \frac{k_n(u_{n+1}(x)u_n(y) - u_n(x)u_{n+1}(y))}{k_{n+1}(x-y)} \ \textit{if}\ x \neq y. \tag{1.152}$$

Here, $k_n > 0$ is the coefficient of x^n in $u_n(x)$. We also have

$$K(x,x) = \frac{k_n}{k_{n+1}}\left(u_{n+1}'(x)u_n(x) - u_n'(x)u_{n+1}(x)\right). \tag{1.153}$$

Remark that (1.153) is obtained from a passage of (1.152) to the limit.

Proof. Let $x, y \in I$. We set

$$v_{j+1}(x, y) \equiv u_{j+1}(x)u_j(y) - u_j(x)u_{j+1}(y) \quad (j = 0, 1, 2, \ldots). \tag{1.154}$$

From (1.141), we have

$$\begin{aligned} v_{j+1}(x, y) &= \{(a_{j+1}x + b_{j+1})u_j(x) - c_{j+1}u_{j-1}(x)\}u_j(y) \\ &\quad - u_j(x)\{(a_{j+1}y + b_{j+1})u_j(y) - c_{j+1}u_{j-1}(y)\} \\ &= a_{j+1}(x - y)u_j(x)u_j(y) + c_{j+1}\{u_j(x)u_{j-1}(y) - u_{j-1}(x)u_j(y)\} \\ &= a_{j+1}(x - y)u_j(x)u_j(y) + \frac{a_{j+1}}{a_j}v_j(x, y) \end{aligned}$$

and hence

$$(x - y)u_j(x)u_j(y) = \frac{v_{j+1}(x, y)}{a_{j+1}} - \frac{v_j(x, y)}{a_j} \tag{1.155}$$

for $j = 1, 2, \ldots$. We will show that (1.155) is valid also for $j = 0$. By the definition of k_n

$$\begin{cases} u_0(x) = k_0 > 0 \\ u_1(x) - u_1(y) = k_1(x - y), \end{cases} \tag{1.156}$$

and so, from (1.138), (1.154) and (1.156), we have

$$(x - y)u_0(x)u_0(y) = k_0^2(x - y),$$

$$\frac{v_1(x, y)}{a_1} = \frac{u_1(x)u_0(y) - u_0(x)u_1(y)}{a_1} = \frac{k_0}{k_1}k_0\,(u_1(x) - u_1(y)) = k_0^2(x - y)$$

and so if we set $v_0(x, y) = 0$, (1.155) is valid for $j = 0$.

Therefore, for $x - y \neq 0$, from (1.138) and (1.155), we have

$$\begin{aligned} K(x, y) &= \sum_{j=0}^{n} u_j(x)u_j(y) \\ &= \frac{1}{x - y}\sum_{j=0}^{n}\left(\frac{v_{j+1}(x, y)}{a_{j+1}} - \frac{v_j(x, y)}{a_j}\right) \\ &= \frac{v_{n+1}(x, y)}{a_{n+1}(x - y)} \\ &= \frac{k_n(u_{n+1}(x)u_n(y) - u_n(x)u_{n+1}(y))}{k_{n+1}(x - y)}. \end{aligned}$$

The formula (1.153) is derived from (1.152). Thus the proof is complete.

Zero Points

In order to refer to the sampling theorem, we now examine the structure of the zero points of u_n on I.

Theorem 1.25. *Let $n \in \mathbb{Z}_+$. Then any zero point of the function u_n is simple and lies in I.*

Proof. When $n = 0$, (1.156) shows that the theorem is true. From (1.156) again, we have $u_0 \equiv k_0$. Hence we have

$$\int_a^b u_n(x)w(x)dx = \int_a^b u_n(x) \cdot 1 \cdot w(x)dx = \frac{1}{k_0}\langle u_n, u_0\rangle_{L^2_w(I)} = 0 \quad (n = 1, 2, \ldots).$$

Therefore $u_n(x)$ has at least one zero point with some odd degree on I. So we set their zero distinct points as $\{x_j\}_{j=1}^k$ of odd order. We set

$$p_k(x) \equiv (x - x_1)(x - x_2) \cdots (x - x_k). \tag{1.157}$$

Observe that $p_k(x)$ is a polynomial with degree k. We assume first that $k < n$. Then, of course,

$$\langle u_n, p_k\rangle_{L^2_w(I)} = 0. \tag{1.158}$$

However, in this case all the orders of zeros of $u_n(x)p_k(x)$ on (a, b) are even. And so, $u_n(x)p_k(x)$ does not change the signs on (a, b). So, without loss of generality, we assume

$$u_n(x)p_k(x) \geq 0. \tag{1.159}$$

Then, since $u_n \cdot p_k$ is not zero, we have

$$\langle u_n, p_k\rangle_{L^2_w} = \int_a^b u_n(x)p_k(x)w(x)dx > 0 \tag{1.160}$$

which is a contradiction to (1.158). Hence, $k \geq n$. Meanwhile, $u_n(x)$ has precisely n-zero points and so, we have the desired result $k = n$.

The next theorem describes the relation of zero points of u_n and u_{n+1}.

Theorem 1.26. *The zero points of u_n are separated by the zero points of u_{n+1} on I.*

Proof. At first, we assume that u_n and u_{n+1} vanish simultaneously at $x = a$. From the recurrence formula (1.141), $u_{n-1}(a) = 0$. By repeating this, we have

$$u_{n+1}(a) = u_n(a) = u_{n-1}(a) = \cdots = u_0(a) = 0. \tag{1.161}$$

This is a contradiction to the fact $u_0(a) \neq 0$.

Next we denote some sequence of the consecutive zeros of u_n by $x_r, x_{r+1}(x_r < x_{r+1})$. By (1.153), we have

$$0 \le K(x_r, x_r) = \frac{k_n}{k_{n+1}} \left(-u_n'(x_r)u_{n+1}(x_r)\right), \tag{1.162}$$

and so

$$u_n'(x_r)u_{n+1}(x_r) \le 0. \tag{1.163}$$

Similarly, we deduce

$$u_n'(x_{r+1})u_{n+1}(x_{r+1}) \le 0. \tag{1.164}$$

By the definition of x_r and x_{r+1}, there is no zero point of u_n between x_r and x_{r+1}. Hence we have

$$u_n'(x_r)u_n'(x_{r+1}) < 0. \tag{1.165}$$

Since x_r and x_{r+1} are not zero points of u_{n+1}, from (1.163)–(1.165) we learn

$$u_{n+1}(x_r)u_{n+1}(x_{r+1}) < 0. \tag{1.166}$$

Hence, if we combine (1.164), (1.165) and (1.166), then we see that u_{n+1} has odd numbers of zero points on (x_r, x_{r+1}). Similarly, we can see that for any two sequential zero points of u_{n+1}, there exists one zero point of u_n.

1.5.2 Examples of Orthonormal Systems

We refer to [288] for the proof of the propositions in this section in principle. Let $w : I = (a, b) \to (0, \infty)$ be a weight, that is, a positive measurable function. Let $L^2_w(a, b)$ be a Hilbert space equipped with inner product

$$\langle f, g\rangle_{L^2_w(a,b)} \equiv \int_a^b f(x)\overline{g(x)}w(x)dx. \tag{1.167}$$

Here, w satisfies

$$\int_a^b x^n w(x)dx < \infty \quad (n \in \mathbb{Z}_+). \tag{1.168}$$

For various intervals $[a, b]$ and weight functions w, when we apply the Gram Schmidt orthogonalization for the polynomials $\{u_n(x) \equiv x^n; n = 1, 2, \ldots\}$ we have orthonormal systems. We list typical ones.

Theorem 1.27 (Legendre polynomials). *Let $I = (-1,1)$ and $w(x) \equiv 1$. Define the Legendre polynomola of order n by:*

$$P_n(x) \equiv \frac{1}{2^n n!}\frac{d^n}{dx^n}(x^2-1)^n \quad (x \in I) \tag{1.169}$$

for $n \in \mathbb{Z}_+$. Then we have

$$u_n(x) = \sqrt{\frac{2n+1}{2}}P_n(x) = \frac{1}{2^n n!}\frac{d^n}{dx^n}(x^2-1)^n \quad (x \in I). \tag{1.170}$$

See [288, p. 228, 231] for the proof of (1.170).
Let $n \in \mathbb{N} \cup \{0\}$. Then we have another expression:

$$P_n(x) \equiv \left(n+\frac{1}{2}\right)^{1/2}\left(\frac{x}{2}\right)^n \sum_{\nu=0}^{[n/2]}(-1)^\nu \binom{n}{\nu}\binom{2n-2\nu}{n} x^{-2\nu} \quad (x \in \mathbb{R}). \tag{1.171}$$

Theorem 1.28 (Chebyshev polynomials). *We let $I \equiv (-1,1)$ and $w(x) \equiv \dfrac{1}{\sqrt{1-x^2}}$ for $x \in I$. Set*

$$T_0(x) \equiv 1, \quad T_n(x) \equiv \frac{(-1)^n}{(2n-1)!!}\sqrt{1-x^2}\frac{d^n}{dx^n}\sqrt{(1-x^2)^{2n-1}} \tag{1.172}$$

for $n = 1, 2, \ldots$ Then

$$\begin{cases} u_0(x) = \dfrac{1}{\sqrt{\pi}}T_0(x) \\ u_n(x) = \sqrt{\dfrac{2}{\pi}}T_n(x) \quad (n = 1, 2, \ldots) \end{cases} \tag{1.173}$$

for $x \in I$.

The polynomials $T_0, T_1, \ldots$ are Chebyshev polynomials; see [288, Section 5.7] for the proof of (1.173).

Theorem 1.29 (Hermite polynomials 1). *Let $I = (-\infty, \infty)$ and define the weight by:*

$$w(x) \equiv \exp\left(-\frac{x^2}{2}\right) \quad (x \in I).$$

Define

$$H_n(x) \equiv (-1)^n \exp\left(\frac{x^2}{2}\right)\frac{d^n}{dx^n}\exp\left(-\frac{x^2}{2}\right) \quad x \in I \tag{1.174}$$

for $n \in \mathbb{Z}_+$. *Then*

$$u_n(x) = \frac{1}{\sqrt{n!\sqrt{2\pi}}} H_n(x) \quad x \in I. \tag{1.175}$$

See [288, pp. 250–252] for the proof of (1.175).

The polynomials $H_0^*, H_1^*, \ldots$ are called Hermite polynomials.

Theorem 1.30 (Hermite polynomials 2). *Let* $I = (-\infty, \infty)$ *and define the weight by*

$$w(x) \equiv \frac{\exp(-x^2)}{\sqrt{\pi}} \quad (x \in I).$$

Define

$$H_n^*(x) \equiv (-1)^n \exp(x^2) \frac{d^n}{dx^n} \exp(-x^2) \tag{1.176}$$

for $n \in \mathbb{Z}_+$. *Then we have*

$$u_n(x) = \frac{1}{\sqrt{2^n n!}} H_n^*(x). \tag{1.177}$$

See [288, pp. 250–252] for a proof of (1.177).

The polynomials $H_0^*, H_1^*, \ldots$ are called Hermitian polynomials as well.

The Hermite polynomial of degree n can be expressed differently:

$$H_n^*(x) \equiv (2x)^n \sqrt{\frac{n!}{2^n}} \sum_{\nu=0}^{[n/2]} \frac{(-1)^\nu}{(n-2\nu)!} \frac{(2x)^{-2\nu}}{\nu!} \quad (x \in \mathbb{R}). \tag{1.178}$$

Recall $\Gamma(z)$ is the Gamma function defined by

$$\Gamma(z) \equiv \int_0^\infty e^{-t} t^{z-1} dt \quad (\mathrm{Re}(z) > 0). \tag{1.179}$$

Theorem 1.31 (Laguerre polynomials). *Let* $I = (0, \infty)$ *and define the weight by:*

$$w(x) \equiv \frac{\exp(-x)x^\alpha}{\Gamma(1+\alpha)} \quad (x \in I),$$

where $\alpha > -1$ *is a fixed parameter and* Γ *denotes the Gamma function. Define the Laguerre polynomial* $L_n^{(\alpha)}$ *of order n by:*

$$L_n^{(\alpha)}(x) \equiv \frac{e^x x^{-\alpha}}{n!} \frac{d^n}{dx^n} \left(e^{-x} x^{n+\alpha}\right) \quad (x \in I) \tag{1.180}$$

for $n \in \mathbb{Z}_+$. Then we have

$$u_n(x) = \sqrt{\frac{n!\Gamma(\alpha+1)}{\Gamma(\alpha+n+1)}} L_n^{(\alpha)}(x) \quad (x \in I) \tag{1.181}$$

for $n \in \mathbb{Z}_+$.

See [288, pp. 239–241] for a proof of (1.181).

The polynomials $L_0^{(\alpha)}, L_1^{(\alpha)}, \ldots$ are Laguerre bi-polynomials. When $\alpha = 0$, they are represented by $L_n(x)$ that are called simply as Laguerre polynomials.

Let $n \in \mathbb{Z}_+$ and $\alpha > -1$. Then we have another expression of $L_n^{(\alpha)}$:

$$L_n^{(\alpha)}(x) \equiv \sqrt{\frac{n!\Gamma(\alpha+1)}{\Gamma(\alpha+n+1)}} \sum_{\nu=0}^{n} (-1)^\nu \binom{n+\alpha}{n-\nu} \frac{x^\nu}{\nu!}, \quad (x \in (0,\infty)). \tag{1.182}$$

This system is an orthogonal basis for $L^2(\mathbb{R}^+, \mu)$.

1.5.3 Polynomial Reproducing Kernel Hilbert Spaces

Zero Points and RKHS

We denote the $N+1$ simple zero points of u_{N+1} on (a,b) by $\{x_n\}_{n=0}^N$. From (1.152),

$$K(x_m, x_n) = \frac{k_N(u_{N+1}(x_m)u_N(x_n) - u_N(x_m)u_{N+1}(x_n))}{k_{N+1}(x_m - x_n)} = 0. \tag{1.183}$$

Meanwhile, if $m = n$, from (1.153)

$$K(x_n, x_n) = \frac{k_N}{k_{N+1}} \left(u'_{N+1}(x_n)u_N(x_n) - u'_N(x_n)u_{N+1}(x_n)\right) = \frac{k_N}{k_{N+1}} u'_{N+1}(x_n)u_N(x_n).$$

Let $H_K(a,b)$ be the (finite-dimensional) Hilbert space with the reproducing kernel K. We consider the linear transform of $H_K(a,b)$

$$f(x_n) = \langle f, K_{x_n} \rangle_{H_K(a,b)}. \tag{1.184}$$

Then, if we set

$$\omega_n = \|K_{x_n}\|_{H_K(a,b)}^2 = K(x_n, x_n) > 0,$$

we have

$$\langle K_{x_m}, K_{x_n} \rangle_{H_K(a,b)} = K(x_n, x_m) = \delta_{m,n}\omega_n. \tag{1.185}$$

Of course, since $\{K_{x_n}\}_{n=0}^{N+1}$ are linearly independent, the linear mapping is isometric and we have the identity

$$\|f\|_{H_K(a,b)} = \sqrt{\sum_{n=0}^{N} \frac{|f(x_n)|^2}{\omega_n}}. \tag{1.186}$$

We thus have the sampling theorem

$$f(x) = \langle f, K_x \rangle_{H_K(a,b)} = \sum_{n=0}^{N} \frac{f(x_n)\overline{K(x_n, x)}}{\omega_n} = \sum_{n=0}^{N} f(x_n) \frac{K(x, x_n)}{K(x_n, x_n)}. \tag{1.187}$$

Note that one orthonormal system is given by

$$\varphi_n(x) = \sqrt{\frac{k_N u_N(x_n)}{k_{N+1} u'_{N+1}(x_n)}} \cdot \frac{u_{N+1}(x)}{x - x_n}, \tag{1.188}$$

since $\{K_{x_n}/\omega_n^{1/2}\}_{n=0}^{N+1}$ is an orthonormal system.

We consider the sampling theory in detail in Sect. 8.4. In particular, see a recent related article [204] for information.

Scaling Limits of Christoffel Darboux Kernels

For the case of Hermitian polynomials 2 (see Theorem 1.30), we know more. In Theorem 1.30, set

$$K^{(n+1)}(x, y) = \sum_{j=0}^{n} u_j(x) u_j(y). \tag{1.189}$$

Then from (1.152) we have

$$K^{(n+1)}(x, y) = \left(\frac{n+1}{2}\right)^{1/2} \frac{u_{n+1}(x)u_n(y) - u_n(x)u_{n+1}(y)}{x - y} \tag{1.190}$$

when $x \neq y$ and if $x = y$, then a passage of (1.190) to the limit $x \to y$ yields

$$K^{(n+1)}(x, x) = \left(\frac{n+1}{2}\right)^{1/2} \left(u'_{n+1}(x)u_n(x) - u'_n(x)u_{n+1}(x)\right) \tag{1.191}$$

similar to (1.153). By Adamov [6], the function H_n^*, which is given by (1.176), satisfies

$$H_n^*(x) \approx 2^{(n+1)/2} n^{n/2} e^{-n/2} \exp\left(\frac{x^2}{2}\right) \cos\left(\sqrt{2n+1}x - \frac{n\pi}{2}\right) \tag{1.192}$$

as $n \to \infty$. From (1.192), it is easy to see that

$$\lim_{N\to\infty} \frac{K^{(N)}(x,y)}{N} = 0. \tag{1.193}$$

Moreover, according to [430],

$$\lim_{N\to\infty} \frac{\pi}{\sqrt{2N}} K^{(N)}\left(\frac{\pi x}{\sqrt{2N}}, \frac{\pi y}{\sqrt{2N}}\right) = \frac{\sin \pi(x-y)}{\pi(x-y)}. \tag{1.194}$$

In general, for a positive continuous function $w(x)$ on the interval $[-1,1]$, we define similarly $K^{(n+1)}(x,y)$. Then, as in [461], we have

$$\lim_{N\to\infty} \frac{K^{(N)}(x,x)}{N} = \frac{1}{\pi w(x)\sqrt{1-x^2}} \quad \text{for almost every} \quad (-1,1). \tag{1.195}$$

For any $x \in (-1,1)$,

$$\lim_{N\to\infty} K^{(N)}\left(x + \frac{a}{M^{(N)}(x,x)}, x + \frac{b}{M^{(N)}(x,x)}\right) = \frac{\sin \pi(a-b)}{\pi(a-b)} K^{(N)}(x,x), \tag{1.196}$$

for

$$M^{(N)}(x,y) = w(x)^{1/2} w(y)^{1/2} K^{(N)}(x,y) \tag{1.197}$$

holds for a, b on any compact subset of $\mathbb{C}$ [286, 287, 432]. The relationship of these formulas to random matrix theory and nuclear physics is described in [430]. As for the proof of the universality limit formula, according to [287, 433] it is based on the theory of entire functions of exponential type, as in the case of the celebrated sampling theorem of Whittaker Ogura Kotelnikov Shannon Someya; see [74] for a detailed account of why this theorem deserves the name of 5 people. There is another approach using the Riemann Hilbert method for the proof of (1.196) [270, Theorem 1.1]. See also [346] for the case of complex manifolds of higher dimensions.

This paragraph was introduced by T. Ohsawa.

1.5.4 RKHS Constructed by Meixner-Type Polynomials

The taxonomy of Meixner polynomials does not seem to be consistent in the literature [106, Section 5.4] and [187, Section 9.8]. The Meixner class will consist of the polynomial systems of Hermite, Laguerre, Poisson Charlier, Krawtchouk and Pollaczek (the last system is typified by the Legendre polynomials). All five systems are orthogonal with respect to a probability measure and each is complete in an appropriate L^2 space (these facts are proved in many places; see, e.g., [451, p. 33, 39, 105]. The Hermite case has been treated in [205]. For the case of Meixner-type polynomials, see [206] for details. See also [342] as well.

Example 1.4 (Poisson Carlier Polynomials). Let $a > 0$ and define

$$PC_n(x) \equiv \frac{(-1)^n a^{n/2}}{\sqrt{n!}} \sum_{\nu=0}^{\min(x,n)} (-1)^\nu \binom{n}{\nu}\binom{x}{\nu} \nu! a^{-\nu} \quad (x \in \mathbb{R}). \tag{1.198}$$

In (1.198) we could replace $\min(x, n)$ with just n. This system is an orthogonal basis for $L^2((0,\infty),\mu)$, where μ is the discrete probability measure determined by the step function with jumps $j_\nu = \exp(-a)a^\nu/\nu!$ at $\nu = 0, 1, \ldots$; see [426].

Example 1.5 (Krawtchouk Polynomials). Let $n = 0, 1, \ldots, N$ and let p and q satisfy $p > 0$, $q > 0$, and $p + q = 1$. Then define

$$K_n(x) \equiv \left[\binom{N}{n}(pq)^n\right]^{-1/2} \sum_{\nu=0}^{n} (-1)^{n-\nu} \binom{N-x}{n-\nu}\binom{x}{\nu} p^{n-\nu} q^\nu \quad (x \in \mathbb{R}). \tag{1.199}$$

This system is an orthogonal basis for $L^2(\mathbb{R},\mu)$, where μ is the discrete probability measure determined by the step function with jumps

$$j_x = \binom{N}{x} p^x q^{N-x}, \quad (x = 0, 1, \ldots, N).$$

We refer to [106, p. 176] and [451, pp. 35–36] for this information.

For general orthogonal polynomials, see the reference [23, Chapters 5–7].

1.5.5 RKHS for Trigonometric Polynomials

On the interval $[-\ell, \ell]$, a direct calculation shows that the functions

$$\varphi_n(x) \equiv \exp\left(i\frac{n\pi}{\ell}x\right), \tag{1.200}$$

where $n = 0, \pm 1, \pm 2, \ldots, \pm N$, are an orthonormal system with inner product

$$\langle f, g\rangle = \frac{1}{2\ell}\int_{-\ell}^{\ell} f(x)\overline{g(x)}dx. \tag{1.201}$$

Hence, the reproducing kernel of the space spanned by the system $\{\varphi_n\}_{n=-N}^{N}$ is given for $x \neq y$, and we have

$$K(x,y) = \sum_{n=-N}^{N} \varphi_n(x)\overline{\varphi_n(y)} = \sum_{n=-N}^{N} \exp\left(\frac{n\pi}{\ell}(x-y)i\right) = \frac{\sin[(2N+1)\pi(x-y)/2\ell]}{\sin[\pi(x-y)/2\ell]}.$$

For $x = y$, a passage to the limit yields

$$K(x,x) = 2N+1. \tag{1.202}$$

For $c \in \left(-\frac{l}{2N+1}, \frac{l}{2N+1}\right)$ and $n = 0, \pm 1, \pm 2, \ldots, \pm N$, we set

$$x_n = c + \frac{2ln}{2N+1}. \tag{1.203}$$

Then we have

$$\langle K_{x_m}, K_{x_n}\rangle_{H_K(a,b)} = \delta_{m,n}\omega_n, \quad \omega_n = \|K_{x_n}\|^2_{H_K(a,b)} = K(x_n,x_n) > 0. \tag{1.204}$$

Hence we have the sampling theorem, similarly

$$f(x) = \langle f, K_x\rangle_{H_K(a,b)} = \sum_{n=-N}^{N} \frac{f(x_n)\overline{K(x_n,x)}}{\omega_n} = \sum_{n=-N}^{N} f(x_n)\frac{K(x,x_n)}{K(x_n,x_n)}. \tag{1.205}$$

Note that one orthonormal system is

$$\varphi_n(x) = \frac{1}{\sqrt{2N+1}}\operatorname{cosec}\left[\frac{\pi}{2\ell}\left(x-c-\frac{2\ell n}{2N+1}\right)\right]\sin\left[\frac{(2N+1)\pi}{2\ell}\left(x-c-\frac{2\ell n}{2N+1}\right)\right].$$

See also [37, 174, 204] for some general sampling theorems and converse sampling theorems. For example, in [37, Theorem 2], the sampling theorem for the class $F^2 \equiv \{f \in L^2(\mathbb{R}) \cap C(\mathbb{R}) : \mathscr{F}f \in L^1(\mathbb{R})\}$.

The book [342] by H. Ogawa was very instructive to write this section.

1.6 Graphs and Reproducing Kernels

Following [453], we connect graph theory and the theory of reproducing kernels.

Here we recall the definitions for graphs.

Definition 1.5.

1. A graph $G = (V(G), E(G))$ is a pair of a nonempty finite set $V(G)$, called the vertex set, and a subset $E(G)$ in $\{\{x,y\} : x, y \in V(G) \text{ and } x \neq y\}$, called the edge set. We further assume that a graph always means simple, namely it has neither loops nor multiedges, and has a finite number of vertices. We set $V = V(G)$ and $E = E(G)$ for simplicity.

2. For any vertex $i \in V$, let d_i or $\deg_G(i)$ denote the degree of i, that is, d_i is the cardinality of the set $\{j \in V : \{i,j\} \in E\}$.
3. A graph is said to be complete, if $E(G) = \{\{x,y\} : x,y \in V \text{ and } x \neq y\}$.
4. A path from x to y in the graph G is a sequence of $x_0 = x, x_1, x_2, \ldots, x_l = y$ such that $\{x_i, x_{i+1}\} \in E(G)$ for $0 \leq i \leq l-1$ and the vertices $x_0, \ldots, x_l$ are all distinct. The graph G is said to be connected if, for any two distinct vertices in G, there exists a path from one to the other.
5. A graph is said to be a tree if, for any two distinct vertices, there exists the unique path between them.
6. Let d_G denote the path-length distance for a connected graph G.
7. Graphs $G_1 = (V(G_1), E(G_1))$ and $G_2 = (V(G_2), E(G_2))$ are isomorphic if there exists a bijection $\Phi : V(G_1) \to V(G_2)$ such that $\{x,y\} \in E(G_1)$ if and only if $\{\Phi(x), \Phi(y)\} \in E(G_2)$.

A multiedge between points means that there is more than one way to go from one point to another.

Now we recall some notions on matrices.

Definition 1.6.

1. *The adjacency matrix* of G is a square matrix $A = A^G = \{A^G_{xy}\}_{x,y \in G}$ whose rows and columns are indexed by V with (x,y)-entry 1 if $\{x,y\} \in E$ and 0 otherwise.
2. *The degree matrix* D of G is a diagonal matrix with (x,x)-entry equal to its degree.
3. *The Laplacian matrix* L of G is defined as $L = D - A$. Let $\lambda_j (1 \leq j \leq s)$ be all distinct eigenvalues of L with increasing order $\lambda_1 < \cdots < \lambda_s$. Let m_i be the multiplicity of λ_i.
4. Let E_j denote the projection onto the eigenspace corresponding to λ_j for $1 \leq j \leq s$.
5. Since L is symmetric, L has the following spectral decomposition; $L = \sum_{j=1}^{s} \lambda_j E_j$ with $E_j^T = E_j$, $E_i E_j = \delta_{ij} E_i$ and $\sum_{j=1}^{s} E_j = I$, where I denotes the identity matrix and δ_{ij} denotes the Kronecker delta.

The matrix L has the smallest eigenvalue 0 with multiplicity 1 and its eigenvector is the all-ones vector $\mathbf{1}$ provided that G is connected. If G is connected, then $E_1 = \frac{1}{n} J$ holds, where n denotes the number of vertices and J denotes the all-ones matrix.

1.6.1 RKHS $\mathscr{H}_G$

Let G be a connected graph with adjacency matrix A. The set of all real-valued functions on V will be denoted by $\mathscr{F} = \mathscr{F}(G)$. We define a bilinear form on $\mathscr{F}$ as follows:

$$\mathscr{E}(u,v) = \frac{1}{2} \sum_{x,y \in V} A_{xy}(u(x) - u(y))(v(x) - v(y)) \quad (u, v \in \mathscr{F}). \tag{1.206}$$

Then, we see the elementary facts on the bilinear form $\mathscr{E}$:

Lemma 1.6. *Assume that G is connected. Then*

1. *The Cauchy Schwarz inequality for $\mathscr{E}$ holds:*

$$|\mathscr{E}(u,v)|^2 \le \mathscr{E}(u,u)\mathscr{E}(v,v) \tag{1.207}$$

   for all u, v in $\mathscr{F}$.
2. *$\mathscr{E}(u,u) = 0$ if and only if u is constant on V.*

Proof.

1. Observe that $J(tu+v, tu+v) \ge 0$ for all $t \in \mathbb{R}$ from the definition (1.206).
2. Let $x_0 \in V(G)$ be fixed. Then since G is connected, for any x we can find a path from x_0 to x in the graph G; we can find a sequence of $x_0, x_1, \dots, x_l = x$ such that $\{x_i, x_{i+1}\} \in E(G)$ for $0 \le i \le l-1$ and the vertices $x_0, \dots, x_l$ are all distinct. Since $A_{x_i x_{i+1}} = 1$, $\mathscr{E}(u,u) = 0$ implies $u(x_i) = u(u_{i+1})$ for $0 \le i \le l-1$. Thus, $u(x_0) = u(x)$, which implies that u is a constant function.

   Next, we define another bilinear form on $\mathscr{F}$ as follows:

$$\langle u, v\rangle = \left(\sum_{x\in V} u(x)\right)\left(\sum_{x\in V} v(x)\right) + \mathscr{E}(u,v) \quad (u, v \in \mathscr{F}). \tag{1.208}$$

Then, we have:

Lemma 1.7. *Assume that G is connected. The mapping $\langle\cdot,\cdot\rangle$ is an inner product invariant under graph isomorphisms in* (1.208).

Proof. Let us start by proving that $\langle\cdot,\cdot\rangle$ is an inner product. What is unclear is the strict positivity of the bilinear form $\langle\cdot,\cdot\rangle$. To this end, we suppose $\langle u,u\rangle = 0$ for u in $\mathscr{F}$. Set $c \equiv \sum_{x\in V} u(x)$. Then $\mathscr{E}(u,u) = 0$ from the definition of the inner product and (1.207). Thus, by Lemma 1.6, u is constant. Meanwhile, $\langle u,u\rangle = 0$ also implies $|c|^2 = 0$ and hence $c = 0$. Therefore we have that $u = 0$. Thus $\langle\cdot,\cdot\rangle$ is an inner product.

Next, let Φ be a graph isomorphism from G_1 onto G_2, and let $\langle\cdot,\cdot\rangle_1$ and $\langle\cdot,\cdot\rangle_2$ be the corresponding inner products defined by G_1 and G_2, respectively. Let us prove that the inner product is invariant under Φ. Since $A_{xy} = A_{\Phi^{-1}(x)\Phi^{-1}(y)}$, we have

$$\begin{aligned}
&\langle u\circ\Phi^{-1}, u\circ\Phi^{-1}\rangle_2\\
&= \left|\sum_{x\in V(G_2)} u(\Phi^{-1}(x))\right|^2 + \frac{1}{2}\sum_{x,y\in V(G_2)} A_{xy}|u(\Phi^{-1}(x)) - u(\Phi^{-1}(y))|^2\\
&= \left|\sum_{x\in V(G_2)} u(\Phi^{-1}(x))\right|^2 + \frac{1}{2}\sum_{x,y\in V(G_2)} A_{\Phi^{-1}(x)\Phi^{-1}(y)}|u(\Phi^{-1}(x)) - u(\Phi^{-1}(y))|^2
\end{aligned}$$

$$= \left| \sum_{x' \in V(G_1)} u(x') \right|^2 + \frac{1}{2} \sum_{x',y' \in V(G_1)} A_{x'y'} |u(x') - u(y')|^2$$
$$= \langle u, u \rangle_1.$$

By the polarization identity, it follows that inner product $\langle \cdot, \cdot \rangle$ is invariant under graph isomorphisms. This concludes the proof.

Now we can introduce the Hilbert space $\mathscr{H}_G$:

Definition 1.7. The space $\mathscr{H}_G$ will denote the Hilbert space $(\mathscr{F}, \langle \cdot, \cdot \rangle)$. The norm induced by $\langle \cdot, \cdot \rangle$ will be denoted by $\| \cdot \|$, that is, we set

$$\|u\|^2 \equiv \left| \sum_{x \in V} u(x) \right|^2 + \mathscr{E}(u, u).$$

The space $\mathscr{H}_G$ will be written as $\mathscr{H}$ if there is no confusion.

Since $\dim(\mathscr{H}) < \infty$, the point evaluation on $\mathscr{H}$ is norm-continuous. By the Riesz representation theorem, there exists the reproducing kernel function $k(y, x)$. Before we go further, a helpful remark may be in order.

Remark 1.4. In (1.208), $k(x, y) = k(y, x)$ for $x, y \in G$. In fact,

$$k(y, x) = \langle k_x, k_y \rangle = \langle k_y, k_x \rangle = k(x, y),$$

where (1.208) is used for the second equality.

Definition 1.8. Let G_1 and G_2 be connected graphs, and let k and j be reproducing kernel functions of G_1 and G_2, respectively. We will say that $\mathscr{H}_{G_1}$ and $\mathscr{H}_{G_2}$ are *isomorphic* as reproducing kernel Hilbert spaces if there exists a unitary operator U from $\mathscr{H}_{G_1}$ onto $\mathscr{H}_{G_2}$, a graph isomorphism Φ from G_1 onto G_2, and a function φ on $V(G_2)$ such that the following diagram commutes:

$$\begin{array}{ccc} \mathscr{H}_{G_1} & \xrightarrow{\ U\ } & \mathscr{H}_{G_2} \\ {\scriptstyle \iota_1}\uparrow & & \uparrow{\scriptstyle \frac{1}{\varphi}\iota_2} \\ G_1 & \xrightarrow[\ \Phi\ ]{} & G_2, \end{array}$$

where ι_1 denotes the mapping $x \mapsto k_x$ from $V(G_1)$ into $\mathscr{H}_{G_1}$ and ι_2 denotes the mapping $x \mapsto j_x$ from $V(G_2)$ into $\mathscr{H}_{G_2}$.

Lemma 1.8.

1. *In Definition* 1.8, *we have;*

   (a) *for all* $x \in G_1$ *and* $y \in G_2$,

$$j_{\Phi(x)}(y) = \varphi(\Phi(x))\varphi(y)k_x(\Phi^{-1}(y)), \tag{1.209}$$

(b) *for all* $x \in G_1$ *and* $y \in G_2$

$$Uk_x(y) = \varphi(y)k_x(\Phi^{-1}(y)), \tag{1.210}$$

(c) $Uf(y) = \varphi(y)f(\Phi^{-1}(y))$ for all $f \in \mathscr{H}_{G_1}$.

2. *Conversely, given* Φ *and* φ *satisfying* (1.209), $\mathscr{H}_{G_1}$ *and* $\mathscr{H}_{G_2}$ *are isomorphic by the unitary operator* U *defined by*

$$Uk_x = j_{\Phi(x)}/\varphi(\Phi(x)). \tag{1.211}$$

Proof.

1. (a) The above commutative diagram reads as

$$Uk_x = j_{\Phi(x)}/\varphi(\Phi(x)). \tag{1.212}$$

By the reproducing property of j_y and this relation, we obtain

$$j_{\Phi(x)}(y) = \langle j_{\Phi(x)}, j_y \rangle_{\mathscr{H}_{G_2}} = \varphi(\Phi(x))\langle Uk_x, j_y \rangle_{\mathscr{H}_{G_2}}.$$

Since U is a unitary operator, we deduce $\varphi(\Phi(x))k_x = U^* j_{\Phi(x)}$ from (1.212) for all $x \in G_1$, or equivalently $\varphi(y)k_{\Phi^{-1}(y)} = U^* j_y$ for all $y \in G_2$. Therefore,

$$j_{\Phi(x)}(y) = \varphi(\Phi(x))\langle k_x, U^* j_y \rangle_{\mathscr{H}_{G_2}} = \varphi(\Phi(x))\langle k_x, \varphi(y)k_{\Phi^{-1}(y)} \rangle_{\mathscr{H}_{G_2}},$$

proving (1.209).

(b) This is a consequence of (1.209) and (1.212).

(c) This is a consequence of (1.210) and $\mathscr{H}_{G_1} = \mathrm{Span}(\{k_x\}_{x \in G_1})$.

2. Observe that (1.209) preserves the inner product and that the inverse is given by $U^* j_y = \varphi(y)k_{\Phi^{-1}(y)}$.

Let δ_x denote the delta function for a vertex x in V. It follows from the definition of inner product that δ_x is orthogonal to δ_y in $\mathscr{H}_G$ if and only if $\{x, y\}$ belongs to E. Hence, if G is complete, then $\{\delta_x\}_{x \in V}$ is an orthogonal set in $\mathscr{H}_G$.

Then we obtain the fundamental theorem:

Theorem 1.32. *If* G_1 *and* G_2 *are isomorphic, then* $\mathscr{H}_{G_1}$ *and* $\mathscr{H}_{G_2}$ *are isomorphic.*

Proof. Let Φ be a graph isomorphism from G_1 onto G_2. Setting $Uu = u \circ \Phi^{-1}$, we learn U is a unitary operator from $\mathscr{H}_{G_1}$ onto $\mathscr{H}_{G_2}$ by Lemma 1.7, and then it is obvious that $U^* v = v \circ \Phi$. In order to show the statement, we shall see that $Uk_x = j_{\Phi(x)}$ for any x in $V(G_1)$. Let v be in $\mathscr{H}_{G_2}$. Then

$$\langle v, Uk_x \rangle_{\mathscr{H}_{G_2}} = \langle U^* v, k_x \rangle_{\mathscr{H}_{G_1}} = \langle v \circ \Phi, k_x \rangle_{\mathscr{H}_{G_1}} = v(\Phi(x)) = \langle v, j_{\Phi(x)} \rangle_{\mathscr{H}_{G_2}}.$$

This concludes the proof.

1.6.2 Gram Matrices

We introduce the definition:

Definition 1.9. We set $K = (k(x,y))_{x,y \in V}$, where we use the notation $k(y,x) = k_x(y)$. K will be called *the Gram matrix* of G. The (i,j)-entry of K is also denoted by $K_{i,j}$.

Then we obtain

Theorem 1.33. *Let G_1 and G_2 be connected graphs, and let K_1 and K_2 be Gram matrices of G_1 and G_2, respectively. Then G_1 and G_2 are isomorphic if and only if $K_1 = K_2$ up to permutation.*

Proof. Let k and j be reproducing kernels of $\mathcal{H}_{G_1}$ and $\mathcal{H}_{G_2}$, respectively. First, we show "the only if" part. Suppose that G_1 and G_2 are isomorphic. Let U, j, Φ be as in Definition 1.8. Then, by Theorem 1.32, we have that

$$k(y,x) = \langle k_x, k_y \rangle_{\mathcal{H}_{G_1}} = \langle Uk_x, Uk_y \rangle_{\mathcal{H}_{G_2}} = \langle j_{\Phi(x)}, j_{\Phi(y)} \rangle_{\mathcal{H}_{G_2}} = j(\Phi(y), \Phi(x)).$$

This concludes "the only if" part. Next, we show "the if" part. Assume that $K_1 = K_2$ up to permutation. Then there exists a bijective mapping Φ from $V(G_1)$ onto $V(G_2)$ such that $\langle k_x, k_y \rangle_{\mathcal{H}_{G_1}} = \langle j_{\Phi(x)}, j_{\Phi(y)} \rangle_{\mathcal{H}_{G_2}}$ for any x, y in $V(G_1)$. Let U be the operator defined by $U : k_x \mapsto j_{\Phi(x)}$. Then, since $\{k_x\}_{x \in G_1}$ is linearly independent, U is a unitary operator from $\mathcal{H}_{G_1}$ onto $\mathcal{H}_{G_2}$. We show that Φ is a graph isomorphism from G_1 onto G_2. Since

$$\begin{aligned}
\langle \delta_x, \delta_y \rangle_{\mathcal{H}_{G_1}} &= \sum_{z \in V(G_1)} \delta_x(z) \sum_{z \in V(G_1)} \delta_y(z) + \mathcal{E}(\delta_x, \delta_y) \\
&= 1 + \frac{1}{2} \sum_{z,w \in V(G_1)} A^{G_1}_{zw} (\delta_x(z) - \delta_x(w))(\delta_y(z) - \delta_y(w)) \\
&= 1 + \frac{1}{2} \sum_{(z,w) \in \{(x,y),(y,x)\}} A^{G_1}_{zw} (\delta_x(z) - \delta_x(w))(\delta_y(z) - \delta_y(w)) \\
&= 1 - A^{G_1}_{xy},
\end{aligned}$$

if $x \neq y$, we have that

$$\langle \delta_x, \delta_y \rangle_{\mathcal{H}_{G_1}} = \langle U^* \delta_{\Phi(x)}, U^* \delta_{\Phi(y)} \rangle_{\mathcal{H}_{G_1}} = \langle \delta_{\Phi(x)}, \delta_{\Phi(y)} \rangle_{\mathcal{H}_{G_2}}.$$

Therefore $\{\Phi(x), \Phi(y)\}$ belongs to $E(G_2)$ if and only if $\{x, y\}$ belongs to $E(G_1)$. This concludes the proof.

By Theorems 1.32 and 1.33, the following three conditions are mutually equivalent:

1. G_1 and G_2 are isomorphic,
2. $\mathscr{H}_{G_1}$ and $\mathscr{H}_{G_2}$ are isomorphic,
3. $K_1 = K_2$, up to permutation.

We set $\|u\|_2 \equiv (\sum_{x\in V} |u(x)|^2)^{1/2}$ for every u in $\mathscr{F}$, and $l^2(G)$ will denote the Hilbert space over G with the norm $\|\cdot\|_2$. We set $\|u\|_\infty = \max_{x\in V} |u(x)|$ for every u in $\mathscr{F}$, and $l^\infty(G)$ will denote the Banach space over G with the norm $\|\cdot\|_\infty$.

Then we see:

Lemma 1.9. *Let G be a connected graph with n vertices, and let T be the operator from $\mathscr{H}_G$ into $l^2(G)$ defined by $Tu \equiv (u(x))_{x\in G}$. Then*

1. $TT^* = K$, more precisely,

$$TT^*\mathbf{a} = \left(\sum_{y\in G} \mathbf{a}(y)K(y,x)\right)_{x\in G}, \tag{1.213}$$

2. $T^*T = \sum_{x\in G} k_x \otimes k_x$, where $(u \otimes v)w \equiv \langle w, v\rangle u$.

Proof.

1. Let us now calculate the adjoint of T. Let $\mathbf{a} = (a_x)_{x\in G} \in \ell^2(G)$. Then we have

$$T^*\mathbf{a}(x) = \langle T^*\mathbf{a}, k_x\rangle = \langle \mathbf{a}, Tk_x\rangle = \sum_{y\in G} a_y Tk_x(y) = \sum_{y\in G} a_y k_x(y) = \sum_{y\in G} a_y k(y,x).$$

   Thus (1.213) follows.
2. From the above calculation,

$$T^*Tu = T^*(u(x))_{x\in G}) = \sum_{y\in G} u(y)k(y,\cdot).$$

   Meanwhile, $(k_y \otimes k_y)u = \langle u, k_y\rangle k_y = u(y)k(y,\cdot)$ from Remark 1.4. Thus, we obtain the desired result.

From the general property of reproducing kernel Hilbert spaces, we note that:

Lemma 1.10. *Let G be a connected graph. Then*

$$\|u\|_\infty^2 \le \{\max_{x\in V} k(x,x)\}\|u\|_{\mathscr{H}_G}^2.$$

Furthermore, we have:

Theorem 1.34. *Let G be a connected graph, and let $\sigma(K)$ denote the set of all eigenvalues of K. Then*

$$\max \sigma(K) \le |V| \max_{x\in V} k(x,x).$$

Proof. Let T be as in Lemma 1.9. Let u be an eigenvector with respect to an eigenvalue λ of T^*T. Then we have that

$$\lambda \|u\|_{\mathscr{H}}^2 = \langle \lambda u, u \rangle_{\mathscr{H}} = \langle T^* Tu, u \rangle_{\mathscr{H}} = \langle Tu, Tu \rangle_{l^2} = \|u\|_2^2.$$

It follows that $\sigma(T^*T) \subset \{\|u\|_2^2/\|u\|_{\mathscr{H}}^2 : u \neq 0\}$. Since $\sigma(TT^*) = \sigma(T^*T)$, we have that $\sigma(K) \subset \{\|u\|_2^2/\|u\|_{\mathscr{H}}^2 : u \neq 0\}$. By Lemma 1.10, we have that

$$\|u\|_2^2 \leq |V| \|u\|_\infty^2 \leq |V| \|u\|_{\mathscr{H}}^2 \max_{x \in V} k(x, x).$$

This concludes the proof.

Following the above settings we now connect graph theory and reproducing kernels and we will have further developments from this viewpoint. See [453] for further detailed results with the references [36, 309, 501]. We may consider the topics as a very interesting new research topics with connections to matrix theory and reproducing kernels.

1.7 Green's Functions and Reproducing Kernels

1.7.1 Basic Concepts of Dirac's Delta Function, Green's Functions and Reproducing Kernels

In the final part, we refer to fundamental concepts of Dirac's delta function, Green's functions and reproducing kernels. These concepts are fundamental, deep and have involved backgrounds. In order to look simply at their fundamental concepts, rough representations will be suitable.

For simplicity, we place ourselves in the setting of a bounded regular domain D whose boundary is made up of a finite number of analytic Jordan curves.

Consider a unit point mass distribution $\delta(p, q)$ (called *Dirac's delta function*) at $q \in D$, this will mean that, for example, for any continuous function f on D, we have

$$f(q) = \int_D f(p)\delta(p, q)dp. \tag{1.214}$$

We will be able to consider such a representation for various function spaces. This basic concept was found by a physician Paul Adrien Maurice Dirac and by field medalist L. Schwartz, who expanded this theory known as *distribution theory*. The representation will show that $\delta(p, q)$ is not the usual function, but a distribution.

Meanwhile, suppose that a solution $G(p,q)$ for some linear (differential) operator L on some function space on D is given by the equation, symbolically, for any fixed $q \in D$

$$LG(p,q) = \delta(p,q) \tag{1.215}$$

whose identity is valid on D except for the point $q \in D$. When G depends only on the distance $|p-q|$, then the function $G(p,q)$ will be called a *fundamental solution* for the operator L and further when some boundary conditions are satisfied, then the function $G(p,q)$ will be called a *Green's function* for the operator L satisfying the imposed boundary conditions.

In general, the function $G(p,q)$ may be understood physically as the *impulse response* for the system L.

1.7.2 Relationship Between Dirac's Delta Function, Green's Functions and Reproducing Kernels

For the adjoint operator L^* of L, we shall consider the self-adjoint operator L^*L and its Green's function $G(p,q)$ satisfying

$$L^*LG(p,q) = \delta(p,q) \tag{1.216}$$

at the formal level, whose identity is valid on D except for the point $q \in D$. Then, from (1.214) we obtain the representation

$$f(q) = \int_D f(p)L^*LG(p,q)dp. \tag{1.217}$$

Then, we can obtain the representation symbolically, using the Green Stokes formula,

$$f(q) = \int_D Lf(p)LG(p,q)dp + \text{ some boundary integrals.} \tag{1.218}$$

If the boundary integrals are zero, then we have

$$f(q) = \int_D Lf(p)LG(p,q)dp. \tag{1.219}$$

If the function space is made up of all f satisfying

$$\int_D |Lf(p)|^2 dp < \infty, \tag{1.220}$$

then the space forms a Hilbert space, and if the function $G(p, q)$ is the usual function on D belonging to this Hilbert space, then the function $G(p, q)$ will represent the reproducing property for the Hilbert space.

Indeed, we can find many cases satisfying these properties. In these frameworks, we can see the basic relationship between Dirac's delta function, Green's functions and reproducing kernels.

We will be able to consider Dirac's delta function, Green's functions and reproducing kernels as a family in the above sense.

1.7.3 The Simplest Example

In Theorem 1.7, recall the following.

For fixed $a, b > 0$,

$$G(s,t) \equiv \frac{1}{2ab}\exp\left(-\frac{b}{a}|s-t|\right) \quad (s,t \in \mathbb{R}) \tag{1.221}$$

is the reproducing kernel for the Hilbert space $H_G(\mathbb{R}) = W^{1,2}(\mathbb{R})$ equipped with the norm

$$\|f\|_{W^{1,2}(\mathbb{R})} = \sqrt{\int_{\mathbb{R}} (a^2|f'(x)|^2 + b^2|f(x)|^2)\,dx}. \tag{1.222}$$

The reproducing property is verified as follows: using (1.30) we have

$$\begin{aligned}
&\int_{\mathbb{R}} \left(a^2 f'(s)\frac{dG_t}{ds}(s) + b^2 f(s)G(s,t)\right) ds \\
&= \int_{-\infty}^{t} f'(s)\frac{dG_t}{ds}(s) + b^2 f(s)G(s,t)\,ds + \int_{t}^{\infty} a^2 f'(s)\frac{dG_t}{ds}(s) + b^2 f(s)G(s,t)\,ds \\
&= \frac{1}{2}\int_{-\infty}^{t} \left(\frac{b}{a}f(s) - f'(s)\right)\exp\left(\frac{b}{a}(s-t)\right) ds \\
&\quad + \frac{1}{2}\int_{t}^{\infty} \left(f'(s) + \frac{b}{a}f(s)\right)\exp\left(\frac{b}{a}(t-s)\right) ds \\
&= \frac{1}{2}\left[f(s)\exp\left(\frac{b}{a}(s-t)\right)\right]_{-\infty}^{t} - \frac{1}{2}\left[f(s)\exp\left(\frac{b}{a}(t-s)\right)\right]_{t}^{\infty} = f(t).
\end{aligned}$$

However, this will mean that

$$\int_{\mathbb{R}} f(s) \left(b^2 G_t(s) - a^2 \frac{d^2 G_t}{ds^2}(s) \right) ds = \int_{\mathbb{R}} \left(a^2 f'(s) \frac{dG_t}{ds}(s) + b^2 f(s) G(s,t) \right) ds$$
$$= f(t);$$

that is, $G(s,t)$ is the Green's function satisfying the differential equation

$$-a^2 y''(s) + b^2 y(s) = \delta(s,t)$$

on $\mathbb{R}$ satisfying the null property at ∞.

For some general and deep relationship between the Green's functions and the reproducing kernels, see [29]. We refer to [366, Chapter 4] for a beautiful discussion of the Green kernel on the complex plane.

Berlinet and C. Thomas-Agnan dealt with many concrete reproducing kernels in probability and statistics theory [48, Chapter 7]. For many other concrete reproducing kernels of the Sobolev type on intervals and rectangle domains, we refer to the books [118, 388] and references therein.

Chapter 2
Fundamental Properties of RKHS

In Chap. 2, which is at the heart of this book, we develop the general theory of reproducing kernels and its fundamental properties. First, we establish that any positive definite kernel corresponds uniquely to an RKHS. Then we discuss operations for RKHS. These operations will be fundamental for applications. After clarifying the basic properties, we will present the methods to construct RKHSs. One of the ideas is to use complete orthonormal systems or the Fourier integral. Finally, we discuss the relation between RKHS and linear mappings. One of the powerful properties that RKHSs enjoy is that we can characterize completely the image of linear mappings. But the proof of the results presented earlier than Sect. 2.4 is independent from Sect. 2.4.

We list [209, 347, 403] for the introduction of theory of reproducing kernel Hilbert spaces.

2.1 Basic Properties of RKHS

2.1.1 Elementary Properties of RKHS

As we will see when we consider a linear mapping in the framework of Hilbert spaces, its image space forms in a natural way a reproducing kernel Hilbert space. Therefore, we will be interested in a natural way in the properties of reproducing kernel Hilbert spaces. Here and below we denote by ev_p the *evaluation operator* (mapping) at $p \in E$. Also, we denote by $\mathscr{F}(E)$ the set of all functions defined on E which assume their value in $\mathbb{R}$ or $\mathbb{C}$. We put a topology on $\mathscr{F}(E)$ by the family $\{\mathrm{ev}_p\}_{p \in E}$. Thus, a mapping Φ from a topological space X to $\mathscr{F}(E)$ is continuous, if and only if $x \in X \mapsto \Phi(x)(p) \in \mathbb{R}$ (or $\mathbb{C}$) is continuous for all $p \in E$. Theorem 2.1 shows that a reproducing kernel Hilbert space is a very good and natural function space.

S. Saitoh, Y. Sawano, *Theory of Reproducing Kernels and Applications*,
Developments in Mathematics 44, DOI 10.1007/978-981-10-0530-5_2

Denote by $\mathbb{K}$ either $\mathbb{R}$ or $\mathbb{C}$.

Lemma 2.1. *Suppose that H is a Hilbert space which is made up of $\mathbb{K}$-valued functions defined on E. The embedding $H \subset \mathscr{F}(E)$ is continuous if and only if*

$$\|\mathrm{ev}_p|_H\|_{H\to\mathbb{C}} = \sup\{|f(p)| \,:\, f \in H,\ \|f\|_H = 1\} < \infty \tag{2.1}$$

for all $p \in E$.

Proof. Suppose $H \hookrightarrow \mathscr{F}(E)$. Then the mapping $f \in H \to \mathrm{ev}_p(f) = f(p) \in \mathbb{K}$ is a continuous mapping. Therefore (2.1) holds.

Suppose instead that (2.1) holds. Then if $\{f_k\}_{k=1}^{\infty}$ is a sequence in H that converges to $f \in H$, then

$$|f(p) - f_k(p)| \le \sup\{|g(p)| \,:\, f \in H,\ \|g\|_H = 1\} \times \|f - f_k\|_H$$

implies $f_k(p) \to f(p)$. Thus $H \hookrightarrow \mathscr{F}(E)$.

Theorem 2.1 ([185, Theorem 5]). *Suppose that H is a Hilbert space consisting of functions on a set E.*

1. *The Hilbert space H is realized as a reproducing kernel Hilbert space $H_K(E)$ with a positive definite quadratic form function $K : E \times E \to \mathbb{C}$ if and only if the embedding $H \subset \mathscr{F}(E)$ is continuous.*
2. *If a sequence $\{f_j\}_{j=1}^{\infty}$ in $H_K(E)$ converges to f in $H_K(E)$, then*

$$\lim_{j\to\infty} f_j(p) = f(p) \tag{2.2}$$

for all $p \in E$. Furthermore, on any subset of E on which $p \mapsto K(p,p)$ is bounded, its convergence is uniform on there.

As is written in [316, p. 71], one reason for using RKHSs in approximation theory and numerical analysis is that strong convergence in RKHS implies pointwise convergence. This fact can be quantified as in the theorem above.

Proof.

1. If $H = H_K(E)$ for some positive definite function $K : E \times E \to \mathbb{C}$, then we have a reproducing formula

$$f(p) = \langle f, K_p \rangle_H \quad (p \in E),$$

which together with the Cauchy Schwarz inequality implies the continuity of the embedding $H = H_K(E) \hookrightarrow \mathscr{F}(E)$.

We need to show sufficiency. Let $p \in E$. Then by the Riesz representation theorem, we have

$$f(p) = \mathrm{ev}_p(f) = \langle f, h_p \rangle_H, f \in H \tag{2.3}$$

for some $h_p \in H$. Now let us consider

$$K(p,q) \equiv h_q(p),\ p,q \in E. \tag{2.4}$$

We need to check $H = H_K(E)$ with norm-coincidence.

Let us check first that $\{h_q\}_{q\in E}$ spans a dense subspace of H. To this end, let $f \in H$ be orthogonal to h_p for all $p \in E$. Then by (2.1), $f(p) = \langle f, h_p \rangle_H = 0$ for all $p \in E$. Thus, $f \equiv 0$.

Next, let us verify that the inner products of H and $H_K(E)$ coincide for the restriction to $\mathrm{Span}(\{K_q\}_{q\in E})$. We have

$$K(p,q) = h_q(p) = \langle h_q, h_p \rangle_H \tag{2.5}$$

from (2.3) and (2.4). Meanwhile since $K_q = h_q$, we have

$$K(p,q) = \langle K_q, K_p \rangle_{H_K(E)} = \langle h_q, h_p \rangle_{H_K(E)}. \tag{2.6}$$

Therefore it follows from (2.5) and (2.6) that the inner products of $H_K(E)$ and H coincide in the linear space $\mathrm{Span}(\{K_q\}_{q\in E})$.

Since $\mathrm{Span}(\{K_q\}_{q\in E})$ is dense both in $H_K(E)$ and H, it follows that $H_K(E) = H$ with norm-coincidence.

2. (2.2) as well as its uniform convergence on the set where $K(p,p)$ is bounded is a direct consequence of the Hölder inequality and the reproducing property.

2.1.2 *Positive Definite Quadratic form Functions*

Let us start by recalling the definition of positive definite quadratic form functions. Related to this terminology, we have similar ones, like *positive matrix* (N. Aronszajn), *positive definite function* (S. Bochner) – in this case, in Fourier integral forms – *positive type function* (A. Berlinet and C. Thomas-Agnan), and *positive definite matrix* in the sense of Moore (B. Okutmustur and A. Gheondea).

Definition 2.1. A complex-valued function $k : E \times E \to \mathbb{C}$ is called a *positive definite quadratic form function* on the set E, or *positive definite function*, when it satisfies the property that for an arbitrary function $X : E \to \mathbb{C}$ and any finite subset F of E,

$$\sum_{p,q\in F} \overline{X(p)} X(q) k(p,q) \geq 0. \tag{2.7}$$

Example 2.1. Assume that $\rho : \mathbb{R} \to [0,\infty)$ is integrable. Then $K : \mathbb{R} \times \mathbb{R} \to \mathbb{C}$, which is given by

$$K(x,y) \equiv \frac{1}{(2\pi)^n} \int_{\mathbb{R}} \exp(-it \cdot (x-y))\rho(t)\, dt \ (x,y \in \mathbb{R}), \tag{2.8}$$

is a positive definite quadratic form function. Just check (2.7).

See [447] for more examples of positive definite functions including the history.

As we see easily from (2.7), a reproducing kernel $K(p,q)$ on E is a positive definite quadratic form function on E, however, its converse statement is valid and important:

Theorem 2.2. *For any positive definite quadratic form function $K : E \times E \to \mathbb{C}$, there exists a uniquely determined reproducing kernel Hilbert space $H_K = H_K(E)$ admitting the reproducing kernel K on E.*

To prove Theorem 2.2, we recall that the evaluation mapping is defined by

$$\mathrm{ev}_p(f) \equiv f(p) \quad p \in E. \tag{2.9}$$

We put a topologize on $\mathscr{F}(E)$ with the family $\{\mathrm{ev}_p\}_{p\in E}$.

Lemma 2.2. *Define a linear subspace $H_{0,K}(E) \subset \mathscr{F}(E)$ by*

$$H_{0,K}(E) \equiv \mathrm{Span}\{K_p \ : \ p \in E\}, \tag{2.10}$$

that is,

$$H_{0,K}(E) = \bigcup_{N=1}^{\infty} \left\{ \sum_{j=1}^{N} \alpha_j K_{p_j} \ : \ p_1, p_2 \ldots, p_N \in E,\ \alpha_1, \alpha_2, \ldots, \alpha_N \in \mathbb{C} \right\}.$$

Then the function

$$\left(\sum_{j=1}^{N} \alpha_j K_{p_j}, \sum_{k=1}^{M} \beta_k K_{q_k} \right) \in H_{0,K}(E) \times H_{0,K}(E) \mapsto \sum_{j=1}^{N} \sum_{k=1}^{M} \alpha_j \overline{\beta_k} K(q_k, p_j) \in \mathbb{C} \tag{2.11}$$

defines a well-defined inner product. In particular, via this inner product (2.11), *we can equip $H_{0,K}(E)$ with the structure of a pre-Hilbert space.*

Proof. What is difficult about the proof is that the definition makes sense. That is, we have to show that the definition does not depend on the presentation of

$$\sum_{j=1}^{N} \alpha_j K_{p_j}.$$

It can happen that $\{K_p\}_{p\in E}$ is linearly dependent. In this case, we have to be more careful. See Example 2.2 below.

Let us consider the function

$$F = \sum_{j=1}^{N} \alpha_j K_{p_j} \in H_{0,K}(E) \mapsto \left\| \sum_{j=1}^{N} \alpha_j K_{p_j} \right\|_{H_{0,K}(E)} = \sqrt{\sum_{j,k=1}^{N} \alpha_j \overline{\alpha_k} K(p_k, p_j)} \in [0, \infty).$$

As for the proof, it is reduced to proving that

$$\sum_{j,k=1}^{N} \alpha_j \overline{\alpha_k} K(p_k, p_j) = \sum_{j,k=1}^{M} \beta_j \overline{\beta_k} K(q_k, q_j), \tag{2.12}$$

if we assume that $F \in H_{0,K}(E)$ admits two different expressions:

$$F = \sum_{j=1}^{N} \alpha_j K_{p_j} = \sum_{j=1}^{M} \beta_j K_{q_j}. \tag{2.13}$$

This can be achieved as follows: First, we compute that

$$\begin{aligned} \sum_{j,k=1}^{N} \alpha_j \overline{\alpha_k} K(p_k, p_j) &= \sum_{k=1}^{N} \overline{\alpha_k} \sum_{j=1}^{N} \alpha_j K(p_k, p_j) \\ &= \sum_{k=1}^{N} \overline{\alpha_k} F(p_k) \\ &= \sum_{k=1}^{N} \overline{\alpha_k} \sum_{j=1}^{M} \beta_j K(p_k, q_j). \end{aligned}$$

Now that $K(*_1, *_2) = \overline{K(*_2, *_1)}$, we have

$$\sum_{j,k=1}^{N} \alpha_j \overline{\alpha_k} K(p_k, p_j) = \sum_{k=1}^{N} \overline{\alpha_k} \left(\sum_{j=1}^{M} \beta_j \overline{K(q_j, p_k)} \right) = \sum_{j=1}^{M} \beta_j \overline{\left(\sum_{k=1}^{N} \alpha_k K(q_j, p_k) \right)}.$$

Inserting (2.13) once more, we obtain

$$\sum_{j,k=1}^{N} \alpha_j \overline{\alpha_k} K(p_k, p_j) = \sum_{j=1}^{M} \beta_j \overline{\left(\sum_{k=1}^{M} \beta_k K(q_j, q_k) \right)} = \sum_{j,k=1}^{M} \beta_j \overline{\beta_k} K(q_k, q_j),$$

which yields (2.12).

Now we prove Lemma 2.2. Note that

$$\mathrm{I} \equiv \left\langle \sum_{j=1}^{N} \alpha_j K_{p_j}, \sum_{k=1}^{M} \beta_k K_{q_k} \right\rangle_{H_{0,K}(E)} \tag{2.14}$$

can be obtained by polarization, that is, we have

$$\begin{aligned}\mathrm{I} = &\frac{1}{4}\left\| \sum_{j=1}^{N} \alpha_j K_{p_j} + \sum_{k=1}^{M} \beta_k K_{q_k} \right\|_{H_{0,K}(E)}^2 - \frac{1}{4}\left\| \sum_{j=1}^{N} \alpha_j K_{p_j} - \sum_{k=1}^{M} \beta_k K_{q_k} \right\|_{H_{0,K}(E)}^2 \\ &+ \frac{i}{4}\left\| \sum_{j=1}^{N} \alpha_j K_{p_j} + i\sum_{k=1}^{M} \beta_k K_{q_k} \right\|_{H_{0,K}(E)}^2 - \frac{i}{4}\left\| \sum_{j=1}^{N} \alpha_j K_{p_j} - i\sum_{k=1}^{M} \beta_k K_{q_k} \right\|_{H_{0,K}(E)}^2 .\end{aligned}$$

This identity shows that the definition of the inner product is meaningful.

Example 2.2. If $K \equiv 1$, then $\{K_p\}_{p \in E}$ is linearly dependent.

Lemma 2.3 below shows that $H_{0,K}(E)$ enjoys the reproducing property and that ev_p is continuous.

Lemma 2.3. *Let $f \in H_{0,K}(E)$. Then we have the reproducing identity*

$$f(p) = \langle f, K_p \rangle_{H_{0,K}(E)} \quad (p \in E). \tag{2.15}$$

Furthermore,

$$|f(p)| \le \sqrt{K(p,p)}\|f\|_{H_{0,K}(E)} \quad (p \in E). \tag{2.16}$$

Proof. Equality (2.15) is an immediate consequence of the definition of the inner product. Estimate (2.16) is obtained from the first equality (2.15) and the Cauchy Schwarz inequality.

We define

$$H_K(E) \equiv \{f \in \mathscr{F}(E) : f \text{ is a pointwise limit of a Cauchy sequence in } H_{0,K}(E)\}$$

and the norm is by

$$\|f\|_{H_K(E)} = \lim_{k \to \infty} \|f_k\| \tag{2.17}$$

for $f \in H_K(E)$, where $\{f_k\}_{k=1}^{\infty}$ is a Cauchy sequence in $H_{0,K}(E)$ converging to f pointwise. The space $H_K(E)$ carries the natural structure of the Hilbert space induced by $H_{0,K}(E)$.

Proposition 2.1. *The Hilbert space $H_K(E)$ is a unique Hilbert space such that the following conditions are fulfilled:*

1. *$H_K(E)$ is a linear subspace of $\mathscr{F}(E)$ and the original topology of $H_K(E)$ is stronger than the one via the embedding $H_K(E) \subset \mathscr{F}(E)$,*
2. *$K_q \in H_K(E)$ for all $q \in E$ and $\{K_q\}_{q \in E}$ spans a dense subspace of $H_K(E)$,*
3. *$f(p) = \langle f, K_p \rangle_{H_K(E)}$ for all $f \in H_K(E)$ and $p \in E$.*

In particular, we have: $K(p,q) = \langle K_q, K_p \rangle_{H_K(E)}$ for all $p, q \in E$.

The third property is the reproducing property for the function $K(\cdot, p)$ in $H_K(E)$.

Proof. It is easy to see that $H_K(E)$ defined above satisfies the above properties. Let H be another Hilbert space satisfying the above conditions. Then we have

$$\langle K_q, K_p \rangle_H = \langle K_q, K_p \rangle_{H_K(E)}. \tag{2.18}$$

Now that $H_{0,K}(E)$ is dense in H and $H_K(E)$ respectively, in view of Proposition 1.1, we see that H and $H_K(E)$ coincide as a Hilbert space.

We can guarantee the unique correspondence of the space $H_K(E)$ and the kernel K. The function $K : E \times E \to \mathbb{C}$ is unique in the following sense:

Proposition 2.2. *Let $L : E \times E \to \mathbb{C}$ be a function enjoying the following two properties:*

$$L_p \in H_K(E), \tag{2.19}$$

$$f(p) = \langle f, L_p \rangle_{H_K(E)} \text{ for all } f \in H_K(E). \tag{2.20}$$

Then $L \equiv K$.

Before we come to the proof, we remark that K, constructed above, satisfies the above properties; see Lemma 2.3.

Proof. Let $p, q \in E$ be fixed. Then we have $L(p,q) = \langle L_q, K_p \rangle_{H_K(E)}$ by the reproducing property (2.19) of K. Now that L satisfies the reproducing property (2.20), we obtain

$$L(p,q) = \langle L_q, K_p \rangle_{H_K(E)} = \overline{\langle K_p, L_q \rangle_{H_K(E)}} = \overline{K(q,p)}. \tag{2.21}$$

Since $K(p,q) = \overline{K(q,p)}$, we obtain the desired result.

Example 2.3. For any complex-valued function $f \in \mathscr{F}(E) \setminus \{0\}$, we consider a positive definite quadratic function $K = f \otimes \overline{f}$. Then, the reproducing kernel Hilbert space $H_K(E)$ is given by

$$H_K(E) = \{af : a \in \mathbb{C}\}, \ \langle af, bf \rangle_{H_K(E)} = a\overline{b} \quad a, b \in \mathbb{C}. \tag{2.22}$$

Example 2.4. For any set E, the function K given by

$$K(p,q) \equiv \delta(p,q) = \begin{cases} 1 & p = q, \\ 0 & \text{otherwise} \end{cases}$$

is positive definite. The corresponding space $H_K(E)$ is given by:

$$H_K(E) = \left\{ f : E \to \mathbb{C} \,:\, f \text{ assumes zero with finite exception and} \sum_{x \in E} |f(x)|^2 < \infty \right\}.$$

from the construction above. The inner product is

$$\langle f, g \rangle = \sum_{x \in E} f(x)\overline{g(x)}$$

for $f, g \in H_K(E)$. It matters that E can be a set containing uncountably many points. The identity $\langle f, \delta_p \rangle_{\ell^2(E)} = f(p)$ implies that the function δ_p is the reproducing kernel of $\ell^2(E)$ at $p \in E$ where δ_p is the function defined by $\delta_p(x) = \delta_{xp}$ (Kronecker's delta). Hence, $\ell^2(E)$ is an RKHS on E.

We see that this is a prototype of separable reproducing kernel Hilbert spaces. Indeed, as we will establish in Theorem 2.27, for any separable reproducing kernel Hilbert space $H_K(E)$, we have

$$K(p,q) = \sum_{j \in J} f_j(p)\overline{f_j(q)} \tag{2.23}$$

for some $\{f_j\}_{j \in J} \subset H_K(E)$ indexed by an at most countable set J.

Finally, in Sect. 2.5.2 we present certain fundamental property of $H_K(E)$. We give a criterion for any function f to vanish at a fixed point p_0.

Proposition 2.3. *Suppose that $K : E \times E \to \mathbb{C}$ is a positive definite quadratic form function on a set E. Let $p_0 \in E$. Then*

$$f(p_0) = 0 \tag{2.24}$$

for all $f \in H_K(E)$ if and only if $K(p_0, p_0) = 0$.

Proof. If $K(p_0, p_0) = 0$, then, as we have verified before, $K_{p_0} \equiv 0$. As a consequence, we have

$$f(p_0) = \langle f, K_{p_0} \rangle_{H_K(E)} = 0 \tag{2.25}$$

for all $f \in H_K(E)$.

Suppose conversely that $f(p_0) = 0$ for all $f \in H_K(E)$. Then, letting $f = K_{p_0} \in H_K(E)$, we have

$$K(p_0, p_0) = f(p_0) = 0. \tag{2.26}$$

In view of (2.25) and (2.26), we have the desired result.

Example 2.5. Let

$$K(s,t) \equiv \int_0^\infty \frac{\sin(s\,u)\sin(t\,u)}{u^2+1}\,du = \frac{\pi}{4}\,(\exp(-|s-t|) - \exp(-s-t))$$

for $s, t > 0$ as in Theorem 1.5. We extend in a natural way any function in $H_K(0,\infty)$ to $\{0\}$ by letting $f(0) = 0$ for any $f \in H_K(0,\infty)$. Then $H_K[0,\infty)$ obtained in this way is an example of Proposition 2.3.

2.1.3 Smoothness of Functions and Kernels of RKHS

We suppose here that E carries the structure of a topological space in discussing the topological property of the kernel function K.

Theorem 2.3 (Continuity of functions in $H_K(E)$). *Suppose that E is a topological space and that $K : E \times E \to \mathbb{C}$ is a positive definite quadratic form function. Then K is continuous if and only if the mapping $p \in E \mapsto K_p \in H_K(E)$ is continuous. If this is the case, then all the functions in $H_K(E)$ are continuous in E.*

Proof. Continuity of K is trivially equivalent to continuity of $p \in E \mapsto K_p \in H_K(E)$ because $K(p,q) = \langle K_q, K_p \rangle_{H_K(E)}$ and

$$\|K_p - K_q\|_{H_K(E)} = \sqrt{K(p,p) - K(p,q) - K(q,p) + K(q,q)}$$

for all $p, q \in E$.

Assume that K is continuous and hence the mapping $p \in E \mapsto K_p \in H_K(E)$ is continuous. Then every function in $H_{0,K}(E)$ is continuous. Since

$$|f(p)| = |\langle f, K_p \rangle_{H_K(E)}| \le \sqrt{K(p,p)}\|f\|_{H_K(E)}, \tag{2.27}$$

the norm convergence in $H_K(E)$ implies uniform convergence on $\{p \in E : K(p,p) < M\}$ for every $M > 0$. Hence it follows that every function in $H_K(E)$ is continuous.

Next, we turn to the differentiability of the functions. The most fundamental theorem is given by the following

Theorem 2.4. *Let $H_K(E)$ be a separable reproducing kernel Hilbert space. Choose an orthonormal basis $\{e_j\}_{j=1}^{\infty} \subset H_K(E)$. If $\ell : H_K(E) \to \mathbb{C}$ is a bounded linear operator, then*

$$\ell \bowtie K \equiv \sum_{j=1}^{\infty} \overline{\ell(e_j)} e_j \tag{2.28}$$

converges in $H_K(E)$ and does not depend on the choice of $\{e_j\}_{j=1}^{\infty}$.

Proof. To prove that the right-hand side of (2.28) actually converges, it suffices to prove that $\{\ell(e_j)\}_{j=1}^{\infty}$ is an $\ell^2(\mathbb{N})$-sequence. To this end, we take $\{a_j\}_{j=1}^{\infty} \subset \ell^2(\mathbb{C})$ such that $a_j = 0$ with finite exception. Then we have

$$\left|\sum_{j=1}^{\infty} a_j \ell(e_j)\right| = \left|\ell\left(\sum_{j=1}^{\infty} a_j e_j\right)\right| \leq \|\ell\| \left|\sum_{j=1}^{\infty} a_j e_j\right| = \|\ell\| \sqrt{\sum_{j=1}^{\infty} |a_j|^2}.$$

Thus $\{\ell(e_j)\}_{j=1}^{\infty} \in \ell^2(\mathbb{N})$ by the (converse) of the Cauchy Schwarz inequality. To prove that the limit does not depend the choice of $\{e_j\}_{j=1}^{\infty}$, we choose another orthonormal basis $\{f_j\}_{j=1}^{\infty} \subset H_K(E)$. Write

$$f_k \equiv \sum_{i=1}^{\infty} a_{ik} e_i,$$

where $\{a_{ik}\}_{i,k=1}^{\infty}$ satisfies

$$\sum_{i=1}^{\infty} a_{ik_1} \overline{a_{ik_2}} = \delta_{k_1 k_2} \quad 1 \leq k_1, k_2 < \infty.$$

Then for all $j \in \mathbb{N}$,

$$\left\langle \sum_{k=1}^{\infty} \overline{\ell(e_k)} e_k, f_j \right\rangle = \sum_{k=1}^{\infty} \overline{\ell(e_k)} \overline{a_{kj}} = \sum_{k=1}^{\infty} \overline{\ell(a_{kj} e_k)} = \overline{\ell(f_j)}.$$

Hence, the right-hand side of (2.28) does not depend on the choice of $\{e_j\}_{j=1}^{\infty}$.

Remark 2.1. When we are given $f \in H_K(E)$, we regard f as the linear operator by $g \in H_K(E) \mapsto \langle g, f \rangle_{H_K(E)}$ and denote it by $f \bowtie K$.

The differential properties of functions can be inherited the kernel functions nicely as the following theorems assert:

Define

$$C^{1,1}(O \times O) \equiv \{f \in C(O \times O) \,:\, \partial_x f, \partial_y f, \partial_x \partial_y f, \partial_y \partial_x f \text{ exists and continuous in } O\}.$$

For any $M \in \mathbb{N}$ $C^{M,M}(O \times O)$ is defined analogously.

Theorem 2.5. *Let O be an open set in $\mathbb{R}^n$ and let $1 \leq j \leq n$ be an integer. If a positive definite $C^{1,1}(O \times O)$-function $K : (p,q) \in O \times O \mapsto K(p,q) \in \mathbb{C}$, then $\partial p_j \partial q_j K$ is a positive definite kernel and we have*

$$\|\partial_j f\|_{H_{\partial p_j \partial q_j K(O)}} \leq \|f\|_{H_K(O)}. \tag{2.29}$$

If $H_K(O)$ contains 0 as the unique constant function, then we have

$$\|\partial_j f\|_{H_{\partial p_j \partial q_j K(O)}} = \|f\|_{H_K(O)}. \tag{2.30}$$

Proof. We can show that $\partial_{q_j} K \in H_K(O)$ by considering

$$\left\| \frac{K(\cdot, q + h\mathbf{e}_j) - K(\cdot, q)}{h} - \frac{K(\cdot, q + h'\mathbf{e}_j) - K(\cdot, q)}{h'} \right\|_{H_K(O)},$$

where $\mathbf{e}_j$ is the j-th elementary vector. That $\partial p_j \partial q_j K$ is a positive definite kernel can be proved by a simple limit argument and Theorem 2.4. Define $P : H_K(O) \to H_K(O)$ as the projection into the space of all functions that are perpendicular to any constant function in $H_K(O)$. Let us set

$$\mathscr{H} \equiv \{\partial_j f \,:\, f \in H_K(O)\}, \tag{2.31}$$

and equip $\mathscr{H}$ with the inner product given by

$$\langle \partial_j f, \partial_j g \rangle_{\mathscr{H}} = \langle P(f), P(g) \rangle_{H_K(O)}. \tag{2.32}$$

Let us establish that $\mathscr{H} = H_{\partial p_j \partial q_j K}$ with inner product coincidence.

Since

$$\partial_{p_j} \partial_{q_j} K(p,q) = \langle (\partial_{q_j} K)_q, (\partial_{q_j} K)_p \rangle_{H_K(O)}, \tag{2.33}$$

we have

$$\partial_{p_j} \partial_{q_j} K(p,q) = \frac{\partial}{\partial p_j} \left[\langle \partial_{q_j} K_q, K_p \rangle_{H_K(O)} \right] = \frac{\partial [\partial_{jq} K_q]}{\partial p_j}(p) \tag{2.34}$$

by Theorem 2.4. Hence it follows that $\partial p_j \partial q_j K \in \mathscr{H}$.

Next we observe that

$$\langle f, \partial_{qj} K_q \rangle_{H_K(O)} = \partial_{q_j} f(q) = 0, \tag{2.35}$$

where $f \equiv c$ is a constant function. Hence we obtain

$$P(\partial_{q_j} K_q) = \partial_{q_j} K_q. \tag{2.36}$$

Hence it follows that

$$\begin{aligned}\langle \partial_{p_j} \partial_{q_j} K_q, \partial_{p_j} \partial_{q_j} K_p \rangle_{\mathscr{H}} &= \langle P((\partial_{q_j} K)_q), P((\partial_{q_j} K)_p) \rangle_{H_K(O)} \\ &= \langle (\partial_{q_j} K)_q, (\partial_{q_j} K)_p \rangle_{H_K(O)} \\ &= \partial_{p_j} \partial_{q_j} K(p, q)\end{aligned}$$

for all $p, q \in E$. Therefore, in view of these two observations, we obtain $\mathscr{H} = H_{K_{p_j q_j}}(E)$. Once this is established, it is easy from the definition of P to deduce the conclusions of the theorem.

Theorem 2.6 (Differential properties of functions in $H_K(O)$). *Suppose that O is an open set in $\mathbb{R}^n$ and that $K \in C^{M,M}(O \times O)$, $M \in \mathbb{N}$. Then:*

1. *$\partial_q^r K(\cdot, q) \in H_K(O)$ for all $q \in O$ and all $r \in \mathbf{Z}_+{}^n$ with length less than or equal to M,*
2. *Any function in $H_K(O)$ belongs to $C^M(O)$, More precisely,*

$$\partial^r f(q) = \langle f, \partial_q^r K(\cdot, q) \rangle_{H_K(O)} \quad (q \in O). \tag{2.37}$$

Here and below, for the proof we use

$$e_k = (\overbrace{0, 0, \ldots, 0}^{k-1 \text{ times}}, 1, \overbrace{0, 0, \ldots, 0}^{n-k \text{ times}})$$

to denote the k-th fundamental vector.

Proof. If $M = 0, 1$, then the assertion is trivial. Assume that the assertion is true for $M = M_0$.

Now suppose that $K \in C^{M_0+1, M_0+1}(O)$. It then follows from induction the assumption that $\partial_q^r K(p, q) \in H_K(O)$, if $|r| = M_0$. Suppose that $1 \le k \le n$ is given. Then we have

$$\partial_p^r \partial_q^r K(p, q) = \langle \partial_q^r K_q, \partial_q^r K_p \rangle_{H_K(O)}. \tag{2.38}$$

From this we can readily deduce that

$$\|\Delta_{h_1,q} - \Delta_{h_2,q}\|_{H_K(O)} = \sqrt{\sum_{i,j=1}^{2}(-1)^{i+j}\int_0^1\int_0^1 \partial_p^\alpha \partial_q^\alpha K(q+s\,h_i, q+t\,h_j)\,ds\,dt} \to 0$$

as $h_1, h_2 \to 0$, where

$$\Delta_{h,q} \equiv \frac{\partial^\alpha K_{q+h\,e_k} - \partial^\alpha K_q}{h}, \; h \neq 0. \tag{2.39}$$

As a result, we see that

$$\lim_{h\to 0} \frac{\partial^r K_{q+h\,e_k} - \partial^r K_q}{h} \tag{2.40}$$

exists in the topology of $H_K(O)$. However, convergence in $H_K(O)$ implies pointwise convergence. Therefore, it follows that

$$\partial^{r+e_j} K_q = \lim_{h\to 0} \frac{\partial^r K_{q+h\,e_k} - \partial^r K_q}{h} \tag{2.41}$$

takes place in the topology of $H_K(O)$.

With (2.41) established, we obtain

$$\partial^r f(p) = \langle f, \partial_q^r K_p \rangle_{H_K(O)} \tag{2.42}$$

for all multi-indexes r with length less than $M_0 + 1$. It remains to proving that $\partial^r f$ is continuous on O, which amounts to proving that $p \mapsto \partial_q^r K_p$ is continuous in the topology of $H_K(O)$. From what we have just proved we see that

$$\begin{aligned}
&\|\partial_q^r K_p - \partial_q^r K_{p'}\|_{H_K(O)}^2 \\
&= \partial_p^r \partial_q^r K(p,p) - \partial_p^r \partial_q^r K(p',p) - \partial_p^r \partial_q^r K(p,p') + \partial_p^r \partial_q^r K(p',p').
\end{aligned}$$

Letting $p' \to p$, we obtain that $f \in C^{M_0+1}(O)$.

Remark 2.2. We can use (2.37) to deduce that $\partial_p^r \partial_q^r K$ is a positive definite form function under the same assumption as in Theorem 2.6.

Even the similar property of analyticity can be transferred easily:

Theorem 2.7. *Let $\Omega \subset \mathbb{C}^n$ be a domain. Suppose that $K : \Omega \times \Omega \to \mathbb{C}$ is a continuous positive definite kernel. Assume in addition that K_q is analytic for all $q \in \Omega$. Then any function in $H_K(\Omega)$ is analytic in Ω.*

Proof. This fact follows from the Cauchy Riemann equation and Hartog's theorem of separation of holomorphy.

As in this corollary, we can derive many properties for the expansion functions g_j from the properties of the kernel K, such interesting properties were derived by M. G. Krein [261] independent of the theory of reproducing kernels.

See [272, Theorem 1] and [280, 301] for some applications of the continuity of the reproducing kernel Hilbert spaces to continuous stochastic process.

Example 2.6. In Theorem 2.4, we let $E = \{1, 2, \ldots, N\}$ and $K(i,j) \equiv \delta(i,j)$ for $i, j \in E$. Let $a_1, a_2, \ldots, a_N \in \mathbb{C}$ and define

$$\ell(f) \equiv \sum_{j=1}^{N} a_j f(j) \quad (f \in \mathscr{F}(E)).$$

Let us set $f_j(i) \equiv \delta(i,j)$ for $j \in E$. Then

$$\ell \bowtie K = \sum_{j=1}^{N} \overline{\ell(f_j)} f_j = \sum_{j=1}^{N} \overline{a_j} f_j.$$

For any positive definite quadratic form function K on a set E, there exists a uniquely determined reproducing kernel Hilbert space $H_K(E)$ admitting the kernel K. We can consider intuitively the relations induced from relations among positive definite quadratic form functions. This problem will propose still up-to-date new problems. We will examine such problems induced from the sum, restriction, product and other transforms of positive definite quadratic functions and their basic general applications.

2.2 Operations of One Reproducing Kernel Hilbert Spaces

2.2.1 *Restriction of a Reproducing Kernel*

Now suppose that we are given a positive definite quadratic form function $K : E \times E \to \mathbb{C}$. The main concern here is to consider the restriction of K to $E_0 \times E_0$, where E_0 is a subset of E. Of course, the restriction is again a positive definite quadratic form function on the subset $E_0 \times E_0$. We will consider the relation between the two reproducing kernel Hilbert spaces.

In Sect. 2.2.1 we aim to prove the following result:

Theorem 2.8 (Restriction of RKHS). *Suppose that $K : E \times E \to \mathbb{C}$ is a positive definite quadratic form function on a set E. Let E_0 be a subset of E. Then the Hilbert space that $K|E_0 \times E_0 : E_0 \times E_0 \to \mathbb{C}$ defines is given by*

$$H_{K|E_0 \times E_0}(E_0) = \{f \in \mathscr{F}(E_0) : f = \tilde{f}|E_0 \text{ for some } \tilde{f} \in H_K(E)\}. \tag{2.43}$$

Furthermore, the norm is expressed in terms of $H_K(E)$:

$$\|f\|_{H_{K|E_0 \times E_0}(E_0)} = \min\{\|\tilde{f}\|_{H_K(E)} : \tilde{f} \in H_K(E), f = \tilde{f}|E_0\}. \tag{2.44}$$

For the proof, let us set

$$H_0(E_0) \equiv \{f \in \mathscr{F}(E_0) : \text{ there exists } \tilde{f} \in H_K(E) \text{ such that } f = \tilde{f}|E_0\}. \tag{2.45}$$

The proof of Theorem 2.8 is made up of Propositions 2.4 and 2.5.

Proposition 2.4.

1. *Let $f \in H_0(E_0)$. Then the set $\left\{\|\tilde{f}\|_{H_K(E)} : \tilde{f} \in H_K(E), \tilde{f}|E_0 = f\right\} \subset [0, \infty)$ has a minimum.*
2. *If we define*

$$\|f\|_{H_0(E_0)} \equiv \min\left\{\|\tilde{f}\|_{H_K(E)} : \tilde{f} \in H_K(E), \tilde{f}|E_0 = f\right\} \tag{2.46}$$

for $f \in H_0$, then $(H_0(E_0), \|\cdot\|_{H_0(E_0)})$ is a Hilbert space.

Proof.

1. Choose a minimizing sequence $\{f_j\}_{j=1}^{\infty} \subset H_K(E)$. Then since $H_K(E)$ is a Hilbert space and $\{f_j\}_{j=1}^{\infty} \subset H_K(E)$ is a bounded sequence, it follows from the Banach Alaoglu theorem that we can assume that $\{f_j\}_{j=1}^{\infty} \subset H_K(E)$ is weakly convergent. Due to the reproducing property, weak convergence implies pointwise convergence. Therefore, the limit f of the weakly convergent sequence $\{f_j\}_{j=1}^{\infty} \subset H_K(E)$ attains the minimum.
2. Just recall the parallelogram law.

We use the Banach Alaoglu theorem, see [304, 372, 406, 458, 483] for example for the statement.

Proposition 2.5. *Let $f \in H_0(E_0)$. Then we have*

$$f(p) = \langle f, (K|E_0 \times E_0)_p \rangle_{H_0(E_0)} \tag{2.47}$$

for all $p \in H_0(E_0)$.

In particular $H_0(E_0) \subset H_{K|E_0 \times E_0}(E_0)$.

Proof. Let $\tilde{f} \in H_K(E)$ be chosen so that $f = \tilde{f}|E_0$ and $\|f\|_{H_0(E_0)} = \|\tilde{f}\|_{H_K(E)}$. Notice that

$$\|(K|E_0 \times E_0)_p\|_{H_0(E_0)} = \|K_p\|_{H_K(E)}$$

for all $p \in E_0$. Indeed, for any $g \in H_K(E)$ such that $g|E_0 \equiv 0$,

$$\begin{aligned}\|K_p + g\|_{H_K(E)}^2 &= \|K_p\|_{H_K(E)}^2 + \|g\|_{H_K(E)}^2 + 2\mathrm{Re}(\langle g, K_p \rangle_{H_K(E)}) \\ &= \|K_p\|_{H_K(E)}^2 + \|g\|_{H_K(E)}^2.\end{aligned}$$

Thus we have

$$\langle f, (K|E_0 \times E_0)_p \rangle_{H_0(E)} = \langle \tilde{f}, K_p \rangle_{H_K(E)} = \tilde{f}(p) \tag{2.48}$$

for all $p \in E_0$. Here for the first equality we have used the projection property of $H_0(E_0)$.

Proof (of Theorem 2.8).

1. Equation (2.43) follows from Proposition 2.5.
2. Notice that (2.47) implies that $H_0(E_0) = H_{K|E_0 \times E_0}(E_0)$ with inner product coincidence.

It may be interesting to compare Theorem 2.8 with [207, Theorem 4].

Remark 2.3. Let E_0 be a subset of E. Suppose that $f \in H_K(E_0)$. Then we can find $\tilde{f} \in H_K(E)$ such that $\tilde{f}|E_0 = f$ and $\|f\|_{H_{K|E_0 \times E_0}(E_0)} = \|\tilde{f}\|_{H_K(E)}$. Observe that

$$\{f^\dagger \in H_K(E) \,:\, f^\dagger|E_0 = f\} = \{\tilde{f} + g \,:\, g \in H_K(E), \,:\, g|E_0 = 0\}.$$

Let $h \in H_K(E)$. Then $h \in H_K(E)$ satisfies $\|h\|_{H_K(E)} = \|h|E\|_{H_{K|E_0 \otimes E_0}(E_0)}$ if and only if h is perpendicular to $\{g \in H_K(E) \,:\, g|E_0 = 0\}$.

In general, for a reproducing kernel K defined on a set E and a subset F of E, the relation between H_K and $H_{K|F \times F}$ is very complicated.

Example 2.7. Theorem 1.7 shows that the function $K : \mathbb{R} \times \mathbb{R} \to \mathbb{C}$ given by

$$K(x, y) = \frac{1}{2} \exp(-|x - y|) \quad (x, y \in \mathbb{R}) \tag{2.49}$$

is the reproducing kernel for the Hilbert space $W^{1,2}(\mathbb{R})$ with the norm

$$\|f\|_{W^{1,2}(\mathbb{R})} \equiv \sqrt{\int_{\mathbb{R}} (|f'(x)|^2 + |f(x)|^2)\, dx}. \tag{2.50}$$

Meanwhile, by Theorem 1.3, its restriction to the finite interval $[a, b]$ is the reproducing kernel for the Sobolev space $W^{1,2}[a, b]$, where the inner product is given by

$$\langle f_1, f_2 \rangle_{W^{1,2}[a,b]} \equiv f_1(a) f_2(a) + f_1(b) f_2(b) + \int_a^b (f_1(x) f_2(x) + f_1'(x) f_2'(x))\, dx. \tag{2.51}$$

When reproducing kernels are analytic for the first variable z and are anti-analytic for the second variable u, their restrictions to some subsets will have mysteriously deep structures, see [388, Sections 2.6–2.12] for many concrete examples. Of course, these results may be related to analytic extension problems.

2.2.2 *Pullback of a Reproducing Kernel by Any Mapping*

Here we generalize Sect. 1.4.7. Suppose that we have a positive definite quadratic form function $K : E \times E \to \mathbb{C}$ and an arbitrary mapping $\varphi : F \to E$. Then define $\varphi^*K : F \times F \to \mathbb{C}$ by:

$$\varphi^*K(p,q) \equiv K(\varphi(p),\varphi(q)) \quad (p,q \in F), \tag{2.52}$$

which is a positive definite quadratic form function.

Let us recall again that ev_p was defined by (2.9). As for this positive definite quadratic form function, we have the following:

Theorem 2.9 (Pullback of RKHS). *Set*

$$\mathscr{H}(E) \equiv \bigcap_{p \in F} \ker(\mathrm{ev}_{\varphi(p)}) \subset H_K(E).$$

Denote by $\mathscr{H}^{\perp}(E)$ the orthogonal complement of $\mathscr{H}(E)$ in $H_K(E)$ and by P the projection from $H_K(E)$ to $\mathscr{H}^{\perp}(E) \subset H_K(E)$. Then the pullback $H_{\varphi^*K}(F)$ is described as follows:

$$H_{\varphi^*K}(F) = \{f \circ \varphi \,:\, f \in H_K(E)\} \tag{2.53}$$

as a set and the inner product is

$$\langle f \circ \varphi, g \circ \varphi \rangle_{H_{\varphi^*K}(F)} = \langle Pf, Pg \rangle_{H_K(E)} \tag{2.54}$$

for all $f, g \in H_K(E)$.

Proof. Let $f \in \mathscr{H}(E)$. Then $\langle f, K_{\varphi(p)} \rangle_{H_K(E)} = f(\varphi(p)) = 0$ for all $p \in F$. Therefore, since f is arbitrary we have $K_{\varphi(p)} \in \mathscr{H}^{\perp}(E)$ and hence

$$P\left[K_{\varphi(p)}\right] = K_{\varphi(p)}. \tag{2.55}$$

With (2.55) in mind, we let $\mathscr{H}$ to be the pre-Hilbert space given by the right-hand side of (2.53) equipped with the inner product (2.54). It is easy to check that $\mathscr{H}$ is a Hilbert space and that (2.54) is well defined. Since $(\varphi^*K)_q(p) = \varphi^*K(p,q) = K(\varphi(p),\varphi(q)) = K_{\varphi(q)}(\varphi(p))$, for all $p \in E$, it follows that $(\varphi^*K)_q = K_{\varphi_q} \circ \varphi \in \mathscr{H}$ for all $q \in E$. In view of (2.55), we have $\langle f \circ \varphi, (\varphi^*K)_p \rangle_{\mathscr{H}} = \langle f \circ \varphi, K_{\varphi_p} \circ \varphi \rangle_{\mathscr{H}} = \langle f, K_{\varphi_p} \rangle_{H_K(E)} = f \circ \varphi(p)$ for all $p \in E$. Hence it follows that $\mathscr{H} = H_{\varphi^*K}(F)$ with coincidence of the inner product.

Before we consider some applications, it may be interesting to reconsider the restriction using the inclusion mapping $\varphi = i : F \hookrightarrow E$.

Example 2.8. Let F be a subset of E. Set

$$\mathscr{H}(E) \equiv \bigcap_{p \in F} \ker(\mathrm{ev}_p) \subset H_K(E).$$

Denote by P the projection from $H_K(E)$ to $\mathscr{H}^{\perp}(E) \subset H_K(E)$. Then the pullback $H_{K|F\times F}(F)$ is described as follows: $H_{K|F\times F}(F) = \{f|F : f \in H_K(E)\}$ as a set of functions and the inner product is $\langle f|F, g|F\rangle_{H_{K|F\times F}(F)} = \langle Pf, Pg\rangle_{H_K(E)}$ for all $f, g \in H_K(E)$.

2.2.3 Balloon of Reproducing Kernel Hilbert Spaces

Suppose that $K : E\times E \to \mathbb{C}$ is a positive definite quadratic form function and $\mathscr{H}$ is a Hilbert space. Let $T \in \mathscr{B}(H_K(E), \mathscr{H})$. We define another inner product on $H_K(E)$ by:

$$\langle f, g\rangle_{H_K(E:\mathscr{H})} \equiv \langle f, g\rangle_{H_K(E)} + \langle Tf, Tg\rangle_{\mathscr{H}} \quad (f, g \in H_K(E)). \tag{2.56}$$

Let us consider a linear space $H_K(E)$ having the inner product $\langle\cdot,\cdot\rangle_{H_K(E:\mathscr{H})}$. Now that the evaluation is trivially continuous, we see that the Hilbert space $(H_K(E), \langle\cdot,\cdot\rangle_{H_K(E:\mathscr{H})})$ admits a reproducing kernel $K_{\mathscr{H}}$.

Theorem 2.10. *The function $K_{\mathscr{H}} : E\times E \to \mathbb{C}$ satisfies the following equation and the solution is unique*:

$$\begin{cases} (K_{\mathscr{H}})_q \in H_K(E) \text{ for all } q \in E, \\ K(p,q) = K_{\mathscr{H}}(p,q) + \langle T[(K_{\mathscr{H}})_q], TK_p\rangle_{\mathscr{H}} \text{ for all } p, q \in E. \end{cases} \tag{2.57}$$

Proof. The fact that $(K_{\mathscr{H}})_q \in H_K(E)$ for all $q \in E$ follows from the definition of $H_K(E)$ with the inner product $\langle\cdot,\cdot\rangle_{H_K(E:\mathscr{H})}$; $(K_{\mathscr{H}})_q \in H_K(E;\mathscr{H}) = H_K(E)$. The equation in (2.57) follows because

$$\begin{aligned} K_{\mathscr{H}}(p,q) + \langle T(K_{\mathscr{H}})_q, TK_p\rangle_{\mathscr{H}} &= \langle (K_{\mathscr{H}})_q, K_p\rangle_{H_K(E)} + \langle T[(K_{\mathscr{H}})_q], TK_p\rangle_{\mathscr{H}} \\ &= K_{\mathscr{H}}(p,q) + \langle T[(K_{\mathscr{H}})_q], TK_p\rangle_{\mathscr{H}} \\ &= \langle (K_{\mathscr{H}})_q, K_p\rangle_{H_K(E)} = \overline{\langle K_p, (K_{\mathscr{H}})_q\rangle_{H_K(E)}} \\ &= \overline{K(q,p)} = K(p,q). \end{aligned}$$

This is the desired result.

Let us prove the uniqueness. Observe that $K_q = (K_{\mathscr{H}})_q + T^*T[(K_{\mathscr{H}})_q]$. Suppose that $\overline{K}$ is another solution to (2.57). Then we have $K_q = (\overline{K})_q + T^*T[(\overline{K})_q]$ and hence

$$
\begin{aligned}
&\|(K_{\mathscr{H}})_q - \overline{K}_q\|_{H_K(E)}^2 + \|T(K_{\mathscr{H}})_q - T\overline{K}_q\|_{H_K(E)}^2 \\
&= \langle (K_{\mathscr{H}})_q - \overline{K}_q, (1 + T^*T)((K_{\mathscr{H}})_q - \overline{K}_q)\rangle_{H_K(E)} \\
&= \langle (K_{\mathscr{H}})_q - \overline{K}_q, K_q - K_q\rangle_{H_K(E)} \\
&= 0,
\end{aligned}
$$

proving that $\overline{K} = K_{\mathscr{H}}$.

The next theorem will appear several times in this book.

Theorem 2.11 (Highlight of several points). *Suppose that we are given a finite number of points $\Theta = \{\theta_j\}_{j=1}^N \subset E$ and a positive sequence $\{\lambda_j\}_{j=1}^N$. If we set*

$$
A_\Theta \equiv \{A_{\Theta j,j'}\}_{j,j'=1}^N \equiv ({}^t\{\delta_{j,j'} + \lambda_j K(\theta_{j'}, \theta_j)\}_{j,j'=1}^N)^{-1}
$$

$$
K_\Theta(p,q) \equiv K(p,q) - \sum_{j,j'=1}^N \lambda_j K(p,\theta_j) A_{\Theta j,j'} K(\theta_{j'}, q)
$$

for $p, q \in E$. Then $H_{K_\Theta}(E) = H_K(E)$ as a set and the inner product of $H_{K_\Theta}(E)$ is

$$
\langle f, g\rangle_{H_{K_\Theta}(E)} = \langle f, g\rangle_{H_K(E)} + \sum_{j=1}^N \lambda_j f(\theta_j)\overline{g(\theta_j)} \quad f, g \in H_K(E). \tag{2.58}
$$

Proof. Let us define an inner product $\langle\cdot,\cdot\rangle_*$ on $H_K(E)$ by the right-hand side of (2.58).
Let $f \in H_K(E)$. Then we have

$$
\langle f, (K_\Theta)_p\rangle_* - f(p) = \langle f, (K_\Theta)_p\rangle_{H_K(E)} - f(p) + \sum_{j=1}^n \lambda_j f(\theta_j)\overline{K_\Theta(\theta_j, p)}
$$

and hence by the definition of K_θ, we have

$$
\begin{aligned}
&\langle f, (K_\Theta)_p\rangle_* - f(p) \\
&= -\sum_{j,j'=1}^N \lambda_j \cdot \overline{A_{\Theta j,j'} \cdot K(\theta_{j'}, p)} \cdot \langle f, K_{\theta_j}\rangle_{H_K(E)} + \sum_{j=1}^N \lambda_j f(\theta_j)\overline{K_\Theta(\theta_j, p)} \\
&= -\sum_{j,j'=1}^N \lambda_j \cdot \overline{A_{\Theta j,j'} \cdot K(\theta_{j'}, p)} \cdot f(\theta_j) + \sum_{j=1}^N \lambda_j f(\theta_j)\overline{K_\Theta(\theta_j, p)}.
\end{aligned}
$$

Since

$$
\begin{aligned}
&K_\Theta(\theta_m,p)\\
&\quad= K(\theta_m,p)-\sum_{j,j'=1}^{N}\lambda_j\,K(\theta_m,\theta_j)\,A_{\Theta,j,j'}K(\theta_{j'},p)\\
&\quad= K(\theta_m,p)-\sum_{j,j'=1}^{N}(\delta_{mj}+\lambda_j\,K(\theta_m,\theta_j))\,A_{\Theta,j,j'}K(\theta_{j'},p)+\sum_{j'=1}^{N}A_{\Theta,m,j'}K(\theta_{j'},p)\\
&\quad=\sum_{j'=1}^{N}A_{\Theta,m,j'}K(\theta_{j'},p),
\end{aligned}
$$

we see $f(p)=\langle f,(K_\theta)_p\rangle_*$ and that $H_K(E)=H_{K_\Theta}(E)$ as a set of functions and that the inner product is given as in the above.

Example 2.9. Let us consider the case when $N=1$; $\Theta=\{\theta_1\}$. Then

$$
K_\Theta(p,q)=K(p,q)-\frac{\lambda_1}{1+\lambda_1K(\theta_1,\theta_1)}K(p,\theta_1)K(\theta_1,q)\quad (p,q\in E).
$$

Example 2.10. Let us consider the case when $N=2$. If $N=2$ and $\xi_1,\xi_2\in E$, then we have

$$
\begin{aligned}
&M_{\{\theta_1,\theta_2\},\lambda_1,\lambda_2}(\xi_1,\xi_2)\\
&\quad\equiv \det\begin{pmatrix}1+\lambda_1\,K(\xi_1,\xi_1) & \lambda_1\,K(\xi_2,\xi_1)\\ \lambda_2\,K(\xi_1,\xi_2) & 1+\lambda_2\,K(\xi_2,\xi_2)\end{pmatrix}\\
&\quad=1+\lambda_1\,K(\xi_1,\xi_1)+\lambda_2\,K(\xi_2,\xi_2)+\lambda_1\,\lambda_2\,(K(\xi_1,\xi_1)K(\xi_2,\xi_2)-|K(\xi_1,\xi_2)|^2)
\end{aligned}
$$

and

$$
\begin{pmatrix}A_{\{\theta_1,\theta_2\},1,1} & A_{\{\theta_1,\theta_2\},1,2}\\ A_{\{\theta_1,\theta_2\},2,1} & A_{\{\theta_1,\theta_2\},2,2}\end{pmatrix}=\frac{1}{M_{\lambda_1,\lambda_2}(\xi_1,\xi_2)}\begin{pmatrix}1+\lambda_2\,K(\xi_2,\xi_2) & -\lambda_1\,K(\xi_2,\xi_1)\\ -\lambda_2\,K(\xi_1,\xi_2) & 1+\lambda_1\,K(\xi_1,\xi_1)\end{pmatrix}.
$$

As a result, we obtain

$$
\begin{aligned}
&K_{\lambda_1,\lambda_2,\xi_1,\xi_2}(x,y)\\
&\quad=K(x,y)+\frac{\lambda_1\,\lambda_2(K(\xi_1,\xi_2)K(x,\xi_1)K(\xi_2,y)+K(\xi_2,\xi_1)K(x,\xi_2)K(\xi_1,y))}{M_{\lambda_1,\lambda_2}(\xi_1,\xi_2)}\\
&\qquad-\frac{\lambda_1(1+\lambda_2\,K(\xi_2,\xi_2))K(x,\xi_1)K(\xi_1,y)+\lambda_2(1+\lambda_1K(\xi_1,\xi_1))K(x,\xi_2)K(\xi_2,y)}{M_{\lambda_1,\lambda_2}(\xi_1,\xi_2)}.
\end{aligned}
$$

Example 2.11. Consider the case when $E = \mathbb{R}$ and $H_K(\mathbb{R}) = W^{2,2}(\mathbb{R})$. Let $T : H_K(\mathbb{R}) \to \mathbb{C}$ be $Tf = f'(0)$. Equip $H_K(\mathbb{R})$ with an inner product

$$\langle f, g\rangle^* = \langle f, g\rangle_{H_K(\mathbb{R})} + f'(0)\overline{g'(0)} \quad (f, g \in H_K(\mathbb{R})).$$

Here we calculate the solution $K_{\mathscr{H}}(p, q)$ to

$$K(p, q) = K_{\mathscr{H}}(p, q) + \langle T(K_{\mathscr{H}})_q, TK_p\rangle_{\mathscr{H}} \text{ for all } p, q \in E = \mathbb{R}. \tag{2.59}$$

Let $p, q \in E$. In our case this Eq. (2.59) reads as

$$K(p, q) = K_{\mathscr{H}}(p, q) + \partial_p K_{\mathscr{H}}(0, q)\overline{\partial_p K(0, p)}. \tag{2.60}$$

Thus we see that $K_{\mathscr{H}}(p, q) = K(p, q) - a(q)\overline{\partial_p K(0, p)}$ for some function $a : \mathbb{R} \to \mathbb{C}$. Observe also that $K_{\mathscr{H}}(p, q) = \overline{K_{\mathscr{H}}(q, p)} = K(p, q) - \overline{a(p)}\partial_p K(0, q)$. This implies $a(q)\overline{\partial_p K(0, p)} = \overline{a(p)}\partial_p K(0, q)$. Note that this is equivalent to $a(q) = \kappa \partial_p K(0, q)$ for some constant $\kappa \in \mathbb{C}$. Hence,

$$K_{\mathscr{H}}(p, q) = K(p, q) - \kappa\overline{\partial_p K(0, p)}\partial_p K(0, q). \tag{2.61}$$

Remark that $\partial_p \partial_q K(0, 0) \in \mathbb{R}$. By inserting (2.61) into (2.60), we obtain

$$\begin{aligned} K(p, q) &= K(p, q) - \kappa\overline{\partial_p K(0, p)}\partial_p K(0, q) \\ &\quad + \partial_p K(0, q)\overline{\partial_p K(0, p)} - \kappa\partial_p K(0, q)\partial_q\partial_p K(0, 0)\overline{\partial_p K(0, p)}. \end{aligned}$$

Consequently, we obtain

$$\kappa = \frac{1}{1 + \partial_q \partial_p K(0, 0)}$$

and hence

$$K_{\mathscr{H}}(p, q) = K(p, q) - \frac{\overline{\partial_p K(0, q)}\partial_p K(0, p)}{1 + \partial_q \partial_p K(0, 0)}.$$

Example 2.12. Suppose that we are given a continuous operator $\ell : H_K(E) \to \mathbb{C} = \mathscr{H}$ and $\lambda > 0$. Equip $H_K(E)$ with an inner product

$$\langle f, g\rangle_\dagger \equiv \langle f, g\rangle_{H_K(E)} + \ell(f)\overline{\ell(g)} \quad (f, g \in H_K(E)).$$

Observe that the evaluation mapping $f \in H_K(E) \mapsto f(p) \in \mathbb{C}$ is continuous from $(H_K(E), \langle\cdot,\cdot\rangle_\dagger)$ to $\mathbb{C}$. Let us find the reproducing kernel $K_{\mathscr{H}}$ of the Hilbert space $(H_K(E), \langle\cdot,\cdot\rangle_\dagger)$.

If $\ell = 0$, then we know that $K_{\mathscr{H}} = K$. So let us suppose otherwise. Fix $p, q \in E$. According to (2.57), we need to solve

$$K(p,q) = K_{\mathscr{H}}(p,q) + \ell((K_{\mathscr{H}})_q)\overline{\ell(K_p)}. \tag{2.62}$$

Similar to (2.61), we know that $K_{\mathscr{H}}(p,q) = K(p,q) - \kappa\ell(K_q)\overline{\ell(K_p)}$ and that

$$K_{\mathscr{H}}(p,q) = K(p,q) - \overline{\ell((K_{\mathscr{H}})_p)}\ell(K_q) \quad (p, q \in E).$$

Since $\ell \neq 0$, $\ell((K_{\mathscr{H}})_q) = \overline{\kappa}\ell(K_q)$ for some complex number κ. Thus, $K_{\mathscr{H}}(p,q) = K(p,q) - \kappa\ell(K_q)\overline{\ell(K_p)}$. Recall that we have defined an operation $\ell \bowtie K$ in Theorem 2.4. Note that in Theorem 2.4 we needed to assume that $H_K(E)$ is separable but later in Example 2.18, we can pass to the general case. If we insert (4.27) into (2.62), then we have $K(p,q) = K(p,q) - \kappa\ell(K_q)\overline{\ell(K_p)} - \ell(\{K_q - \kappa\ell(K_q)\ell \bowtie K\})\overline{\ell(K_p)}$. Thus it follows that

$$\kappa = \frac{1}{1 + \ell(\ell \bowtie K)}$$

and

$$K_{\mathscr{H}}(p,q) = K(p,q) - \frac{\ell(K_q)\overline{\ell(K_p)}}{1 + \ell(\ell \bowtie K)} \quad (p, q \in E). \tag{2.63}$$

Theorem 2.10 with $\|L\|_{H_K \to \mathscr{H}} \ll 1$ is taken up in [410, Theorem 2.4], where a geometric series is employed.

2.2.4 *Squeezing of RKHS*

Suppose that $K : E \times E \to \mathbb{C}$ is a positive definite quadratic function and that $\mathscr{H}$ is a Hilbert space, as before. Assume that $T \in \mathscr{B}(H_K(E), \mathscr{H})$ satisfies

$$\|Tf\|_{\mathscr{H}} < \|f\|_{H_K(E)}, \text{ for all } f \in H_K(E) \setminus \{0\}. \tag{2.64}$$

Then define an inner product $\langle \cdot, \cdot \rangle_{H_K^-(E:\mathscr{H})}$ by

$$\langle f, g \rangle_{H_K^-(E:\mathscr{H})} \equiv \langle f, g \rangle_{H_K(E)} - \langle Tf, Tg \rangle_{\mathscr{H}} \quad (f, g \in H_K(E)). \tag{2.65}$$

We shall denote by $H_K^-(E : \mathscr{H})_0$ the pre-Hilbert space $H_K(E)$ equipped with the inner product $\langle \cdot, \cdot \rangle_{H_K^-(E:\mathscr{H})}$ above. Since $\|f\|_{H_K^-(E:\mathscr{H})} \leq \|f\|_{H_K(E)}$ for all $f \in H_K(E)$, it then follows that $H_K(E) \hookrightarrow H_K^-(E : \mathscr{H})_0$. Denote by $H_K^-(E : \mathscr{H})$ the completion of $H_K(E)$ with respect to the inner product $\langle \cdot, \cdot \rangle_{H_K^-(E:\mathscr{H})}$.

About the Hilbert space $H_K^-(E : \mathscr{H})$, we have the following assertion:

Theorem 2.12 (Squeezing of RKHS). *Let $T \in \mathscr{B}(H_K(E), \mathscr{H})$ satisfy (2.64) and define an inner product $\langle \cdot, \cdot \rangle_{H_K^-(E:\mathscr{H})}$ by (2.65). Then $H_K^-(E : \mathscr{H})$ is realized as a RKHS for some positive definite quadratic function $L : E \times E \to \mathbb{C}$, if and only if*

$$\sup_{t \in [0,1)} [(1 - tT^*T)^{-1} K_p](p) < \infty \tag{2.66}$$

for all $p \in E$. In this case,

$$L(p, q) = \lim_{t \uparrow 1} [(1 - tT^*T)^{-1} K_q](p) \quad (p, q \in E). \tag{2.67}$$

Proof. We write $K_t(p, q) = [(1 - tT^*T)^{-1} K_q](p)$ for $p, q \in E$.

1. Suppose that $H_K^-(E : \mathscr{H})$ is realized as $H_L(E)$ for some positive definite quadratic function $L : E \times E \to \mathbb{C}$. Let $t \in [0, 1)$. If we define the inner product

$$\langle f, g \rangle_{H_K^-(E:\mathscr{H};t)} \equiv \langle f, g \rangle_{H_K(E)} - t \langle Tf, Tg \rangle_{\mathscr{H}}$$

for $f, g \in H_K(E)$, then we have that $(H_K(E), \langle \cdot, \cdot \rangle_{H_K^-(E:\mathscr{H};t)})$ is embedded into $H_L(E)$ with the embedding constant less than 1. Since the kernel of the Hilbert space $(H_K(E), \langle \cdot, \cdot \rangle_{H_K^-(E:\mathscr{H};t)})$ is given by the function

$$(p, q) \in E \times E \mapsto [(1 - tT^*T)^{-1} K_q](p),$$

similar to Theorem 2.10, we have

$$\begin{aligned} [(1 - tT^*T)^{-1} K_p](p) &= \langle (1 - tT^*T)^{-1} K_p, L_p \rangle_{H_K^-(E:\mathscr{H})} \\ &\le \sqrt{L(p, p)} \|(1 - tT^*T)^{-1} K_p\|_{H_K^-(E:\mathscr{H})} \\ &\le \sqrt{L(p, p)} \|(1 - tT^*T)^{-1} K_p\|_{H_K^-(E:\mathscr{H};t)} \\ &\le \sqrt{L(p, p)[(1 - tT^*T)^{-1} K_p](p)}. \end{aligned}$$

If we arrange this inequality, we obtain (2.66).

2. Suppose instead (2.66) holds. Let us check that $H_K^-(E : \mathscr{H})$ is realized by a reproducing kernel Hilbert space and that the kernel is given by (2.67). First, since $K_t - K_s$ is positive definite for $0 < s < t < 1$, the limit

$$L(p, q) \equiv \lim_{t \uparrow 1} K_t(p, q)$$

exists. We need to check (1.1) and (1.2), which amounts to showing that

$$L_q \in H_K^-(E : \mathscr{H}) \tag{2.68}$$

and

$$f(q) = \langle f, L_q \rangle_{H_K^-(E:\mathscr{H})} \quad (f \in H_K^-(E:\mathscr{H})), \tag{2.69}$$

respectively.

Let us check (2.68). Note that $H_K(E) \subset H_L(E)$ because $K \ll L$ and hence we obtain that

$$H_K^-(E:\mathscr{H}) \hookrightarrow H_L(E) \hookrightarrow \mathscr{F}(E). \tag{2.70}$$

Meanwhile, we calculate that

$$\begin{aligned}
&\|(K_t)_q - (K_s)_q\|_{H_K^-(E:\mathscr{H})} \\
&= (t-s)\|(1-T^*T)(1-tT^*T)^{-1}(1-sT^*T)^{-1}K_q\|_{H_K(E)} \\
&\le (t-s)\|(t^{-1}-T^*T)(1-tT^*T)^{-1}(1-sT^*T)^{-1}K_q\|_{H_K(E)} \\
&= (t-s)t^{-1}\|(1-sT^*T)^{-1}K_q\|_{H_K(E)} \\
&\le (t-s)t^{-1} \sup_{u\in[0,1)} [(1-uT^*T)^{-1}K_q](q).
\end{aligned}$$

Consequently, by the completeness of $H_K^-(E:\mathscr{H})$, the limit

$$A_{t,q} = \lim_{t\uparrow 1}(K_t)_q$$

exists in $H_K^-(E:\mathscr{H})$. Since the pointwise topology is shown to be weaker than the $H_K^-(E:\mathscr{H})$-topology, we see that $A_{t,q}(p) = L(p,q)$ and hence (2.68) holds.

To check (2.69), by the fact that $\mathrm{Span}(\{K_p\}_{p\in E})$ is dense in $H_K(E)$, we can assume that $f \in H_K(E)$. In this case, we have

$$\begin{aligned}
\langle f, L_q \rangle_{H_K^-(E:\mathscr{H})} &= \lim_{t\uparrow 1}\langle f, (1-tT^*T)^{-1}K_q\rangle_{H_K^-(E:\mathscr{H})} \\
&= \lim_{t\uparrow 1}\langle f, (1-tT^*T)^{-1}(1-T^*T)K_q\rangle_{H_K(E)} \\
&= \langle f, K_q\rangle_{H_K(E)} \\
&= f(p).
\end{aligned}$$

Thus we obtain (2.69).

Several helpful examples may be in order.

Example 2.13. Let $E = \mathbb{N}$ and $K(p,q) = \delta(p,q)$, $p,q \in E$ as usual. Define a bounded linear operator $T(\{a_j\}_{j=1}^\infty) \equiv \{\sqrt{1-j^{-1}}a_j\}_{j=1}^\infty$. Then

$$H_K^-(E:\mathscr{H}) = \left\{ \{b_j\}_{j=1}^\infty \,:\, \|\{b_j\}_{j=1}^\infty\|_{H_K^-(E:\mathscr{H})} = \sqrt{\sum_{j=1}^\infty \frac{|b_j|^2}{j}} < \infty \right\}.$$

We give an example of T satisfying (2.64) and having norm 1.

Example 2.14. Let $\{a_{ij}\}_{i,j=1}^\infty$ be a doubly indexed sequence such that

$$\sum_{j=1}^\infty a_{ij}\overline{a_{jk}} = \delta_{ik},\ a_{1j} = 2^{-j/2}$$

for all $j, k \in \mathbb{N}$. Define a unitary transform U on $\ell^2(\mathbb{N})$ by

$$U(\{b_j\}_{j=1}^\infty) \equiv \left\{ \sum_{k=1}^\infty a_{jk} b_k \right\}_{j=1}^\infty, \quad \{b_j\}_{j=1}^\infty \in \ell^2(\mathbb{N}).$$

Let T_0 be a bounded linear operator given by

$$T_0(\{b_j\}_{j=1}^\infty) \equiv \left\{ \sqrt{1-3^{-j}}\, b_j \right\}_{j=1}^\infty, \quad \{b_j\}_{j=1}^\infty \in \ell^2(\mathbb{N}).$$

Let $K : \mathbb{N} \times \mathbb{N} \to \mathbb{R}$ be the Dirac delta. Then $T = U^* T_0 U$ fails (2.66). Indeed, as $t \uparrow 1$,

$$(1 - tT^*T)^{-1} K_1(1) = \sum_{j=1}^\infty \frac{1}{2^j(1 - t + t3^{-j})} \uparrow \infty.$$

2.2.5 Transforms of RKHS by Operators

Here we suppose that we have a Hilbert space $\mathscr{H}$ and a mapping h from E to $\mathscr{H}$. Assume that $\mathscr{H}$ is densely spanned by $\{h(p) : p \in E\}$. Let $f \in \mathscr{H}$. Then define

$$Lf(p) \equiv \langle f, h(p) \rangle_{\mathscr{H}} \quad (p \in E). \tag{2.71}$$

If we set $K(p,q) \equiv \langle h(q), h(p) \rangle_{\mathscr{H}}$ for $p, q \in E$, then we can quantify the above observation as follows:

$$H_K(E) = \{ Lf \in \mathscr{F}(E) : f \in \mathscr{H} \} \tag{2.72}$$

and

$$\|Lf\|_{H_K(E)} = \|f\|_{\mathscr{H}} \quad (f \in \mathscr{H}) \tag{2.73}$$

according to Theorem 2.37 to follow. Note that $L : \mathscr{H} \to H_K(E)$ is bijective in view of Theorem 2.37, since $\{h(p) : p \in E\}$ spans a dense subspace of $\mathscr{H}$. Now let $A \in \mathscr{B}(\mathscr{H})$. Then define a function $K_A : E \times E \to \mathbb{C}$ by

$$K_A(p,q) \equiv \langle A[h(q)], A[h(p)]\rangle_{\mathscr{H}}. \tag{2.74}$$

The aim of Sect. 2.2.5 is to investigate the relation between $H_K(E)$ and $H_{K_A}(E)$.

We use Theorem 2.17, which is proved independently. Note that $H_{K_A}(E) \hookrightarrow H_K(E)$ because $K_A \ll \|A\|^2 K$, that is

$$\{\|A\|^2 K(p_i,p_j) - K_A(p_i,p_j)\}_{i,j=1}^N$$

is positive semidefinite in the sense of matrices. Given $F \in H_K(E)$, we define a function $F_A \in \mathscr{F}(E)$ by

$$F_A(p) \equiv \langle L^{-1}F, A[h(p)]\rangle_{\mathscr{H}}, \ (p \in E). \tag{2.75}$$

We can check that $F_A \otimes F_A \ll \gamma^2 K_A$ with $\gamma \equiv \|L^{-1}F\|_{\mathscr{H}}$. Therefore, $F_A \in H_{K_A}(E)$ by Corollary 2.1 to follow. Let $f \in \mathscr{H}$. Set

$$L_A f(p) \equiv \langle f, A[h(p)]\rangle_{\mathscr{H}}, \ (p \in E). \tag{2.76}$$

Remark that $F_A = L_A[L^{-1}F]$ for all $F \in H_K(E)$ from (2.75) and (2.76).

The following theorem is of fundamental importance:

Theorem 2.13 (Transforms of RKHS by operators). *Under the notations* (2.71), (2.74) *and* (2.75), *we have:*

1. *for all* $f \in H_K(E)$

$$\|F_A\|_{H_{K_A}(E)} \leq \|f\|_{H_K(E)}, \tag{2.77}$$

2. *the mapping*

$$F \in H_K(E) \mapsto F_A = L_A[L^{-1}F] \in H_{K_A}(E) \tag{2.78}$$

is isometric if and only if $\{A[h(p)] : p \in E\}$ *spans a dense subspace of* $\mathscr{H}$. *If* (2.78) *is isometric, then the mapping* (2.78) *is bijective.*

Proof.

1. Let $F = Lf$ for some $f \in \mathscr{H}$ and write $P \equiv P_{\mathscr{H} \to \ker(L_A)^\perp}$. Then we have

$$\|F_A\|_{H_{K_A}(E)} = \|L_A[L^{-1}F]\|_{H_{K_A}(E)} = \|PL^{-1}F\|_{\mathscr{H}} \leq \|L^{-1}F\|_{\mathscr{H}} = \|F\|_{H_K(E)}, \tag{2.79}$$

which proves (2.77).

2. Equality in (2.79) holds for all $f \in H_K(E)$ if and only if $\ker(L_A) = 0$. Note that

$$\begin{aligned}\ker(L_A) = 0 &\iff L_A f(p) = 0 \text{ for all } p \in E \text{ implies } f = 0\\ &\iff \langle f, A[h(p)]\rangle_{\mathscr{H}} = 0 \text{ for all } p \in E \text{ implies } f = 0\\ &\iff \{A[h(p)] \,:\, p \in E\} \text{ spans a dense subspace in } \mathscr{H}.\end{aligned}$$

As a consequence we see that the mapping $f \in H_K(E) \mapsto f_A \in H_{K_A}(E)$ is isometric if and only if $\{A[h(p)] \,:\, p \in E\}$ spans a dense subspace.

The surjectivity of the mapping (2.78) is clear if we assume it is isometry, because we know that $f_A = L_A[L^{-1}f], f \in H_K(E)$ and $L_A : \mathscr{H} \mapsto H_{K_A}(E)$ and $L : \mathscr{H} \to H_K(E)$ are both isomorphic.

The next theorem explains how to obtain the inverse.

Theorem 2.14. *Keep to the same notation above. Assume $F \in H_K(E) \mapsto F_A \in H_{K_A}(E)$ is isometry. Then the inverse mapping of* (2.78) *is*

$$G \in H_{K_A}(E) \mapsto [p \in E \mapsto \langle G, (K_A)_p\rangle_{H_{K_A}(E)} \in \mathbb{C}] \in H_K(E), \tag{2.80}$$

namely, if we set

$$F(p) \equiv \langle G, (K_A)_p\rangle_{H_{K_A}(E)} \quad (p \in E),$$

then $F \in H_K(E)$ and $L_A[L^{-1}F] = G$.

Proof. In view of Theorem 2.13 we have only to show that

$$F(p) = \langle F_A, (K_A)_p\rangle_{H_{K_A}(E)}, \, p \in E \tag{2.81}$$

for all F of the form $F = K_q$ for some $q \in E$. Indeed, we are assuming that $F \in H_K(E) \mapsto F_A \in H_{K_A}(E)$ is isomorphic and $\{K_p \,:\, p \in E\}$ spans a dense subspace of $H_K(E)$.

From the definition we obtain

$$\langle (K_A)_q, (K_A)_p\rangle_{H_{K_A}(E)} = \langle L_A[L^{-1}K_q], L_A[L^{-1}K_p]\rangle_{H_{K_A}(E)} = \langle L^{-1}K_q, L^{-1}K_p\rangle_{\mathscr{H}}.$$

The reproducing property yields

$$\langle (K_A)_q, (K_A)_p\rangle_{H_{K_A}(E)} = \langle K_q, K_p\rangle_{H_K(E)} = K(p,q),$$

which establishes (2.81).

The idea here is important when we consider some general non linear mappings in connection with linear mappings, because, for example, for $f \in H_K(E)$ we have the natural space containing f^2 as the space H_{K^2} and furthermore, we can derive the norm inequality between the two norms f in $H_K(E)$ and f^2 in $H_{K^2}(E)$ as in Corollary 2.4 to follow.

Example 2.15. In comparison with Theorems 1.4 and 1.5, about the kernels, we have

$$K_1(s,t) \equiv \frac{\pi}{4}\left(\exp(-|s-t|) + \exp(-s-t)\right)$$
$$\gg K_2(s,t) \equiv \frac{\pi}{4}\left(\exp(-|s-t|) - \exp(-s-t)\right)$$

for $s,t \in (0,\infty)$, which reflects $H_{K_1}(0,\infty) \supset H_{K_2}(0,\infty)$.

2.2.6 Pullback of Reproducing Kernels by a Hilbert Space-Valued Function

Let $\hat{E}$ be a set and $\hat{\phi} : \hat{E} \to E$ a mapping. Assume that we are given a Hilbert space $\mathscr{H}$ and another mapping $h : E \to \mathscr{H}$. Consider $K(p,q) = \langle h(q), h(p)\rangle_{\mathscr{H}}$ and $K_{\hat{\phi}}(\hat{p},\hat{q}) = \langle h(\hat{\phi}(\hat{p})), h(\hat{\phi}(\hat{q}))\rangle_{\mathscr{H}}$. Denote by L the mapping $Lf = \langle f, h(\cdot)\rangle_{\mathscr{H}}$ as before. Also, we set $L_{\hat{\phi}}f \equiv \langle f, \hat{\phi}^* h(\cdot)\rangle_{\mathscr{H}}$. We remark that $\ker(L_{\hat{\phi}}) \supset \ker(L)$ because $L_{\hat{\phi}} = \hat{\phi}^* L$. Let us consider the restriction of

$$\hat{\phi}^* : f \in \mathscr{F}(E) \mapsto f \circ \hat{\phi} \in \mathscr{F}(\hat{E}) \tag{2.82}$$

to $H_K(E)$. That is, we aim to investigate the mapping $\hat{\phi}^*$ given by

$$[\hat{\phi}^* Lf](\hat{p}) = \langle f, h(\hat{\phi}(\hat{p}))\rangle_{\mathscr{H}} \quad \text{for} \quad f \in \mathscr{H}. \tag{2.83}$$

We claim that $\hat{\phi}^* : H_K(E) \to H_{K_{\hat{\phi}}}(\hat{E})$ is a contraction mapping. Indeed, let $f \in \ker(L)^{\perp}$. Then we have

$$\|\hat{\phi}^* Lf\|_{H_{K_{\hat{\phi}}}(\hat{E})} = \|P_{\mathscr{H} \to \ker(L_{\hat{\phi}})^{\perp}} f\|_{\mathscr{H}} \leq \|f\|_{\mathscr{H}} = \|Lf\|_{H_K(E)}. \tag{2.84}$$

using Theorem 2.37.

The next theorem shows when $\hat{\phi}^* : H_K(E) \to H_{K_{\hat{\phi}}}(\hat{E})$ is an isometry. Note that ev denotes the evaluation mapping (see (2.9) above).

Theorem 2.15. *The mapping $\hat{\phi}^* : H_K(E) \to H_{K_{\hat{\phi}}}(\hat{E})$ is an isometry if and only if*

$$\bigcap_{p \in E} \ker(\mathrm{ev}_p \circ L) = \bigcap_{\hat{p} \in \hat{E}} \ker(\mathrm{ev}_{\hat{\phi}(\hat{p})} \circ L_{\hat{\phi}}) \subset \mathscr{H}. \tag{2.85}$$

If (2.85) is the case, then we have $\hat{\phi}^* : H_K(E) \to H_{K_{\hat{\phi}}}(\hat{E})$ is a bijection and the inverse is given by

$$[\hat{\phi}^{*-1}(\hat{f})](p) = \langle \hat{f}, \hat{\phi}^*[K_p]\rangle_{H_{K_{\hat{\phi}}}(\hat{E})} \quad \hat{f} \in H_{K_{\hat{\phi}}}(\hat{E}). \tag{2.86}$$

Proof. According to Theorem 2.37, we know that $L : \mathscr{H} \to H_K(E)$ is a surjective partial isometry. Likewise $L_{\hat{\phi}} : \mathscr{H} \to H_{K_{\hat{\phi}}}(\hat{E})$ is a surjective partial isometry. Note that $\ker(L) = \bigcap_{p \in E} \ker(\mathrm{ev}_p)$, $\ker(L_{\hat{\phi}}) = \bigcap_{\hat{p} \in \hat{E}} \ker(\mathrm{ev}_{\hat{\phi}(\hat{p})} \circ L_{\hat{\phi}})$. As a result, $\hat{\phi}^* : H_K \to H_{K_{\hat{\phi}}}$ is an isometry if and only if

$$\ker(L) = \ker(L_{\hat{\phi}}) \Longleftrightarrow \bigcap_{p \in E} \ker(\mathrm{ev}_p \circ L) = \bigcap_{\hat{p} \in \hat{E}} \ker(\mathrm{ev}_{\hat{\phi}(\hat{p})} \circ L_{\hat{\phi}}).$$

Assume $\ker(L) = \ker(L_{\hat{\phi}})$ here and below. Then, from the above, we have $L|\ker(L) : \ker(L)^{\perp} \to H_K(E)$ and $L_{\hat{\phi}}|\ker(L) : \ker(L_{\hat{\phi}}) = \ker(L) \to H_{K_{\hat{\phi}}}(\hat{E})$ are both isomorphisms and $\hat{\phi}^*$ can be written as $\hat{\phi}^* = (L_{\hat{\phi}}|\ker(L)) \circ (L|\ker(L))^{-1}$. As a result, $\hat{\phi}^* : H_K(E) \to H_{K_{\hat{\phi}}}(\hat{E})$ is a bijection.

Note that for $f \in H_K(E)$

$$f(p) = \langle f, K_p \rangle_{H_K(E)} = \langle \phi^* f, \phi^*[K_p] \rangle_{H_{K_{\hat{\phi}}}(\hat{E})}. \tag{2.87}$$

Thus we see that the inverse can be written as above.

Example 2.16. Let E' be a subset of E. Then according to Theorem 2.15, the restriction mapping $f \in H_K(E) \to f|E' \in H_{K|E' \times E'}(E')$ is an isometry if and only if $f(p) = 0$ for all $p \in E'$ implies $f = 0$.

2.3 Operations of More Than One Reproducing Kernel Hilbert Space

2.3.1 Sum of Reproducing Kernels

Suppose that we are given two positive definite quadratic form functions $K_1, K_2 : E \times E \to \mathbb{C}$. The usual sum $K(p,q) = K_1(p,q) + K_2(p,q)$ on $E \times E$ is also a positive definite quadratic form function on E. We will consider the relation among the corresponding reproducing kernel Hilbert spaces $H_{K_1}(E)$, $H_{K_2}(E)$ and $H_K(E)$. We prove:

Theorem 2.16. *Let $K_1, K_2 : E \times E \to \mathbb{C}$ be positive definite. Set $K \equiv K_1 + K_2$.*

1. *We have*

$$H_K(E) = \{f_1 + f_2 \in \mathscr{F}(E) : f_1 \in H_{K_1}(E), f_2 \in H_{K_2}(E)\}. \tag{2.88}$$

   So, as a linear space, we have $H_{K_1+K_2}(E) = H_{K_1}(E) + H_{K_2}(E)$.
2. *The norm of $H_K(E)$ is represented in terms of norms of $H_{K_1}(E)$ and $H_{K_2}(E)$ as follows:*

$$\|f\|_{H_K} = \min_{f_1 \in H_{K_1}(E), f_2 \in H_{K_2}(E), f = f_1 + f_2} \sqrt{\|f_1\|_{H_{K_1}(E)}^2 + \|f_2\|_{H_{K_2}(E)}^2}. \tag{2.89}$$

We prove this theorem in several steps. Let us set

$$H \equiv \{f \in \mathscr{F}(E) : \text{ there exist } f_1 \in H_{K_1}(E) \text{ and } f_2 \in H_{K_2}(E) \text{ such that } f = f_1 + f_2\}$$

in the sequel.

Proposition 2.6.

1. Let $f \in H$. Then

$$\left\{ \sqrt{\|f_1\|_{H_{K_1}(E)}^2 + \|f_2\|_{H_{K_2}(E)}^2} : f_1 \in H_{K_1}(E), f_2 \in H_{K_2}(E), f = f_1 + f_2 \right\} \tag{2.90}$$

has a minimum.

2. If we define

$$\|f\|_H \equiv \min_{f_1 \in H_{K_1}(E), f_2 \in H_{K_2}(E), f = f_1 + f_2} \sqrt{\|f_1\|_{H_{K_1}(E)}^2 + \|f_2\|_{H_{K_2}(E)}^2} \tag{2.91}$$

for $f \in H$, then $(H, \|\cdot\|_H)$ is a Hilbert space.

Proof. Let us set

$$M \equiv \inf \left\{ \sqrt{\|f_1\|_{H_{K_1}(E)}^2 + \|f_2\|_{H_{K_2}(E)}^2} : f_1 \in H_{K_1}(E), f_2 \in H_{K_2}(E), f = f_1 + f_2 \right\}$$

for $f \in H_{K_1}(E) + H_{K_2}(E)$. Then, in view of the definition of M, there exist sequences $\{f_{j,1}\}_{j=1}^\infty \subset H_{K_1}(E)$ and $\{f_{j,2}\}_{j=1}^\infty \subset H_{K_2}(E)$ such that

$$\lim_{j \to \infty} \sqrt{\|f_{j,1}\|_{H_{K_1}(E)}^2 + \|f_{j,2}\|_{H_{K_2}(E)}^2} = M. \tag{2.92}$$

Note that (2.92) implies that $\{f_{j,1}\}_{j=1}^\infty \subset H_{K_1}(E)$ and $\{f_{j,2}\}_{j=1}^\infty \subset H_{K_2}(E)$ are both bounded. Hence in view of the Banach Alaoglu theorem, if we pass to the subsequence from the start, we can assume that $\{f_{j,1}\}_{j=1}^\infty \subset H_{K_1}(E)$ and $\{f_{j,2}\}_{j=1}^\infty \subset H_{K_2}(E)$ converge to $f_1 \in H_{K_1}(E)$ and $f_2 \in H_{K_2}(E)$ in weak topology on the Hilbert spaces $H_{K_1}(E)$ and $H_{K_2}(E)$, respectively. Hence, from the weak convergence, we have

$$\begin{aligned} f(p) &= \lim_{j \to \infty} \left(\langle f_{j,1}, (K_1)_p \rangle_{H_{K_1}(E)} + \langle f_{j,2}, (K_2)_p \rangle_{H_{K_2}(E)} \right) \\ &= \langle f_1, (K_1)_p \rangle_{H_{K_1}(E)} + \langle f_2, (K_2)_p \rangle_{H_{K_2}(E)} \\ &= f_1(p) + f_2(p) \end{aligned}$$

for all $p \in E$. Since the weak convergence also implies

$$\sqrt{\|f_1\|_{H_{K_1}(E)}^2 + \|f_2\|_{H_{K_2}(E)}^2} \le \liminf_{j\to\infty} \sqrt{\|f_{j,1}\|_{H_{K_1}(E)}^2 + \|f_{j,2}\|_{H_{K_2}(E)}^2},$$

we see that f_1 and f_2 give the minimum.

To prove that $(H, \|\cdot\|_H)$ is a Hilbert space, it suffices to prove the parallelogram law:

$$2(\|f\|_H^2 + \|g\|_H^2) = \|f+g\|_H^2 + \|f-g\|_H^2 \quad (f, g \in H).$$

Let $f = f_1 + f_2$ and $g = g_1 + g_2$. Then we have

$$2(\|f_1\|_{H_{K_1}(E)}^2 + \|g_1\|_{H_{K_1}(E)}^2) = \|f_1 + g_1\|_{H_{K_1}(E)}^2 + \|f_1 - g_1\|_{H_{K_1}(E)}^2$$

and

$$2(\|f_2\|_{H_{K_2}(E)}^2 + \|g_2\|_{H_{K_2}(E)}^2) = \|f_2 + g_2\|_{H_{K_2}(E)}^2 + \|f_2 - g_2\|_{H_{K_2}(E)}^2.$$

Thus we have

$$\begin{aligned}
&2(\|f\|_H^2 + \|g\|_H^2) \\
&\le 2(\|f_1\|_{H_{K_1}(E)}^2 + \|g_1\|_{H_{K_1}(E)}^2) + 2(\|f_2\|_{H_{K_2}(E)}^2 + \|g_2\|_{H_{K_2}(E)}^2) \\
&= \|f_1 + g_1\|_{H_{K_1}(E)}^2 + \|f_1 - g_1\|_{H_{K_1}(E)}^2 + \|f_2 + g_2\|_{H_{K_2}(E)}^2 + \|f_2 - g_2\|_{H_{K_2}(E)}^2 \\
&= \|f_1 + g_1\|_{H_{K_1}(E)}^2 + \|f_2 + g_2\|_{H_{K_2}(E)}^2 + \|f_1 - g_1\|_{H_{K_1}(E)}^2 + \|f_2 - g_2\|_{H_{K_2}(E)}^2.
\end{aligned}$$

Note that in view of (2.90), we can choose f_1, f_2, g_1, g_2 so that

$$\|f+g\|_H^2 = \|f_1 + g_1\|_{H_{K_1}(E)}^2 + \|f_2 + g_2\|_{H_{K_2}(E)}^2$$

and

$$\|f-g\|_H^2 = \|f_1 - g_1\|_{H_{K_1}(E)}^2 + \|f_2 - g_2\|_{H_{K_2}(E)}^2.$$

Thus we obtain

$$2(\|f\|_H^2 + \|g\|_H^2) \le \|f+g\|_H^2 + \|f-g\|_H^2.$$

The reverse inequality can be proved analogously.

Remark 2.4. One has

$$\|(K_1)_p + (K_2)_p\|_H^2 = \|(K_1)_p\|_{H_{K_1}(E)}^2 + \|(K_2)_p\|_{H_{K_2}(E)}^2 \tag{2.93}$$

for $p \in E$. Indeed, let $(K_1)_p + (K_2)_p = f_1 + f_2$ with $f_1 \in H_{K_1}(E)$ and $f_2 \in H_{K_2}(E)$. Then $g = f_1 - (K_1)_p = -f_2 + (K_2)_p \in H_{K_1}(E) \cap H_{K_2}(E)$. Since

$$2\mathrm{Re}(g(p)) = \langle g, (K_i)_p \rangle + \langle (K_i)_p, g \rangle$$

for $i = 1, 2$, we obtain

$$\begin{aligned} &\|f_1\|^2_{H_{K_1}(E)} + \|f_2\|^2_{H_{K_2}(E)} \\ &= \|(K_1)_p + g\|^2_{H_{K_1}(E)} + \|(K_2)_p - g\|^2_{H_{K_2}(E)} \\ &= \|(K_1)_p\|^2_{H_{K_1}(E)} + \|g\|^2_{H_{K_1}(E)} + \|(K_2)_p\|^2_{H_{K_2}(E)} + \|g\|^2_{H_{K_2}(E)}. \end{aligned}$$

Therefore (2.93) follows from (2.91).

Proposition 2.7. *Let $f \in H$. Then we have $f(p) = \langle f, K_p \rangle_H$ for all $p \in E$.*

Proof. Let $f = f_1 + f_2, f_1 \in H_{K_1}(E), f_2 \in H_{K_2}(E)$ be the decomposition satisfying

$$\|f\|^2_H = \|f_1\|^2_{H_{K_1}(E)} + \|f_2\|^2_{H_{K_2}(E)}. \tag{2.94}$$

Then the polarization formula and Remark 2.4 together with an analogous argument give

$$\langle f, K_p \rangle_H = \langle f_1, (K_1)_p \rangle_{H_{K_1}(E)} + \langle f_2, (K_2)_p \rangle_{H_{K_2}(E)} = f_1(p) + f_2(p) = f(p).$$

This is the desired result.

In view of the uniqueness of $H_K(E)$, we obtain the desired relation and Theorem 2.16 is proved.

It is worth noting that by (2.89), in general we have the inequality

$$\|f_1 + f_2\|^2_{H_{K_1+K_2}(E)} \le \|f_1\|^2_{H_{K_1}(E)} + \|f_2\|^2_{H_{K_2}(E)}. \tag{2.95}$$

In particular, if $K = K_1 = K_2$, we learn from (2.99)

$$\|f_1 + f_2\|^2_{H_K(E)} = 2\|f_1 + f_2\|^2_{H_{2K}(E)} \le 2(\|f_1\|^2_{H_K(E)} + \|f_2\|^2_{H_K(E)}). \tag{2.96}$$

Let $K : E \times E \to \mathbb{C}$ and $K_0 : E \times E \to \mathbb{C}$ be positive definite quadratic form functions. As an application of the sum of two reproducing kernels, let us now consider the condition for which the embedding $H_{K_0}(E) \subset H_K(E)$ holds as a set. Recall that for positive definite kernels K_1 and K_2, the notation $K_1 \ll K_2$ means that $K_2 - K_1$ is a positive definite quadratic form function on the set E.

Theorem 2.17. *Let $K_0, K : E \times E \to \mathbb{C}$ be positive definite quadratic form functions. Then the following are equivalent:*

1. *The Hilbert space $H_{K_0}(E)$ is a subset of $H_K(E)$;*

2. *there exists* $\gamma > 0$ *such that*

$$K_0 \ll \gamma^2 K. \tag{2.97}$$

If these conditions hold, then the embedding $H_{K_0}(E) \subset H_K(E)$ *is actually continuous and the embedding norm* M *is given by*

$$M = \inf\{\gamma > 0 : K_0 \ll \gamma^2 K\} = \min\{\gamma > 0 : K_0 \ll \gamma^2 K\}.$$

Proof. Suppose that $K_0 \ll \gamma^2 K$. Then from Theorem 2.16 we deduce that

$$H_{K_0}(E) \hookrightarrow H_{\gamma^2 K - K_0}(E) + H_{K_0}(E) = H_{\gamma^2 K}(E) = H_K(E). \tag{2.98}$$

Here the inequalities and inclusion are understood in the set-theoretical sense. In view of the fact that

$$\|f\|_{H_{\gamma^2 K}(E)} = \gamma^{-1}\|f\|_{H_K(E)}, \tag{2.99}$$

we see that the embedding constant is less than or equal to γ.

Suppose instead that $H_{K_0}(E) \subset H_K(E)$. Then we see from the closed graph theorem that the embedding is continuous. Let us denote the embedding constant by M. Then we have

$$\begin{aligned}\sum_{j,k=1}^m a_j\overline{a_k}\, K_0(p_j,p_k) &= \sum_{j,k=1}^m a_j\overline{a_k}\langle (K_0)_{p_k}, K_{p_j}\rangle_{H_K(E)} \\ &= \left\langle \sum_{k=1}^m \overline{a_k}\,(K_0)_{p_k}, \sum_{k=1}^m \overline{a_k}\,K_{p_k}\right\rangle_{H_K(E)}.\end{aligned}$$

By the Cauchy Schwarz inequality, we have

$$\begin{aligned}\sum_{j,k=1}^m a_j\overline{a_k}\, K_0(p_j,p_k) &\le \left\|\sum_{k=1}^m \overline{a_k}\,(K_0)_{p_k}\right\|_{H_K(E)} \cdot \left\|\sum_{k=1}^m \overline{a_k}\,K_{p_k}\right\|_{H_K(E)} \\ &\le M\left\|\sum_{k=1}^m \overline{a_k}\,(K_0)_{p_k}\right\|_{H_{K_0}(E)} \cdot \left\|\sum_{k=1}^m \overline{a_k}\,K_{p_k}\right\|_{H_K(E)} \\ &\le M\sqrt{\sum_{j,k=1}^m a_j\overline{a_k}\,K_0(p_j,p_k)}\cdot\sqrt{\sum_{j,k=1}^m a_j\overline{a_k}\,K(p_j,p_k)}.\end{aligned}$$

If we arrange the above inequality, then we obtain

$$\sqrt{\sum_{j,k=1}^{m} a_j\overline{a_k}\, K_0(p_j, p_k)} \le M \sqrt{\sum_{j,k=1}^{m} a_j\overline{a_k}\, K(p_j, p_k)} \tag{2.100}$$

and γ in condition (8.51) can be taken as M.

Theorem 2.17 yields the following interesting corollary: Here $f \otimes \overline{f}$ is a function defined on $E \times E$ by

$$f \otimes \overline{f}(p, q) \equiv f(p)\overline{f}(q) \quad (p, q \in E).$$

Corollary 2.1. *Let $K : E \times E \to \mathbb{C}$ be a positive definite quadratic function. Then $f \in \mathscr{F}(E)$ belongs to $H_K(E)$ if and only if $f \otimes \overline{f} \ll \gamma^2 K$ for some $\gamma > 0$. If this is the case, then the norm of f is given by*

$$\|f\|_{H_K(E)} = \inf\{\gamma > 0 : f \otimes \overline{f} \ll \gamma^2 K\}. \tag{2.101}$$

Proof. This is because

$$\|f\|_{H_{f\otimes\overline{f}}(E)} = 1. \tag{2.102}$$

Indeed, (2.102) is a special case of Example 1.1 with $N = 1$ and $a_{11} = 1$.

From the above arguments, we see, in particular, the following interesting result:

Corollary 2.2. *Suppose K is continuous on $E \times E$ and that $\{g_j\}_{j=1}^{\infty} \subset \mathscr{F}(E)$ such that*

$$K(p, q) = \sum_{j=1}^{\infty} g_j(p)\overline{g_j(q)} \quad (p, q \in E). \tag{2.103}$$

In particular, assume that

$$\sum_{j=1}^{\infty} |g_j(p)|^2 < \infty.$$

1. *All the functions g_j are continuous on E and belong to $H_K(E)$.*
2. *The convergence of (2.103) is uniform on $E_0 \times E_0$, for any compact subset E_0 of E.*

Proof.

1. Let j be fixed. Then by Theorem 2.17, we have $g_j \in H_K(E)$ because $g_j \otimes \overline{g_j} \ll K$.

2. Uniform convergence can be deduced from the formula

$$K(p,p) = \sum_{j=1}^{\infty} |g_j(p)|^2 \tag{2.104}$$

and from Dini's theorem. Another proof of uniform convergence is to resort to the latter part of Theorem 2.1 using (2.104).

Example 2.17. Here we revisit Theorem 2.4. Let $H_K(E)$ be a reproducing kernel Hilbert space. Let $\ell \in H_K(E)^*$. Then $\ell \bowtie K$ can be extended to the case when $H_K(E)$ is not separable. Indeed, if we set

$$\ell \bowtie K(p) = \overline{\ell(K_p)},$$

then $\ell \bowtie K$ agrees with the definition (2.28). Even when $H_K(E)$ is not separable, we can check that $\ell \bowtie K \in H_K(E)$. In fact, according to Corollary 2.1, this amounts to the inequality

$$\ell \bowtie K \otimes \overline{\ell \bowtie K} \ll K. \tag{2.105}$$

Note that

$$\ell \bowtie K \otimes \overline{\ell \bowtie K}(p,q) = \overline{\ell(K_p)}\ell(K_q) \quad (p,q \in E).$$

Keeping this in mind, let us take $p_1, p_2, \ldots, p_N \in E$ and $\alpha_1, \alpha_2, \ldots, \alpha_N \in E$. Then

$$\sum_{j,k=1}^{N} \alpha_j \overline{\alpha_k} \ell \bowtie K \otimes \overline{\ell \bowtie K}(p_j, p_k) = \sum_{j,k=1}^{N} \alpha_j \overline{\alpha_k} \overline{\ell(K_{p_j})} \ell(K_{p_k}) = \left| \ell \left(\sum_{k=1}^{N} \overline{\alpha_k} K_{p_k} \right) \right|^2.$$

By the Riesz representation theorem, we can suppose that ℓ is realized by $f \in H_K(E)$. Thus,

$$\sum_{j,k=1}^{N} \alpha_j \overline{\alpha_k} \ell \bowtie K \otimes \overline{\ell \bowtie K}(p_j, p_k) = \left| \sum_{k=1}^{N} \overline{\alpha_k} f(p_k) \right|^2 = \left| \left\langle f, \sum_{k=1}^{N} \alpha_k K_{p_k} \right\rangle_{H_K(E)} \right|^2.$$

If we use the Schwarz inequality, then we have

$$\sum_{j,k=1}^{N} \alpha_j \overline{\alpha_k} \ell \bowtie K \otimes \overline{\ell \bowtie K}(p_j, p_k) \le \left\| \sum_{k=1}^{N} \alpha_k K_{p_k} \right\|_{H_K(E)}^2 = \sum_{j,k=1}^{N} \alpha_j \overline{\alpha_k} K(p_j, p_k),$$

proving (2.105).

2.3.2 Sum of Arbitrary Abstract Hilbert Spaces

Let $\mathscr{S}$ be a Hilbert space which is equipped with a Hilbert space $\mathscr{S}$-valued function $\mathbf{h}_{\mathscr{S}}$ on E. We assume further that

$$\mathscr{V} \equiv \{\mathbf{h}_{\mathscr{S}}(p) : p \in E\} \text{ is complete in } \mathscr{S}, \text{ that is, } \mathscr{S} = \overline{\mathscr{V}}^{\mathscr{S}}. \tag{2.106}$$

We will introduce the sum $\mathscr{H}_1[+]\mathscr{H}_2$ of abstract Hilbert spaces $\mathscr{H}_1$ and $\mathscr{H}_2$, when we are given two mappings $\mathbf{h}_1 : E \to \mathscr{H}_1$ and $\mathbf{h}_2 : E \to \mathscr{H}_2$ satisfying

$$\langle \mathbf{h}_1(q), \mathbf{h}_1(p)\rangle_{\mathscr{H}_1} + \langle \mathbf{h}_2(q), \mathbf{h}_2(p)\rangle_{\mathscr{H}_2} = \langle \mathbf{h}_{\mathscr{S}}(q), \mathbf{h}_{\mathscr{S}}(p)\rangle_{\mathscr{S}} \quad \text{on} \quad E \times E. \tag{2.107}$$

Let us set

$$K_1(p,q) \equiv \langle \mathbf{h}_1(q), \mathbf{h}_1(p)\rangle_{\mathscr{H}_1}, \quad K_2(p,q) \equiv \langle \mathbf{h}_2(q), \mathbf{h}_2(p)\rangle_{\mathscr{H}_2} \quad (p, q \in E).$$

Then, we can consider the linear mapping from $\mathscr{S}$ to $H_{K_1+K_2}(E)$

$$f_{\mathscr{S}}(p) = \langle \mathbf{f}_{\mathscr{S}}, \mathbf{h}_{\mathscr{S}}(p)\rangle_{\mathscr{S}}, \quad (\mathbf{f}_{\mathscr{S}} \in \mathscr{S}). \tag{2.108}$$

Theorem 2.37 and (2.106) yield

$$\|f_{\mathscr{S}}\|_{H_{K_1+K_2}(E)} = \|\mathbf{f}_{\mathscr{S}}\|_{\mathscr{S}}. \tag{2.109}$$

To define an element $\mathbf{f}_1[+]\mathbf{f}_2$ as a "sum", we will consider two systems

$$f_j(p) = \langle \mathbf{f}_j, \mathbf{h}_j(p)\rangle_{\mathscr{H}_j}, \quad (p \in E) \tag{2.110}$$

for $\mathbf{f}_j \in \mathscr{H}_j\, (j = 1, 2)$. Then, by Theorems 2.1 and 2.37, we have $f_1 \in H_{K_1}(E) \subset H_{K_1+K_2}(E)$ and $f_2 \in H_{K_2}(E) \subset H_{K_1+K_2}(E)$ and hence $f_1 + f_2 \in H_{K_1+K_2}(E)$.

Now, by the density (2.106) and Theorem 2.37 for $f_1 + f_2 \in H_{K_1+K_2}(E)$, there exists a uniquely determined $\mathbf{f}_{\mathscr{S}} \in \mathscr{S}$ satisfying

$$f_1(p) + f_2(p) = \langle \mathbf{f}_{\mathscr{S}}, \mathbf{h}_{\mathscr{S}}(p)\rangle_{\mathscr{S}} \text{ on } E. \tag{2.111}$$

Then $\mathbf{f}_{\mathscr{S}}$ will be considered as the sum of $\mathbf{f}_1$ and $\mathbf{f}_2$ through these mappings and so we introduce the notation

$$\mathbf{f}_{\mathscr{S}} = \mathbf{f}_1[+]\mathbf{f}_2 \in \mathscr{S}. \tag{2.112}$$

This sum for the members $\mathbf{f}_1 \in \mathscr{H}_1$ and $\mathbf{f}_2 \in \mathscr{H}_2$ is introduced through the three mappings induced by $\{\mathscr{H}_j,\, E,\, \mathbf{h}_j\}$ $(j = 1, 2)$ and $\{\mathscr{S},\, E,\, \mathbf{h}_{\mathscr{S}}\}$.

The operator $\mathbf{f}_1[+]\mathbf{f}_2$ is represented in terms of $\mathbf{f}_1$ and $\mathbf{f}_2$ by the inversion formula

$$\langle \mathbf{f}_1, \mathbf{h}_1(\cdot)\rangle_{\mathscr{H}_1} + \langle \mathbf{f}_2, \mathbf{h}_2(\cdot)\rangle_{\mathscr{H}_2} \mapsto \mathbf{f}_1[+]\mathbf{f}_2 \tag{2.113}$$

from $H_{K_1+K_2}(E)$ onto $\mathscr{S}$ via (2.111). Then we obtain the triangle inequality.

Theorem 2.18 ([390, Theorem 2.1], [83, Proposition 2.4]). *We have the triangle inequality*

$$\|\mathbf{f}_1[+]\mathbf{f}_2\|_{\mathscr{S}}^2 \le \|\mathbf{f}_1\|_{\mathscr{H}_1}^2 + \|\mathbf{f}_2\|_{\mathscr{H}_2}^2. \tag{2.114}$$

Proof. By Theorem 2.37, (2.106) and (2.112), we have

$$\|\mathbf{f}_1[+]\mathbf{f}_2\|_{\mathscr{S}} = \|\mathbf{f}_{\mathscr{S}}\|_{\mathscr{S}} = \|f_1 + f_2\|_{H_{K_1+K_2}(E)}. \tag{2.115}$$

By (2.88), we have

$$\|f_1 + f_2\|_{H_{K_1+K_2}(E)}^2 \le \|f_1\|_{H_{K_1}(E)}^2 + \|f_2\|_{H_{K_2}(E)}^2. \tag{2.116}$$

Finally, by (2.241) below, we have

$$\|f_1\|_{H_{K_1}} \le \|\mathbf{f}_1\|_{\mathscr{H}_1}, \quad \|f_2\|_{H_{K_2}} \le \|\mathbf{f}_1\|_{\mathscr{H}_2}. \tag{2.117}$$

Combining (2.115), (2.116) and (2.117), we obtain (2.114).

If $\{\mathbf{h}_j(p);\ p \in E\}$ are complete in $\mathscr{H}_j$ $(j = 1, 2)$, then $\mathscr{H}_j$ and $H_{K_j}(E)$ are isometric for each $j = 1, 2$ again by Theorem 2.37 2. Using the isometric mappings induced by Hilbert space valued functions $\mathbf{h}_j$ $(j = 1, 2)$ and $\mathbf{h}_{\mathscr{S}}$, we can introduce the sum space of $\mathscr{H}_1$ and $\mathscr{H}_2$ in the form $\mathscr{H}_1[+]\mathscr{H}_2$ through the mappings. Of course, the sum carries the structure of a Hilbert space. Furthermore, such spaces are determined in the framework of isometric relation (2.109).

Example 2.18 ([390, Section 7]). Let $(T, \mathscr{T}, dm)$ and $(E, \mathscr{E}, d\mu)$ be σ-finite measure spaces. Let $\rho_1 : T \to (0, \infty)$ and $\rho_2 : T \to (0, \infty)$ be measurable functions such that $\rho_1 + \rho_2 > 0$. For a measurable function $\rho : T \to (0, \infty)$, define

$$L^2(\rho, m) \equiv \left\{f : T \to \mathbb{C} : f \text{ is } m\text{-measurable and } \|f\|_{L^2(\rho,m)} < \infty\right\},$$

where

$$\|f\|_{L^2(\rho,m)} \equiv \sqrt{\int_T |f(t)|^2 \rho(t) dm(t)}.$$

We assume that $\{h(\cdot, p); p \in E\}$ is complete in the spaces $L^2(\rho_1, m)$ and $L^2(\rho_2, m)$ and that

$$\int_T |h(t,p)|^2(\rho_1(t)+\rho_2(t))dm(t) < \infty \quad (p \in E). \tag{2.118}$$

Here the convergence of the integral is guaranteed by (2.118). As we learn from the decomposition,

$$F = F\chi_{\{\rho_1<\rho_2\}} + F\chi_{\{\rho_2\le\rho_1\}} \quad F \in L^2(\rho_1+\rho_2),$$

we see that $\{h(\cdot,p); p \in E\}$ is complete in the space $L^2(\rho_1+\rho_2, m)$ as well, which will yield (2.106). For simplicity we assume that $\rho_1+\rho_2 > 0$ on E.

Here we claim that $L^2(\rho_1)[+]L^2(\rho_2) \subset L^2(\rho_1+\rho_2)$ and that regarding the operation (2.112),

$$(F_1[+]F_2)(t) = \frac{F_1(t)\rho_1(t)+F_2(t)\rho_2(t)}{\rho_1(t)+\rho_2(t)} \quad (t \in E) \tag{2.119}$$

holds.

Let us start by specifying that $\mathscr{S} = L^2(\rho_1+\rho_2, m)$. As we mentioned, (2.106) is already guaranteed. Next, we define $\mathbf{h}_1 : E \to L^2(\rho_1)$, $\mathbf{h}_2 : E \to L^2(\rho_2)$ and $\mathbf{h}_{\mathscr{S}} : E \to L^2(\rho_1+\rho_2) = \mathscr{S}$ by

$$\mathbf{h}_1(p) = h(\cdot,p),\ \mathbf{h}_2(p) = h(\cdot,p),\ \mathbf{h}_{\mathscr{S}}(p) = h(\cdot,p) \quad (p \in E).$$

Note that these definitions make sense by (2.118). Let $p, q \in E$. As is seen from

$$K_1(p,q) \equiv \langle \mathbf{h}_1(q), \mathbf{h}_1(p)\rangle_{L^2(\rho_1)} = \int_T h(t,q)\overline{h(t,p)}\rho_1(t)\,dm(t),$$

$$K_2(p,q) \equiv \langle \mathbf{h}_2(q), \mathbf{h}_2(p)\rangle_{L^2(\rho_2)} = \int_T h(t,q)\overline{h(t,p)}\rho_2(t)\,dm(t),$$

$$K(p,q) \equiv \langle \mathbf{h}(q), \mathbf{h}(p)\rangle_{L^2(\rho_1+\rho_2)} = \int_T h(t,q)\overline{h(t,p)}(\rho_1(t)+\rho_2(t))\,dm(t),$$

we have (2.107). Therefore Theorem 2.18 is applicable. Let us remark that we are able to characterize $H_{K_1}(E)$, $H_{K_2}(E)$ and $H_K(E)$ by Theorem 2.36.

Let $F_1 \in L^1(\rho_1)$ and $F_2 \in L^2(\rho_2)$ be given. We will consider two linear transforms

$$f_j(p) = \int_T F_j(t)\overline{h(t,p)}\rho_j(t)dm(t), \quad p \in E. \tag{2.120}$$

So we need to look for an element $F \in L^2(\rho_1+\rho_2)$ such that

$$f_1(p)+f_2(p) = \int_T F(t)\overline{h(t,p)}(\rho_1(t)+\rho_2(t))dm(t). \tag{2.121}$$

If we define F by (2.119), then from (2.120) we conclude that (2.121) follows and that $F \in L^2(\rho_1 + \rho_2)$. Also (2.114) reads

$$\int_E \frac{|F_1(t)\rho_1(t) + F_2(t)\rho_2(t)|^2}{\rho_1(t) + \rho_2(t)}\, dm(t) \le \|F_1\|_{L^2(\rho_1)} + \|F_2\|_{L^2(\rho_2)}. \tag{2.122}$$

Observe also that (2.122) can be deduced from the Hölder inequality. Inequality (2.122) dates back to [390, (7.1.6)].

The main idea of Sect. 2.3.2 is [390, Section 2].

For the generalized triangle inequality (2.114), some interesting discussions and remarks may be found in [408, Section 4], and some fully general triangle inequalities for Banach space versions were given in [454].

2.3.3 *Pasting of Reproducing Kernel Hilbert Spaces*

Following [421], we introduce new operations for reproducing kernel Hilbert spaces.

Theorem 2.19. *Let $\{E_1, E_2\}$ be a partition of a set E. Suppose that we are given a reproducing kernel K on E. Denote by K_1, K_2 the restrictions of K to $E_1 \times E_1$ and $E_2 \times E_2$ respectively. Then the following are equivalent*:

1. $K|E_1 \times E_2 \equiv 0$;
2. $f \in H_K(E) \mapsto (f|E_1, f|E_2) \in H_{K_1}(E_1) \oplus H_{K_2}(E_2)$ *is an isomorphism.*

If one of these conditions is fulfilled, then we have

$$K(x,y) = \begin{cases} K_1(x,y) & x, y \in E_1, \\ K_2(x,y) & x, y \in E_2, \\ 0 & \text{otherwise}. \end{cases} \tag{2.123}$$

Before we come to the proof, we need a setup. We define

$$H_j(E) \equiv \overline{\operatorname{Span}\{K_p \,:\, p \in E_j\}}$$

for $j = 1, 2$, where the closure is considered in $H_K(E)$. We also denote by p_j the projection from $H_K(E)$ to $H_j(E)$ for $j = 1, 2$.

Proof. Assume that (1) holds. Observe by assumption (1) that $H_{K_1}(E)$ and $H_{K_2}(E)$ are mutually orthogonal. Let $f \in H_K(E)$. Then this orthogonality shows that

$$p_1(f)|E_2 \equiv 0,\ p_2(f)|E_1 \equiv 0, \tag{2.124}$$

where p_1 and p_2 denote the projections from $H_K(E)$ onto $H_{K_1}(E)$ and $H_{K_2}(E)$, respectively.

We next claim that $f \in H_K(E)$, if

$$\|p_1(f)\|_{H_K(E)} = \min\{\|h\|_{H_K(E)} \,:\, h|E_1 = f|E_1\}.$$

Indeed, if $x \in E_1$ and if f and h agree on E_1, then the equality (2.124) shows that $p_1(f) = p_1(h) = \chi_{E_1} f$. Consequently

$$\|h\|_{H_K(E)}^2 = \|p_1(h)\|_{H_K(E)}^2 + \|p_2(h)\|_{H_K(E)}^2 \geq \|p_1(f)\|_{H_K(E)}^2$$

and our claim is justified.

In general, we have

$$\|f|E_j\|_{H_{K_j}(E_j)} = \min\{\|h\|_{H_K(E)} \,:\, h|E_j = f|E_j\}$$

and hence, with what we have just proved,

$$\|f|E_j\|_{H_{K_j}(E)} = \|p_j(f)\|_{H_K(E)}.$$

This implies that $f \in H_K(E) \mapsto (f|E_1, f|E_2) \in H_{K_1}(E) \oplus H_{K_2}(E)$ is an isometry. The mapping is also surjective, since we can deduce from the equality (2.124) that

$$p_1(f_1) + p_2(f_2) \mapsto (f_1|E_1, f_2|E_2).$$

Assume instead (2). Let $x \in E_1$ be frozen and denote $f_x = K_x$. Then we have $f_x \in H_K(E)$ and $f_x|E_1 \in H_{K_1}(E)$. Their norm squares are both $K(x,x)$. By the assumption (2), this means $f_x|E_2 \equiv 0$. Consequently, $K(x,y) = \overline{K(y,x)} = 0$ for all $x \in E_1$ and $y \in E_2$.

Finally let us prove (2.123). If $(x,y) \in E \times E \setminus (E_1 \times E_1 \cup E_2 \times E_2)$, then this is clear from the assumption (1) in Theorem 2.19. If $(x,y) \in E_1 \times E_1$, then, since $K_1 = K|E_1 \times E_1$, we have $K_1(x,y) = K(x,y)$. Likewise if $(x,y) \in E_1 \times E_1$, then we have $K_2(x,y) = K(x,y)$.

Now we present some examples.

Example 2.19. Let E' be a subset of E. Then we have a partition of E: $\{E', E \setminus E'\}$. In view of Theorem 2.19, $f \in H_K(E) \mapsto f|E' \in H_{K|E' \otimes E'}(E')$ is an isometry if and only if $K'|(E \times E \setminus (E' \times E \cup E \times E')) \equiv 0$.

Example 2.20. Let $H_{K_1}[0,\infty)$ be the set of all absolutely continuous functions f on $(0,\infty)$ such that f and its derivative f' satisfy

$$\lim_{x \downarrow 0} f(x) = 0, \quad \int_0^\infty |f'(x)|^2 e^x \, dx < \infty.$$

Then a direct calculation shows that $H_{K_1}[0,\infty)$ is a reproducing kernel Hilbert space with kernel $K_1(x,y) = 1 - e^{-\min(x,y)}$. Likewise let $H_{K_2}(-\infty,0]$ be the set of all absolutely continuous functions g on $(-\infty,0)$ such that g and its derivative g' satisfy

$$\lim_{x \uparrow 0} g(x) = 0, \quad \int_{-\infty}^{0} |g'(x)|^2 e^{-x} \, dx < \infty.$$

Then $H_{K_2}(-\infty, 0]$ is also a reproducing kernel Hilbert space with kernel $K_2(x, y) = 1 - e^{\max(x,y)}$. Consequently, (2.123) shows that

$$K(x, y) = \begin{cases} 1 - e^{-\min(|x|,|y|)} & xy \geq 0, \\ 0 & \text{otherwise} \end{cases}$$

is a reproducing kernel on $\mathbb{R}$. The space $H_K[0, \infty)$ is given as the set of all absolutely continuous function h, such that h and its derivative h' satisfy

$$h(0) = 0, \quad \int_{-\infty}^{\infty} |h'(x)|^2 e^{|x|} \, dx < \infty$$

according to Theorem 2.19. We refer to Sect. 4.1 and [160, Section 4] for an example of an application of reproducing kernel Hilbert space $H_{K_1}[0, \infty)$ to the real inversion formula of the Laplace transform.

2.3.4 *Tensor Product of Hilbert Spaces*

Given two Hilbert spaces H_1 and H_2, we can make the tensor product $H_1 \otimes H_2$ from a pre-Hilbert space via the following inner product:

For example, we have

$$\|f_1 \otimes f_2\|_{H_{K_1 \otimes K_2}(E_1 \times E_2)} = \|f_1\|_{H_{K_1}(E_1)} \|f_2\|_{H_{K_2}(E_2)} \tag{2.125}$$

for $f_1 \in H_{K_1}(E_1)$ and $f_2 \in H_{K_2}(E_2)$.

$$\langle h_1 \otimes h_2, h_1' \otimes h_2' \rangle_{H_1 \otimes H_2} = \langle h_1, h_1' \rangle_{H_1} \langle h_2, h_2' \rangle_{H_2} \quad (h_1, h_1' \in H_1, h_2, h_2' \in H_2).$$

When we are given two complex-valued functions f and g defined on E_1 and E_2, an operation is possible: $(p_1, p_2) \in E_1 \times E_2 \to f \otimes g(p_1, p_2) \equiv f(p_1)g(p_2) \in \mathbb{C}$. Here we consider what happens when we are given positive definite quadratic form functions $K_1 : E_1 \times E_1 \to \mathbb{C}$ and $K_2 : E_2 \times E_2 \to \mathbb{C}$.

Theorem 2.20. *Let $K_1 : E_1 \times E_1 \to \mathbb{C}$ and $K_2 : E_2 \times E_2 \to \mathbb{C}$ be positive definite quadratic form functions. Then $K_1 \otimes K_2 : E_1 \times E_2 \times E_1 \times E_2 \to \mathbb{C}$ is a positive definite quadratic form function and*

$$H_{K_1}(E_1) \otimes H_{K_2}(E_2) = H_{K_1 \otimes K_2}(E_1 \times E_2). \tag{2.126}$$

To prove this theorem, let us set

$$\widetilde{H_{K_1} \otimes H_{K_2}}(E_1 \times E_2) = \operatorname{Span}\{f_1 \otimes f_2 \,:\, f_1 \in \mathscr{F}(E_1), f_2 \in \mathscr{F}(E_2)\}.$$

Thus if $f \in \widetilde{H_{K_1} \otimes H_{K_2}}(E_1 \times E_2)$, then we can find a finite set Λ and collections $\{f_{1;\lambda}\}_{\lambda\in\Lambda} \subset \mathscr{F}(E_1)$ and $\{f_{2;\lambda}\}_{\lambda\in\Lambda} \subset \mathscr{F}(E_2)$ such that

$$f = \sum_{\lambda\in\Lambda} f_{1;\lambda} \otimes f_{2;\lambda}$$

and the inner product is given by way of the tensor.

Lemma 2.4. *Let $p_1 \in E_1$ and $p_2 \in E_2$. Then we have*

$$\left|\sum_{\lambda\in\Lambda} f_{1;\lambda} \otimes f_{2;\lambda}(p_1, p_2)\right| \\ \leq \sqrt{K_1(p_1, p_1)K_2(p_2, p_2)} \left\|\sum_{\lambda\in\Lambda} f_{1;\lambda} \otimes f_{2;\lambda},\right\|_{\widetilde{H_{K_1} \otimes H_{K_2}}(E_1\times E_2)}.$$

Proof. Let us remark that

$$\sum_{\lambda\in\Lambda} f_{1;\lambda} \otimes f_{2;\lambda}(p_1, p_2) = \sum_{\lambda\in\Lambda} f_{1;\lambda}(p_1) f_{2;\lambda}(p_2) \\ = \sum_{\lambda\in\Lambda} \langle f_{1;\lambda}, (K_1)_{p_1}\rangle_{H_{K_1}(E_1)} \cdot \langle f_{2;\lambda}, (K_2)_{p_2}\rangle_{H_{K_2}(E_2)}.$$

Therefore

$$\left|\sum_{\lambda\in\Lambda} f_{1;\lambda} \otimes f_{2;\lambda}(p_0, p_1)\right| \\ = \left|\sum_{\lambda\in\Lambda} \langle f_{1;\lambda} \otimes f_{2;\lambda}, (K_1)_{p_1} \otimes (K_2)_{p_2}\rangle_{H_{K_1}(E_1)\otimes H_{K_2}(E_2)}\right| \\ = \left|\left\langle \sum_{\lambda\in\Lambda} f_{1;\lambda} \otimes f_{2;\lambda}, (K_1)_{p_1} \otimes (K_2)_{p_2}\right\rangle_{\widetilde{H_{K_1} \otimes H_{K_2}}(E_1\times E_2)}\right| \\ \leq \sqrt{K_1(p_1, p_1)K_2(p_2, p_2)} \left\|\sum_{\lambda\in\Lambda} f_{1;\lambda} \otimes f_{2;\lambda},\right\|_{\widetilde{H_{K_1} \otimes H_{K_2}}(E_1\times E_2)}.$$

This is the desired result.

In view of Lemma 2.4 we obtain Theorem 2.20.

Assume that $E_1 = E_2 = E$. Now we consider its restriction to the diagonal set. Let us set

$$D = \{(p, q, p, q) \, : \, p, q \in E\}. \tag{2.127}$$

Then we see that $K_1 \otimes K_2|D$ is a positive definite quadratic form function. An immediate consequence of Proposition 2.4 is the following:

Theorem 2.21. *Suppose that $K_1, K_2 : E \times E \to \mathbb{C}$ are positive definite quadratic form functions. Then so is the pointwise product $K \equiv K_1 \cdot K_2 : E \times E \to \mathbb{C}$.*

This theorem, whose proof is simple, contains important corollaries. The first concerns linear algebra.

Corollary 2.3. *Let $N \in \mathbb{N}$ and let $\{a_{ij}\}_{i,j=1}^N$ and $\{b_{ij}\}_{i,j=1}^N$ be $N \times N$ positive definite matrices. Then the matrix $C \equiv \{a_{ij}\, b_{ij}\}_{i,j=1,2,\ldots,N}$ is positive definite.*

The next results are used frequently in applications.

Corollary 2.4. *Let $K_1, K_2 \, : \, E \times E \to \mathbb{C}$ be positive definite quadratic from functions. Then*

$$\|f_1 + f_2\|_{H_{K_1+K_2}(E)} \leq \|f_1\|_{H_{K_1}(E)} + \|f_2\|_{H_{K_2}(E)} \tag{2.128}$$

and

$$\|f_1 \cdot f_2\|_{H_{K_1K_2}(E)} \leq \|f_1\|_{H_{K_1}(E)} \|f_2\|_{H_{K_2}(E)} \tag{2.129}$$

for $f_1 \in H_{K_1}(E)$ and $f_2 \in H_{K_2}(E)$.

Proof. Equation (2.128) is a direct consequence of (2.89). Let $f_1 \in H_{K_1}(E)$ and $f_2 \in H_{K_2}(E)$. Then $K_1 \cdot K_2$ is a positive definite quadratic form function by Theorem 2.21. Let $D \equiv \{(p, p) \, : \, p \in E\}$. By (2.46) and (2.125), we have

$$\|f_1 \cdot f_2\|_{H_{K_1K_2}(D)} \leq \|f_1 \otimes f_2\|_{H_{K_1 \otimes K_2}(E)} = \|f_1\|_{H_{K_1}(E_1)} \|f_2\|_{H_{K_2}(E_2)},$$

using $f_1 \cdot f_2 = f_1 \otimes f_2|D$ which proves (2.129).

Now we consider a special case of the above situation. Denote by $\mathbb{C}^*$ the set of all complex values other than 0.

Corollary 2.5. *Let $K : E \times E \to \mathbb{C}$ be a positive definite quadratic form function and let s be a mapping from E to $\mathbb{C}^*$. Define $K_s(p, q) \equiv s(p)\overline{s(q)}K(p, q)$ for $p, q \in E$. Then one has*

$$H_{K_s}(E) = \{F \in \mathscr{F}(E) \, : \, F = f \cdot s \text{ for some } f \in H_K(E)\}. \tag{2.130}$$

Furthermore, one has $\langle f \cdot s, g \cdot s\rangle_{H_{K_s}(E)} = \langle f, g\rangle_{H_K(E)}$ for all $f, g \in H_K(E)$.

Proof. Let us define a Hilbert space H by

$$H = \{F \in \mathscr{F}(E) \,:\, F = f \cdot s \text{ for some } f \in H_K(E)\}, \tag{2.131}$$

where the inner product is given by

$$\langle f \cdot s, g \cdot s \rangle_{H_{K_s}(E)} = \langle f, g \rangle_{H_K(E)} \tag{2.132}$$

for $f, g \in H_K(E)$. Our task is to establish that $H = H_{K_s}(E)$.

First, we observe

$$K_s(\cdot, q) = s(\cdot)\overline{s(q)}K_q = \overline{s(q)} \cdot [s \cdot K_q] \in H. \tag{2.133}$$

From the definition of the inner product, we obtain

$$\begin{aligned}
\langle K_s(\cdot, q), K_s(\cdot, p) \rangle_{H_{K_s}(E)} &= \langle \overline{s(q)} \cdot [s \cdot K_q], \overline{s(p)} \cdot [s \cdot K_p] \rangle_{H_{K_s}(E)} \\
&= s(p)\overline{s(q)} \langle s \cdot K_q, s \cdot K_p \rangle_{H_{K_s}(E)} \\
&= s(p)\overline{s(q)} \langle K_q, K_p \rangle_{H_K(E)} \\
&= s(p)\overline{s(q)} K(p, q).
\end{aligned}$$

As a result we obtain

$$\langle K_s(\cdot, q), K_s(\cdot, p) \rangle_{H_{K_s}(E)} = s(p)\overline{s(q)} K(p, q). \tag{2.134}$$

Finally, if $F = f \cdot s \in H$ with $f \in H_K(E)$, then, from $F(p) = \langle f, K_p \rangle_{H_K(E)} s(p)$, we have

$$F(p) = \langle f \cdot s, K_p \cdot s \rangle_{H_{K_s}(E)} s(p) = \langle f \cdot s, K_p \cdot s \cdot \overline{s(p)} \rangle_{H_{K_s}(E)} = \langle F, (K_s)_p \rangle_{H_{K_s}(E)}.$$

Hence the reproducing property

$$F(p) = \langle F, (K_s)_p \rangle_{H_{K_s}(E)} \tag{2.135}$$

is established.

From (2.133), (2.134) and (2.135) we conclude $H = H_{K_s}(E)$.

A trivial inquality (2.129) is a very strong technique which produces many inequalities. For example, in [83, Section 2.2], we can find an integral inequality using the operator L_j given by (7.75). Furthermore, by Corollary 2.5, we can realize many reproducing kernel Hilbert spaces from known reproducing kernel Hilbert spaces, and therefore the technique is a very important in the related norm inequalities. See, for example, [388].

Meanwhile, the product of a reproducing kernel will give the basic relation between the related linear mapping and nonlinear mappings, as we shall see in Theorem 8.1. See also, for example, [388, 389].

2.3.5 *Increasing Sequences of RKHS*

We shall consider a positive semi-definite quadratic form function $\{K_t\}_{t>0}$ on E such that $K_{t_2} \ll K_{t_1}$ for all $0 < t_1 < t_2$. We wish to introduce a pre-Hilbert space by

$$H_{K_0} \equiv \bigcup_{t>0} H_{K_t}(E).$$

For any $f \in H_{K_0}$, there exists $t > 0$ such that $f \in H_{K_t}(E)$. Then, for any $t' \in (0, t)$, $H_{K_t}(E) \subset H_{K_{t'}}(E)$ and

$$\|f\|_{H_{K_t}(E)} \geq \|f\|_{K_{t'}(E)}$$

for all $f \in H_{K_t}$. Therefore the limit exists for all $f \in H_{K_0}$:

$$\|f\|_{H_{K_0}} \equiv \lim_{t \downarrow 0} \|f\|_{H_{K_t}(E)}.$$

Assume $\|f\|_{H_{K_0}} \neq 0$ unless $f = 0$.

Denote by H_0 the completion of H_{K_0} with respect to this norm.

We state two theorems.

Theorem 2.22. *Suppose that we are given a decreasing sequence $\{K_t\}_{t>0}$ of positive definite quadratic form functions satisfying*

$$\langle f, g\rangle_{H_{K_{t_1}}(E)} = \langle f, g\rangle_{H_{K_{t_2}}(E)} \tag{2.136}$$

for all $t_2 > t_1 > 0$ and $f, g \in H_{K_{t_2}}(E)$.

1. *Let $f \in H_0$. If we define*

$$f_t^*(x) = \langle f, K_t(\cdot, x)\rangle_{H_0} \quad (x \in E),$$

then $f_t^* \in H_{K_t}(E)$ for all $t > 0$, and as $t \downarrow 0$, $f_t^* \to f$ in the topology of H_0.

2. *The space $H_{K_t}(E)$ is a closed subspace of H_0.*

Proof. 1. As before,

$$
\begin{aligned}
&\|f\|_{H_0}^2 \sum_j \sum_{j'} C_j \overline{C_{j'}} K_t(x_{j'}, x_j) - \left| \sum_j C_j f_t^*(x_j) \right|^2 \\
&= \|f\|_{H_0}^2 \sum_j \sum_{j'} C_j \overline{C_{j'}} K_t(x_{j'}, x_j) - \left| \left\langle f, \sum_j \overline{C_j} K_t(\cdot, x_j) \right\rangle_{H_0} \right|^2 \\
&\geq \|f\|_{H_0}^2 \sum_j \sum_{j'} C_j \overline{C_{j'}} K_t(x_{j'}, x_j) - \|f\|_{H_0}^2 \left\| \sum_j \overline{C_j} K_t(\cdot, x_j) \right\|_{H_0}^2 \\
&\geq \|f\|_{H_0}^2 \sum_j \sum_{j'} C_j \overline{C_{j'}} K_t(x_{j'}, x_j) - \|f\|_{H_0}^2 \left\| \sum_j \overline{C_j} K_t(\cdot, x_j) \right\|_{H_{K_t}(E)}^2 = 0
\end{aligned}
$$

for any finite number of points $\{x_j\}$ of the set E and for any complex numbers $\{C_j\}$. Therefore, $f_t^* \in H_{K_t}(E)$.

The mapping $f \mapsto f_t^*$ being uniformly bounded, we can therefore assume that $f \in H_{K_r}(E)$ for some fixed $r > 0$. Therefore we see that $f_t^* \in H_{K_t}(E)$ and that $\|f_t^*\|_{H_{K_t}(E)} \leq \|f\|_{H_0}$ for all $t \in (0, r)$.

However, in this case, the result is clear, since, $f \in H_{K_t}(E)$ for $0 < t < r$

$$\lim_{t \downarrow 0} f_t^*(x) = \lim_{t \downarrow 0} \langle f, K_t(\cdot, x) \rangle_{H_0} = \langle f, K_t(\cdot, x) \rangle_{H_{K_t}(E)} = f(x).$$

Here we used (2.136) for the second equality.

2. Let $\{f_j\}_{j=1}^{\infty}$ be a sequence in $H_{K_t}(E)$ convergent to f in H_0. Since we have

$$f_j(x) = \langle f_j, K_t(\cdot, x) \rangle_{H_{K_t}(E)} = \langle f_j, K_t(\cdot, x) \rangle_{H_0} \quad (x \in E).$$

for all $f \in H_{K_t}$, we have

$$f(x) = \langle f, K_t(\cdot, x) \rangle_{H_0} \quad (x \in E).$$

Therefore $f \in H_{K_t}(E)$.

Next we discuss what happens if we drop (2.136).

Theorem 2.23. *Let E be a set and suppose that we have a family of positive definite quadratic form functions $\{K_t\}_{t>0}$ such that $K_{t_1} \leq K_{t_2}$ for all $0 < t_2 < t_1$. Then for all $f \in H_0$, we have*

$$\lim_{t \to +0} u_f(\cdot, t) \equiv f \tag{2.137}$$

in the space H_0, for the functions

$$u_f(x,t) \equiv \langle f, K_t(\cdot,x)\rangle_{H_0} \quad (x \in H_0, t > 0). \tag{2.138}$$

Remark that the right-hand side does not depend on (2.138).

Proof. Let us check whether $f_t^* = u_f(\cdot,t) \in H_{K_t}(E)$ for $f \in H_0$. We can check

$$f_t^* \otimes \overline{f_t^*} \ll \|f\|_{H_0}^2 K_t$$

using (2.138). Indeed, as before,

$$\begin{aligned}
&\|f\|_{H_0}^2 \sum_j \sum_{j'} C_j \overline{C_{j'}} K_t(x_{j'},x_j) - \left| \sum_j C_j f_t^*(x_j) \right|^2 \\
&= \|f\|_{H_0}^2 \sum_j \sum_{j'} C_j \overline{C_{j'}} K_t(x_{j'},x_j) - \left| \left\langle f, \sum_j \overline{C_j} K_t(\cdot,x_j) \right\rangle_{H_0} \right|^2 \\
&\geq \|f\|_{H_0}^2 \sum_j \sum_{j'} C_j \overline{C_{j'}} K_t(x_{j'},x_j) - \|f\|_{H_0}^2 \left\| \sum_j \overline{C_j} K_t(\cdot,x_j) \right\|_{H_0}^2 \\
&\geq \|f\|_{H_0}^2 \sum_j \sum_{j'} C_j \overline{C_{j'}} K_t(x_{j'},x_j) - \|f\|_{H_0}^2 \left\| \sum_j \overline{C_j} K_t(\cdot,x_j) \right\|_{H_{K_t}(E)}^2 = 0
\end{aligned}$$

for any finite number of points $\{x_j\}$ of the set E and any complex numbers $\{C_j\}$. Therefore, $f_t^* \in H_{K_t}(E)$. From this calculation we see that $f_t^* \in H_{K_t}(E)$ and that

$$\|f_t^*\|_{H_{K_t}(E)} \leq \|f\|_{H_0}. \tag{2.139}$$

The mapping $f \mapsto f_t$ being uniformly bounded, we can assume that $f \in H_{K_r}(E)$ for some $r > 0$. Since $\{K_r(\cdot,q)\}_{q\in E}$ spans a dense subspace of $H_{K_r}(E)$, we may assume that $f = K_r(\cdot,q)$ for some $q \in E$. Let $0 < t < s < r$. Then we have

$$f_t^*(x) = \langle K_r(\cdot,q), K_t(\cdot,x)\rangle_{H_0} \quad (x \in E)$$

and hence

$$\|f_t^*\|_{H_{K_s}(E)} \leq \|f_t^*\|_{H_{K_r}(E)} \leq \|K_r(\cdot,q)\|_{H_0} \leq \|K_r(\cdot,q)\|_{H_{K_s}(E)},$$

where we used (2.139) for the second inequality.

Let $\{\varphi_\lambda^{(t)}\}_{\lambda\in\Lambda_t}$ be a CONS of $H_{K_t}(E)$. Then we have

$$K_t(\cdot,x) = \sum_{\lambda\in\Lambda_t} \overline{\varphi_\lambda^{(t)}(x)}\varphi_\lambda^{(t)} \tag{2.140}$$

in $H_{K_t}(E)$ for any fixed $x \in E$. Therefore, for $t \in (0,r)$,

$$f_t^*(x) = \sum_{\lambda\in\Lambda_t} \langle K_r(\cdot,q), \varphi_\lambda^{(t)}\rangle_{H_0}\varphi_\lambda^{(t)}(x) \quad (x \in E).$$

We claim this pointwise convergence takes place in $H_{K_0}(E)$ as well. Note that

$$\sum_{\lambda\in\Lambda_t} |\langle K_r(\cdot,q), \varphi_\lambda^{(t)}\rangle_{H_0}|^2 = (\|f_t^*\|_{H_{K_t(E)}})^2 \le (\|f\|_{H_0})^2 \le (\|K_r(\cdot,q)\|_{H_{K_r(E)}})^2 < \infty$$

thanks to Parseval's inequality and (2.139).

For any finite set $F \subset \Lambda_t$, we define

$$f_F(x) \equiv \sum_{\lambda\in\Lambda_t\setminus F} \langle K_r(\cdot,q), \varphi_\lambda^{(t)}\rangle_{H_{K_t(E)}}\varphi_\lambda^{(t)}(x) \quad (x \in E).$$

Then

$$(f_F)_t^*(x) = \sum_{\lambda\in\Lambda_t\setminus F} \langle K_r(\cdot,q), \varphi_\lambda^{(t)}\rangle_{H_t}\langle \varphi_\lambda^{(t)}, K_t(\cdot,x)\rangle_{H_0} = \sum_{\lambda\in\Lambda_t\setminus F} \langle K_r(\cdot,q), \varphi_\lambda^{(t)}\rangle_{H_t}\varphi_\lambda^{(t)}(x).$$

Arguing as before, we have

$$\sqrt{\sum_{\lambda\in\Lambda_t\setminus F} |\langle K_r(\cdot,q), \varphi_\lambda^{(t)}\rangle_{H_0}|^2} \le \|f_F\|_{H_t}. \tag{2.141}$$

(2.141) implies that

$$f_t^* = \sum_{\lambda\in\Lambda_t} \langle K_r(\cdot,q), \varphi_\lambda^{(t)}\rangle_{H_0}\varphi_\lambda^{(t)}, \tag{2.142}$$

where the convergence takes place in the topology of $H_{K_t}(E)$ for any $q \in E$. Using this formula, we obtain

$$\langle K_r(\cdot,q), f_t^*\rangle_{H_{K_t}(E)} = \sum_{\lambda\in\Lambda_t} \overline{\langle K_r(\cdot,q), \varphi_\lambda^{(t)}\rangle_{H_0}}\langle K_r(\cdot,q), \varphi_\lambda^{(t)}\rangle_{H_{K_t}(E)} \tag{2.143}$$

and

$$\|f_t^*\|_{H_{K_t}(E)} = \sqrt{\sum_{\lambda\in\Lambda_t} |\langle K_r(\cdot,q), \varphi_\lambda^{(t)}\rangle_{H_0}|^2}. \tag{2.144}$$

We also have

$$\|K_r(\cdot,q)\|_{H_{K_t}(E)} = \sqrt{\sum_{\lambda\in\Lambda_t} |\langle K_r(\cdot,q),\varphi_\lambda^{(t)}\rangle_{H_{K_t}(E)}|^2} (< \infty) \tag{2.145}$$

for all $0 < t \leq r$. Meanwhile,

$$f = \sum_{\lambda\in\Lambda_t} \langle K_r(\cdot,q),\varphi_\lambda^{(t)}\rangle_{H_t}\varphi_\lambda^{(t)}.$$

By inequality (2.141), we obtain

$$0 = \limsup_{t\to 0} \|f - f_t^*\|_{H_{K_t}(E)} \geq \limsup_{t\to 0} \|f - f_t^*\|_{H_0} = 0.$$

Thus the proof of Theorem 2.23 is complete.

The source of this subsection is [417, 418] and see the applications to initial value problems for some general linear operator equations.

2.3.6 Wedge Product of RKHS

Let n be a natural number and let $H_K(E)$ be a reproducing kernel Hilbert space on E. Then write $\otimes^n H_K(E)$ for the n-fold tensor product of $H_K(E)$. We define the n-fold wedge product $\wedge^n H_K(E)$ by the closed subspace of $\otimes^n H_K(E)$ which is spanned by

$$\left\{ \sum_{\sigma\in S_n} \operatorname{sgn}(\sigma) x_{\sigma(1)} \otimes x_{\sigma(2)} \otimes \ldots \otimes x_{\sigma(n)} \ : \ x_1, x_2, \ldots, x_n \in H_K(E) \right\}.$$

Theorem 2.24. *Let n be a natural number and let $H_K(E)$ be a reproducing kernel Hilbert space on E. Then the function $\wedge^n K$, given by*

$$\wedge^n K(x_1, x_2, \ldots, x_n, y_1, y_2, \ldots, y_n) \equiv \frac{1}{n!} \det\{K(x_i, y_j)\}_{i,j=1,2,\ldots,n}$$

for $x_1, x_2, \ldots, x_n, y_1, y_2, \ldots, y_n \in E$, is positive definite and a reproducing kernel of the space $\wedge^n H_K(E)$.

Proof. We establish the reproducing property solely.

We shall calculate

$$\mathrm{I} = \left\langle \sum_{\sigma\in S_n} \operatorname{sgn}(\sigma) \otimes_{j=1}^n K(\cdot_j, z_{\sigma(j)}), \wedge^n K(\cdot_1, \cdot_2, \ldots, \cdot_n, y_1, y_2, \ldots, y_n) \right\rangle_{\otimes^n H_K(E)}$$

to show the reproducing property. Notice that

$$\wedge^n K(\cdot_1, \cdot_2, \ldots, \cdot_n, y_1, y_2, \ldots, y_n) = \frac{1}{n!} \sum_{\sigma \in S_n} \mathrm{sgn}(\sigma) \otimes_{j=1}^n K(\cdot_j, y_{\sigma(j)}).$$

Consequently,

$$\begin{aligned}
\mathrm{I} &= \sum_{\sigma \in S_n} \mathrm{sgn}(\sigma) \left\langle \otimes_{j=1}^n K(\cdot_j, z_{\sigma(j)}), \wedge^n K(\cdot_1, \cdot_2, \ldots, \cdot_n, y_1, y_2, \ldots, y_n) \right\rangle_{\otimes^n H_K(E)} \\
&= \frac{1}{n!} \sum_{\sigma \in S_n} \mathrm{sgn}(\sigma) \left\langle \otimes_{j=1}^n K(\cdot_j, z_{\sigma(j)}), \sum_{\tau \in S_n} \mathrm{sgn}(\tau) \otimes_{j=1}^n K(\cdot_j, y_{\tau(j)}) \right\rangle_{\otimes^n H_K(E)} \\
&= \frac{1}{n!} \sum_{\sigma \in S_n} \sum_{\tau \in S_n} \mathrm{sgn}(\tau)\mathrm{sgn}(\sigma) \left\langle \otimes_{j=1}^n K(\cdot_j, z_{\sigma(j)}), \otimes_{j=1}^n K(\cdot_j, y_{\tau(j)}) \right\rangle_{\otimes^n H_K(E)}.
\end{aligned}$$

By the reproducing property of K, we have

$$\mathrm{I} = \frac{1}{n!} \sum_{\sigma \in S_n} \sum_{\tau \in S_n} \mathrm{sgn}(\tau)\mathrm{sgn}(\sigma) \prod_{j=1}^n K(y_{\tau(j)}, z_{\sigma(j)}) = \sum_{\sigma \in S_n} \mathrm{sgn}(\sigma) \prod_{j=1}^n K(y_j, z_{\sigma(j)}).$$

This equality shows the reproducing property (1.2).

Example 2.21. We can readily identify the vectors in $\mathbb{R}^m$ with the functions from $\{1, 2, \ldots, m\}$ to $\mathbb{C}$. If we write $K(i,j) \equiv \delta_{ij}$, for $i, j = 1, 2, \ldots, m$, then $\mathbb{R}^m$, which carries the usual inner product, becomes a reproducing kernel Hilbert space.

Let us consider $\wedge^n \mathbb{R}^m$ for $n \leq m$ and suppose that we have an $m \times n$ matrix L. We write

$$L = (\mathbf{L}_1, \mathbf{L}_2, \ldots, \mathbf{L}_n),$$

where each $\mathbf{L}_j$ is an $m \times 1$ matrix, or equivalently, an m-column vector. Then a calculation similar to the proof of Theorem 2.24 shows that

$$\begin{aligned}
\det({}^t LL) &= \det(\{\langle \mathbf{L}_i, \mathbf{L}_j \rangle\}_{i,j=1,2,\ldots,n}) \\
&= \frac{1}{n!} \langle \mathbf{L}_1 \wedge \mathbf{L}_2 \wedge \ldots \wedge \mathbf{L}_n, \mathbf{L}_1 \wedge \mathbf{L}_2 \wedge \ldots \wedge \mathbf{L}_n \rangle_{\otimes^n \mathbb{R}^m} \\
&= \frac{1}{n!} \|\mathbf{L}_1 \wedge \mathbf{L}_2 \wedge \ldots \wedge \mathbf{L}_n\|^2_{\otimes^n \mathbb{R}^m},
\end{aligned}$$

where ${}^t L$ stands for the transpose of L. Let us say that two n-tuples of $\{1, 2, \ldots, m\}$ are equivalent if they are equal modulo a permutation. Let us choose a representative $\{I_\lambda\}_{\lambda \in \Lambda}$ with respect to this equivalence. If the two elements $(a_1, a_2, \ldots, a_n)$ and $(b_1, b_2, \ldots, b_n)$ are not equivalent, then

$$K_{\wedge^n \mathbb{R}^m}((a_1, a_2, \ldots, a_n), (b_1, b_2, \ldots, b_n)) = 0. \tag{2.146}$$

Therefore, by Theorem 2.19 we have

$$\|\mathbf{L}_1 \wedge \mathbf{L}_2 \wedge \ldots \wedge \mathbf{L}_n\|^2_{\otimes^n \mathbb{R}^m} = \sum_{\lambda \in \Lambda} \|(\mathbf{L}_1 \wedge \mathbf{L}_2 \wedge \ldots \wedge \mathbf{L}_n)|I_\lambda\|^2_{\otimes^n \mathbb{R}^m}. \tag{2.147}$$

Observe also that if $\sharp I_\lambda < n$, then

$$K_{\wedge^n \mathbb{R}^m}|(I_\lambda \times \wedge^n \mathbb{R}^m) \equiv 0.$$

Consequently,

$$\|\mathbf{L}_1 \wedge \mathbf{L}_2 \wedge \ldots \wedge \mathbf{L}_n\|^2_{\otimes^n \mathbb{R}^m} = \sum_{\lambda \in \Lambda,\, \sharp I_\lambda = n} \|(\mathbf{L}_1 \wedge \mathbf{L}_2 \wedge \ldots \wedge \mathbf{L}_n)|I_\lambda\|^2_{\otimes^n \mathbb{R}^m}. \tag{2.148}$$

If we write down of (2.148), then we have

$$\det({}^t LL) = \sum_{\lambda \in \Lambda,\, \sharp I_\lambda = n} \det({}^t L_\lambda L_\lambda), \tag{2.149}$$

where L_λ is a square matrix obtained by deleting the rows whose indexes do not belong to I_λ. The equality (2.149) is well known as the Cauchy Binet formula appearing in vector-analysis. When $m = 2$ and $n = 3$, then (2.149) dates back to Lagrange [271, pp. 662–663]. When $m = 2$, then (2.149) dates back to Cauchy [98, pp. 373–374].

2.4 Construction of Reproducing Kernel Hilbert Spaces

We meet in many situations a positive definite quadratic form function (kernel form). One basic important problem is its realization as the reproducing kernel Hilbert space. By considering various realizations, we can see the function space property in more detail ways. Numerical realizations are also important problems for some practical realization in order to obtain the function property using computers.

By combining with some operator, we will exhibit how we construct reproducing kernel Hilbert spaces.

2.4.1 Use Linear Mappings and CONS

As we have seen, given a positive definite quadratic form function $K : E \times E \to \mathbb{C}$, we can construct $H_K(E)$ in a canonical way. However, this does not always suffice. We sometimes need a concrete structure like Theorem 1.8. Here, we will see another typical method. The spirit in Sect. 2.4.1 seems to run counter to the one in the last

Sect. 2.1.1. However, once getting out of the canonical construction, we see a lot of new things. Let $\mathscr{H}$ be a separable Hilbert space and let $\mathbf{h} : E \to \mathscr{H}$ be a mapping. Here we do not assume that $\ker(L) = 0$ so far. Let $K(p,q) \equiv \langle \mathbf{h}(q), \mathbf{h}(p) \rangle$ for $p, q \in E$.

We define $Lf(p) \equiv \langle f, \mathbf{h}(p) \rangle_{\mathscr{H}}$ for $f \in \mathscr{H}$.

Then as we have seen, this couple $(\mathscr{H}, \mathbf{h})$ gives us a reproducing kernel Hilbert space $H_K(E)$. However, let us investigate it in more detail via a CONS $\{V_j\}_{j=1}^{\infty}$ of $\mathscr{H}$. A modification is readily made if $\mathscr{H}$ is of finite dimensional. Let us define a function v_j on E by

$$v_j(\cdot) \equiv LV_j(\cdot) = \langle V_j, \mathbf{h}(\cdot) \rangle_{\mathscr{H}} \in H_K(E). \tag{2.150}$$

Let $p \in E$. Now that $\{V_j\}_{j=1}^{\infty}$ forms a CONS, we have

$$h(p) = \sum_{j=1}^{\infty} \langle \mathbf{h}(p), V_j \rangle V_j = \sum_{j=1}^{\infty} \overline{v_j(p)} V_j \tag{2.151}$$

as well as

$$\sum_{j=1}^{\infty} |v_j(p)|^2 = \sum_{j=1}^{\infty} |\langle V_j, \mathbf{h}(p) \rangle_{\mathscr{H}}|^2 = \|\mathbf{h}(p)\|_{\mathscr{H}}^2 < \infty. \tag{2.152}$$

Now we define

$$\overline{h}(p) \equiv \sum_{j=1}^{\infty} v_j(p) V_j \quad (p \in E), \tag{2.153}$$

whose convergence in $\mathscr{H}$ will be guaranteed by (2.152).

If $\{\mathbf{h}(p) : p \in E\}$ spans a dense subspace of $\mathscr{H}$, then $H_K(E)$ has the following beautiful expression:

Theorem 2.25. *Assume that L is injective. Let v_j be given by* (2.150)*. Assume that $\{\mathbf{h}(p) : p \in E\}$ spans a dense subspace of $\mathscr{H}$. Then we have*

$$H_K(E) = \left\{ \sum_{j=1}^{\infty} a_j v_j : \{a_j\}_{j=1}^{\infty} \in \ell^2(\mathbb{N}) \right\} \tag{2.154}$$

and the inner product is

$$\left\| \sum_{j=1}^{\infty} a_j v_j \right\|_{H_K(E)} = \sqrt{\sum_{j=1}^{\infty} |a_j|^2}$$

for all $\{a_j\}_{j=1}^{\infty} \in \ell^2(\mathbb{N})$.

We now prove Theorem 2.25 after some remarks: The series

$$S = \sum_{j=1}^{\infty} a_j\, v_j = L\left(\sum_{j=1}^{\infty} a_j V_j\right) \tag{2.155}$$

converges absolutely by the Bessel inequality provided $\sum_{j=1}^{\infty} |a_j|^2 < \infty$. Furthermore,

$$\langle v_j, v_k\rangle_{H_K(E)} = \langle L\, V_j, L\, V_k\rangle_{H_K(E)} = \langle V_j, V_k\rangle_{\mathscr{H}} = \delta_{jk}, \tag{2.156}$$

which implies that the inner product on the right-hand side makes sense.

Proof (of Theorem 2.25). The reproducing property of H with the inner product two lines below (1.54) is clear. Indeed, it suffices to verify this property for $f = v_j$ for $j \in \mathbb{N}$.

Define

$$H \equiv \left\{\sum_{j=1}^{\infty} a_j\, v_j \in H_K(E) \;:\; \sum_{j=1}^{\infty} |a_j|^2 < \infty\right\}. \tag{2.157}$$

Then, for $p \in E$, we have

$$K_p = \langle \mathbf{h}(p), \mathbf{h}(\cdot)\rangle_{\mathscr{H}} = \sum_{j=1}^{\infty} \langle \mathbf{h}(p), V_j\rangle_{\mathscr{H}} \cdot \langle V_j, \mathbf{h}(\cdot)\rangle_{\mathscr{H}} = \sum_{j=1}^{\infty} \overline{v_j(p)} \cdot v_j \in H. \tag{2.158}$$

as was to be shown.

If $f \in H_K(E)$, we have the following expression of $f(p)$ by using

$$T_{H_K(E)}(f) \equiv \sum_{j=1}^{\infty} \langle f, v_j\rangle_{H_K(E)} V_j.$$

Proposition 2.8. *Let $f \in H_K(E)$. Then we have*

$$f(p) = \langle f, L[\mathbf{h}(p)]\rangle_{H_K(E)} = \langle T_{H_K(E)}(f), \mathbf{h}(p)\rangle_{\mathscr{H}} \quad (p \in E). \tag{2.159}$$

Proof. Let $f = LF$ with $F \in \mathscr{H}$ as usual. Then, for $p \in E$, we have

$$\begin{aligned}\langle f, L[\mathbf{h}(p)]\rangle_{H_K(E)} &= \langle F, P_{\mathscr{H} \to \ker(L)^{\perp}} \mathbf{h}(p)\rangle_{\mathscr{H}} \\ &= \sum_{j=1}^{\infty} \langle P_{\mathscr{H} \to \ker(L)^{\perp}} F, V_j\rangle_{\mathscr{H}} \langle V_j, \mathbf{h}(p)\rangle_{\mathscr{H}}\end{aligned}$$

$$= \sum_{j=1}^{\infty} \langle f, v_j \rangle_{\mathscr{H}} \langle V_j, \mathbf{h}(p) \rangle_{\mathscr{H}}$$
$$= \langle T_{H_K(E)}(f), \mathbf{h}(p) \rangle_{\mathscr{H}}.$$

This proves (2.159).

Lemma 2.5. *Keep the same notation as above. Then* $T_{H_K(E)}$ *is an isometry from* $H_K(E)$ *to* $\mathscr{H}$.

Proof. We need to prove

$$\|f\|_{H_K(E)} = \|T_{H_K(E)}(f)\|_{\mathscr{H}} \tag{2.160}$$

for all $f \in H_K(E)$. Let $f = LF$ with $F \in \mathscr{H}$ as usual using Theorem 2.37. Then we have

$$\begin{aligned}
\|f\|_{H_K(E)} &= \sqrt{\langle P_{\mathscr{H} \to \ker(L)^\perp} F, P_{\mathscr{H} \to \ker(L)^\perp} F \rangle_{\mathscr{H}}} \\
&= \sqrt{\sum_{j=1}^{\infty} \langle P_{\mathscr{H} \to \ker(L)^\perp} F, V_j \rangle_{\mathscr{H}} \cdot \langle V_j, P_{\mathscr{H} \to \ker(L)^\perp} F \rangle_{\mathscr{H}}} \\
&= \sqrt{\sum_{j=1}^{\infty} |\langle f, v_j \rangle_{H_K(E)}|^2} = \|T_{H_K(E)}(f)\|_{\mathscr{H}}.
\end{aligned}$$

This is the desired result.

The next theorem characterizes $T_{H_K(E)}(f)$.

Theorem 2.26. *Let* $f \in H_K(E)$. *The function* $T_{H_K(E)}(f)$ *is a unique element in* $\ker(L)^\perp$ *such that*

$$f(p) = \langle T_{H_K(E)}(f), \mathbf{h}(p) \rangle \quad (p \in E). \tag{2.161}$$

Proof. Everything in Theorem 2.26 was proved, except for the fact that $T_{H_K(E)}(f)$ belongs to $\ker(L)^\perp$. However, this is a direct consequence of the following equality:

$$\begin{aligned}
\langle T_{H_K(E)}(f), a \rangle_{\mathscr{H}} &= \sum_{j=1}^{\infty} \langle f, v_j \rangle_{H_K(E)} \langle V_j, a \rangle_{\mathscr{H}} \\
&= \sum_{j=1}^{\infty} \langle L^* f, V_j \rangle_{\mathscr{H}} \langle V_j, a \rangle_{\mathscr{H}} \\
&= \langle L^* f, a \rangle_{\mathscr{H}} \\
&= \langle f, La \rangle_{H_K(E)} = 0
\end{aligned} \tag{2.162}$$

for all $a \in \ker(L)$.

So we are done.

The following is a fundamental equality in the theory of reproducing kernel Hilbert spaces:

Theorem 2.27. *With the notation above, we have*

$$K(p,q) = \sum_{j=1}^{\infty} v_j(p)\overline{v_j(q)} \quad (p, q \in E). \tag{2.163}$$

Proof. Let $p, q \in E$. By Parseval's theorem, we have

$$K(p,q) = \langle \mathbf{h}(q), \mathbf{h}(p)\rangle_{\mathscr{H}} = \sum_{j=1}^{\infty} \langle \mathbf{h}(q), V_j\rangle_{\mathscr{H}} \cdot \langle V_j, \mathbf{h}(p)\rangle_{\mathscr{H}} = \sum_{j=1}^{\infty} v_j(p)\overline{v_j(q)},$$

proving (2.163).

Orthonormal systems can be constructed by the Gram Schmidt orthogonalization procedure from elementary linearly independent functions. Therefore, we consider that we can calculate reproducing kernels. This idea was fundamental due to S. Bergman. See [46, Chapter 2] and [279].

2.4.2 Use the Fourier Integral

In Sect. 2.4.2 we assume that $E = \mathbb{R}$.

Here and below we assume that ρ is non-negative and integrable and that K is given by (2.8):

$$K(x,y) \equiv \frac{1}{2\pi} \int_{\mathbb{R}} \exp(-it \cdot (x-y))\rho(t)\, dt \; (x, y \in \mathbb{R}).$$

Definition 2.2. Define

$$L^2(\mathbb{R}, \rho) \equiv \left\{ F : \mathbb{R} \to \mathbb{C} \; : \; \|F\|_{L^2(\mathbb{R},\rho)} \equiv \sqrt{\frac{1}{2\pi} \int_{\mathbb{R}} |F(t)|^2 \rho(t)\, dt} < \infty \right\}. \tag{2.164}$$

Also define

$$I_\rho : L^2(\mathbb{R}, \rho) \to \mathscr{B} \equiv \{f : \mathbb{R} \to \mathbb{C} \; : f \text{ is bounded and continuous }\} \tag{2.165}$$

by

$$I_\rho F(x) \equiv \frac{1}{2\pi} \int_{\mathbb{R}} F(t) \exp(-it \cdot x)\rho(t)\, dt. \tag{2.166}$$

Lemma 2.6. *The mapping $I_\rho : L^2(\mathbb{R}, \rho) \to \mathscr{B}$ is an injection.*

Proof. This is clear from the fact that the Fourier transform is an injection.

The next theorem characterizes the range of I_ρ.

Theorem 2.28. *With the notation above,*

$$H_K(\mathbb{R}) \equiv \left\{ I_\rho F \,:\, F \in L^2(\mathbb{R}, \rho) \right\} \tag{2.167}$$

and the norm is given by

$$\|I_\rho F\|_{H_K(\mathbb{R})} = \|F\|_{L^2(\mathbb{R},\rho)},\ F \in L^2(\mathbb{R}, \rho). \tag{2.168}$$

Proof. Let us set

$$H = \left\{ I_\rho F \,:\, F \in L^2(\mathbb{R}, \rho) \right\},\ \|I_\rho F\|_H = \|F\|_{L^2(\mathbb{R},\rho)}. \tag{2.169}$$

Note that

$$K_y = I_\rho[E_y], \tag{2.170}$$

where $E_y(t) = \exp(it \cdot y)$. Next, observe that

$$\langle K_y, K_x \rangle = \langle E_y, E_x \rangle_{L^2(\mathbb{R},\rho)} = \frac{1}{2\pi} \int_{\mathbb{R}} E_y(t)\overline{E_x(t)}\rho(t)\, dt = K(x, y). \tag{2.171}$$

Also it is trivial that $H_K(\mathbb{R}) \subset \mathscr{F}(\mathbb{R})$ because we have $H_K(\mathbb{R}) \subset \mathscr{B}$.

As a consequence we have $H = H_K(\mathbb{R})$ with norm-coincidence.

Now that

$$I_\rho f(t) = \frac{1}{\rho(t)} \int_{\mathbb{R}} f(x) \exp(it \cdot x)\, dx, \tag{2.172}$$

(2.168) read as

$$\|f\|_{H_K(\mathbb{R})} = \sqrt{\frac{1}{2\pi} \int_{\mathbb{R}} \left| \int_{\mathbb{R}} f(x) \exp(it \cdot x)\, dx \right|^2 \frac{dt}{\rho(t)}}. \tag{2.173}$$

2.4.3 Mercer Expansion: A General Approach

For a positive definite quadratic form function, in order to construct its reproducing kernel Hilbert space, we will consider its expansion in terms of the related eigenfunctions and eigenvalues. This expansion extends the notion of CONS. In Sect. 2.4.3 we assume that X is a compact Hausdorff space coming with a Radon measure μ.

Assume that $K : X \times X \to \mathbb{R}$ is a function in $\mathscr{B}(X \times X)$, where $\mathscr{B}(X \times X)$ is a function space obtained by replacing X with $X \times X$ in (0.1). Define

$$I_K f(x) \equiv \int_X K(x, y) f(y)\, d\mu(y). \tag{2.174}$$

The operator I_K is a compact operator as the following theorem asserts:

Theorem 2.29. *Under the above notation* (2.174), *I_K is an $L^2(\mu)$-compact operator.*

Proof. We may assume that K is a positive function. We also remark that

$$\|I_K\|_{L^2(\mu)\to L^2(\mu)} \le \mu(X)\|K\|_{\mathscr{B}(X\times X)} \tag{2.175}$$

by the Hölder inequality. By the Weierstrass approximation theorem, we see that

$$\{\phi \otimes \eta \,:\, \phi, \eta \in C_c(X)\} \tag{2.176}$$

spans a dense subspace of $\mathscr{B}(X \times X)$. As a result the matter is reduced to the case

$$K = \sum_{j=1}^{N} \phi_j \otimes \eta_j,\ \phi_j,\ \eta_j \in C_c(X). \tag{2.177}$$

However, if K takes such a form, I_K is of finite rank, hence compact.

The next theorem is fundamental and is called the Mercer expansion.

Theorem 2.30 (Mercer expansion). *Let μ be a Radon measure on a topological space X. Assume that $K : X \times X \to \mathbb{R}$ is a continuous function that satisfies the following conditions*:

1. $K(x, y) = K(y, x)$ *for all* $x, y \in \mathrm{supp}(\mu)$;
2. $\iint_{X\times X} K(x, y) f(x)\overline{f(y)}\, d\mu(x)\, d\mu(y) \ge 0$ *for all* $f \in L^2(\mu)$.

Let $\{\lambda_j\}_{j\in J}$ be the set of positive eigenvalues of I_K and $\{\varphi_j\}_{j\in J}$ the corresponding orthonormal eigenfunctions. Then we have

$$K = \sum_{j=1}^{\infty} \lambda_j \varphi_j \otimes \varphi_j$$

on $\mathrm{supp}(\mu \times \mu)$, *where the convergence takes place in* $C(\mathrm{supp}(\mu \times \mu))$.

Proof. By replacing X with $\mathrm{supp}(\mu)$ we may assume that $X = \mathrm{supp}(\mu)$. If J is a finite set, then there is nothing to prove. Therefore, we shall assume otherwise. In this case J is countable because $L^2(X)$ is separable. Set $J = \mathbb{N}$.

Let us begin with the expansion. To do this, we consider

$$\mathrm{I} \equiv \int_{X\times X}\left(K(x,y)-\sum_{j=1}^{N}\lambda_j\varphi_j(x)\varphi_j(y)\right)\overline{f(x)}f(y)\,d\mu(x)\,d\mu(y)$$

for $f \in C(X)$. Then, by the Fubini theorem,

$$\begin{aligned}\mathrm{I} &= \int_X\left(\int_X K(x,y)f(y)\,d\mu(y)-\sum_{j=1}^{N}\int_X\lambda_j\varphi_j(x)\varphi_j(y)f(y)\,d\mu(y)\right)\overline{f(x)}\,d\mu(y)\\ &= \int_X\left(\int_X\sum_{j=N+1}^{\infty}\int_X\lambda_j\varphi_j(x)\varphi_j(y)f(y)\,d\mu(y)\right)\overline{f(x)}\,d\mu(y)\\ &\geq 0.\end{aligned}$$

Hence we conclude that

$$I \geq 0 \tag{2.178}$$

for all $f \in C(X)$. Now assume that

$$K(x,x) < \sum_{j=1}^{\infty}\lambda_j\varphi_j(x)^2. \tag{2.179}$$

Then we choose N large enough to have

$$K(x,x) < \sum_{j=1}^{N}\lambda_j\varphi_j(x)^2. \tag{2.180}$$

Now that K is assumed to be continuous,

$$\varphi_j(x) = \frac{1}{\lambda_j}\int_X K(x,y)\varphi_j(y)\,d\mu(y) \tag{2.181}$$

is continuous as well. Therefore, there exists a neighborhood U of x such that

$$K(y,z) < \sum_{j=1}^{N}\lambda_j\varphi_j(y)\varphi_j(z),\ y,z \in U. \tag{2.182}$$

Now that X is assumed to be a compact Hausdorff space, there exists a nontrivial non-negative continuous function f supported on U.

In view of the definition of I and (2.182) we see that $I < 0$, which contradicts (2.178). As a result, we obtain

$$K(x,x) \geq \sum_{j=1}^{\infty} \lambda_j \varphi_j(x)^2. \tag{2.183}$$

Assume that (2.183) is strict. Choose $\varepsilon = \varepsilon(x) > 0$ so that

$$K(x,x) > \varepsilon + \sum_{j=1}^{\infty} \lambda_j \varphi_j(x)^2. \tag{2.184}$$

Then, for all $N \in \mathbb{N}$, we have

$$K(x,x) > \varepsilon + \sum_{j=1}^{N} \lambda_j \varphi_j(x)^2. \tag{2.185}$$

By the continuity of K and φ_j, there exists a neighborhood U_N of x such that

$$K(y,z) > \varepsilon + \sum_{j=1}^{N} \lambda_j \varphi_j(y) \varphi_j(z) \tag{2.186}$$

for all $y, z \in U_N$. Pick a non zero continuous function f_N so that $0 \leq f_N \leq \chi_U$. Then we have

$$\int_{X \times X} \left(K(y,z) - \sum_{j=1}^{N} \lambda_j \varphi_j(y) \varphi_j(z) \right) f_N(y) f_N(z) \, d\mu(y) \, d\mu(z) \geq \varepsilon \, \|\varphi_j\|_{L^2(\mu)}^2.$$

This contradicts the fact that

$$I_N = \sum_{j=1}^{\infty} \lambda_j \, \langle \cdot, \varphi_j \rangle \varphi_j \tag{2.187}$$

takes place in the operator topology of $L^2(\mu)$.

Once we establish that

$$K(x,x) = \sum_{j=1}^{\infty} \lambda_j \, \varphi_j(x)^2, \tag{2.188}$$

it follows from the Dini convergence theorem that the convergence is uniform.

From the uniform convergence of (2.188), we deduce that

$$\sum_{j=1}^{\infty} \lambda_j \, \varphi_j \otimes \varphi_j \tag{2.189}$$

converges uniformly and hence defines a continuous function. This function agrees with K as an integral kernel, and it follows that

$$K = \sum_{j=1}^{\infty} \lambda_j \, \varphi_j \otimes \varphi_j \tag{2.190}$$

for $\mu \times \mu$-almost every on $X \times X$. However, both sides are continuous functions. Hence it follows that

$$K = \sum_{j=1}^{\infty} \lambda_j \, \varphi_j \otimes \varphi_j \tag{2.191}$$

and the convergence takes place in the uniform topology.

By Theorem 2.30, the reproducing kernel Hilbert space $H_K(X)$ is represented as follows:

Theorem 2.31. *Under the same notation, we have*

$$H_K(X) = \left\{ \sum_{j=1}^{\infty} \frac{c_j}{\lambda_j} \varphi_j \, : \, (c_1, c_2, \ldots, c_j, \ldots) \in \mathbb{C}^{\mathbb{N}}, \quad \sum_{j=1}^{\infty} \frac{|c_j|^2}{\lambda_j} < \infty \right\} \tag{2.192}$$

and the norm is given by

$$\left\| \sum_{j=1}^{\infty} \frac{c_j}{\lambda_j} \varphi_j \right\|_{H_K(X)} = \sqrt{\sum_{j=1}^{\infty} \frac{|c_j|^2}{\lambda_j}}, \tag{2.193}$$

where $(c_1, c_2, \ldots, c_j, \ldots) \in \mathbb{C}^{\mathbb{N}}$ *satisfies*

$$\sum_{j=1}^{\infty} \frac{|c_j|^2}{\lambda_j} < \infty.$$

Proof. Let us set the Hilbert space given by (2.192) as H with the inner product defined by (2.192). Now that

$$K(x,x) = \sum_{j=1}^{\infty} \lambda_j \, \varphi_j(x)^2, \tag{2.194}$$

it is easy to deduce that

$$H \subset \mathscr{F}(X). \tag{2.195}$$

Also,

$$K_x = \sum_{j=1}^{\infty} \lambda_j \varphi_j(x)\varphi_j \tag{2.196}$$

with

$$\sum_{j=1}^{\infty} \frac{(\lambda_j \varphi_j(x))^2}{\lambda_j} = \sum_{j=1}^{\infty} \lambda_j \varphi_j(x)^2 < \infty \tag{2.197}$$

gives $K_x \in H$. Also note that

$$\langle K_y, K_x \rangle_H = \sum_{j=1}^{\infty} \frac{\lambda_j^2 \varphi_j(x)\varphi_j(y)}{\lambda_j} = \sum_{j=1}^{\infty} \lambda_j \varphi_j(x)\varphi_j(y) = K(x, y). \tag{2.198}$$

As a consequence we deduce that $H_K(X) = H$ with norm-coincidence.

As a standard textbook, we list K. Yoshida [496] and an interesting connection with the theory of reproducing kernels and learning theory, see F. Cucker and S. Smell [117].

2.4.4 *Realization by Discrete Points*

We let H be a Hilbert space such that $H \subset \mathscr{F}(E)$ in the sense of continuous embedding. Suppose that K is a positive definite quadratic form function K on a set E. We have two points of view about a realization of RKHS $H_K(E)$. One is a discretization of the RKHS $H_K(E)$, and the other is the representation of the functions in $H_K(E)$ in terms of the values on some discrete point of E. We will work on separable Hilbert spaces consisting of functions on E and this idea is based on [262, Theorems 1–3]. The results will give nice relationships between *finite* and *infinite* and the ones between *discrete* and *continuous*. They are basic ideas in computer sciences and analysis. The result may be considered as a general sampling theory in that the functions can be represented by discrete point data.

Definition 2.3. A countable sequence $\{e_j\}_{j=1}^{\infty}$ in a Hilbert space is said to be *a basis*, if each $x \in H$ admits a unique expansion

$$x = \sum_{j=1}^{\infty} \lambda_j(x) e_j, \tag{2.199}$$

where each $\lambda_j : H \to \mathbb{C}$ is a continuous linear functional.

We define a matrix $\Gamma = \{\Gamma^n_{ij}\}_{1 \le i,j \le n}$ to be an inverse matrix of $\{\langle e_i, e_j \rangle_H\}_{1 \le i,j \le n}$. Let

$$K_n(p,q) \equiv \sum_{i,j=1}^{n} \Gamma^n_{ij} e_j(p) \overline{e_i(q)} \quad (p,q \in E). \tag{2.200}$$

Lemma 2.7. *Let us define*

$$e'_j \equiv \sum_{k=1}^{n} a_{jk} e_k, \quad \Theta = \{\theta^n_{ij}\}_{1 \le i,j \le n} \equiv \left(\{\langle e'_i, e'_j \rangle\}_{1 \le i,j \le n}\right)^{-1} \tag{2.201}$$

for an $n \times n$ *matrix* $A = \{a_{jk}\}_{1 \le j,k \le n}$. *Then we have*

$$K_n(p,q) = \sum_{i,j=1}^{n} \theta^n_{ij} e'_i(p) \overline{e'_j(q)} \quad (p,q \in E). \tag{2.202}$$

Proof. Note that

$$\Theta^{-1} = \left\{\left\langle \sum_{k=1}^{n} a_{ik} e_k, \sum_{l=1}^{n} a_{jl} e_l \right\rangle\right\}_{1 \le i,j \le n} = \left\{\sum_{k,l=1}^{n} a_{ik} \overline{a_{jl}} \langle e_k, e_l \rangle\right\}_{1 \le i,j \le n} = A\Gamma^{-1}A^*.$$

Thus we obtain $\Theta = (A^*)^{-1}\Gamma A^{-1}$. Therefore, if we write $A^{-1} \equiv \{a^{ij}\}_{1 \le i,j \le n}$, we have

$$\begin{aligned}
\sum_{i,j=1}^{n} \theta^n_{ij} e'_j(p) \overline{e'_i(q)} &= \sum_{i,j,k,l=1}^{n} \theta^n_{ij} a_{jk} e_k(p) \, \overline{a_{il} e_l(q)} \\
&= \sum_{i,j,k,l,\mu,\nu=1}^{n} \overline{a^{\mu i}} \, \overline{a_{il}} a^{\nu j} a_{jk} \Gamma^n_{\mu\nu} \, e_k(p) \overline{e_l(q)} \\
&= \sum_{i,j,k,l=1}^{n} \Gamma^n_{lk} \, e_k(p) \overline{e_l(q)} = K_n(p,q).
\end{aligned}$$

Thus the proof is complete.

In view of Lemma 2.7, we may assume that the e_j are orthonormal.

Theorem 2.32. *Define* K_n *by* (2.200). *Then a Hilbert space* H *admits a reproducing kernel if and only if* $\lim_{n \to \infty} K_n(q,q)$ *exists for all* $q \in E$ *and in this case the norm convergence implies the pointwise convergence.*

Proof. Assume that $\lim_{n \to \infty} K_n(q,q)$ exists for all $q \in E$. Then we have

$$\sum_{j=1}^{\infty} |e_j(q)|^2 < \infty. \tag{2.203}$$

As a result, if we fix $q \in E$, then the limit

$$K_q = \sum_{j=1}^{\infty} \overline{e_j(q)} e_j \tag{2.204}$$

exists in $H_K(E)$. With this in mind, let us set

$$K(p,q) \equiv K_q(p) = \sum_{j=1}^{\infty} e_j(p)\overline{e_j(q)}.$$

Let $f \in H$ be chosen arbitrarily. Since $\{e_j\}_{j=1}^{\infty}$ is a CONS, it follows that

$$f = \sum_{j=1}^{\infty} \langle f, e_j \rangle_H e_j \tag{2.205}$$

and in H and hence pointwise. As a result, from (2.204) and (2.205), we obtain

$$\langle f, K_q \rangle_{\mathscr{H}} = \sum_{j=1}^{\infty} \langle f, e_j \rangle_{\mathscr{H}} e_j(q) = f(q). \tag{2.206}$$

Thus H admits a reproducing kernel K.

The convergence of (2.206) is exactly what we have been proved before.

Let us now consider $H_K(E)$. Assume that $H_K(E)$ is separable. Then there exists a countable subset $\{q_j\}_{j=1}^{\infty}$ such that $\{K_{q_j}\}_{j=1}^{\infty}$ spans a dense subspace in $H_K(E)$. This observation justifies the following definition:

Definition 2.4. A set $\{p_j\}_{j=1}^{\infty}$ is said to be *a uniqueness set* for the function space $H_K(E)$ if, for any function $f \in H_K(E) \setminus \{0\}, f(p_j) = 0$ for all j, then f is identically zero.

For a uniqueness set $S = \{q_j\}_{j=1}^{\infty}$, we write $S_n \equiv \{q_1, q_2, \ldots, q_n\} \subset S$. Let $\Gamma^n \equiv \{\, \Gamma^n_{ij} \,\}_{1 \le i,j \le n}$ be the inverse matrix of $\{K(q_j, q_i)\}_{1 \le i,j \le n} = \{\, \langle K_{q_i}, K_{q_j} \rangle_{H_K(E)} \,\}_{1 \le i,j \le n}$.

The following result is fundamental about the result of the convergence:

Theorem 2.33 (Ultimate realization of reproducing kernel Hilbert spaces). *Let $f \in H_K(E)$ and Γ be given by* (2.200). *We assume that $\{q_j\}_{j=1}^{\infty}$ is a uniqueness set of $H_K(E)$ and that $\{K_{q_j}\}_{j=1}^{\infty}$ is linearly independent. Then*

$$f = \lim_{n \to \infty} \sum_{i,j=1}^{n} f(q_i) \Gamma^n_{ij} K_{q_j} \tag{2.207}$$

in the topology of $H_K(E)$. In particular,

$$\langle f, g \rangle_{H_K(E)} = \lim_{n \to \infty} \sum_{i,j=1}^{n} f(q_i) \Gamma^n_{ij} \overline{g(q_j)}$$

for $f, g \in H_K(E)$ *and*

$$f(p) = \lim_{n\to\infty} \sum_{j=1}^{n} \sum_{j'=1}^{n} f(q_j)\Gamma_{ij}^n K(p, q_{j'}) \tag{2.208}$$

for $f \in H_K(E)$ *and* $p \in E$.

Equation (2.208) may be considered as the ultimate sampling theory. The important view points are the norm non decreasing property and the norm may be calculated by computers for the above norm realization. See [91–93] for details.

Proof. Let us set

$$A_n(f) \equiv \sum_{i,j=1}^{n} f(q_i)\Gamma_{ij}^n K_{q_j}. \tag{2.209}$$

Let us first establish that $\{A_n\}_{n=1}^{\infty}$ is uniformly bounded, which will allow us to assume that $f = K_{q_l}$.

$$\begin{aligned}\left\| \sum_{i,j=1}^{n} f(q_i)\Gamma_{ij}^n K_{q_j} \right\|_{H_K(E)} &= \sqrt{\sum_{i,j,k,l=1}^{n} K(q_l, q_j) f(q_i)\overline{f(q_k)}\Gamma_{ij}^n \overline{\Gamma_{kl}^n}} \\ &= \sqrt{\sum_{i,k=1}^{n} \overline{f(q_k)} f(q_i)\overline{\Gamma_{ki}^n}}.\end{aligned}$$

Note that

$$\sum_{i,k=1}^{n} f(q_i)\overline{f(q_k)\Gamma_{ki}^n} = \left\langle f, \sum_{i,k=1}^{n} f(q_k)\Gamma_{ki}^n K_{q_i} \right\rangle_{H_K(E)} \le \|f\|_{H_K(E)} \left\| \sum_{i,j=1}^{n} f(q_i)\Gamma_{ij}^n K_{q_j} \right\|_{H_K(E)}.$$

As a consequence, it follows that

$$\left\| \sum_{i,j=1}^{n} f(q_i)\Gamma_{ij}^n K_{q_j} \right\|_{H_K(E)} \le \|f\|_{H_K(E)}. \tag{2.210}$$

With (2.210) established, as we have remarked we can assume that $f = K_{q_l}$. Let $n \ge l$. Then we have

$$\sum_{i,j=1}^{n} K(q_i, q_l)\Gamma_{ij}^n K(p, q_j) = K(p, q_l), \quad (p \in E) \tag{2.211}$$

hence

$$f = \lim_{n\to\infty} \sum_{i,j=1}^{n} f(q_i)\Gamma_{ij}^{n} K_{q_j} \text{ in } H_K(E) \tag{2.212}$$

is trivial if $f = K_{q_l}$. A passage to the limit allows us to have (2.212) for general f.

The next corollary can be applied to the interpolation problem.

Corollary 2.6. *Assume that $f \in \mathcal{F}(E)$ satisfies*

$$\mathrm{I} \equiv \sup_{n\in\mathbb{N}} \left(\sum_{i,j=1}^{n} f(q_i)\Gamma_{ij}^{n}\overline{f(q_j)} \right) < \infty. \tag{2.213}$$

Then there exists $F \in H_K(E)$ such that $F(q_j) = f(q_j)$ for all $j \in \mathbb{N}$.

Proof. Suppose that I is finite. Let us consider

$$f_n \equiv \sum_{i,j=1}^{n} f(q_i)\Gamma_{ij}^{n} K_{q_j}. \tag{2.214}$$

Then we obtain

$$\|f_n\|_{H_K(E)} \le \sup_{m\in\mathbb{N}} \sqrt{\sum_{i,j=1}^{m} f(q_i)\Gamma_{ij}^{m}\overline{f(q_j)}} < \infty. \tag{2.215}$$

The Banach Alaoglu theorem gives us a subsequence $\{f_{n_j}\}_{j=1}^{\infty}$ convergent to $F \in H_K(E)$ in weak-topology. Let us prove that $F(q_l) = f(q_l)$ for all $l \in \mathbb{N}$. To do this, we take $l \in \mathbb{N}$. Then we have

$$F(q_l) = \langle F, K_{q_l}\rangle = \lim_{j\to\infty} \langle f_{n_j}, K_{q_l}\rangle,$$

from weak convergence of $\{f_{n_j}\}_{j=1}^{\infty}$ and hence

$$F(q_l) = \lim_{j\to\infty} \sum_{i,m=1}^{n_j} f(q_i)\Gamma_{im}^{n_j}\langle K_{q_m}, K_{q_l}\rangle = \lim_{j\to\infty} \sum_{i,m=1}^{n_j} f(q_i)\Gamma_{im}^{n_j} K(q_l, q_m) = f(q_l).$$

Thus $F \in H_K(E)$ is an element that agrees with f on $\{q_j\}_{j=1}^{\infty}$.

This result illustrates that the function f in $H_K(E)$ is represented in terms of $f(q_j)$ on the discrete point set $\{q_j\}$ on the set E, that is, a very general sampling theorem. However, the numerical treatment of this algorithm will be very complicated. However, H. Fujiwara gave numerical experiments through his powerful method based on his infinite precision arithmetic algorithm *exflib*. Furthermore, the background properties of Theorem 2.33 were given in [92, 93] with many concrete applications and numerical experiments. Furthermore, for the sampling theory and many numerical experiments with detailed discussions, see [166].

Example 2.22. In this example we will clarify the structure of the reproducing kernel Hilbert spaces of finite dimension. Let $H = H_K(E)$ be the Hilbert kernel space of dimension n.

Let $\{W_j\}_{j=1}^n$ be a basis for H which is not necessarily orthonormal and write

$$\beta = \{\beta_{ij}\}_{1\le i,j\le n} \equiv \{\langle W_i, W_j\rangle_H\}_{1\le i,j\le n}^{-1}.$$

Then as we have seen, K is given by

$$K(p,q) = \sum_{i,j=1}^{n} \beta_{ij}\overline{W_i(p)}W_j(q) \quad (p,q \in E). \tag{2.216}$$

Positive Definite Hermitian Matrices

Example 2.22 may be stated clearly, for positive definite Hermitian matrices, as follows: Let $E = \{1,2,\ldots,N\}$. For any positive definite Hermitian matrix $A = \{a_{\nu\mu}\}_{\nu,\mu=1}^N$ we set $\tilde{A} = \{\tilde{a}_{\nu\mu}\}_{\nu,\mu=1}^N \equiv \overline{A^{-1}}$. We think of $\mathbb{C}^N$ as the vectors composed of all functions on E. We also regard $\mathbb{C}^n$ as the space of $N \times 1$ matrices and denote the space by $H[A]$ when it is equipped with the inner product

$$\langle \mathbf{x}, \mathbf{y}\rangle_{H[A]} = \mathbf{y}^* A\mathbf{x} \quad (\mathbf{x}, \mathbf{y} \in \mathbb{C}^n). \tag{2.217}$$

We consider the function K defined on $E \times E$ defined by:

$$K(\nu,\mu) \equiv \tilde{a}_{\nu\mu} \quad (\nu,\mu \in E). \tag{2.218}$$

This function K is the reproducing kernel for the space $H[A] = H_K(\{1,2,\ldots,N\})$. Namely, if we consider (2.218) as a row of column vectors

$$\mathbb{K}_\mu \equiv {}^t(\tilde{a}_{1\mu},\ldots,\tilde{a}_{N\mu}),$$

then for any vector $\mathbf{x} = (x_1, x_2, \ldots, x_N)^T$, we have

$$\langle \mathbf{x}, \mathbb{K}_\mu\rangle_{H[A]} = x_\mu \quad (\mu = 1,2,\ldots,N). \tag{2.219}$$

We can connect the theory of reproducing kernels to that of positive definite Hermitian matrices in this setting. See [388, pp. 11–14] and [416] for more details with many related references. We refer to [369] for systematic applications of finite dimensional reproducing kernels to a constructive theory of approximations to multivariate functions by polynomials restrictions. There we find several concrete reproducing kernels in terms of special functions.

2.4.5 Aveiro Discretization Method

Discretization by using reproducing kernels is established as the *Aveiro discretization method* globally with many concrete applications and numerical experiments in [91–93].

In particular, the representation by (2.208) and matrix representations are numerically computable in a practical way using Fujiwara's method. In the discretization method we will need precision in a deep way with huge computer resources. However, both these requirements were already prepared by Fujiwara (e.g., consider the case of the Laplace transform). Meanwhile, the method relies on general linear problems which may be dealt with as shown for arbitrary linear partial differential equations with general domains. Furthermore, boundary values and initial boundary values can also be given with freedom. Of course, the essential new restriction of the method is that at only a finite number of points of time and space the equations can be considered.

We shall see a prototype result using a prototype differential operator:

$$Ly \equiv \alpha y'' + \beta y' + \gamma y. \tag{2.220}$$

Here we consider a general situation in which coefficients are *arbitrary* functions on a general interval I.

We wish to construct some natural solution of

$$Ly = g \tag{2.221}$$

for a general function g on a general interval I.

For a practical construction of the solution, we obtain the following formula.

Theorem 2.34. *Let us fix a positive number h and take a finite number of points $\{t_j\}_{j=1}^n$ of I such that*

$$(\alpha(t_j), \beta(t_j), \gamma(t_j)) \neq \mathbf{0} \in \mathbb{R}^3$$

for each $j = 1, 2, \ldots, n$. Then the optimal solution y_h^A of the Eq. (2.221) is

$$y_h^A(t) = \frac{1}{2\pi} \int_{-\frac{\pi}{h}}^{\frac{\pi}{h}} F_h^A(\xi) e^{-it\xi} d\xi \quad (t \in I)$$

in terms of the function $F_h^A \in L^2\left(-\frac{\pi}{h}, \frac{\pi}{h}\right)$ in the sense that F_h^A has the minimum norm in $L^2\left(-\frac{\pi}{h}, \frac{\pi}{h}\right)$ among the functions $F \in L^2\left(-\frac{\pi}{h}, \frac{\pi}{h}\right)$ that satisfy, for the characteristic function $\chi_h(t)$ of the interval $\left(-\frac{\pi}{h}, \frac{\pi}{h}\right)$:

$$\frac{1}{2\pi} \int_{\mathbb{R}} F(\xi)[\alpha(t)(-\xi^2) + \beta(t)(-i\xi) + \gamma(t)]\chi_h(\xi) \exp(-it\xi) d\xi = g(t) \tag{2.222}$$

for all $t = t_j$.

The best extremal function F_h^A *is given by*

$$F_h^A(\xi) = \sum_{j,j'=1}^{n} g(t_j)\widetilde{a_{jj'}}\overline{(\alpha(t_{j'})(-\xi^2) + \beta(t_{j'})(-i\xi) + \gamma(t_{j'}))}\exp(it_{j'}\xi). \tag{2.223}$$

Here, the matrix $A = \{a_{jj'}\}_{j,j'=1}^{n}$, *formed by the elements*

$$a_{jj'} = K_{hh}(t_j, t_{j'})$$

with

$$2\pi K_{hh}(t,t') = \int_{\mathbb{R}} \chi_h(\xi)\exp(-i(t-t')\xi)(-\alpha(t)\xi^2 - i\beta(t)\xi + \gamma(t)) \tag{2.224}$$
$$\cdot \overline{(-\alpha(t')\xi^2 - i\beta(t')\xi + \gamma(t'))}d\xi,$$

is positive definite and the $\widetilde{a_{jj'}}$ *are the* (j,j')*-elements of the inverse of* $\overline{A}$*, the complex conjugate of* A.

Therefore the optimal solution y_h^A *of Eq.* (2.221) *is given by*

$$y_h^A(t) = \frac{1}{2\pi}\sum_{j,j'=1}^{n} g(t_j)\widetilde{a_{jj'}} \cdot \left[\overline{\gamma(t_{j'})}\int_{-\frac{\pi}{h}}^{\frac{\pi}{h}} e^{-i(t-t_{j'})\xi}d\xi - \overline{\alpha(t_{j'})}\int_{-\frac{\pi}{h}}^{\frac{\pi}{h}} \xi^2 e^{-i(t-t_{j'})\xi}d\xi\right.$$
$$\left. + i\overline{\beta(t_{j'})}\int_{-\frac{\pi}{h}}^{\frac{\pi}{h}} \xi e^{-i(t-t_{j'})\xi}d\xi\right].$$

We will refer to the content of the theorem. At first, we consider approximate solutions of the differential equation (2.221) and we consider the Paley Wiener function spaces with parameter h as approximating function spaces; the function spaces are formed by analytic functions of all the functions of exponential type that are decreasing to zero exponential order. Next, using the Fourier inversion, the differential equation (2.221) may be transformed into (2.222). However, to solve the integral equation (2.222) is very difficult for general coefficient functions. Therefore we assume that (2.222) is valid on some finite number of points t_j. This assumption will be reasonable for the discretization of the integral equation. By this assumption we can obtain optimal approximate solutions in a very simple way.

Here we assume that Eq. (2.221) is valid on I and so, as a practical case we consider the equation in (2.221) on I subject to some boundary conditions. In the present case, the boundary conditions are given as zero at infinity for $I = \mathbb{R}$.

However, the result approximates general solutions subject to boundary values. For example, for a finite interval (a,b), we consider $t_1 = a$ and $t_n = b$ and $\alpha(t_1) = \beta(t_1) = \alpha(t_n) = \beta(t_n) = 0$. Then we can obtain the approximate solution having the given arbitrary boundary values $y_h^A(t_1)$ and $y_h^A(t_n)$. In addition, by a simple modification we may approximate the general solutions satisfying the corresponding boundary values.

For a finite interval case I, following the boundary conditions, we can consider the corresponding reproducing kernels by the Sobolev Hilbert spaces. However, concrete representations of the reproducing kernels are involved depending on the boundary conditions. However, we can still consider them and use them.

Of course, the smaller parameter $h > 0$ is, the better approximate solutions we can obtain. However, analytic functions will be too strong in numerical complexes, and therefore, sometimes, we will need more general function spaces such as the Sobolev spaces.

For the representation (2.224) of the reproducing kernel K_{hh}, we can calculate it for some simple cases.

The important facts are: for linear differential equations with variable coefficients, we can represent their approximate solutions satisfying their boundary conditions without integrals.

In [92, 93], the principal and many examples such as real and numerical inversion formulas of the Laplace transform, many linear type partial differential equations, convolution integral equations and singular integral equations are examined.

We are looking for some optimal solutions satisfying the differential equations at certain discrete points, and so we are free from important restrictions on the domains that occur in using ordinary methods. For instance, this is not the case of the *Finite Element Method* (FEM) and the *Difference Method* (DM) that depend seriously on the domains. In our case, we can consider the problems on any domains. Furthermore, we can consider the error estimates for exact solutions and our approximate solutions over the global space and time spaces depending on the numbers of points. However, these error estimates depend case-by-case on the contexts. These problems will create a new field of interesting inequalities. Here, the error estimates mean that our approximate solutions of the differential equations satisfying the equations for some discrete points and the solutions of the equations are close in some sense. For the one-dimensional case we will easily be able to image its meaning. However, on some higher-dimensional cases, its detail and exact meanings will be delicate.

Anyhow, error estimates for the present approximate solutions are entirely new open problems.

E. M. Rocha presented a scheme to extend the Aveiro Discretization Method in Mathematics (ADMM) to non linear PDEs [371].

2.5 RKHS and Linear Mappings

We show clearly what reproducing kernels are and why the theory of reproducing kernels is fundamental, and then we will introduce a basic context for the fundamental general theory of reproducing kernels from the viewpoints of general and broad applications. In particular, in the last parts of this section, we refer to basic relations between the theory of reproducing kernels and other research topics that are expected to yield further developments.

2.5.1 Identification of the Images of Linear Transforms in Terms of the Adjoint

Thanks to Theorem 2.36, we see that general linear transforms in the framework of Hilbert spaces map from Hilbert spaces onto some naturally determined reproducing kernel Hilbert spaces. First, we recall the well-known theorem in connection with the images of linear mappings. Suppose that we know that a linear transform maps from $\mathcal{H}$ into some Hilbert space H consisting of functions on E. Then, for $f \in H$ we wish to know whether f is an image of $\mathcal{H}$ or not. For this fundamental problem, there exists the following simple, however, very important theorem which characterizes the members of the image of unbounded linear operators $L : \mathcal{H} \to H$ in terms of the adjoint operator L^*. Recall that L, whose domain is $\mathrm{Dom}(L)$, is closed in H if $\{(x, Lx) : x \in \mathrm{Dom}(L)\}$ is closed. In this case, by the von-Neumann theorem $L^{**} = L$. The following Theorem 2.35 is powerful, in particular, for L^2-solvability of the $\overline{\partial}$-equations; see [219, 220, 264, 265, 345]. In particular, in [264], the next theorem is effectively used to characterize the regular domains.

Theorem 2.35. *Assume that L is a densely defined closed linear operator from $\mathcal{H}$ into H. Let $M > 0$. For any $f \in H$, f is an image of $\mathbf{f} \in \mathrm{Dom}(L)$ with $\|\mathbf{f}\|_{\mathcal{H}} \leq M$ by L if and only if*

$$|\langle g, f\rangle_H| \leq M\|L^* g\|_{\mathcal{H}} \text{ for all } g \in \mathrm{Dom}(L^*). \tag{2.225}$$

Proof. If inequality (2.225) holds, then the linear map

$$u = L^* g \mapsto \langle g, f\rangle_H \tag{2.226}$$

is defined without ambiguity for the choice of $g \in \mathrm{Dom}(L^*) \subset H$ satisfying $u = L^* g$. Also it is bounded with norm M. Hence, by the Riesz representation theorem, there exists $\mathbf{f} \in \mathcal{H}$ such that

$$\langle L^* g, \mathbf{f}\rangle_{\mathcal{H}} = \langle g, f\rangle_H, \quad \|\mathbf{f}\|_{\mathcal{H}} \leq M \tag{2.227}$$

for all $g \in \mathrm{Dom}(L^*)$, which implies $f \in \mathrm{Dom}(L^{**}) = \mathrm{Dom}(L)$ and hence the desired result $L\mathbf{f} = f$. The reverse implication is trivial.

2.5.2 Identification of the Image of a Linear Mapping

We prove the theorem we have been using in this book many times. Let $\mathcal{F}(E)$ be the linear space consisting of all complex-valued functions on any fixed abstract set E. Let $\mathcal{H}$ be a Hilbert space with the inner product $\langle \cdot, \cdot \rangle_{\mathcal{H}}$. Let $\mathbf{h} : E \to \mathcal{H}$ be a fixed $\mathbf{h}$ valued mapping on E. Then, we consider the linear mapping L from $\mathbf{f} \in \mathcal{H}$ into $\mathcal{F}(E)$ defined by

$$L\mathbf{f}(p) \equiv \langle \mathbf{f}, \mathbf{h}(p)\rangle_{\mathcal{H}}. \tag{2.228}$$

The most fundamental problem in the linear mapping (2.228) will be to characterize the image $L\mathbf{f}(p)$ of L and our next concern will be to grasp how the input $\mathbf{f}$ and the output $L\mathbf{f}(p)$ are related.

The key to consider these fundamental problems is to form the two variables complex-valued function; that is, a positive definite quadratic form function

$$K(p,q) \equiv \langle \mathbf{h}(q), \mathbf{h}(p) \rangle_{\mathcal{H}} \tag{2.229}$$

defined on $E \times E$. We denote by $\mathcal{R}(L)$ the linear function space consisting of all complex-valued functions of the images of $\mathbf{h}$ by L defined on E. In the image space $\mathcal{R}(L)$, we will introduce the norm by

$$\|f\|_{\mathcal{R}(L)} \equiv \inf\{\|\mathbf{f}\|_{\mathcal{H}} \, : \, \mathbf{f} \in \mathcal{H}, \quad f = L\mathbf{f}\}, \tag{2.230}$$

and then the image space will form a Hilbert space; indeed, we obtain precisely

Theorem 2.36.

1. *By* (2.230) *we can define a norm which makes $\mathcal{R}(L)$ into a Hilbert space.*
2. *The function K defined by* (2.229) *enjoys three properties*:
   (a) *For any $q \in E$,*

$$K_q = K(\cdot, q) \in \mathcal{R}(L). \tag{2.231}$$

   (b) *The function K has the reproducing property*;

$$f(q) = \langle f, K_q \rangle_{\mathcal{R}(L)} \tag{2.232}$$

   for any function $f \in \mathcal{R}(L)$ and for any point $q \in E$.
   (c) *The mapping L is isomorphic from $\mathcal{H}$ to $(\mathcal{R}(L), \langle \cdot, \cdot \rangle_{\mathcal{R}(L)})$, that is,*

$$\|L\mathbf{f}\|_{\mathcal{R}(L)} = \|\mathbf{f}\|_{\mathcal{H}} \quad (\mathbf{f} \in \mathcal{H}), \tag{2.233}$$

   if and only if $\{\mathbf{h}(p) \, : \, p \in E\}$ spans a dense subspace of $\mathcal{H}$.
3. *The function K satisfying* (2.231) *and* (2.232) *is unique.*

Definition 2.5. The component of $\ker(L)^{\perp}$ is called *the visible component* for the linear mapping (2.228).

Proof (of Theorem 2.36*).* Let $P_{\ker(L)^{\perp}}$ be the orthogonal projection from $\mathcal{H}$ to $\ker(L)^{\perp}$.

That means that we cannot obtain the $\ker(L)$-part from the output of the linear mapping.

1. First, (2.228) yields the following representation of $\ker(L)$, the null space or the kernel of L :

$$\ker(L)(\equiv \{\mathbf{f} \in \mathcal{H} \, : \, L\mathbf{f} = 0\}) = \bigcap_{p \in E} \{\mathbf{f} \in \mathcal{H} \, : \, \langle \mathbf{f}, \mathbf{h}(p) \rangle_{\mathcal{H}} = 0\}. \tag{2.234}$$

So we see that $\ker(L)$ is a closed subspace in $\mathscr{H}$. Hence we have

$$\|f\|_{\mathscr{R}(L)} = \|L\mathbf{f}\|_{\mathscr{R}(L)} = \inf_{\mathbf{g}\in\ker(L)} \|\mathbf{f}-\mathbf{g}\|_{\mathscr{H}} = \|P_{\ker(L)^\perp}\mathbf{f}\|_{\mathscr{H}} \tag{2.235}$$

from (2.230) for $f = L\mathbf{f}$ with $\mathbf{f} \in \mathscr{H}$. The restriction $L|_{\ker(L)^\perp}$ of L to $\ker(L)^\perp$ is an isometry between the Hilbert spaces $(\ker(L)^\perp, \langle\cdot,\cdot\rangle_{\mathscr{H}})$ and $(\mathscr{R}(L), \langle\cdot,\cdot\rangle_{\mathscr{R}(L)})$. Thus $(\mathscr{R}(L), \langle\cdot,\cdot\rangle_{\mathscr{R}(L)})$ is a Hilbert space.

2. (a) Clearly, from (2.229), K satisfies the property (2.231).
   (b) In order to verify the property (2.232) of K, note that for $\mathbf{f}_0 \in \ker(L)$,

$$\langle \mathbf{f}_0, \mathbf{h}(p)\rangle_{\mathscr{H}} = 0 \quad \text{on} \quad E. \tag{2.236}$$

From (2.236) we conclude that

$$\mathbf{h}(q) \in \ker(L)^\perp \quad (q \in E). \tag{2.237}$$

If $\mathbf{f} \in \mathscr{H}$ and $f = L\mathbf{f} \in \mathscr{F}(E)$, then we can deduce (2.232) as follows:

$$\begin{aligned}\langle f, K_q\rangle_{\mathscr{R}(L)} &= \langle L\mathbf{f}, L[\mathbf{h}(q)]\rangle_{\mathscr{R}(L)} = \langle P_{\ker(L)^\perp}\mathbf{f}, P_{\ker(L)^\perp}[\mathbf{h}(q)]\rangle_{\mathbf{h}} \\ &= \langle \mathbf{f}, \mathbf{h}(q)\rangle_{\mathbf{h}} = f(q).\end{aligned} \tag{2.238}$$

Here for the third equality we refer to (2.235).

   (c) First, observe from (2.235) that $L : \mathscr{H} \to R(L)$ is isomorphic if and only if $\mathscr{H} = \ker(L)^\perp$. Suppose now that $\mathscr{H} = \ker(L)^\perp$. Let $\mathbf{f} \in \mathscr{H}$ be perpendicular to $\mathbf{h}(p)$ for all $p \in E$. Then from (2.236) we have $L\mathbf{f}(p) = \langle \mathbf{f}, \mathbf{h}(p)\rangle_{\mathscr{H}} = 0$ for all $p \in E$. Since L is assumed to be injective, we have $\mathbf{f} = 0$. This shows that $\{h(p) : p \in E\}$ spans a dense subspace of $\mathscr{H}$. Suppose instead that it spans a dense subspace of $\mathscr{H}$. From (2.237) we conclude that $\{h(p) : p \in E\} \subset \ker(L)^\perp$ and hence $\mathscr{H} = \ker(L)^\perp$.

3. Suppose that K^+ satisfies the properties (2.231) and (2.232). Then we have for any $q \in E$,

$$\begin{aligned}\|K_q - K_q^+\|^2_{\mathscr{R}(L)} &= \langle K_q - K_q^+, K_q\rangle_{\mathscr{R}(L)} - \langle K_q - K_q^+, K_q^+\rangle_{\mathscr{R}(L)} \\ &= K(q,q) - K^+(q,q) - K(q,q) + K^+(q,q) \\ &= 0,\end{aligned}$$

using the property (2.232) for both K_q and K_q^+. Hence, K and K^+ are identical.

Let us summarize again what we have obtained. Here the function $K(p,q)$ satisfying (2.231) and (2.232) is determined uniquely in the space $\mathscr{R}(L)$. Furthermore L is an isometry between $\mathscr{H}$ onto $\mathscr{R}(L)$ if and only if $\{\mathbf{h}(p); p \in E\}$ is complete in $\mathscr{H}$. The properties (2.231) and (2.232) of K are referred to as the reproducing property in $\mathscr{R}(L)$ of K and we call K a *reproducing kernel*. This is why the function f in the inner product in (2.232) is produced after taking the inner product in the same way as in (2.232). In this case, we call such a Hilbert space admitting a reproducing kernel *a reproducing kernel Hilbert space*, and we shall abbreviate it to RKHS.

In fact, the reproducing kernel Hilbert spaces will be determined from some positive definite quadratic form kernel K intuitively and independently of the linear mapping as in Theorem 2.36. That is, the Hilbert space is determined by K and so we will denote its Hilbert space by $H_K(E)$. Then, the image space of $\mathcal{H}$ by (2.228) is uniquely determined by the function K as a reproducing kernel Hilbert space $\mathcal{R}(L) = H_K(E)$. From this viewpoint, Theorem 2.36 may be restated as follows:

Theorem 2.37. *The image space* $\{L\mathbf{f}\}_{\mathbf{f}\in\mathcal{H}}$ *by* (2.228) *of* $\mathcal{H}$ *is characterized as the reproducing kernel Hilbert space* $H_K(E)$ *determined by* K *in* (2.229) *and we have the inequality*

$$\|L\mathbf{f}\|_{H_K(E)} \leq \|\mathbf{f}\|_{\mathcal{H}} \quad (\mathbf{f} \in \mathcal{H}). \tag{2.239}$$

Furthermore, for any $f \in H_K(E)$*, there exists a uniquely determined vector* $\mathbf{f}^* \in \mathcal{H}$ *satisfying*

$$f \equiv \langle \mathbf{f}^*, \mathbf{h}(\cdot)\rangle_{\mathcal{H}} \quad \text{on} \quad E \quad \text{and} \quad \|f\|_{H_K(E)} = \|\mathbf{f}^*\|_{\mathcal{H}}. \tag{2.240}$$

Namely $\mathcal{R}(L)$ in Theorem 2.36 is equal to $H_K(E)$ and $L : \mathcal{H} \to H_K(E)$ is a surjective partial isometry.

Theorem 2.37 gives us a method of how to construct a reproducing kernel Hilbert space for the image space that is isometric to $\mathcal{H}$. Therefore, the main point in Theorem 2.37 is to construct practically the space $H_K(E)$ from K defined by (2.229). For the construction, we have many methods depending on the kernels K and their representations, as we have seen already.

See [123, Theorem 4] for some special case.

Theorem 2.38. *Let* $\{p_j\}_{j=1}^n$ *be a sequence of points in* E *such that*

$$a_{jj'} = K(p_j, p_{j'}), \quad \det(\{a_{jj'}\}_{j,j'=1}^n) \neq 0. \tag{2.241}$$

Let $\{d_j\}_{j=1}^n$ *be a finite sequence of complex numbers. Set* $\tilde{a_{jj'}}$ *is the element of the complex conjugate inverse of the matrix* $\{a_{jj'}\}_{j,j'=1,\ldots,n}$.

1. *Let us set*

$$g \equiv \sum_{j,j'=1}^{n} d_j \tilde{a_{jj'}} \mathbf{h}(p_{j'}).$$

Then we have $\langle g, \mathbf{h}(p_j)\rangle_{\mathcal{H}} = d_j$ *for* $j = 1, 2, \ldots, n$.

2. *If* $f \in \mathcal{H}$ *satisfies* $\langle f, \mathbf{h}(p_j)\rangle_{\mathcal{H}} = d_j$ *for* $j = 1, 2, \ldots, n$*, then* $\|g\|_{\mathcal{H}} \leq \|f\|_{\mathcal{H}}$.

Proof. 1. This is a direct consequence of the related definitions.
2. Set $h \equiv g - f$. Then

$$\langle h, g \rangle_{\mathscr{H}} = \sum_{j,j'=1}^{n} d_j \tilde{a_{jj'}} \langle h, \mathbf{h}(p_{j'}) \rangle_{\mathscr{H}} = 0.$$

Thus, $\|f\|_{\mathscr{H}}^2 = \|g\|_{\mathscr{H}}^2 + \|h\|_{\mathscr{H}}^2 \geq \|g\|_{\mathscr{H}}^2$.

To conclude Sect. 2.5.2 let us present two examples that have a strong connection with harmonic analysis.

Example 2.23. Let us take a bump function $g \in C_c^\infty(\mathbb{R})$ so that $\chi_{[-a,a]} \leq g \leq \chi_{[-b,b]}$ for $0 < a < b$. The windowed Fourier transform in time-frequency analysis is given by

$$T_g f(\omega, t) \equiv \int_{\mathbb{R}} f(s) g(s-t) \exp(-i\omega \cdot s)\, ds. \tag{2.242}$$

Here f is usually assumed to be an $L^2(\mathbb{R})$-function.

Our aim here is to specify the range of T_g.

Let $\mathscr{H} = L^2(\mathbb{R})$ and $E = \mathbb{R} \times \mathbb{R}$. Define

$$\mathbf{h}_{\omega,t} \equiv \overline{g(\cdot - t)} \exp(i\omega \cdot) \in L^2(\mathbb{R}).$$

Since $T_g f(\omega, t) = \langle f, \mathbf{h}_{\omega,t} \rangle_{\mathscr{H}}$, the image of T_g is exactly $H_K(\mathbb{R} \times \mathbb{R})$. The reproducing kernel $K : \mathbb{R} \times \mathbb{R} \times \mathbb{R} \times \mathbb{R} \to \mathbb{C}$ is given by

$$K(\omega, t, \omega', t') = \langle \mathbf{h}_{\omega',t'}, \mathbf{h}_{\omega,t} \rangle_{L^2(\mathbb{R})} = \int_{\mathbb{R}} g(s-t) \overline{g(s-t')} \exp(i(\omega' - \omega)s)\, ds. \tag{2.243}$$

The windowed Fourier transform is important in that it will lead us to the theory of modulation spaces initiated by Feichtinger and Gröchenig [148–150, 154, 155].

Example 2.24. Let us take $\psi \in L^1(\mathbb{R}) \cap L^\infty(\mathbb{R})$ satisfying

$$\int_{\mathbb{R}} \psi(t)\, dt = 0, \int_{\mathbb{R}} |\xi|^{-1} |\mathscr{F}\psi(\xi)|^2\, d\xi < \infty. \tag{2.244}$$

Then define

$$Tf(a, b) \equiv \frac{1}{\sqrt{a}} \int_{\mathbb{R}} f(x) \overline{\psi\left(\frac{x-b}{a}\right)}\, dx \tag{2.245}$$

for $a > 0$ and $b \in \mathbb{R}$. Again we assume that $f \in L^2(\mathbb{R})$.

To see that the operator T falls under the scope of Theorem 2.36, we let

$$E \equiv (0, \infty) \times \mathbb{R},\ \mathscr{H} \equiv L^2(\mathbb{R}),\ \mathbf{h}_{a,b} \equiv \frac{1}{\sqrt{a}} \psi\left(\frac{\cdot - b}{a}\right).$$

Then we have

$$Tf(a,b) = \langle f, \mathbf{h}_{a,b} \rangle_{\mathscr{H}}.$$

Therefore, from Theorem 2.36, the image is a reproducing kernel Hilbert space given by

$$K(a,b;a',b') = \frac{1}{\sqrt{a\,a'}} \int_{\mathbb{R}} \psi\left(\frac{x-b}{a}\right) \overline{\psi\left(\frac{x-b'}{a'}\right)}\, dx. \tag{2.246}$$

For this example, we have the interesting formula (2.248).

The following theorem is a supplement on the transform T given by (2.245).

Theorem 2.39. *Suppose that T is given by* (2.245).

1. *Let $f \in L^2(\mathbb{R})$. Then we have*

$$\int_0^\infty \int_{-\infty}^\infty |Tf(a,b)|^2 \frac{da\, db}{a^2} < \infty. \tag{2.247}$$

2. *Let $f, g \in L^2(\mathbb{R})$. Then*

$$\int_0^\infty \int_{-\infty}^\infty Tf(a,b)\overline{Tg(a,b)} \frac{da\, db}{a^2} = 2\pi \left(\int_0^\infty |\mathscr{F}\psi(a)|^2 \frac{da}{a} \right) \langle f, g \rangle_{L^2(\mathbb{R})}. \tag{2.248}$$

Proof. A polarization allows us to assume that $f = g$ when we prove (2.248). Also, we may suppose that $f \in \mathscr{S}(\mathbb{R})$ because $\mathscr{S}(\mathbb{R})$ is dense in $L^2(\mathbb{R})$. Then by the Plancherel formula, we have

$$\begin{aligned}
\int_0^\infty \int_{-\infty}^\infty |Tf(a,b)|^2 \frac{da\, db}{a^2} &= \int_0^\infty \int_{-\infty}^\infty \left| \int_{-\infty}^\infty \mathscr{F}f(\xi) \exp(-ib\xi) \overline{\psi(a\,\xi)}\, d\xi \right|^2 \frac{da\, db}{a} \\
&= \int_0^\infty \lim_{S\to\infty} \int_{-S}^S \left| \int_{-\infty}^\infty \mathscr{F}f(\xi) \exp(-ib\xi) \overline{\psi(a\,\xi)}\, d\xi \right|^2 \frac{da\, db}{a}
\end{aligned}$$

by the monotone convergence theorem. The Fubini theorem in turn gives

$$\begin{aligned}
&\int_0^\infty \int_{-\infty}^\infty |Tf(a,b)|^2 \frac{da\, db}{a^2} = 2 \int_0^\infty \lim_{S\to\infty} \\
&\quad \cdot \left(\int_{-\infty}^\infty \int_{-\infty}^\infty \mathscr{F}f(\xi)\overline{\mathscr{F}\psi(a\,\xi)} \cdot \overline{\mathscr{F}f(\xi')}\mathscr{F}\psi(a\,\xi') \frac{\sin(S(\xi-\xi'))}{\xi-\xi'}\, d\xi\, d\xi' \right) \cdot \frac{da}{a} \\
&= \int_0^\infty \lim_{S\to\infty} \cdot \left(\int_{-\infty}^\infty \mathscr{F}f(\xi)\overline{\mathscr{F}\psi(a\,\xi)} \right. \\
&\quad \left. \cdot \left(\int_{-\infty}^\infty \overline{\mathscr{F}f(\xi')}\mathscr{F}\psi(a\,\xi') \frac{2\sin(S(\xi-\xi'))}{\xi-\xi'}\, d\xi' \right) d\xi \right) \cdot \frac{da}{a}.
\end{aligned}$$

Note that the inner integral has the following expression:

$$\int_{-\infty}^{\infty} |\mathscr{F}^{-1}[\mathscr{F}f\overline{\mathscr{F}\psi(a\,\cdot)}](\xi)|^2 \chi_{(-S,S)}(\xi)\,d\xi$$
$$= \int_{-\infty}^{\infty} \mathscr{F}f(\xi)\overline{\mathscr{F}\psi(a\,\xi)} \cdot \left(\int_{-\infty}^{\infty} \overline{\mathscr{F}f(\xi')}\mathscr{F}\psi(a\,\xi')\frac{2\sin(S(\xi-\xi'))}{\xi-\xi'}\,d\xi'\right)d\xi.$$

Therefore, it follows again from the Fubini theorem that

$$\begin{aligned}\int_0^{\infty}\int_{-\infty}^{\infty} |Tf(a,b)|^2 \frac{da\,db}{a^2} &= 2\pi \int_0^{\infty}\left(\int_{-\infty}^{\infty} |\mathscr{F}^{-1}[\mathscr{F}f\overline{\mathscr{F}\psi(a\,\cdot)}](x)|^2\,dx\right)\frac{da}{a}\\ &= 2\pi \int_0^{\infty}\left(\int_{-\infty}^{\infty} |\mathscr{F}f(\xi)\overline{\mathscr{F}\psi(a\xi)}|^2\,d\xi\right)\frac{da}{a}\\ &= 2\pi\left(\int_0^{\infty} |\mathscr{F}\psi(a)|^2\frac{da}{a}\right)\left(\int_{-\infty}^{\infty} |\mathscr{F}f(\xi)|^2\,d\xi\right).\end{aligned}$$

This yields (2.247) and (2.248).

See [439] for a similar formula using the Dunkl transform.

Identity (2.248) gives the fundamental isometric identity for the wavelet analysis. For more detailed results and the wavelet theory, see, for example, I. Daubechies [121].

Here, however, in particular, note that the isometric identity (2.248) does not give any inversion formula of the wavelet transform and does not give the image identification, because the isometric mapping shows up only into a mapping. For the image identification and inversion mapping for typical Meyer wavelets, see, [388, Chapter 3, Section 4].

For any positive definite quadratic form function K on E, we can construct a Hilbert space $\mathscr{H}$ and a mapping $\mathbf{h}$ from E into $\mathscr{H}$ satisfying (2.229) by Gaussian probability distribution with n th order co-distributions and zero mean which are given by $n \times n$ matrices $\{K(p_j, p_{j'})\}_{j,j'=1,2,\ldots,n}$. For this construction, we use the *Kolmogorov theorem on measures* whose proof is not elementary [353]. This theorem is very important, from which we can derive simply Theorem 2.41 from Theorems 2.36 or 2.37. Furthermore, this Kolmogorov factorization theorem [11, 260, 314] will essentially be indispensable when we introduce various operators on a Hilbert space and among Hilbert spaces.

This important result was interestingly derived from the theory of stochastic theory independent of the theory of reproducing kernels. Furthermore, such factorization representation of a positive definite quadratic form function may be used conversely to realize reproducing kernel Hilbert spaces admitting reproducing kernels.

For some deep theory on the Kolmogorov factorization theorem, see, for example, [11, 111–113]. Here, note that for a positive definite quadratic form function K on E, when we know its reproducing kernel Hilbert space, it is easy to show that K can be written in the form (2.229). Indeed. it is given simply by

$$K(p,q) = \langle K_q, K_p \rangle_{H_K(E)} \quad ((p,q) \in E) \tag{2.249}$$

and such a factorization is determined among isometric Hilbert spaces.

2.5.3 Image Identification by Reproducing Kernels

In order to see how we use Theorem 2.37, by way of Theorem 2.37 we shall refer to a typical example that succeeds in identifying the image for a linear transform using the theory of reproducing kernels. Theorem 2.40 was derived simply, first however the result was very surprising and had a great impact as we see from the book [388] and many references therein.

We consider the simple heat equation

$$u_t(x,t) = u_{xx}(x,t) \quad \text{on} \quad \mathbb{R} \times T_+ \quad (T_+ \equiv \{t > 0\}) \tag{2.250}$$

subject to the initial condition

$$u_F(\cdot, 0) = F \in L^2(\mathbb{R}) \quad \text{on} \quad \mathbb{R}. \tag{2.251}$$

Using the Fourier transform, we obtain a representation of the solution $u_F(x,t)$

$$\begin{aligned} u_F(x,t) &= \frac{1}{\sqrt{4\pi t}} \int_{\mathbb{R}} F(\xi) \exp\left(-\frac{(x-\xi)^2}{4t}\right) d\xi \\ &= \frac{1}{\sqrt{4\pi t}} \int_{\mathbb{R}} F(x-\xi) \exp\left(-\frac{\xi^2}{4t}\right) d\xi \end{aligned} \tag{2.252}$$

at least in the formal sense. Apart from classical and educative arguments about the properties of this transform, let us content ourselves with seeing that the heat kernel is nice enough to consider u_F for $F \in L^2(\mathbb{R})$. For any fixed $t > 0$, we first examine the integral transform $F \mapsto u_F$ and then characterize the image function $u_F(x,t)$.

Note here that M. Nishio generalized the heat kernels. They are related to the parabolic Bergman and Bloch spaces. There are great works in connection with operator theory; see, for example, [332–336] as well as a series of works [213–216] concerning function spaces.

Let us write

$$k(x;t) \equiv \frac{1}{\sqrt{4\pi t}} \exp\left(-\frac{x^2}{4t}\right) \quad (x \in \mathbb{R},\ t > 0). \tag{2.253}$$

The highlight of Sect. 2.5.3 is to prove the following theorems, which identify the RKHS $H_k(\mathbb{R})$:

Theorem 2.40. *Let $t > 0$. A function f takes the form $u_F(\cdot, t)$ for some $F \in L^2(\mathbb{R})$ if and only if f admits analytic extension $\tilde{f}$ to $\mathbb{C}$ and satisfies*

$$\sqrt{\frac{1}{\sqrt{2\pi t}} \iint_{\mathbb{C}} |\tilde{f}(x+iy)|^2 \exp\left(\frac{-y^2}{2t}\right) dxdy} < \infty. \tag{2.254}$$

In this case, $f \in H_k(\mathbb{R})$, where k is given by (2.253), and the norm is

$$\|f\|_{H_k(\mathbb{R})} = \sqrt{\frac{1}{\sqrt{2\pi t}} \iint_{\mathbb{C}} |\tilde{f}(x+iy)|^2 \exp\left(-\frac{y^2}{2t}\right) dxdy}$$

for $f \in H_k(\mathbb{R})$.

Theorem 2.41. *Let $t > 0$. In the integral transform $F \mapsto u_F(\cdot, t)$ of $L^2(\mathbb{R})$ functions F, the images $u_F(\cdot, t)$ extend analytically onto $\mathbb{C}$ to a function, which we still write $u_F(\cdot, t)$. Furthermore, we have the isometrical identities*

$$\|F\|^2_{L^2(\mathbb{R})} = \frac{1}{\sqrt{2\pi t}} \iint_{\mathbb{C}} |u_F(z,t)|^2 \exp\left(-\frac{y^2}{2t}\right) dxdy = \sum_{j=0}^{\infty} \frac{(2t)^j}{j!} \|u_F(\cdot, t)\|^2_{L^2(\mathbb{R})} \tag{2.255}$$

for any fixed $t > 0$.

Corollary 2.7. *If a C^∞-function $f : \mathbb{R}^n \to \mathbb{C}$ has a finite integral on the most right-hand side in (2.255), then f is extended analytically onto $\mathbb{C}$ and*

$$\sum_{j=0}^{\infty} \frac{(2t)^j}{j!} \int_{\mathbb{R}} |\partial_x^j f(x)|^2 dx = \frac{1}{\sqrt{2\pi t}} \iint_{\mathbb{C}} |f(x+iy)|^2 \exp\left(-\frac{y^2}{2t}\right) dxdy. \tag{2.256}$$

Before the proof of the theorems and its corollary, we recall briefly the properties of $u_F(\cdot, t)$.

For $F \in C^1(\mathbb{R})$ such that F and F' are bounded continuous functions, we have the classical solution $u(x, t)$ satisfying

$$\lim_{t \downarrow 0} u_F(x, t) = F(x) \quad \text{on} \quad \mathbb{R} \tag{2.257}$$

uniformly in view of Remark 1.3. If in addition F is compactly supported, then

$$\lim_{t \downarrow 0} u_F(x, t) = F(x) \quad \text{on} \quad \mathbb{R} \tag{2.258}$$

in the topology of $L^2(\mathbb{R})$. Since u is given (2.252), then we have

$$\|u_F(\cdot, t)\|_{L^2(\mathbb{R})} \leq \|F\|_{L^2(\mathbb{R})} \tag{2.259}$$

by the Plancherel theorem or the Young inequality. Again by the Plancherel theorem, for $F \in L^2(\mathbb{R})$, we have

$$\lim_{t \downarrow 0} \|u_F(\cdot, t) - F\|_{L^2(\mathbb{R})} = 0. \tag{2.260}$$

We define the kernel

$$K(x, x'; t) \equiv \int_{\mathbb{R}} k(x - \xi; t) k(x' - \xi; t) d\xi = \frac{1}{2\sqrt{2\pi t}} \exp\left(-\frac{x^2}{8t} - \frac{x'^2}{8t} + \frac{xx'}{4t}\right) \tag{2.261}$$

according to (2.229).

With $t > 0$ frozen, the solution operator $F \mapsto u_F$ can be regarded as

$$u_F(x, t) = \langle F, k(x - \cdot; t) \rangle_{L^2(\mathbb{R})}.$$

Therefore we apply (2.228) with

$$E = \mathbb{R}, \quad \mathscr{H} = L^2(\mathbb{R}), \quad \mathbf{h}(x) \equiv k(x - \cdot; t), \quad L = [F \in \mathscr{H} \mapsto u_F(\cdot, t) \in \mathscr{F}(E)].$$

According to (2.230), the reproducing Hilbert kernel space $\mathscr{R}(L)$ is

$$\mathscr{R}(L) = \{u_F(\cdot, t) \, : \, F \in L^2(\mathbb{R})\}.$$

Below we write h_k for $\mathscr{R}(L)$.

Lemma 2.8. *For any fixed $t > 0$, the system*

$$\{k(\cdot - \xi, t); \xi \in \mathbb{R}\} \tag{2.262}$$

spans a dense subspace in $L^2(\mathbb{R})$.

Proof. Recall that the representation (2.253) of $k(x - \xi, t)$ in terms of the Fourier integral and the uniqueness of the Fourier integral (the completeness of the exponential functions). We omit the further detail.

A direct consequence of Lemma 2.8 is that L is isometric.

Now let us view this mapping from the point of complex analysis. Note that the kernel $K(x, x'; t)$ extends analytically to $\mathbb{C} \times \overline{\mathbb{C}}$;

$$K(z, \overline{u}; t) = \frac{1}{2\sqrt{2\pi t}} \exp\left(-\frac{z^2}{8t} - \frac{\overline{u}^2}{8t} + \frac{z\overline{u}}{4t}\right). \tag{2.263}$$

Observe that (2.263) stands for

$$K(z,\overline{u};t) = \int_{\mathbb{R}} K(z-\xi;t)\overline{K(u-\xi;t)}\,d\xi. \tag{2.264}$$

Consequently, the extended kernel K is again positive semidefinite. Denote by $H_K(\mathbb{C})$ the RKHS associated with K. The following is a description of $H_K(\mathbb{C})$; recall that in Theorem 1.17, we proved

$$H_K(\mathbb{C}) = \left\{ f \in \mathcal{O}(\mathbb{C}) \,:\, \|f\|_{H_K(\mathbb{C})} = \sqrt{\iint_{\mathbb{C}} \frac{|\tilde{f}(x+iy)|^2}{\sqrt{2\pi t}} \exp\left(\frac{-y^2}{2t}\right) dxdy} < \infty \right\}. \tag{2.265}$$

However, (2.265) was derived in a natural way and simply from the reproducing structure of (2.252). See the original paper [379, Section 3].

Lemma 2.9. *Let $t > 0$. Then, the set $\{K(\cdot,u;t); u \in \mathbb{C}\}$ spans a dense subspace in $L^2(\mathbb{R})$.*

Proof. This is a consequence of Lemma 2.8.

Proof (of Theorem 2.40). If f has such a property for some $\tilde{f} \in \mathcal{O}(\mathbb{C})$, then we see that

$$\tilde{f}(z) = \langle F, K_z\rangle = \iint_{\mathbb{C}} \frac{\tilde{f}(x+iy)\overline{K_z(x+iy)}}{\sqrt{2\pi t}} \exp\left(\frac{-y^2}{2t}\right) dxdy$$

for all $z \in \mathbb{C}$. Since $F \to \tilde{f}$ is an isometry, we see that $f = u_F$. For the converse, it just suffices to set $\tilde{f}(z) = \langle F, K_z\rangle$ if we have $f = u_F(\cdot,t)$ for some $F \in L^2(\mathbb{R})$.

Proof (of Theorem 2.41). The norm (2.254) can also be expressed in terms of the trace $f(x)$ of $\tilde{f}(z)$ to the real line.

Using the identity

$$k(x-\xi,t) = \frac{1}{2\pi}\int_{\mathbb{R}} \exp\{-p^2t + ip(x-\xi)\}dp \quad (x,\xi \in \mathbb{R}, t > 0), \tag{2.266}$$

we have

$$K(x,x';t) = \frac{1}{2\pi}\int_{\mathbb{R}} \exp\{-2p^2t + ip(x-x')\}dp \quad (x,x' \in \mathbb{R}, t > 0). \tag{2.267}$$

This implies that any member $f(x)$ of $H_k(\mathbb{R})$ can be expressed in the form

$$f(x) = \frac{1}{2\pi}\int_{\mathbb{R}} g(p)\exp(ipx - 2p^2t)dp = \frac{1}{\sqrt{2\pi}}\mathscr{F}^{-1}[g\cdot\exp(-2p^2t)](x) \tag{2.268}$$

for a function g satisfying

$$\int_{\mathbb{R}} |g(p)|^2 \exp(-2p^2 t)dp < \infty \tag{2.269}$$

and we have the isometric identity

$$\|f\|_{h_k} = \sqrt{\frac{1}{2\pi}\int_{\mathbb{R}} |g(p)|^2 \exp(-2p^2 t)dp}. \tag{2.270}$$

Meanwhile, by the Fourier transform and (2.268), we have

$$g(p) = \sqrt{2\pi}\,\mathscr{F}f(p)\exp(2p^2 t) \quad (p \in \mathbb{R}) \tag{2.271}$$

in $L^2(\mathbb{R})$. Hence, we obtain

$$\|f\|_{h_k}^2 = \int_{\mathbb{R}} |\mathscr{F}f(p)|^2 \exp(-2p^2 t)dp = \sum_{j=0}^{\infty} \frac{(2t)^j}{j!}\int_{\mathbb{R}} |f^{(j)}(x)|^2 dx. \tag{2.272}$$

by the monotone convergence theorem and the Parseval–Plancherel identity. Hence, we have (2.255) by Theorems 2.37 and 2.40. The first equality follows from Theorems 2.41 and 2.36. Meanwhile, the second equality follows from (2.272).

Proof (of Corollary 2.7*).* This is a consequence of Theorem 2.41 and (2.255).

In this way, we can derive many isometric identities for various integral transforms and furthermore, as the starting point from Corollary 2.7, we can consider many analytic extension formulas. See, for example, [379, Section 3] as well as [388, Section 3.2] and [400]. Other typical results are as follows: For the Bergman–Selberg spaces $H_{K_q}(S_r)$ $(q > 0)$ for the strip $S_r = \{z \in \mathbb{C} : |\mathrm{Im}(z)| < r\}$, stated in the following theorem, we can derive:

Theorem 2.42 ([384, Theorem 1.1]). *Let* $q > \frac{1}{2}$. *For* $f \in H_{K_q}(S_r)$ *we have the identity*

$$\|f\|_{H_{K_q}(S_r)} \equiv \sqrt{\frac{1}{\Gamma(2q-1)\pi^q}\iint_{S_r} |f(z)|^2 K_q(z,\bar{z})^{1-q} dxdy}$$

$$= \sqrt{\sum_{n=0}^{\infty}\left(\frac{2r}{\pi}\right)^{2n+2q-1}\left(\sum_{j_1>\cdots>j_n \geqq 0}\prod_{k=1}^{n}\frac{1}{(q+j_k)^2}\right)\int_{\mathbb{R}} |f^{(n)}(x)|^2 \frac{dx}{\Gamma(q)^2}},$$

where $K(z,\overline{u})$ stands for the usual Bergman kernel on S_r and the summation $\sum_{j_1>j_2>\ldots>j_n\geqq 0}$ *in* (2.273) *is understood as one for $n = 0$.*

Conversely, any $C^\infty(\mathbb{R})$-function f defined on the real line with convergent summation in (2.273) *admits an analytic extension onto the strip S_r, and the analytic extension F belongs to $H_{K_q}(S_r)$ and satisfies the identity* (2.273).

We have the following formulas: For the proof of Theorems 2.43, 2.44 and 2.45, we refer to [8, 384, 388].

Theorem 2.43 ([384, Theorem 1.2]). *Let $q > \frac{1}{2}$. For the right-half plane $R^+ = \{z \in \mathbb{C} : \mathrm{Re}(z) > 0\}$ we have the identity*

$$\|f\|_{H_{K_q}(R^+)} \equiv \sqrt{\frac{1}{\Gamma(2q-1)\pi}\iint_{R+}|f(z)|^2(2x)^{2q-2}dxdy}$$
$$= \sqrt{\sum_{n=0}^{\infty}\frac{1}{n!\Gamma(n+2q+1)}\int_{\mathbb{R}}|(xf'(x))^{(n)}|^2x^{2n+2q-1}dx}. \qquad (2.273)$$

Conversely, any $C^\infty(0,\infty)$-function f with convergent summation in (2.273) *extends analytically onto the right-half plane R^+. The analytic extension $f(z)$ satisfying $\lim_{x\to\infty} f(x) = 0$ belongs to $H_{K_q}(R^+)$ and the identity* (2.273) *is valid.*

We shall write $S(r) \equiv \{z \in \mathbb{C} : 0 < \arg(z) < r\}$ for the open sector and its boundary $\partial S(r) \equiv \{z \in \mathbb{C} : z = 0 \text{ or } \arg(z) = \pm r\}$.

Theorem 2.44 ([8, Theorem 1]). *Let $r \in (0,\pi/2)$. For an analytic function f on the open sector $S(r)$, we have the identity*

$$\iint_{S(r)}|f(x+iy)|^2dxdy = \sin(2r)\sum_{j=0}^{\infty}\frac{(2\sin r)^{2j}}{(2j+1)!}\int_{\mathbb{R}}x^{2j+1}|f^{(j)}(x)|^2dx. \qquad (2.274)$$

Conversely, if any $f \in C^\infty(S(r))$ has a convergent sum on the right-hand side in (2.274)*, then the function $f(x)$ can be extended analytically onto the sector $S(r)$ in the form $f(z)$ and the identity* (2.274) *is valid.*

In the Szegö space, we have the following formula:

Theorem 2.45 ([8, Theorem 2]). *Let $r \in (0,\pi/2)$. For any member f in the Szegö space on the open sector $S(r)$, we have the identity*

$$\oint_{\partial S(r)}|f(z)|^2|dz| = 2\cos r\sum_{j=0}^{\infty}\frac{(2\sin r)^{2j}}{(2j)!}\int_{\mathbb{R}}x^{2j}|f^{(j)}(x)|^2dx, \qquad (2.275)$$

where $f(x)$ means the nontangential Fatou limit on $\partial S(r)$ for $x \in \mathbb{R}$. Conversely, if any $f \in C^{\infty}(0,\infty)$ has a convergent sum on the right-hand side in (2.275)*, then the function $f(x)$ extends analytically onto the open sector $\Delta(r)$ and the identity* (2.275) *is valid.*

These results were applied to deriving analyticity properties of the solutions of nonlinear partial differential equations in [196–200, 400]. In particular, in [199, Lemma 1], an analogue of Theorem 2.44 is obtained and applied to the proof of the unique existence of the Schrödinger equation [199, Theorems 1 and 2]. H. Aikawa considered the class $W(c_j;\mathbb{R})$ by changing $\frac{(2\sin r)^{2j}}{(2j)!}$ with other general positive sequences in [7], where he proved that some function cannot be extended beyond a sector.

2.5.4 Integro-Differential Equations

We can refer to the important connection with a general integro-differential equation. We state its essential idea simply.

We will now consider the following general linear integro-differential equations: For open intervals T and E,

$$\sum_{j=0}^{n} a_j(t)F^{(j)}(t) + \int_T F(\xi)\overline{h(\xi,t)}dm(\xi) = f(t) \quad \text{on} \quad E, \tag{2.276}$$

where

$$h(\cdot,t) \in L^2(T,dm) \quad \text{for} \quad t \in E, \tag{2.277}$$

and $\{a_j\}_{j=0}^{n}$ are *arbitrary* complex-valued functions on E.

Let $(\hat{E}, d\hat{m})$ be a measure space. From the form (2.276), we will assume that F belongs to some reproducing kernel Hilbert space $H_{\mathbb{K}}(E)$ on E, where $\mathbb{K}$ can be written as

$$\mathbb{K}(t,t') = \int_{\hat{E}} \hat{h}(\hat{\xi},t')\overline{\hat{h}(\hat{\xi},t)}d\hat{m}(\hat{\xi}), \quad \text{on} \quad E \times E \tag{2.278}$$

for some complete system $\{\hat{h}(\cdot,t)\}_{t\in E}$ in $L^2(\hat{E},d\hat{m})$. Then any member $F \in H_{\mathbb{K}}(E)$ is represented in the form

$$F(t) = \int_{\hat{E}} \hat{F}(\hat{\xi})\overline{\hat{h}(\hat{\xi},t)}d\hat{m}(\hat{\xi}) \quad (t \in E), \tag{2.279}$$

and we have the isometric identity from the completeness of the system

$$\|F\|_{H_{\mathbb{K}}(E)} = \sqrt{\int_{\hat{E}} |\hat{F}(\hat{\xi})|^2 d\hat{m}(\hat{\xi})}. \tag{2.280}$$

Motivated by (2.276), we assume furthermore that the functions

$$\frac{\partial^{j+j'}\mathbb{K}}{\partial t^j \partial t'^{j'}}(t,t') \quad (j,j' = 0,1,2,\ldots,n) \tag{2.281}$$

are continuous on $E \times E$. Then we see that any member F of $H_{\mathbb{K}}$ belongs to $C^n(T)$ thanks to Theorem 2.6 and we have the expression

$$F^{(j)}(t) = \int_{\hat{E}} \hat{F}(\hat{\xi}) \overline{\frac{\partial^j \hat{h}}{\partial t^j}(\hat{\xi},t)} d\hat{m}(\hat{\xi}), \quad \text{on} \quad E. \tag{2.282}$$

Lemma 2.10. *If* $\displaystyle\int_{\hat{E}\times T} |\hat{F}(\hat{\xi})\hat{h}(\hat{\xi},\eta)h(\eta,t)|\, d\mu(\eta)d\hat{m}(\hat{\xi}) < \infty$ *then we have*

$$\begin{aligned} f(t) &= \int_{\hat{E}} \hat{F}(\hat{\xi}) \left\{ a_0(t)\overline{\hat{h}(\hat{\xi},t)} + \ldots + a_n(t)\overline{\partial_t^n \hat{h}(\hat{\xi},t)} \right\} d\hat{m}(\hat{\xi}) \\ &\quad + \int_{\hat{E}} \hat{F}(\hat{\xi}) \left\{ \int_T \overline{\hat{h}(\hat{\xi},\eta)} \cdot \overline{h(\eta,t)} dm(\eta) \right\} d\hat{m}(\hat{\xi}) \end{aligned}$$

for all $t \in E$.

Proof. From (2.276) and (2.282) the desired result follows. Here we can exchange the order of the integrals by Fubini's theorem.

Our procedure implies that the integro-differential equation (2.276) can be transformed into the Fredholm integral equation of the first kind.

The difficulty of solving the integro-differential equation (2.276) with variable coefficients will be transformed into that of the complicated form in the integral kernel in (2.276). However, we should note that we can deal with integro-differential equations (2.276) with arbitrary functions as the coefficients.

2.5.5 *Inversion Mapping from Many Types of Information Data*

We will give an application to inversion from many types of information data.

We suppose that we are given $\mathscr{H}_\lambda$ for each $\lambda \in \Lambda$, where Λ is an abstract set. We are in addition given a bounded linear operator

$$L_\lambda \in B(\mathscr{H}, \mathscr{H}_\lambda). \tag{2.283}$$

In particular, with λ fixed, we are interested in the inversion formula

$$L_\lambda x \mapsto x, \quad x \in \mathscr{H}. \tag{2.284}$$

Here we consider $\{L_\lambda x; \lambda \in \Lambda\}$ as information obtained from x and we wish to determine x from the data given.

However, the information about $L_\lambda x$ belongs to various Hilbert spaces $\mathscr{H}_\lambda$, and so, in order to unify the data in a sense, we will take fixed elements $\mathbf{b}_{\lambda,\omega} \in \mathscr{H}_\lambda$ and consider the linear mapping from $\mathscr{H}$

$$X_\mathbf{b}(\lambda, \omega) = \langle L_\lambda x, \mathbf{b}_{\lambda,\omega}\rangle_{\mathscr{H}_\lambda} = \langle x, L_\lambda^* \mathbf{b}_{\lambda,\omega}\rangle_{\mathscr{H}}, \quad x \in \mathscr{H} \tag{2.285}$$

into a linear space consisting of functions on $\Lambda \times \Omega$. For the data on $L_\lambda x$, we will consider $X_\mathbf{b}(\lambda, \omega)$ as observations (measurements, in fact) for x depending on λ and ω. For this linear mapping (2.285), we form the positive definite quadratic form function $K_\mathbf{b}(\lambda, \omega; \lambda', \omega')$ on $\Lambda \times \Omega$ defined by

$$K_\mathbf{b}(\lambda, \omega; \lambda', \omega') = \langle L_{\lambda'}^* \mathbf{b}_{\lambda',\omega'}, L_\lambda^* \mathbf{b}_{\lambda,\omega}\rangle_{\mathscr{H}} = \langle L_\lambda L_{\lambda'}^* \mathbf{b}_{\lambda',\omega'}, \mathbf{b}_{\lambda,\omega}\rangle_{\mathscr{H}_\lambda} \quad \text{on} \quad \Lambda \times \Omega. \tag{2.286}$$

Then we can apply the theory. The concept was derived by generalizing the Pythagorean theorem in the following way [367]:

Let $x \in \mathbb{R}^n$ and $\{e_j\}_{j=1}^n$ be linearly independent unit vectors. We consider the linear mappings

$$L : x \mapsto \{x - \langle x, \mathbf{e}_j\rangle \mathbf{e}_j\}_{j=1}^n \tag{2.287}$$

from $\mathbb{R}^n$ into $\mathbb{R}^n$. Then we wish to establish an isometric identity and inversion formula for the operator. Recall the Pythagorean theorem for $n = 2$. By our operator versions, we can establish the desired results.

Note that in (2.287), for $n \geq 3$ if we instead consider

$$\{\|x - \langle x, \mathbf{e}_j\rangle \mathbf{e}_j\|\}_{j=1}^n \tag{2.288}$$

as scalar-valued mappings, then the mappings are no longer linear. So, we must consider the operator-valued mappings (2.287) in order to obtain isometric mappings in the framework of Hilbert spaces.

We see that some related equations were considered as follows [321, p. 128–157]:

Let $H, H_j(E); j = 1, 2, \ldots, p$ be Hilbert spaces and let

$$R_j : H \mapsto H_j(E), \quad j = 1, 2, \ldots, p \tag{2.289}$$

be linear continuous maps from H onto $H_j(E)$. Let $g_j \in H_j(E)$ be given. Then, consider the problem to compute $f \in H$ such that

$$R_j f = g_j, \quad j = 1, 2, \ldots, p. \tag{2.290}$$

These equations are very important in the theory of computerized tomography by discretization. The typical method is *Kaczmarz's method* based on an iterative method using the orthogonal projections P_j in H onto the affine subspaces $R_j f = g_j$. See [321] for the details.

As for our direct solutions for (2.290) it seems that the result is stable for the sake of (2.290) as data.

2.5.6 Random Fields Estimations

We now place ourselves in the following situation:

1. X is a set,
2. $(\Omega, \mathscr{F}, P)$ is a probability space.

We assume that the random field is of the form

$$u(x) = s(x) + n(x) \quad (x \in X), \tag{2.291}$$

where, for each $x \in X$, $s(x) = s(x; \cdot) : \Omega \to \mathbb{R}$ is the useful signal and $n(x) = n(x; \cdot) : \Omega \to \mathbb{R}$ is a noise. Note that $s(x)$ and $n(x)$ are not necessarily independent of each $x \in X$. Without loss of generality, we can assume that the mean values of $u(x)$ and $n(x)$ are zero. We assume that the covariance functions

$$R(x, y) = E[u(x)u(y)] \quad (x, y \in X) \tag{2.292}$$

and the information

$$f(x, y) = E[u(x)s(y)] \quad (x, y \in X) \tag{2.293}$$

are known.

Here and below, we equiv X with the structure of the measure space: let (X, dm) be a measure space.

In addition, we will consider the general form of a linear estimation $\hat{u}$ of u in the form

$$\hat{u}(x) = \int_X u(y)h(x, y)dm(y) = \langle u, h(x, \cdot)\rangle_{L^2(X,dm)} \tag{2.294}$$

for an $L^2(X, dm)$ space and for a function $h(x, \cdot)$ belonging to $L^2(X, dm)$ for any fixed $x \in E$. For the desired information $As : X \times \Omega \to \mathbb{R}$, which satisfies

$$A(ks) = kAs, \; A(s_1 + s_2) = As_1 + As_2$$

for all $s, s_1, s_2 : X \times \Omega \to \mathbb{R}$ and $k \in \mathbb{C}$, we wish to determine the function $h(x, t)$ attaining

$$\inf\{E[(\hat{u}(x) - As(x))^2] \; : \; \hat{u} \text{ is given by (2.294) and } h(x, \cdot) \in L^2(X, dm)\} \tag{2.295}$$

which gives the minimum of the variance by the least squares method.

Many topics in filtering and estimation theory in signal and image processing, underwater acoustics, geophysics, optical filtering, etc., which were initiated by N. Wiener (1894–1964), will be presented in this framework. Then we see that the linear transform $h(x, t)$ is given by the integral equation

$$\int_X R(x', y)h(x, y)dm(y) = f(x', x). \tag{2.296}$$

Therefore, our random fields estimation problems will be reduced to finding the inversion formula

$$f \mapsto h \tag{2.297}$$

in our framework. So, our general method for integral transforms will be applied to these problems. For this situation and other topics and methods for the inversion formulas, see [364] for the details.

2.5.7 Support Vector Machines and Probability Theory for Data Analysis

In connection with the beautiful representation (2.229), we recall fundamental mathematics objects that may be related to the theory of reproducing kernels.

Let T be a set. Let $(\Omega, \mathcal{B}, P)$ be a probability space and $L^2(\Omega, \mathcal{B}, P)$ the Hilbert space composed of the square integrable random variables on Ω with the inner product $E[X\overline{Y}]$. Let $X(t) = X(t, \cdot), t \in T$ be a square integrable stochastic process defined on the probability space $(\Omega, \mathcal{B}, P)$. We set the mean value function as $m(t) = E[X(t)]$. Then the second moment function

$$R(t, s) = E[X(t)\overline{Y(s)}] \quad t, s \in T \tag{2.298}$$

and the covariance function

$$K(t,s) = \mathrm{Cov}[X(t), Y(s)] = E[(X(t) - m(t))\overline{(Y(s) - m(s))}] \quad t, s \in T \tag{2.299}$$

are positive definite quadratic form functions on Ω and so, both theories of stochastic processes and reproducing kernels have fundamental relationship. A typical result is the *Loéve theorem:* The Hilbert space $H(X)$ generated by the process $X(t)$, t on a set T with the covariance function R is *congruent* to the reproducing kernel Hilbert space admitting the kernel R.

The support vector machine is a powerful computational method for solving learning and function estimating problems such as pattern recognition, density and regression estimation and operator inversion. The basic idea can be found in E. Parzen [356], A. Berlinet and C. Thomas-Agnan [48, p. 248], B. Schölkopf, A.J. Smola [427] and Vapnik [471]. See also the recent book Steinwart and Christmann [445].

From some data input space E we consider a general non linear mapping to a feature space F that is a pre-Hilbert space with the inner product $\langle \cdot, \cdot \rangle_F$:

$$\Phi : E \to F; \quad x \mapsto \Phi(x). \tag{2.300}$$

Then we form the positive definite quadratic form function

$$K(x,y) = \langle \Phi(x), \Phi(y) \rangle_F. \tag{2.301}$$

The important point of this method is that we can apply this kernel to the problem of consisting the optimal hyperplanes in the space F by not using the explicit values of the transformed data $\Phi(x)$. See B.E. Boser, I.M. Guyon, and V.N. Vapnik [54] and V.N. Vapnik [471].

A new method appeared as the *kernel method.* We consider the transform of the data in the probability space $(\Omega, \mathscr{B}, P)$ for a reproducing kernel Hilbert space $H_K(\Omega)$ admitting a kernel K on Ω:

$$\Psi : \Omega \to \langle \cdot, \cdot \rangle_F; \quad x \mapsto K_x \tag{2.302}$$

and we can apply the theory of reproducing kernels to the probability problems on the space $(\Omega, \mathscr{B}, P)$.

On the whole space $\mathbb{R}^m$ the following kernels are typical and used:

1. The usual inner product is given by $k(x_1, x_2) = x_1^T x_2$.
2. For $c \geq 0$ and for a positive integer d

$$k_{d,c}^{\mathrm{poly}}(x_1, x_2) = (x_1^T x_2 + c)^d \quad x_1, x_2 \in \mathbb{R}^m.$$

3. The Gauss kernel, for $\sigma > 0$

$$k_\sigma^G(x_1, x_2) = \exp\Big(-\frac{|x_1 - x_2|^2}{2\sigma^2}\Big) \quad x_1, x_2 \in \mathbb{R}^m.$$

In statistical learning theory, reproducing kernel Hilbert spaces are used basically as the hypothesis space in approximating regression functions. See, for example, the books [116, 471]. Here, in connection with a basic formula by F. Cucker and S. Smale [117] which is fundamental in the approximation of error estimates, we will consider a general formula based on the general theory of reproducing kernels. For related results, see also [502].

Meanwhile, the kernel forms (2.229) and (2.301) were considered by K. Kataoka [253–255] in the theory of microlocal energy methods with applications to pseudodifferential operators.

2.5.8 *Inverse Formulas Using CONS*

Let $\mathscr{H}$ be a separable Hilbert space and $\mathbf{h} : E \to \mathscr{H}$ a mapping. Here we do not assume that $\ker(L) = 0$. Then as we have seen, this couple $(\mathscr{H}, \mathbf{h})$ gives us a reproducing kernel Hilbert space $H_K(E)$, where $K(p,q) = \langle \mathbf{h}(q), \mathbf{h}(p) \rangle$. However, let us investigate it in more detail via a CONS $\{V_j\}_{j=1}^{\infty}$ of $\mathscr{H}$. A modification is readily made if $\mathscr{H}$ is finite dimensional. Let us define v_j by (2.150). Then we have (2.151). Now we define a function $\overline{h}$ on E by (2.153). We also recall a linear mapping $T_{H_K(E)}$ given by (2.303). We also define a linear mapping

$$T_{H_K(E)}(f) = \sum_{j=1}^{\infty} \langle f, v_j \rangle_{H_K(E)} V_j. \tag{2.303}$$

The convergence of (2.303) takes place in the topology of $\mathscr{H}$. Indeed, we have the following proposition:

Proposition 2.9. *The convergence of the sum* (2.303) *takes place in the strong topology in* $\mathscr{B}(H_K(E), \mathscr{H})$.

Proof. Let $f = LF$ with $F \in \mathscr{H}$. Then we have

$$\sum_{j=1}^{\infty} |\langle f, v_j \rangle_{H_K(E)}|^2 = \sum_{j=1}^{\infty} |\langle P_{\mathscr{H} \to \ker(L)^{\perp}} F, V_j \rangle_{\mathscr{H}}|^2 = \|P_{\mathscr{H} \to \ker(L)^{\perp}} F\|_{\mathscr{H}}^2 < \infty.$$

Therefore, the sum converges in the strong topology in the space $\mathscr{B}(H_K(E), \mathscr{H})$.

In terms of a complete orthonormal system $\{\mathbf{v}_j\}_{j=1}^{\infty}$ of $\mathscr{H}$, we have the inversion formula

Theorem 2.46. *We have the inversion formula*

$$\mathbf{f}^* = \sum_{j=1}^{\infty} \left\langle f, \langle \mathbf{v}_j, \mathbf{h}(\cdot) \rangle_{\mathscr{H}} \right\rangle_{H_K(E)} \mathbf{v}_j. \tag{2.304}$$

That is, if we define $\mathbf{f} = L^{-1}f$, where L^{-1} stands for $(L|\ker(L)^{\perp})^{-1}$, then

$$\mathbf{f}^*(p) = \sum_{j=1}^{\infty} \left\langle f, \langle \mathbf{v}_j, \mathbf{h}(p)\rangle_{\mathscr{H}} \right\rangle_{H_K(E)} \mathbf{v}_j$$

for $p \in E$.

Proof. For the adjoint L^* of the isometry L between $\ker(L)^{\perp}$ and $H_K(E)$, $L^{-1} = L^*$. Hence, from Parseval's identity we obtain

$$\begin{aligned}
\mathbf{f}^* &= \sum_{j=1}^{\infty} \langle \mathbf{f}^*, \mathbf{v}_j\rangle_{\mathscr{H}} \mathbf{v}_j \\
&= \sum_{j=1}^{\infty} \langle L^* f, \mathbf{v}_j\rangle_{\mathscr{H}} \mathbf{v}_j \\
&= \sum_{j=1}^{\infty} \langle f, L\mathbf{v}_j\rangle_{H_K(E)} \mathbf{v}_j \\
&= \sum_{j=1}^{\infty} \left\langle f, \langle \mathbf{v}_j, \mathbf{h}(\cdot)\rangle_{\mathscr{H}} \right\rangle_{H_K(E)} \mathbf{v}_j.
\end{aligned}$$

Thus proved.

2.5.9 A General Fundamental Theorem on Inversions

Here we consider a concrete case of Theorem 2.36. We will consider the linear mapping defined by (2.228). In order to derive a general inversion formula for (2.228) that is widely applicable in analysis, we assume that $\mathscr{H} = L^2(T, dm)$ and that $H_K(E)$ is a closed subspace of $L^2(E, d\mu)$. Furthermore, below we assume that $(T, \mathscr{T}, dm)$ and $(E, \mathscr{E}, d\mu)$ are both σ-finite measure spaces and that

$$H_K(E) \hookrightarrow L^2(E, d\mu). \tag{2.305}$$

Suppose that we are given a measurable function $h : T \times E \to \mathbb{C}$ satisfying $h_q = h(\cdot, q) \in L^2(T, dm)$ for all $q \in E$. Let us set

$$K(p, q) \equiv \langle h_q, h_p\rangle_{L^2(T,dm)}. \tag{2.306}$$

As we have established in Theorem 2.36, we have

$$H_K(E) \equiv \{f \in \mathscr{F}(E) : f(p) = \langle f, h_p\rangle_{L^2(T,dm)} \text{ for some } f \in \mathscr{H}\}. \tag{2.307}$$

Let us now define

$$L : \mathscr{H} \to H_K(E)(\hookrightarrow L^2(E, d\mu)) \tag{2.308}$$

by

$$LF(p) \equiv \langle F, h_p \rangle_{L^2(T,dm)} = \int_T F(t)\overline{h(t,p)}\, dm(t), \quad p \in E \tag{2.309}$$

for $F \in \mathscr{H} = L^2(T, d\mu)$, keeping in mind (2.305). Observe that $Lf \in H_K(E)$ by Corollary 2.1.

The next theorem, which can be seen in a natural way in view of the proof given below, will lead to the inversion formula.

Theorem 2.47. *Assume that $\{E_N\}_{N=1}^{\infty}$ is an increasing sequence of measurable subsets in E such that*

$$\bigcup_{N=1}^{\infty} E_N = E \tag{2.310}$$

and

$$\iint_{T \times E_N} |h(t,p)|^2\, dm(t)\, d\mu(p) < \infty \tag{2.311}$$

for all $N \in \mathbb{N}$. Then we have

$$L^*f(t) \left(= \lim_{N\to\infty} (L^*[\chi_{E_N} f])(t) \right) = \lim_{N\to\infty} \int_{E_N} f(p)h(t,p)\, d\mu(p) \tag{2.312}$$

for all $f \in L^2(E, d\mu)$ in the topology of $\mathscr{H} = L^2(T, dm)$.

Proof. Let $f \in L^2(E, d\mu)$. The operator L^* being continuous from $L^2(E, d\mu)$ to $\mathscr{H} = L^2(T, dm)$, the first equality in (2.312) in the topology of $\mathscr{H} = L^2(T, dm)$ is trivial. Thus, once we show that

$$L^*[\chi_{E_N} f](t) = \int_{E_N} f(p)h(t,p)\, d\mu(p), \tag{2.313}$$

almost every $t \in T$, (2.313) has been completely established.

To verify (2.312) we choose $g \in L^2(T, dm)$ arbitrarily. Then from the definition of adjoint, the Lebesgue convergence theorem and (2.310), we have

$$\langle L^*[\chi_{E_N} f], g \rangle_{L^2(T,dm)} = \langle \chi_{E_N} f, Lg \rangle_{L^2(E,d\mu)} = \int_{E_N} f(p)\overline{L\,g(p)}\, d\mu(p).$$

By inserting the definition (2.309) of L, we have

$$\langle L^*[\chi_{E_N} f], g\rangle_{L^2(T,dm)} = \int_{E_N} f(p) \overline{\left(\int_T g(t)\overline{h(t,p)}\, dm(t) \right)}\, d\mu(p).$$

Assuming that $f \in L^2(E, d\mu)$ and $g \in L^2(T, dm)$ and that (2.311) holds, we can use the Fubini theorem to obtain

$$\langle L^*[\chi_{E_N} f], g\rangle_{L^2(T,dm)} = \int_T \left(\int_{E_N} f(p)h(t,p)\, d\mu(p) \right) \overline{g(t)}\, dm(t). \tag{2.314}$$

Since g is chosen to be arbitrary, we conclude from (2.314) that (2.313) holds for almost all $t \in T$.

An important observation in Theorem 2.47 is that L, which is regarded as a bounded operator from $\mathscr{H}$ to $H_K(E)$, has the adjoint given by (2.312).

Typical Examples for Inversion Formulas

Practically for many cases, the assumptions in Theorem 2.47, will be satisfied automatically, and so Theorem 2.47 may be applied in many cases. However, the basic assumption (2.305) realizing the reproducing kernnel Hilbert space $H_K(E)$ in terms of the integral form is restrictive. Therefore it is hard to use Theorem 2.47 in general. When the basic assumption (2.305) is satisfied, the formulation in Theorem 2.47 may, however, be considered as a natural inversion formula. Indeed, it was given as strong convergence in the Hilbert space $L^2(T, dm)$ and so, it will be of interest and considered as a natural inversion formula.

Take an example of the Weierstrass transform given by (6.101) to illustrate this. To this end recall that we obtained (2.256) in Corollary 2.7. In the inversion formula in the Weierstrass transform, if the norm in $H_k(\mathbb{R})$ is given by the series

$$\sum_{j=0}^{\infty} \frac{(2t)^j}{j!} \int_{\mathbb{R}} |f^{(j)}(x)|^2 dx, \tag{2.315}$$

then as the integrals on E_N we can take, for example, the N-th sum

$$\sum_{j=0}^{N} \frac{(2t)^j}{j!} \int_{\mathbb{R}} |f^{(j)}(x)|^2 dx. \tag{2.316}$$

Recall that we wrote

$$k(x;t) \equiv \frac{1}{\sqrt{4\pi t}} \exp\left(-\frac{x^2}{4t}\right) \quad (x \in \mathbb{R},\ t > 0)$$

in (2.253). Hence we have the inversion formulas.

Theorem 2.48. *In the Weierstrass integral transform in Theorem* 2.41, *for any fixed* $t > 0$, *we have three expressions*:

$$F(\xi) = \lim_{N\to\infty} \frac{1}{\sqrt{2\pi t}} \iint_{E_N} u_F(z;t)\overline{k(z-\xi,t)} \exp\left(-\frac{y^2}{2t}\right) dxdy, \tag{2.317}$$

$$F(\xi) = \sum_{j=0}^{\infty} \frac{(2t)^j}{j!} \int_{\mathbb{R}} \partial_x^j u(x,t) \partial_x^j k(x-\xi,t) dx, \tag{2.318}$$

$$F(\xi) = \sum_{j=0}^{\infty} \frac{(-2t)^j}{j!} \int_{\mathbb{R}} u(x,t) \partial_x^{2j} k(x-\xi,t) dx, \tag{2.319}$$

where the limits take place in the strong topology in $L^2(\mathbb{R})$ *and* $\{E_N\}_{N=1}^{\infty}$ *is any approximation of* $\mathbb{C}$ *by compact sets in* (2.317).

Proof. Equality (2.319) follows from (2.318) and integration by parts. So we need to establish (2.317) and (2.318).

As for (2.317), we let

$$E = \mathbb{C}, \quad T = \mathbb{R}, \quad \mathscr{H} \equiv L^2(\mathbb{R}, d\xi),$$

and

$$d\mu(x,y) = \exp\left(-\frac{y^2}{2t}\right) dx\,dy, \quad h(z,\xi;t) = k(z-\xi;t).$$

Also define

$$K(z,u;t) = \langle h(\cdot,z;t), h(\cdot,u;t)\rangle_{\mathscr{H}} \quad (z,u \in \mathbb{C},\ t > 0).$$

Then observe that by Lemma 2.8, L, viewed as a mapping from $\mathscr{H} \to H_K(\mathbb{C})$, is an isometry and $L^* = L^{-1}$. As we have seen in Theorem 2.40, $H_K(\mathbb{C}) \hookrightarrow L^2(E, d\mu)$ in the sense of continuous embedding. Noting that $u_F(\cdot;t) = LF$, we have

$$F(\xi) = L^*LF(\xi) = L^*u_F(\xi) = \lim_{N\to\infty} \frac{1}{\sqrt{2\pi t}} \iint_{E_N} u_F(z;t)\overline{k(z-\xi,t)} \exp\left(-\frac{y^2}{2t}\right) dxdy$$

by (2.312).

The equality (2.318) can be dealt with in a similar way. We let

$$\mathscr{H} = L^2(\mathbb{R}, d\xi), \quad T = \mathbb{R}, \quad E = \mathbb{N} \times \mathbb{R}, \quad d\mu = \text{counting measure of } \mathbb{N} \otimes dx.$$

Going through the same argument, we obtain (2.318).

2.5.10 Nonharmonic Transforms

In our general transform (2.309), suppose that φ is close to the integral kernel h in the following sense: For any $F \in L^2(T, dm)$,

$$\left\| \int_T F(t)\overline{(h(\cdot,t) - \varphi(\cdot,t))}dm(t) \right\|^2_{H_K(E)} \leq \omega^2 \int_T |F(t)|^2 dm(t) \tag{2.320}$$

where $0 < \omega < 1$ and ω is independent of $F \in L^2(T, dm)$.

Then we can see that for any $f \in H_K(E)$, there exists a function F^*_φ belonging to the visible component of $L^2(T, dm)$ in (2.238) such that

$$f(p) = \int_T F^*_\varphi(t)\overline{\varphi(p,t)}dm(t) \quad \text{on} \quad E \tag{2.321}$$

and

$$(1-\omega)^2 \int_T |F^*_\varphi(t)|^2 dm(t) \leq \|f\|^2_{H_K(E)} \leq (1+\omega)^2 \int_T |F^*_\varphi(t)|^2 dm(t). \tag{2.322}$$

The integral kernel φ will be considered as a perturbation of the integral kernel h. When we look for the inversion formula of (2.321) following our general method, we must calculate the kernel

$$K_\varphi(p,q) = \int_T \varphi(q,t)\overline{\varphi(p,t)}dm(t) \quad \text{on} \quad E \times E. \tag{2.323}$$

We will, however, in general, not be able to calculate (2.323).

Suppose that the image f of (2.321) belongs to the known space $H_K(E)$. Then, we can construct the inverse F^*_φ using our inversion formula in $H_K(E)$ repeatedly and constructing some approximation of F^*_φ by our inverses.

In particular, for the reproducing kernel K on E we construct (or we obtain, by some other method or directly) the function $\hat{\varphi}(t,p)$ satisfying

$$K(p,q) = \int_T \hat{\varphi}(q,t)\overline{\varphi(p,t)}dm(t) \quad \text{on} \quad E \times E, \tag{2.324}$$

where $\hat{\varphi}(q,\cdot)$ belongs to the *visible component* of $L^2(T, dm)$ for any fixed $q \in E$. Then we have the idea of a '*nonharmonic integral transform*' and we can formulate the inversion formula of (2.321) in terms of the kernel $\hat{\varphi}$ and the space $H_K(E)$, globally [382, Chapter 7].

2.5.11 Determinations of Linear Systems

In Theorem 2.37, consider the mapping $\widetilde{L} : H_K(E) \to \mathscr{H}$, which is given by $\widetilde{L}f = \mathbf{f}$ via (2.240). Note that $\widetilde{L}$ is an isometry from $H_K(E)$ to $\widetilde{L}(H_K(E))$. Indeed, in the statement of Theorem 2.36, we have $\mathscr{R}(L) = H_K(E)$ and we know that L is isomorphic from $\widetilde{L}(H_K(E))$ to $\mathscr{R}(L) = H_K(E)$ from (c) and it is easy to see that $\widetilde{L}$ is the inverse of L. We let

$$\mathbf{g}_{\widetilde{L}}(q) \equiv \widetilde{L}K_q \in \mathscr{H}, \tag{2.325}$$

which is called the *"generating vector"* of $\widetilde{L}$. See [64, 431] for many concrete examples. Then, we have

$$K(p,q) = \langle K_q, K_p \rangle_{H_K(E)} = \langle \widetilde{L}K_q, \widetilde{L}K_p \rangle_{\mathscr{H}} = \langle \mathbf{g}_{\widetilde{L}}(q), \mathbf{g}_{\widetilde{L}}(p) \rangle_{\mathscr{H}}. \tag{2.326}$$

The function $\mathbf{g}_{\widetilde{L}} : E \mapsto \mathscr{H}$ will play the role of $\mathbf{h}$ in (2.229). From (2.326), we obtain

Theorem 2.49 ([378, Theorem 5.1]). *For the linear mapping*

$$\mathbf{f} \in \mathscr{H} \mapsto f \equiv \langle \mathbf{f}, \mathbf{g}_{\widetilde{L}}(\cdot) \rangle_{\mathscr{H}} \in H_K(E), \tag{2.327}$$

we have the following:

1. *Let* $\mathbf{f} \in \mathscr{H}$. *Define* $f(p) \equiv \langle \mathbf{f}, \mathbf{g}_{\tilde{L}}(p) \rangle_{\mathscr{H}}$ *for* $p \in E$. *Then* $f \in H_K(E)$. *In particular, the mapping* (2.327) *makes sense.*
2. *The linear mapping* (2.327) *gives the inverse of the isometry* $\widetilde{L} : H_K(E) \to \widetilde{L}(H_K(E))$.
3. *Let* $f \in H_K(E)$. *We have the isometrical identity* $\|f\|_{H_K(E)} = \|\tilde{L}f\|_{\mathscr{H}}$.
4. *The family of vectors* $\{\mathbf{g}_{\widetilde{L}}(p) : p \in E\}$ *is complete in* $\widetilde{L}(H_K(E))$.

Proof.

1. Just go to Corollary 2.1.
2. This is clear from Theorem 2.36 as we saw before.
3. This is just a consequence of (2.233).
4. This also follows from Theorem 2.36.

Theorem 2.41 will give a new viewpoint and method for the Fredholm integral equation of the first kind that is a fundamental integral transform. The method and solution will have the following properties:

1. use the naturally determined reproducing kernel Hilbert space which is determined by the integral kernel;
2. the solution is given in the sense of $\mathscr{H}$-norm convergence;
3. the solution (inverse) is given by $\mathbf{f}^*$ in Theorem 2.37;
4. for the ill-posed problem in (2.228), our method gives a well-posed solution.

In general, in a linear problem, when we cannot assert its existence of the solutions, or its uniqueness of the solutions, or continuity of its solution, then the linear problem is called an *ill-posed problem* in the sense of Hadamard. The linear problem (2.228) or (2.309) has been considered as a typical ill-posed problem. The method gives a characterization of the image; that is, the solution space. For the inverse or solution, we consider the minimum norm one among many solutions and then we have the isometric relation in Theorem 2.37; that is, the solution has uniqueness and continuity properties. This viewpoint is, however, a mathematical and theoretical one. In practical and physical linear systems, the observation data will be a finite number of data containing error or noises, and so we meet various delicate problems numerically and there exists a field of *ill-posed problems*. In Chap. 3, in order to overcome such cases, we will consider new methods.

Many applications of Theorems 2.36, 2.37 and 2.41 were examined. However, many applications of Theorem 2.35 are not considered. These four theorems are, however, of equivalent importance and therefore we can expect many applications of Theorem 2.35. By Theorem 2.35, we can determine the linear system from the isometric relation; when the generator vector may be considered as a Green's function, it may be applied to determine a physical law of the system. The output space of a linear system in our situation is a reproducing kernel Hilbert space and so by using typical and various reproducing kernel Hilbert spaces (as physical models of output spaces) and corresponding isometric mappings, we will be able to find corresponding laws by the present method.

Chapter 3
Moore Penrose Generalized Inverses and Tikhonov Regularization

By applying the general theory of reproducing kernels, we will consider the practical constructions of approximate solutions for bounded linear operator equations.

3.1 The Best Approximations and the Moore Penrose Generalized Inverses

In Sect. 3.1, we give fundamental properties of the Moore Penrose generalized inverses and reproducing kernels. These results will naturally lead to Tikhonov regularization and the reproducing kernel theory. The applications of Tikhonov regularization and reproducing kernels are the main topics of the book and therefore, in the following chapters, we will gather various concrete applications.

See [317, 318] for the relation between RKHS and generalized inverse.

3.1.1 Frames

Definition 3.1. Let $\mathscr{H}$ be a separable Hilbert space. Let X be a locally compact Hausdorff space endowed with a Radon measure μ such that $\mathrm{supp}(\mu) = X$.

1. A family $\mathscr{F} = \{\psi_x\}_{x\in X}$ of $\mathscr{H}$ is called a *continuous frame* if for all $f \in \mathscr{H}$, the mapping $x \in X \mapsto \langle f, \psi_x\rangle_{\mathscr{H}} \in \mathbb{C}$ is continuous and there exist constants $C_1, C_2 > 0$ such that

$$C_1\|f\|_{\mathscr{H}} \le \sqrt{\int_X |\langle f, \psi_x\rangle_{\mathscr{H}}|^2\, d\mu(x)} \le C_2\|f\|_{\mathscr{H}}$$

   for all $f \in \mathscr{H}$.
2. In defining continuous frame, when $C_1 = C_2$, the family $\mathscr{F}$ is called *tight*.

S. Saitoh, Y. Sawano, *Theory of Reproducing Kernels and Applications*, Developments in Mathematics 44, DOI 10.1007/978-981-10-0530-5_3

Here and below we suppose that a *continuous frame* $\mathscr{F} = \{\psi_x\}_{x \in X}$ is given. Also, X and μ are fixed accordingly.

In view of the inequality

$$\int_X |\langle f, \psi_x \rangle_{\mathscr{H}}| \cdot |\langle \psi_x, g \rangle_{\mathscr{H}}| \, d\mu(x) \leq \sqrt{\int_X |\langle f, \psi_x \rangle_{\mathscr{H}}|^2 \, d\mu(x) \int_X |\langle g, \psi_x \rangle_{\mathscr{H}}|^2 \, d\mu(x)}$$

$$\leq C_2 \|f\|_{\mathscr{H}} \|g\|_{\mathscr{H}}$$

and the Riesz representation theorem, we are led to the following definition:

Definition 3.2. The *frame operator* S is given by

$$Sf \equiv \int_X \langle f, \psi_x \rangle_{\mathscr{H}} \psi_x \, d\mu(x)$$

for $f \in \mathscr{H}$. More precisely, for $f \in \mathscr{H}$, the element $Sf \in \mathscr{H}$ is the unique element in $\mathscr{H}$ such that

$$\langle Sf, g \rangle_{\mathscr{H}} = \int_X \langle f, \psi_x \rangle_{\mathscr{H}} \cdot \langle \psi_x, g \rangle_{\mathscr{H}} \, d\mu(x)$$

for all $g \in \mathscr{H}$.

Lemma 3.1.

1. *The operator S defined in Definition 3.2 is bounded, positive and boundedly invertible.*
2. *If $\mathscr{F}$ is tight, then S is a multiple of the identity.*

Proof.

1. The surjectivity of S is derived from the Riesz representation theorem. To verify positivity, we choose $f \in \mathscr{H}$ arbitrarily. Then we have

$$\langle Sf, f \rangle_{\mathscr{H}} = \int_X \langle f, \psi_x \rangle_{\mathscr{H}} \langle \psi_x, f \rangle_{\mathscr{H}} \, d\mu(x) = \int_X |\langle f, \psi_x \rangle_{\mathscr{H}}|^2 \, d\mu(x) \geq 0. \quad (3.1)$$

Hence S is positive. From (3.1) and the fact that $\mathscr{F}$ is tight, we deduce that

$$\|f\|_{\mathscr{H}} \|Sf\|_{\mathscr{H}} \geq \int_X |\langle f, \psi_x \rangle_{\mathscr{H}}|^2 \, d\mu(x) \geq C_1 \|f\|_{\mathscr{H}}^2,$$

and hence we have $\|Sf\|_{\mathscr{H}} \geq C_1 \|f\|_{\mathscr{H}}$. Consequently, S is boundedly invertible.

2. Since $\mathscr{F}$ is tight, we have

$$\langle Sf, g \rangle_{\mathscr{H}} = \int_X \langle f, \psi_x \rangle_{\mathscr{H}} \cdot \langle \psi_x, g \rangle_{\mathscr{H}} \, d\mu(x) = C_1 \langle f, g \rangle_{\mathscr{H}}$$

by the polarization identity. The element $g \in \mathscr{H}$ being arbitrary, we see that $Sf = C_1 f$ for all $f \in \mathscr{H}$, showing that $S = C_1 \mathrm{id}$.

The next assertion is an immediate consequence of Lemma 3.1.

Lemma 3.2. *The family $\mathscr{F}$ spans a dense subspace of $\mathscr{H}$, that is, $\overline{\mathrm{Span}}(\mathscr{F}) = \mathscr{H}$.*

Now we define two elementary linear operators. See (3.3) again.

Definition 3.3. Keep to the same notation as above. One defines linear operators V, $W \in \mathscr{B}(\mathscr{H}, L^2(X, \mu))$ by

$$Vf(x) \equiv \langle f, \psi_x \rangle_{\mathscr{H}} \quad x \in X,$$

$$Wf(x) \equiv \langle f, S^{-1}\psi_x \rangle_{\mathscr{H}} \quad x \in X$$

for $f \in \mathscr{H}$.

Proposition 3.1. *We have*

$$V^* F = \int_X F(y)\psi_y \, d\mu(y) \in \mathscr{H},$$

$$W^* F = \int_X F(y) S^{-1}\psi_y \, d\mu(y) \in \mathscr{H}$$

for all $F \in L^2(X, \mu)$. Also,

$$S = V^* V, \quad S^{-1} = W^* W, \quad \mathrm{id}_{\mathscr{H}} = V^* W = W^* V.$$

Proof. To verify that V^* is as above, we let

$$\tilde{V} F = \int_X F(y)\psi_y \, d\mu(y).$$

More precisely, we define $\tilde{V}F$ as an element of $\mathscr{H}$ so that

$$\langle \tilde{V}F, g \rangle_{\mathscr{H}} = \int_X F(y) \langle \psi_y, g \rangle_{\mathscr{H}} \, d\mu(y).$$

This definition is justified, since we have

$$\int_X |F(y)\langle \psi_y, g\rangle_{\mathscr{H}}| \, d\mu(y) \le \sqrt{\int_X |F(y)|^2 \, d\mu(y) \int_X |\langle \psi_y, g \rangle_{\mathscr{H}}|^2 \, d\mu(y)}$$
$$\le C_2 \|F\|_{L^2(X,\mu)} \|g\|_{\mathscr{H}}$$

for all $F \in L^2(X, \mu)$ and $g \in \mathscr{H}$. Consequently,

$$\langle \tilde{V}F, g\rangle_{\mathscr{H}} = \int_X F(y)\langle \psi_y, g\rangle_{\mathscr{H}}\, d\mu(y) = \int_X F(y)\overline{Vg(y)}\, d\mu(y) = \langle F, Vg\rangle_{L^2(X,\mu)}.$$

Consequently, $V^* = \tilde{V}$. Since $W = VS^{-1}$ and S is self-adjoint, we have $W^* = S^{-1}V^*$. This means that

$$W^*F = S^{-1}V^*F = S^{-1}\int_X F(y)\psi_y\, d\mu(y) = \int_X F(y)S^{-1}\psi_y\, d\mu(y).$$

The equality $S = V^*V$ is immediate from the expression of V and V^*. Once we prove $S = V^*V$, it immediately follows that $W^*W = S^{-1}V^*VS^{-1} = S^{-1}$. The remaining identity can be proved similarly.

This is a corollary of Proposition 3.1.

Proposition 3.2. *Let $f \in \mathscr{H}$. Then we have*

$$W^*Vf(x) = VW^*f(x) = \int_X F(y)\langle \psi_y, S^{-1}\psi_x\rangle_{\mathscr{H}}\, d\mu(y).$$

Define $R(x, y) \equiv \langle \psi_y, S^{-1}\psi_x\rangle_{\mathscr{H}}$ for $x, y \in X$.

Note that $F \mapsto R(F) = VW^*F = WV^*F$ is a projection from $L^2(X, \mu)$ to $\mathrm{Ran}(W) = \mathrm{Ran}(V)$.

Definition 3.4 (Dual frame). A *dual frame* $\tilde{\mathscr{F}} = \{\tilde{\psi}_x\}_{x\in X}$ is the one satisfying

$$f = \int_X \langle f, \psi_x\rangle_{\mathscr{H}}\tilde{\psi}_x\, d\mu(x)$$

for all $f \in \mathscr{H}$. The canonical dual frame is $\{S^{-1}\psi_x\}_{x\in X}$.

Lemma 3.3. *If we assume $\|\psi_x\|_{\mathscr{H}} \le C$ for some constant $C > 0$ independent of x, then we have*

$$|Vf(x)| \le C\|f\|_{\mathscr{H}}, \quad |Wf(x)| \le C\|S^{-1}\|_{\mathscr{B}(\mathscr{H})}\|f\|_{\mathscr{H}}.$$

Proof. Use Definition 3.3 and Hölder inequality.

See [193] for the relation between sampling and frame.

We have a notion similar to “frame”. Ogawa considered pseudo biorthogonal basis, which we recall now. Let $\{\phi_m, \phi_m^* : 1 \le m \le M\}$ be a set of $2M$ ($M \ge N$) elements in an N-dimensional Hilbert space H_N. If any element f in H_N can be expressed as

$$f = \sum_{m=1}^{M}\langle f, \phi_m^*\rangle_{H_N}\phi_m,$$

then $\{\phi_m, \phi_m^* : 1 \leq m \leq M\}$ is said to be a pseudo biorthogonal basis in H_N. The set $\{\phi_m^* : 1 \leq m \leq M\}$ is called a dual sequence to $\{\phi_m : 1 \leq m \leq M\}$ and $\{\phi_m^* : 1 \leq m \leq M\}$ is a counter dual sequence to $\{\phi_m : 1 \leq m \leq M\}$. See [338–340] for more details.

Proof. Use Definition 3.3 and Hölder inequality.

3.1.2 A General Example of the Moore Penrose Inverses

Before going into the main topics, we will take up a delicate property of the Moore Penrose generalized inverse for some concrete cases. Let X be a topological space, which is equipped with a Borel measure μ. Let $\mathscr{F} = \{\psi_x\}_{x\in X}$ be a continuous frame for $\mathscr{H}$. Define

$$S = \int_X \langle f, \psi_x \rangle \psi_x \, d\mu(x) \tag{3.2}$$

for $f \in \mathscr{H}$. Note that $S \in B(\mathscr{H})$. Define

$$Vf(x) \equiv \langle f, \psi_x \rangle, \quad Wf(x) \equiv \langle f, S^{-1}\psi_x \rangle \quad (x \in X). \tag{3.3}$$

We define

$$A(x, y) = G(\mathscr{F}, \mathscr{F})(x, y) \equiv \langle \psi_y, \psi_x \rangle \tag{3.4}$$

for $x, y \in X$. We introduce an integral operator A by

$$Af(x) \equiv \int_X A(x, y)f(y) \, d\mu(y) \quad (f \in \mathscr{H}) \tag{3.5}$$

and set $A^\dagger \equiv VSW^*$. Observe that $VS = AV$.

Proposition 3.3. *Under the above notation* (3.2) *through* (3.5), *we have the following*:

1. $AA^\dagger f = f$ *for* $f \in \mathrm{Ran}(V)$,
2. $A|\mathrm{Ran}(V) : \mathrm{Ran}(V) \to \mathrm{Ran}(V)$ *is boundedly invertible, and*
3. *the kernels of* $A^\dagger$ *and* A *are given by*

$$\ker(A^\dagger) = \mathrm{Ran}(V)^\perp, \quad \ker(A) = \ker(V^*).$$

Proof. Let $f = VF \in \mathrm{Ran}(V)$ with $F \in V$. Then we have

$$AA^\dagger f = AA^\dagger VF = AVS^{-1}W^*VF = AVS^{-1}F = VSS^{-1}F = VF = f$$

and

$$A^\dagger Af = A^\dagger AVF = A^\dagger VSF = VS^{-1}W^*VSF = VS^{-1}SF = VF = f,$$

proving *1*. and 2. Recall that $A = VV^*$ and that $W^*V = \text{id}$, which shows $\ker(A) = \ker(V^*)$ and that $W = VS^{-1}$. Note that

$$\ker(A^\dagger) = \ker(VSW^*) = \ker(SW^*) = \ker(S^{-1}W^*) = \ker(V^*) = \text{Ran}(V)^\perp.$$

Thus the proof is now complete.

If $F \in \text{Ran}(A)$, then we have the following natural and important conclusion:

Theorem 3.1. *We have* $A^\dagger F = \lim_{\lambda \downarrow 0}(\lambda + A^*A)^{-1}A^*F$ *for all* $F \in \text{Ran}(A)$.

Proof. Note that $A^\dagger A = P_{\mathscr{H} \to \ker(A)^\perp}$. Thus, if $F = Af$ with $f \in \text{Dom}(A)$, then we have

$$\lim_{\lambda \downarrow 0}(\lambda + A^*A)^{-1}F = \lim_{\lambda \downarrow 0}(\lambda + A^*A)^{-1}A^*f = P_{\mathscr{H} \to \ker(A)^\perp}f = A^\dagger Af = A^\dagger F,$$

as was to be shown.

3.1.3 Best Approximation Problems

Before the applications to the Tiknohov regularization, we will examine the concept of the Moore Penrose generalized inverses from the viewpoint of the theory of reproducing kernels. Here we will be able to realize the powerful method of the theory of reproducing kernels for the best approximation problems that lead to Moore Penrose generalized inverses.

Let L be any bounded linear operator from a reproducing kernel Hilbert space $H_K(E)$ admitting a kernel $K : E \times E \to \mathbb{C}$ into a Hilbert space $\mathscr{H}$. Then the following problem is a classical and fundamental problem which is known as the best approximate mean square norm problems: For any member $\mathbf{d}$ of $\mathscr{H}$

$$\inf_{f \in H_K(E)} \|Lf - \mathbf{d}\|_{\mathscr{H}}. \tag{3.6}$$

This problem carries, however, a complicated structure when the Hilbert spaces are infinite dimensional and the problem leads to the generalized inverse in the sense of Moore Penrose. By Theorem 2.37, we are convinced that it has a complicated structure.

Let us set

$$k(p, q) \equiv \langle L^*LK_q, L^*LK_p \rangle_{H_K(E)} = L^*LL^*L[K_q](p) \tag{3.7}$$

and

$$P = P_{\mathscr{H} \to \ker(L)^\perp} = P_{\mathscr{H} \to \overline{\text{Ran}(L^*L)}}. \tag{3.8}$$

Proposition 3.4. *Under the notation* (3.7) *and* (3.8), *we have*

$$H_k(E) = \{L^*Lf \,:\, f \in H_K(E)\} \tag{3.9}$$

and the inner product is

$$\langle L^*Lf, L^*Lg\rangle_{H_k(E)} = \langle Pf, g\rangle_{H_K(E)} \tag{3.10}$$

for $f, g \in H_K(E)$.

To prove Proposition 3.4, we need the following lemma:

Lemma 3.4. *Let* L *be a bounded linear operator on a Hilbert space* H. *Then* $\mathrm{Ran}(L^*L) \cap \ker(L^*L) = \{0\}$.

Proof. Let $a = L^*Lb \in \ker(L^*L)$ with $b \in H$. Then

$$\langle a, a\rangle = \langle a, L^*Lb\rangle = \langle L^*La, b\rangle = \langle 0, b\rangle = 0,$$

proving that $a = 0 \in H$.

Proof (of Proposition 3.4). Equip with $\mathscr{F}(E)$ the weak topology. We make $\mathscr{H}_0 = \mathrm{Ran}(L^*L)$, given by on the right-hand side of (3.9), into the Hilbert space together with the inner product given by (3.10). Then $\mathscr{H}_0 \hookrightarrow \mathscr{F}(E)$ in the sense of continuous embedding. Observe $k_q = L^*L\left[L^*LK_q\right] \in \mathscr{H}_0$ for all $q \in \mathscr{H}_0$. We claim

$$\overline{\{k_q \,:\, q \in E\}}^{\mathscr{H}_0} = \mathscr{H}_0. \tag{3.11}$$

Indeed, let $L^*Lf \in \mathscr{H}_0$ be an element perpendicular to all k_q, $q \in E$. Then we have

$$\langle L^*Lf, k_q\rangle_{\mathscr{H}_0} = \langle Pf, L^*LK_q\rangle_{H_K(E)} = \langle L^*LPf, K_q\rangle_{H_K(E)},$$

which implies $L^*LPf = 0$ and hence $Pf \in \mathrm{Ran}(L^*L) \cap \ker(L^*L) = \{0\}$ by Lemma 3.4. Thus, we conclude that $L^*Lf = 0$ and hence $f = Pf = 0$. We also conclude that $\{k_q : q \in E\}$ spans a dense subspace in $\mathscr{H}_0$.

If $p, q \in \mathscr{H}_0$, then we have

$$\langle k_q, k_p\rangle_{\mathscr{H}_0} = \langle L^*L[L^*LK_q], L^*L[L^*LK_q]\rangle_{\mathscr{H}_0} = \langle L^*LK_q, L^*LK_p\rangle_{H_K(E)} = k(p, q). \tag{3.12}$$

In view of (3.11) and (3.12) we see that $\mathscr{H}_0 = H_K(E)$ with norm coincidence.

The next theorem shows the result of basing a solution in (3.6).

Theorem 3.2. *Equation* (3.6) *admits a solution if and only if* $L^*\mathbf{d} \in H_k(E)$. *If this is the case, then we have* $L^*\mathbf{d} = L^*L\tilde{f}$ *for some* $\tilde{f} \in H_K(E)$ *such that* $\tilde{f}$ *is a solution to* (3.6).

Proof. Suppose that (3.6) admits a solution $\tilde{f}$. Then for all $f \in \mathscr{H}$ we have

$$\|Lf-\mathbf{d}\|^2_{\mathscr{H}} = \|L(f-\tilde{f})\|^2_{\mathscr{H}} + 2\mathrm{Re}(\langle f-\tilde{f}, L^*(L\tilde{f}-\mathbf{d})\rangle_{H_K(E)}) + \|L\tilde{f}-\mathbf{d}\|^2_{\mathscr{H}}. \tag{3.13}$$

In view of (3.13), we see that $\tilde{f}$ is a solution to the minimizing problem (3.6) if and only if $L^*\mathbf{d} = L^*L\tilde{f}$. From Proposition 3.4, we see that (3.6) has a solution if and only if $L^*\mathbf{d} \in H_k(E) = \mathscr{H}_0 = \mathrm{Ran}(L^*L)$.

If $0 \in \rho(L^*L)$, then $\tilde{f}$ can be expressed concretely.

Theorem 3.3. *Keep to the same assumption as above. Suppose further that* $0 \in \rho(L^*L)$, *that is* L^*L *is boundedly invertible. Then we have*

$$\tilde{f} = (L^*L)^{-1}L^*\mathbf{d}, \tag{3.14}$$

which is a unique element that attains the minimum.

Proof. This is a direct consequence of identity $L^*\mathbf{d} = L^*L\tilde{f}$.

Let $f_{\mathbf{d}} \in H_K(E)$ be the element such that

$$L^*\mathbf{d} = L^*Lf_{\mathbf{d}} \tag{3.15}$$

with $f_{\mathbf{d}} \in \ker(L)^{\perp}$.

The extremal function $f_{\mathbf{d}}(p)$ has the following representation:

Theorem 3.4. *Keep to the same assumption as above. Then we have*

$$f_{\mathbf{d}}(p) = \langle L^*\mathbf{d}, L^*LK_p\rangle_{H_k(E)} \quad (p \in E). \tag{3.16}$$

Proof. From the definition of $H_K(E)$, we have

$$f_{\mathbf{d}}(p) = \langle f_{\mathbf{d}}, K_p\rangle_{H_K(E)}. \tag{3.17}$$

Since $f_{\mathbf{d}} \in \ker(L)^{\perp}$, it follows that

$$\langle f_{\mathbf{d}}, K_p\rangle_{H_K(E)} = \langle Pf_{\mathbf{d}}, PK_p\rangle_{H_K(E)}. \tag{3.18}$$

In view of the definition of $H_k(E)$ (see (3.10) above), we see that

$$\langle Pf_{\mathbf{d}}, PK_p\rangle_{H_K(E)} = \langle L^*Lf_{\mathbf{d}}, L^*LK_p\rangle_{H_k(E)}. \tag{3.19}$$

Since $L^*Lf_{\mathbf{d}} = L^*d$, we conclude, from (3.17), (3.18), and (3.19), that

$$f_{\mathbf{d}}(p) = \langle L^*\mathbf{d}, L^*LK_p\rangle_{H_k(E)} \quad (p \in E),$$

which proves (3.16).

Definition 3.5. One defines an unbounded operator $L^\dagger : \mathscr{H} \to H_K(E)$ by

$$L^\dagger \mathbf{d} = f_{\mathbf{d}}, \tag{3.20}$$

where $f_{\mathbf{d}}$ is given in Definition 3.15 and $L^\dagger$ is called the *Moore Penrose generalized inverse* for the operator (equation) $Lf = \mathbf{d}$.

Here, the adjoint operator L^* of L, as we see from equality

$$L^*\mathbf{d}(p) = \langle L^*\mathbf{d}, K_p\rangle_{H_K(E)} = \langle \mathbf{d}, LK_p\rangle_{\mathscr{H}} \quad (p \in E), \tag{3.21}$$

is represented by the known data $\mathbf{d}, L, K(p,q)$, and $\mathscr{H}$. From Theorems 3.2 and 3.4, we see that the problem is well established by the theory of reproducing kernels. That is, the existence, the uniqueness and the representation of the solutions in the problem are well formulated. In particular, note that the adjoint operator is represented in a good way; this fact will turn out to be very important in our framework. The extremal function $f_{\mathbf{d}}$ is the *Moore Penrose generalized inverse* $L^\dagger\mathbf{d}$ of the operator (equation) $Lf = \mathbf{d}$. The criteria in Theorem 3.2 is involved and the Moore Penrose generalized inverse $f_{\mathbf{d}}$ is not good when the data contain the error or noise in some practical cases. Therefore, we will introduce the concept of the Tikhonov regularization later.

However, here note that in Theorem 3.2, if $\{L^*LK_p, p \in E\}$ is complete in $H_K(E)$, then $H_K(E) = H_k(E)$ and so, Theorems 3.2 and 3.4 become fairly simple.

Next, we will give a fundamental theorem for the best approximation by the functions in an RKHS. As a concrete application we discuss the analytic extension problem for functions on the real line to entire functions.

3.1.4 Applications to Best Approximation Problems

Let E be an arbitrary set, and let $H_K(E)$ be an RKHS admitting the reproducing kernel $K(p,q)$. Meanwhile, for any subset X of E we consider a Hilbert space $H(X)$ consisting of functions F on X. In the relationship of two Hilbert spaces $H_K(E)$ and $H(X)$, we assume the following:

1. for the restriction $f|_X$ of the members f of $H_K(E)$ to the set X, $f|_X$ belongs to the Hilbert space $H(X)$, and
2. the restriction operator $Lf = f|X$ is continuous from $H_K(E)$ into $H(X)$.

Then we will consider the fundamental problem

$$\inf_{f \in H_K(E)} \|Lf - F\|_{H(X)} \tag{3.22}$$

for a member F of $H(X)$.

For the sake of the good properties of L and its adjoint L^* in our situation, by Theorem 3.4 we can obtain *algorithms* to decide the best f^* of F in the sense of

$$\inf_{f \in H_K(E)} \|Lf - F\|_{H(X)} = \|Lf^* - F\|_{H(X)}, \tag{3.23}$$

when there exists; that is, the infimum is attained by the functions of $H_K(E)$. Furthermore, when there exist best approximations of f^*, we can obtain the best f^* in a reasonable and constructive way. Indeed, we can obtain intrinsic representations of the best approximation in terms of F and the reproducing kernel $K(p, q)$. By a concrete example, we will see the details.

Example 3.1 (Approximations of functions on $\mathbb{R}$ by entire functions). Following Theorem 3.4, we examine best approximations of functions on the real line by entire functions. Since we need a concrete form of the reproducing kernel in Theorem 3.4, as a typical reproducing kernel Hilbert space, for entire functions we consider the Fischer space $\mathscr{F}_a(\mathbb{R})$ on $\mathbb{C}$ normed by

$$\|f\|_{\mathscr{F}_a(\mathbb{R})} = a\sqrt{\frac{1}{\pi}\iint_{\mathbb{C}} |f(z)|^2 \exp(-a^2|z|^2)\, dx\, dy} \tag{3.24}$$

for fixed $a > 0$, whose reproducing kernel is

$$K_a(z, \overline{u}) = \exp(a^2\overline{u}z) \quad (z, u \in \mathbb{C}). \tag{3.25}$$

Meanwhile, as a function space approximated by the Fischer space $\mathscr{F}_a(\mathbb{R})$ we first determine an $L^2(\mathbb{R}, W(x)dx)$ space with a natural weight $W(x)(\geq 0)$

$$\|F\|_{L^2(W)} = \sqrt{\int_{\mathbb{R}} |F(x)|^2 W(x)\, dx} \tag{3.26}$$

in connection with the Fischer space $\mathscr{F}_a(\mathbb{R})$ and Theorem 3.4. Assume $\mathscr{F}_a(\mathbb{R}) \hookrightarrow L^2(\mathbb{R}, W(x)\, dx)$ and that T is the restriction of $\mathscr{F}_a(\mathbb{R})$ from $\mathbb{C}$ to $\mathbb{R}$. We may assume $W(x) = \exp(-a^2x^2)$. Under these situations we examine the best approximation problem

$$\inf_{f \in \mathscr{F}_a(\mathbb{R})} \|Tf - F\|_{L^2(W)} \tag{3.27}$$

for $F \in L^2(W)$.

In this case, the set $\{Tf; f \in \mathscr{F}_a(\mathbb{R})\}$ will be complete in $L^2(W)$, and so we have

$$\inf_{f \in \mathscr{F}_a(\mathbb{R})} \|Tf - F\|_{L^2(W)} = 0. \tag{3.28}$$

Therefore the condition for the existence of the best approximation f^* in the sense of

$$\|Tf^* - F\|_{L^2(W)} = 0 \tag{3.29}$$

will become the condition that F can be extended analytically to the member $f^* \in \mathscr{F}_a(\mathbb{R})$ except for a null Lebesgue measure set on the real line $\mathbb{R}$.

Furthermore, we can construct a sequence $\{f_n\}_{n=0}^{\infty} \subset \mathscr{F}_a(\mathbb{R})$ such that

$$\lim_{n\to\infty} \|Tf_n - F\|_{L^2(W)} = 0 \tag{3.30}$$

for any function F in $L^2(W)$. We first look for a natural weight W such that the restriction operator T is bounded from $\mathscr{F}_a(\mathbb{R})$ into $L^2(W)$. Note that for any member $f \in \mathscr{F}_a(\mathbb{R})$, the integral exists by Fubini's theorem

$$\int_{-\infty}^{\infty} |f(x+iy)|^2 \exp(-a^2(x^2+y^2))\, dx \tag{3.31}$$

for almost all $y \in \mathbb{R}$. In this setting, we have the following inequality (when $y = 0$):

Theorem 3.5. *Let $\mathscr{F}_a(\mathbb{R})$ be an RKHS normed by* (3.24). *Then we have*

$$\int_{\mathbb{R}} |f(x)|^2 \exp(-a^2x^2)\, dx \leq \frac{\sqrt{2\pi}}{a} \|f\|^2_{\mathscr{F}_a(\mathbb{R})} \tag{3.32}$$

for all $f \in \mathscr{F}_a(\mathbb{R})$. Namely, if we set $W(x) = W_a(x) \equiv \exp(-a^2x^2)$, the restriction operator T from $\mathscr{F}_a(\mathbb{R})$ into $L^2(W_a)$ is bounded with the norm being the square root of $\frac{\sqrt{2\pi}}{a}$.

Proof. The function $f \equiv \exp\left(\frac{a^2}{2}z^2(1-\varepsilon)\right)$, $\varepsilon > 0$ almost attains the norm of T as is easily verified. Recall the identity

$$K_a(z,\overline{u}) = \frac{1}{\sqrt{2\pi}a} \exp\left(-\frac{a^2z^2}{2} - \frac{a^2\overline{u}^2}{2}\right) \int_{\mathbb{R}} \exp\left(z\xi + \overline{u}\xi - \frac{\xi^2}{2a^2}\right) d\xi. \tag{3.33}$$

This representation of $K_a(z,\overline{u})$ implies that any $f \in \mathscr{F}_a(\mathbb{R})$ is represented in the form

$$f(z) = \frac{1}{\sqrt{2\pi}a} \exp\left(-\frac{a^2z^2}{2}\right) \int_{\mathbb{R}} F(\xi) \exp(z\xi) \exp\left(-\frac{\xi^2}{2a^2}\right) d\xi \tag{3.34}$$

for some (of course, uniquely determined) function F satisfying

$$\int_{\mathbb{R}} |F(\xi)|^2 \exp\left(-\frac{\xi^2}{2a^2}\right) d\xi < \infty, \tag{3.35}$$

and we have the isometric identity

$$\|f\|_{\mathscr{F}_a(\mathbb{R})} = \sqrt{\frac{1}{\sqrt{2\pi}a}\int_{\mathbb{R}} |F(\xi)|^2 \exp\left(-\frac{\xi^2}{2a^2}\right) d\xi}. \tag{3.36}$$

Meanwhile, by the Parseval–Plancherel identity, we have, from, (3.34)

$$\begin{aligned}\int_{\mathbb{R}} |f(iy)|^2 \exp(-a^2y^2)dy &= \frac{1}{a^2}\int_{\mathbb{R}} |F(\xi)|^2 \exp\left(-\frac{\xi^2}{a^2}\right) d\xi \\ &\leq \frac{1}{a^2}\int_{\mathbb{R}} |F(\xi)|^2 \exp\left(-\frac{\xi^2}{2a^2}\right) d\xi \\ &= \frac{\sqrt{2\pi}}{a}\|f\|^2_{\mathscr{F}_a(\mathbb{R})}.\end{aligned}$$

For $f \in \mathscr{F}_a(\mathbb{R})$, we set $f_1(z) \equiv f(-iz)$. Then, $f_1 \in \mathscr{F}_a(\mathbb{R})$ and $\|f_1\|_{\mathscr{F}_a(\mathbb{R})} = \|f\|_{\mathscr{F}_a(\mathbb{R})}$. Hence we have the desired result;

$$\int_{\mathbb{R}} |f(x)|^2 \exp(-a^2x^2)dx \leq \frac{\sqrt{2\pi}}{a}\|f_1\|^2_{\mathscr{F}_a(\mathbb{R})} = \frac{\sqrt{2\pi}}{a}\|f\|^2_{\mathscr{F}_a(\mathbb{R})},$$

as was to be shown.

We will determine the condition for the existence of the best approximations $f^* \in \mathscr{F}_a(\mathbb{R})$ of a function $F \in L^2(W_a)$ in the sense of

$$\inf_{f\in\mathscr{F}_a(\mathbb{R})} \|Tf - F\|_{L^2(W_a)} = \|Tf^* - F\|_{L^2(W_a)}. \tag{3.37}$$

By Theorem 3.4, there exist the best approximations f^* in (3.37) if and only if

$$(T^*F)(z) = \int_{\mathbb{R}} F(\xi)\exp(a^2\xi z)\exp(-a^2\xi^2)\, d\xi \in \mathscr{R}(T^*T) \tag{3.38}$$

and $\mathscr{R}(T^*T)$ is characterized as the RKHS $H_k(E)$ admitting the reproducing kernel

$$\begin{aligned}k(z,\overline{u}) &= \langle T^*TK_{\overline{u}}, T^*TK_{\overline{z}}\rangle_{\mathscr{F}_a(\mathbb{R})} \\ &= \langle TK_{\overline{u}}, TT^*TK_{\overline{z}}\rangle_{L^2(W_a)} \\ &= \frac{\sqrt{\pi}}{a}\int_{\mathbb{R}} \exp\left(a^2\overline{u}\xi\right)\exp\left(\frac{a^2(\xi+z)^2}{4}\right)\exp\left(-a^2\xi^2\right)\, d\xi \\ &= \frac{2\pi}{\sqrt{3}a^2}\exp\left(\frac{1}{3}a^2z^2\right)\exp\left(\frac{1}{3}a^2\overline{u}^2\right)\exp\left(\frac{1}{3}a^2\overline{u}z\right).\end{aligned}$$

Note that the RKHS $H_k(E)$ is composed of all entire functions $f(z)$ with finite norms

$$\|f\|_{H_k(E)} = \frac{a^2}{\sqrt[4]{12\pi^4}} \sqrt{\iint_{\mathbb{C}} |f(z)|^2 \exp\left(-a^2x^2 + \frac{a^2y^2}{3}\right) dx\, dy}. \tag{3.39}$$

We can check $\{Tf : f \in \mathscr{F}\}$ and, in particular $\{\exp(a^2\overline{u}\cdot) : u \in \mathbb{C}\}$ are complete in $L^2(W_a)$ and so $Tf^* = F$ in $L^2(W_a)$ in (3.37). Hence, from Theorem 3.2, we have

Theorem 3.6. *For all $F \in L^2(W_a)$, F is realized as an image of $f^* \in \mathscr{F}_a(\mathbb{R})$ by $T : \mathscr{F}_a(\mathbb{R}) \to L^2(W_a)$, if and only if*

$$\iint_{\mathbb{C}} \left| \int_{\mathbb{R}} F(\xi) \exp(a^2\xi z) \exp(-a^2\xi^2)\, d\xi \right|^2 \exp\left(-a^2x^2 + \frac{a^2y^2}{3}\right) dx\, dy < \infty. \tag{3.40}$$

By Theorem 3.4, we can obtain an explicit representation of f^* in terms of F. Of course, f^* is uniquely determined. In order to use Theorem 3.4, note that

$$[T^*T(K_a)_{\overline{u}}](z) = \langle T(K_a)_{\overline{u}}, T(K_a)_{\overline{z}} \rangle_{L^2(W_a)} = \frac{\sqrt{\pi}}{a} \exp\left(\frac{a^2z^2}{4} + \frac{a^2\overline{u}^2}{4} + \frac{a^2\overline{u}z}{2}\right).$$

Therefore, in particular, we have:

Theorem 3.7. *Let $f \in \mathscr{F}_a(\mathbb{R})$ and $z \in \mathbb{C}$. Then we can express $f(z)$ in terms of the trace $f(x), x \in \mathbb{R}$ to the real line in the form*

$$f(z) = \frac{a^3}{\sqrt[4]{144\pi^3}} \iint_{\mathbb{C}} \left(\int_{\mathbb{R}} f(\xi) \exp(a^2 Z\xi - a^2\xi^2)\, d\xi \right)$$
$$\cdot \exp\left(\frac{a^2z^2}{4} + \frac{a^2\overline{Z}}{4} + \frac{a^2z\overline{Z}}{2} - a^2X^2 + \frac{a^2}{3}Y^2\right) dX\, dY,$$

where $Z = X + iY$.

For any $F \in L^2(W_a)$, we now construct a sequence $\{f_n^*\}_{n=0}^{\infty}$ satisfying $f_n^* \in \mathscr{F}_a(\mathbb{R})$ and

$$\lim_{n \to \infty} \|Tf_n^* - F\|_{L^2(W_a)} = 0. \tag{3.41}$$

First, note that the images $T^*F = f$; that is,

$$f(z) = \langle F, TK_{\overline{z}} \rangle_{L^2(W_a)} (z \in \mathbb{C}) \tag{3.42}$$

for $F \in L^2(W_a)$ are characterized as the RKHS $H_{\mathbf{K}}(\mathbb{C})$ admitting the reproducing kernel

$$\mathbf{K}(z,\overline{u}) = \langle TK_{\overline{u}}, TK_{\overline{z}}\rangle_{L^2(W_a)} = \frac{\sqrt{\pi}}{a}\exp\left(\frac{a^2z^2}{4}+\frac{a^2\overline{u}^2}{4}+\frac{a^2\overline{u}z}{2}\right). \tag{3.43}$$

As in the space $H_k(\mathbb{C})$, we see that the space $H_{\mathbf{K}}(\mathbb{C})$ is composed of all entire functions f with finite norms

$$\|f\|^2_{H_{\mathbf{K}}(\mathbb{C})} = \frac{a^3}{2\pi\sqrt{\pi}}\iint_{\mathbb{C}} |f(z)|^2 \exp(-a^2x^2)\,dx\,dy. \tag{3.44}$$

From the representation (3.43) of $\mathbf{K}(z,\overline{u})$ we can verify that the family of functions

$$\varphi_n(z) = \left(\frac{\sqrt{\pi}a^{2n-1}}{2^n n!}\right)^{\frac{1}{2}} \exp\left(\frac{a^2z^2}{4}\right) z^n,\ n = 0,1,2,\ldots$$

is a complete orthonormal system in $H_{\mathbf{K}}(\mathbb{C})$.

For any $F \in L^2(W_a)$, we set

$$a_n = \langle T^*F, \varphi_n\rangle_{H_{\mathbf{K}}(\mathbb{C})} \tag{3.45}$$

and

$$\widetilde{f_N} = \sum_{n=0}^{N} a_n\varphi_n. \tag{3.46}$$

Then, $\widetilde{f_N} \in H_k(E) \subset H_{\mathbf{K}}(\mathbb{C})$ and $\{\widetilde{f_N}\}_{N=0}^{\infty}$ converges to T^*F in $H_{\mathbf{K}}(\mathbb{C})$ as N tends to infinity. By (3.16), we construct the functions f_N^* satisfying $T^*Tf_N^* = T^*\widetilde{f_N}$ from (3.38).

Since the adjoint operator T^* is isometric from $L^2(W_a)$ onto $H_{\mathbf{K}}(\mathbb{C})$, we have

Theorem 3.8 (Approximation of F**).** *For the sequence* $\{f_N^*\}_{N=0}^{\infty}$ *constructed in* (3.47) *by* (3.45) *and* (3.46), *and for any* $F \in L^2(W_a)$, *we have*

$$\lim_{N\to\infty} \|Tf_N^* - F\|_{L^2(W_a)} = 0. \tag{3.47}$$

The above concrete example will show that the Moore Penrose generalized inverses are, in general, a very delicate and complicated structure in analysis.

3.1.5 Applications to Operator Equations with a Parameter

We can also consider the direct sum $\int_X^{\oplus} H_x\,d\mu(x)$ of $\{H_x\}_{x\in X}$ as an extension of $H_0 \oplus H_1$.

Definition 3.6. Let $(X, \mathscr{B}, \mu)$ be a σ-finite measure space.

1. *A measurable family of Hilbert spaces* on (X, μ) is a family $\{H_x\}_{x \in X}$ satisfying

$$X_\infty \equiv \{x \in X : H_x \sim \ell^2(\mathbb{N})\}$$

and

$$X_k \equiv \{x \in X : \dim(H_x) = k\}$$

are measurable for all $k = 0, 1, 2, \ldots$. Below, fix an isomorphism $\varphi_x : X \to \ell^2(\mathbb{N})$ for each $x \in X_\infty$ and an isomorphism $\varphi_x : H_x \to \mathbb{C}^k$ for each $x \in X_k$.
2. *A cross-section* is a family $\{s_x\}_{x \in X}$ such that $s_x \in H_x$ for all $x \in X$. A cross-section is *measurable* if and only if

$$x \in X_\infty \mapsto \varphi_x(s_x) \in \ell^2(\mathbb{N})$$

and

$$x \in X_k \mapsto \varphi_x(s_x) \in \mathbb{C}^k$$

are measurable for all k.
3. We will identify measurable cross-sections s and t that are equal almost everywhere.
4. Given a measurable family $\{H_x\}_{x \in X}$ of Hilbert spaces, the direct sum

$$\int_X^\oplus H_x \, d\mu(x)$$

consists of equivalence classes (with respect to almost everywhere equality) of measurable square integrable cross-sections of $\{H_x\}_{x \in X}$. This is a Hilbert space under the inner product

$$\langle s, t \rangle_{\int_X^\oplus H_x \, d\mu(x)} = \int_X \langle s(x), t(x) \rangle_{H_x} \, d\mu(x); \quad s, t \in \int_X^\oplus H_x \, d\mu(x).$$

Here we work on a σ-finite measure space $(\Lambda, \mathscr{B}, \mu)$. We will consider bounded linear operators on a reproducing kernel Hilbert space $H_K(E)$. We write

$$\mathbb{H} \equiv \int_\Lambda^\oplus \mathbb{H}_\lambda d\mu(\lambda), \tag{3.48}$$

as the direct sum of the Hilbert spaces $\{\mathbb{H}_\lambda\}_{\lambda \in \Lambda}$. We assume that a family $L = \{L_\lambda\}_{\lambda \in \Lambda}$ of bounded linear operators in (2.283) is bounded on $H_K(E)$ into $\mathbb{H}$ and that, for each $f \in H_K(E)$, $\{L_\lambda f\}_{\lambda \in \Lambda}$ is a measurable cross section of $\mathbb{H}$. More precisely, we have

$$\sqrt{\int_{\Lambda} \|L_\lambda f\|_{\mathbb{H}_\lambda}^2 d\mu(\lambda)} \le \sqrt{M}\|f\|_{H_K(E)} \tag{3.49}$$

for some constant $M \ge 0$.

In this setting, given a measurable section $g = \{g_\lambda\}_{\lambda\in\Lambda} \in \mathbb{H}$ we consider the extremal problem

$$\inf\left\{\int_{\Lambda} \| L_\lambda f - g_\lambda \|_{\mathbb{H}_\lambda}^2 \, d\mu(\lambda) \, : f \in H_K(E)\right\}. \tag{3.50}$$

This gives a generalized solution for the equations

$$L_\lambda f = g_\lambda \quad \lambda \in \Lambda, \quad f \in H_K(E) \tag{3.51}$$

when we are given $g_\lambda \in \mathbb{H}_\lambda$ for each λ. Define $L : H_K(E) \to \mathbb{H}$ by $Ld \equiv \{L_\lambda d\}_{\lambda\in\Lambda}$. Let L^* be the adjoint operator of L from $\mathbb{H}$ into $H_K(E)$. We form the positive matrix

$$k(p,q) = \langle L^*LK_q, L^*LK_p\rangle_{H_K(E)} \quad \text{on} \quad E \times E. \tag{3.52}$$

We use the notation (3.8). Then we have the following theorem:

Theorem 3.9 ([407, Theorem 1]). *Let K be a positive definite function on a set E and define k by* (3.52). *Let us set $L = U\sqrt{L^*L}$, where $U : H_K(E) \to \mathbb{H}$ is a partial isometry such that* $\mathrm{Ran}(U) = \mathrm{Ran}(L)$ *and that* $\ker(U) = \mathrm{Ran}(\sqrt{L^*L})^\perp$. *Let $f \in H_K(E)$.*

1. *We have the following identity: for any $g \in \mathbb{H}$*

$$\int_{\Lambda} \| L_\lambda f - g_\lambda \|_{\mathbb{H}_\lambda}^2 \, d\mu(\lambda) = \|\sqrt{L^*L}f - U^*g\|_{H_K(E)}^2 - \|U^*g\|_{\mathbb{H}}^2 + \|g\|_{\mathbb{H}}^2. \tag{3.53}$$

   Consequently, the minimum of the mapping $f \in H_K(E) \mapsto \|Lf - g\|_{\mathbb{H}}$ exists, if and only if $U^*g \in \mathrm{Ran}(\sqrt{L^*L})$.
2. *For a function $g = \{g_\lambda\}_{\lambda\in\Lambda} \in \mathbb{H}$, there exists a function $\overline{f}$ in $H_K(E)$ such that*

$$\inf_{f\in H_K(E)} \int_{\Lambda} \| L_\lambda f - g_\lambda \|_{\mathbb{H}_\lambda}^2 \, d\mu(\lambda) = \int_{\Lambda} \| L_\lambda \overline{f} - g_\lambda \|_{\mathbb{H}_\lambda}^2 \, d\mu(\lambda) \tag{3.54}$$

   if and only if $L^*g \in H_k(E)$.
3. *Furthermore, if there exist the best approximations $\overline{f}$ satisfying* (3.54), *then there exists a unique extremal function $\check{f}$ with the minimum norm in $H_K(E)$, and this function is represented in the form*

$$\check{f}(p) = \langle L^*g, L^*LK_p\rangle_{H_k(E)} \quad \text{on} \quad E. \tag{3.55}$$

Proof. We write $\varphi_\alpha(t) \equiv \dfrac{1}{\sqrt{\max(t,\alpha)}}$ for $\alpha > 0$.

1. We calculate that:

$$\begin{aligned}\int_\Lambda \| L_\lambda f - g_\lambda \|^2_{\mathbb{H}_\lambda}\, d\mu(\lambda) &= \|Lf - g\|^2_{\mathbb{H}} \\ &= \|\sqrt{L^*L}f\|^2_{H_K(E)} - 2\mathrm{Re}(\langle Lf, g\rangle_{\mathbb{H}}) + \|g\|^2_{\mathbb{H}} \\ &= \|\sqrt{L^*L}f\|^2_{H_K(E)} - 2\mathrm{Re}(\langle U\sqrt{L^*L}f, g\rangle_{\mathbb{H}}) + \|g\|^2_{\mathbb{H}} \\ &= \|\sqrt{L^*L}f\|^2_{H_K(E)} - 2\mathrm{Re}(\langle \sqrt{L^*L}f, U^*g\rangle_{\mathbb{H}}) + \|g\|^2_{\mathbb{H}} \\ &= \|\sqrt{L^*L}f - U^*g\|^2_{H_K(E)} - \|U^*g\|^2_{\mathbb{H}} + \|g\|^2_{\mathbb{H}}.\end{aligned}$$

2. Note that $U^*g \in \mathrm{Ran}(U) = \mathrm{Ran}(L)$. From (3.53) it suffices to show:

$$U^*g \in \mathrm{Ran}(\sqrt{L^*L}) \iff L^*g = \sqrt{L^*L}U^*g \in \mathrm{Ran}(L^*L) = H_k(E). \tag{3.56}$$

   The forward implication is trivial from the definition. Let us assume that $g \in \mathbb{H}$ satisfies $L^*g \in \mathrm{Ran}(L^*L)$. Note that $U^*g \in \overline{\mathrm{Ran}(\sqrt{L^*L})}$. Then, since $L^*g \in \mathrm{Ran}(L^*L)$ we can find $h \in H_K(E)$ such that $L^*g = L^*Lh$ and hence

$$\varphi_\alpha(L^*L)\sqrt{L^*L}U^*g = \varphi_\alpha(L^*L)L^*g = \varphi_\alpha(L^*L)L^*Lh \to \sqrt{L^*L}h \tag{3.57}$$

   as $\alpha \downarrow 0$. Since $U^*g \in \overline{\mathrm{Ran}(\sqrt{L^*L})}$ and $\|\varphi_\alpha(L^*L)\sqrt{L^*L}\|_{B(H_K(E))} \le 1$ for all $\alpha > 0$, a routine argument shows that

$$\lim_{\alpha\downarrow 0} \varphi_\alpha(L^*L)\sqrt{L^*L}U^*g = U^*g. \tag{3.58}$$

   Thus we conclude $U^*g \in H_K(E) = \mathrm{Ran}(\sqrt{L^*L})$ from (3.57) and (3.58).
3. Let us suppose that $L^*g \in \mathrm{Ran}(L^*L)$. Set $L^*g = L^*Lh$ with $h \in H_K(E)$ as above. By replacing h suitably, we may assume that h is perpendicular to any element in $\ker(L^*L)$. Observe $U^*g = \sqrt{L^*L}h$ from (3.57) and (3.58). Then we have

$$\int_\Lambda \| L_\lambda f - g_\lambda \|^2_{\mathbb{H}_\lambda}\, d\mu(\lambda) = \|\sqrt{L^*L}(f-h)\|^2_{H_K(E)} - \|\sqrt{L^*L}h\|^2_{H_K(E)} + \|g\|^2_{\mathbb{H}}.$$

   Hence, to minimize the norm of minimizer, we need to choose $\check{f} = \sqrt{L^*L}h$ and this is actually the minimizer.

   Let us prove (3.55). Fix $p \in E$. By the reproducing property of $H_K(E)$ and the fact that $h \in \mathrm{Ran}(L^*L) = \ker(L^*L)^\perp$,

$$\breve{f}(p) = \langle h, K_p \rangle_{H_K(E)} = \langle P_{H_K(E) \to \overline{\mathrm{Ran}(L^*L)}} h, K_p \rangle_{H_K(E)}.$$

By (3.10), we have $\breve{f}(p) = \langle L^*Lh, L^*LK_p \rangle_{H_k(E)} = \langle L^*g, L^*LK_p \rangle_{H_k(E)}$, proving (3.55).

Theorem 3.10. *As for the function given by* (3.52), *the following are equivalent*:

(1) *L is injective*;
(2) *L^*L is injective*;
(3) *$\{L^*LK_x\}_{x \in E}$ is complete in $H_K(E)$*;
(4) *$L^*L : H_K(E) \to H_k(E)$ is an isometry.*

Proof. The equivalence of (1) and (2) and the one of (3) and (4) being elementary, let us prove that (2) and (3) are equivalent. Assume (2) and let f be perpendicular to any element that can be written L^*LK_x for some $x \in E$. Then we have

$$L^*Lf(x) = \langle L^*Lf, K_x \rangle_{H_K(E)} = \langle f, L^*LK_x \rangle_{H_K(E)} = 0$$

for all $x \in E$. Thus, $L^*Lf = 0$. Now that L^*L is injective, it follows that $f = 0 \in H_K(E)$.

Conversely, assume (3) and that $L^*Lf = 0$. Then

$$\langle f, L^*LK_x \rangle_{H_K(E)} = \langle L^*Lf, K_x \rangle_{H_K(E)} = L^*Lf(x) = 0$$

for all $x \in E$. Hence in view of (3) we have $f = 0$.

Example 3.2. As an example, we consider the space $H_K[0, \infty)$ on $[0, \infty)$ given by Theorem 1.6, that is

$$K(x, y) \equiv \min(x, y) \quad (x, y \geq 0).$$

This space is composed of all absolutely continuous real-valued functions on $[0, \infty)$ and $f(0) = 0$ equipped with the norm

$$\|f\|_{H_K(0,\infty)} = \sqrt{\int_0^\infty f'(x)^2 dx}. \tag{3.59}$$

As a space $\mathbb{H}$, we consider the space $L^2((0, \infty), e^{-\lambda} d\lambda)$ and a bounded linear operator L:

$$L_\lambda f = \int_0^\lambda f(\xi) d\xi \tag{3.60}$$

from $H_K(0, \infty)$ into $\mathbb{H}$. Then the adjoint operator L^* from $\mathbb{H}$ into $H_K(0, \infty)$ is given by

$$L^*g(x) = \frac{1}{2}\int_0^x g(\lambda)\lambda^2 e^{-\lambda}\,d\lambda + \int_x^\infty g(\lambda)\left[x\lambda - \frac{1}{2}x^2\right]e^{-\lambda}\,d\lambda \tag{3.61}$$

and we can discuss a problem: Given $g \in \mathbb{H}$, can we find $f \in H_K(0,\infty)$ such that [407, (35)]

$$\int_0^\infty \left|\int_0^\lambda f(\xi)d\xi - g(\lambda)\right|^2 e^{-\lambda}\,d\lambda \leq \int_0^\infty \left|\int_0^\lambda h(\xi)d\xi - g(\lambda)\right|^2 e^{-\lambda}\,d\lambda \tag{3.62}$$

for all $h \in H_K(0,\infty)$? We can give a complete solution for the problem: Such a function f exists if and only if L^*g given by (3.61) belongs to $H_k([0,\infty))$, where

$$k(x,y) = \langle L^*LK_x, L^*LK_y\rangle_{H_K} \quad (x,y \in [0,\infty))$$

according to (3.7).

We followed [413] in this section.

3.2 Spectral Analysis and Tikhonov Regularization

We recall some necessary and important results in spectral theory and then we state the properties about the Tikhonov regularization, where we follow [141]. For the case of compact operators L, see [192].

3.2.1 *Spectral Analysis*

Let $\{E_\lambda\}_{\lambda\in\mathbb{R}}$ be the spectral family for the self-adjoint operator L^*L. When L^*L has a continuous inverse, or equivalently $0 \in \rho(L^*L)$, then we have

$$(L^*L)^{-1} = \int_{\mathbb{R}} \frac{1}{\lambda}dE_\lambda. \tag{3.63}$$

Needless to say, in general there is no guarantee for L to be invertible. So we modify the definition of L^{-1}.

Definition 3.7. Let L be a bounded self-adjoint operator from a Hilbert space $\mathscr{H}_1$ to another Hilbert space $\mathscr{H}_2$. Denote by $\{E_\lambda\}_{\lambda\in\mathbb{R}}$ the spectral decomposition of $\mathbb{R}$ with respect to L^*L. Define $(\mathrm{Dom}(L^\dagger), L^\dagger)$ as follows:

1. The domain $\mathrm{Dom}(L^\dagger)$ is a linear subspace given by

$$\mathrm{Dom}(L^\dagger) \equiv \left\{\mathbf{d} \in \mathscr{H}_2 \,:\, \int_{\mathbb{R}} \lambda^{-2} d(\|E_\lambda L^*\mathbf{d}\|^2_{\mathscr{H}_1}) < \infty\right\},$$

where the integral is understood in the sense of the Stieltjes integral.

2. Let $\mathbf{d} \in \mathrm{Dom}(L^\dagger)$. Then define

$$L^\dagger \mathbf{d} = \int_{\mathbb{R}} \frac{1}{\lambda} d(E_\lambda L^* \mathbf{d}) \equiv \lim_{R\to\infty,\, \varepsilon\downarrow 0} \int_{\lambda\in[\varepsilon,R]\cup[-R,-\varepsilon]} \frac{1}{\lambda} d(E_\lambda L^* \mathbf{d}). \tag{3.64}$$

The limit of (3.64) exists if and only if $\mathbf{d} \in \mathrm{Dom}(L^\dagger)$. Indeed, for any $\mathbf{d} \in \mathcal{H}_2$, we have

$$\left\| \int_F \frac{1}{\lambda} d(E_\lambda L^* \mathbf{d}) \right\|_{\mathcal{H}_1}^2 = \int_F \lambda^{-2} d(\|E_\lambda L^* \mathbf{d}\|_{\mathcal{H}_1}^2)$$

for all Borel measurable sets F. If we use the Cauchy test, then we see that the limit exists if and only if $\mathbf{d} \in \mathrm{Dom}(L^\dagger)$.

Here and below we assume that L maps an RKHS $H_K(E)$ to a Hilbert space $\mathcal{H}$. Then, the Moore Penrose inverse is represented by

$$f_{\mathbf{d}} = \int_{\mathbb{R}} \frac{1}{\lambda} d(E_\lambda L^* \mathbf{d}). \tag{3.65}$$

When $\mathrm{Ran}(L)$ is not closed and $\mathbf{d} \notin \mathrm{Dom}(L^\dagger)$, that is, $Lf = \mathbf{d}$ does not have the Moore Penrose generalized inverse, in this case, by taking $\alpha > 0$, we define

$$f_{\mathbf{d},\alpha} = \int \frac{1}{\lambda+\alpha} d(E_\lambda L^* \mathbf{d}). \tag{3.66}$$

Then, we obtain:

Proposition 3.5. *Let* $\mathbf{d} \in \mathrm{Dom}(L^\dagger)$. *Then we have*

$$\lim_{\alpha\downarrow 0} f_{\mathbf{d},\alpha} = f_{\mathbf{d}} \tag{3.67}$$

in the topology of $H_K(E)$.

Proof. Let $\mathbf{d} \in \mathrm{Dom}(L^\dagger)$. Then $\mathbf{d}$ has an expression; $\mathbf{d} = Lf_{\mathbf{d}}$ for some $f_{\mathbf{d}} \in \mathcal{H}$. Note that

$$f_{\mathbf{d},\alpha} = (L^*L+\alpha)^{-1}L^*\mathbf{d} = (L^*L+\alpha)^{-1}L^*Lf_{\mathbf{d}} = \left(\int_{\mathbb{R}} \frac{\lambda}{\lambda+\alpha} dE_\lambda\right) f_{\mathbf{d}}.$$

As a consequence, we obtain

$$\begin{aligned} \|(L^*L+\alpha)^{-1}L^*\mathbf{d} - f_{\mathbf{d}}\|_{H_K(E)} &= \left\| \left(\int_{\mathbb{R}} \frac{\alpha}{\lambda+\alpha} dE_\lambda\right) f_{\mathbf{d}} \right\|_{H_K(E)} \\ &= \sqrt{\int_{\mathbb{R}} \frac{\alpha^2}{(\lambda+\alpha)^2} d\langle E_\lambda f_{\mathbf{d}}, f_{\mathbf{d}}\rangle_{H_K(E)}}, \end{aligned}$$

where it will be understood that the integral on the far right-hand side is the Stieltjes integral. Now that $d\langle E_\lambda f_{\mathbf{d}}, f_{\mathbf{d}}\rangle$ is supported away from $(-\infty, 0)$ and

$$\int_{\mathbb{R}} d\langle E_\lambda f_{\mathbf{d}}, f_{\mathbf{d}}\rangle_{H_K(E)} = \langle f_{\mathbf{d}}, f_{\mathbf{d}}\rangle_{H_K(E)} < \infty, \tag{3.68}$$

we are in the position of applying the Lebesgue convergence theorem to have

$$\lim_{\alpha\downarrow 0} \sqrt{\int_{\mathbb{R}} \frac{\alpha^2}{(\lambda+\alpha)^2}\, d\langle E_\lambda f_{\mathbf{d}}, f_{\mathbf{d}}\rangle_{H_K(E)}} = 0, \tag{3.69}$$

which yields the desired convergence.

Proposition 3.6. *Under the same notation, for* $\mathbf{d} \in \mathscr{H}$, *we have*

$$\| L f_{\mathbf{d},\alpha}\|_{\mathscr{H}} \le \|\mathbf{d}\|_{\mathscr{H}}. \tag{3.70}$$

Proof. We start with the following elementary identity:

$$L\left(\int_{\mathbb{R}} \frac{dE_\lambda}{\lambda+\alpha}\right) L^* d = L f_{\mathbf{d},\alpha}.$$

Let us use the polar decomposition of L;

$$L = U\sqrt{L^* L}, \tag{3.71}$$

where U is a partial isometry. Then we have

$$L\left(\int_{\mathbb{R}} \frac{dE_\lambda}{\lambda+\alpha}\right) L^* = U\sqrt{L^* L}\left(\int_{\mathbb{R}} \frac{dE_\lambda}{\lambda+\alpha}\right)\sqrt{L^* L}\, U^* = U\left(\int_{\mathbb{R}} \frac{\lambda\, dE_\lambda}{\lambda+\alpha}\right) U^* \le 1.$$

Thus, the result (3.70) is immediate.

Proposition 3.7. *We have*

$$\|f_{\mathbf{d},\alpha}\|_{\mathscr{H}} \le \frac{\|\mathbf{d}\|_{\mathscr{H}}}{2\sqrt{\alpha}}. \tag{3.72}$$

Proof. We go through the same argument as above. We now have to consider

$$\|f_{\mathbf{d},\alpha}\|^2_{\mathscr{H}} = \left\langle \mathbf{d}, L\left(\int_{\mathbb{R}} \frac{dE_\lambda}{(\lambda+\alpha)^2}\right) L^*\mathbf{d}\right\rangle_{\mathscr{H}}. \tag{3.73}$$

We use decomposition (3.71). From a direct calculation

$$L\left(\int_{\mathbb{R}} \frac{dE_\lambda}{(\lambda+\alpha)^2}\right) L^* = U\sqrt{L^*L}\left(\int_{\mathbb{R}} \frac{dE_\lambda}{(\lambda+\alpha)^2}\right)\sqrt{L^*L}\,U^*$$
$$= U\left(\int_{\mathbb{R}} \frac{\lambda\, dE_\lambda}{(\lambda+\alpha)^2}\right) U^* \leq \frac{1}{4\alpha},$$

we obtain the desired result.

The Tikhonov extremal function $f_{\mathbf{d},\alpha}$ is characterized as the following extremal function that makes the Tikhonov functional a minimum:

Theorem 3.11. *Let* $\alpha > 0$. *Then the following minimizing problem admits a unique solution*

$$\min_{f \in H_K(E)} \left(\alpha \|f\|^2_{H_K(E)} + \|\mathbf{d} - Lf\|^2_{\mathcal{H}}\right). \tag{3.74}$$

Furthermore, the minimum is attained by

$$f_{\mathbf{d},\alpha} = (L^*L+\alpha)^{-1}L^*\mathbf{d} = \left(\int_{\mathbb{R}} \frac{1}{\lambda+\alpha} dE_\lambda\right) L^*\mathbf{d} \tag{3.75}$$

by using the spectral decomposition (3.63). *Furthermore,* $\mathbf{d} \mapsto f_{\mathbf{d},\alpha}$ *is almost the inverse of L in the following sense:*

$$\lim_{\alpha \downarrow 0} f_{Lg,\alpha} = g \tag{3.76}$$

in $H_K(E)$ *for all* $g \in H_K(E)$ *and*

$$\lim_{\alpha \downarrow 0} Lf_{\mathbf{d},\alpha} = \mathbf{d} \tag{3.77}$$

in $\mathcal{H}$.

Proof. We complete the square of the formula in question

$$\begin{aligned}
&\alpha \|f\|^2_{H_K(E)} + \|\mathbf{d} - Lf\|^2_{\mathcal{H}} \\
&= \langle(\alpha + L^*L)f, f\rangle_{H_K(E)} - 2\mathrm{Re}\langle L^*\mathbf{d}, f\rangle_{H_K(E)} + \|\mathbf{d}\|^2_{\mathcal{H}} \\
&= \langle(\alpha + L^*L)^{\frac{1}{2}}f, (\alpha + L^*L)^{\frac{1}{2}}f\rangle_{H_K(E)} - 2\mathrm{Re}\langle(\alpha + L^*L)^{-\frac{1}{2}}L^*\mathbf{d}, (\alpha + L^*L)^{\frac{1}{2}}f\rangle_{H_K(E)} \\
&\quad + \|\mathbf{d}\|^2_{\mathcal{H}} \\
&= \left\|(\alpha + L^*L)^{\frac{1}{2}}f - (\alpha + L^*L)^{-\frac{1}{2}}L^*\mathbf{d}\right\|^2_{H_K(E)} + \|\mathbf{d}\|^2_{\mathcal{H}} - \|(\alpha + L^*L)^{-\frac{1}{2}}L^*\mathbf{d}\|^2_{\mathcal{H}}.
\end{aligned}$$

As a result, we see that (3.75) is the unique minimizer. If we proceed as in the calculation used for (3.73), we can readily obtain (3.76) and (3.77).

The next theorem can be used to treat error estimates.

Theorem 3.12. *Suppose that* $\alpha : (0, 1) \to (0, \infty)$ *is a function of* δ *such that*

$$\lim_{\delta \downarrow 0} \alpha(\delta) = 0 \tag{3.78}$$

and

$$\lim_{\delta \downarrow 0} \frac{\delta^2}{\alpha(\delta)} = 0. \tag{3.79}$$

Let $D : (0, 1) \to \mathscr{H}$ *be a function such that*

$$\|D(\delta) - \mathbf{d}\|_{\mathscr{H}} \leq \delta \tag{3.80}$$

for all $\delta \in (0, 1)$. *If* $\mathbf{d} \in \mathrm{Dom}(L^\dagger)$, *then in the topology of* $H_K(E)$ *we have*

$$\lim_{\delta \downarrow 0} f_{D(\delta),\alpha(\delta)} = f_{\mathbf{d}} = L^\dagger \mathbf{d}. \tag{3.81}$$

Proof. Since $f_{\mathbf{d}} = \lim_{\delta \downarrow 0} f_{\mathbf{d},\alpha(\delta)}$ by (3.78), we have only to show that

$$\lim_{\delta \downarrow 0} \left(f_{D(\delta),\alpha(\delta)} - f_{\mathbf{d},\alpha(\delta)} \right) = 0. \tag{3.82}$$

However, as we have seen in Proposition 3.7, the function $f_{D(\delta),\alpha(\delta)} - f_{\mathbf{d},\alpha(\delta)} = f_{D(\delta)-\mathbf{d},\alpha(\delta)}$ does not exceed $\dfrac{\delta}{2\sqrt{\alpha(\delta)}}$ in the $H_K(E)$-norm from (3.80). Thus we obtain (3.81) from (3.79).

We remark that [292, Theorem 1.3] and [293, Theorem 3] can be located as an example of Theorem 3.12.

The relation (3.84) below will be fundamental in practical applications.

We slightly generalize (2.56).

Theorem 3.13. *Let* $L : H_K(E) \to \mathscr{H}$ *be a bounded linear operator. Then define an inner product, for a positive* α

$$\langle f_1, f_2 \rangle_{H_{K_\lambda}(E)} = \alpha \, \langle f_1, f_2 \rangle_{H_K(E)} + \langle L f_1, L f_2 \rangle_{\mathscr{H}} \tag{3.83}$$

for $f_1, f_2 \in H_K(E)$. *Then* $(H_K(E), \langle \cdot, \cdot \rangle_{H_{K_\alpha}(E)})$ *is a reproducing kernel Hilbert space whose reproducing kernel is given by*

$$K_\alpha(p, q) = [(\alpha + L^* L)^{-1} K_q](p) \quad (p, q \in E). \tag{3.84}$$

Here K_α satisfies

$$K_\alpha(p,q) + \frac{1}{\alpha}\langle L[(K_\alpha)_q], L[K_p]\rangle_{\mathscr{H}} = \frac{1}{\alpha}K(p,q) \quad (p,q \in E), \tag{3.85}$$

which is corresponding to the Fredholm integral equation of the second kind for many concrete cases.

Proof. It is easy to check that $(K_\alpha)_q \in H_K(E)$. Furthermore,

$$\begin{aligned}\langle (K_\alpha)_q, (K_\alpha)_p\rangle_{H_{K_\lambda}(E)} &= \alpha\,\langle (K_\alpha)_q, (K_\alpha)_p\rangle_{H_K(E)} + \langle L^*L[(K_\alpha)_q], (K_\alpha)_p\rangle_{H_K(E)}\\ &= \langle (\alpha + L^*L)K_\alpha(\cdot,q), K_\alpha(\cdot,p)\rangle_{H_K(E)}\\ &= \langle K_q, [(\alpha+L^*L)^{-1}[K_p]]\rangle_{H_K(E)}\\ &= \langle [(\alpha+L^*L)^{-1}[K_q]], K_p\rangle_{H_K(E)}\\ &= \left[(\alpha+L^*L)^{-1}[K_q]\right](p) = K_\alpha(p,q).\end{aligned}$$

Thus the proof is complete.

By the Neumann expansion, we obtain

Corollary 3.1. *Assume in addition that $\|L\| < \sqrt{\alpha}$. Then*

$$K_\alpha(p,q) = \sum_{n=0}^{\infty}\left[\left(-\frac{L^*L}{\alpha}\right)^n \frac{K_q}{\alpha}\right](p) \quad (p,q \in E). \tag{3.86}$$

See [410] for a similar approach without using the functional calculus.

3.2.2 Tikhonov Regularization and Reproducing Kernels

Now we wish to represent the Tikhonov extremal function $f_{\mathbf{d},\alpha(\delta)}$ in terms of a reproducing kernel in order to calculate it practically. Furthermore, we will see a very powerful and practical deep application of the theory of reproducing kernels to Tikhonov regularization.

Theorem 3.14. *Under the same assumption as Theorem* 3.13,

$$f \in H_K(E) \mapsto \alpha\,\|f\|^2_{H_K(E)} + \|Lf - \mathbf{d}\|^2_{\mathscr{H}} \in \mathbb{R}$$

attains the minimum only at $f_{\mathbf{d},\alpha} \in H_K(E)$ which satisifes

$$f_{\mathbf{d},\alpha}(p) = \langle \mathbf{d}, L[(K_\alpha)_p]\rangle_{\mathscr{H}}. \tag{3.87}$$

Furthermore, $f_{\mathbf{d},\alpha}(p)$ *satisfies*

$$|f_{\mathbf{d},\alpha}(p)| \leq \|L\|_{H_K(E)\to\mathcal{H}} \sqrt{\frac{K(p,p)}{2\alpha}} \|\mathbf{d}\|_{\mathcal{H}}. \tag{3.88}$$

Proof. The formula (3.87) is a direct consequence of (3.84), using the relation $f_{\mathbf{d},\alpha} = (\alpha + L^*L)^{-1}L^*\mathbf{d}$ and the reproducing property of $H_K(E)$. By the Cauchy–Schwarz inequality we have

$$|f_{\mathbf{d},\alpha}(p)| = |\langle \mathbf{d}, L[(K_\alpha)_p]\rangle_{\mathcal{H}}| \leq \|\mathbf{d}\|_{\mathcal{H}} \; \|L(L^*L+\alpha)^{-1}[K_p]\|_{\mathcal{H}}. \tag{3.89}$$

Note that

$$\begin{aligned} \|L(L^*L+\alpha)^{-1}[K_p]\|_{\mathcal{H}}^2 &= \langle L(L^*L+\alpha)^{-1}[K_p], L(L^*L+\alpha)^{-1}[K_p]\rangle_{\mathcal{H}} \\ &= \langle (L^*L+\alpha)^{-1}[K_p], L^*L(L^*L+\alpha)^{-1}[K_p]\rangle_{H_K} \\ &= \langle K_p, (L^*L+\alpha)^{-1}L^*L(L^*L+\alpha)^{-1}[K_p]\rangle_{H_K}. \end{aligned}$$

Since $4\alpha L^*L \leq (L^*L+\alpha)^2$, we have

$$4\alpha(L^*L+\alpha)^{-1}L^*L(L^*L+\alpha)^{-1} \leq (L^*L+\alpha)^{-1}(L^*L+\alpha)^2(L^*L+\alpha)^{-1} = 1. \tag{3.90}$$

Thus it follows from (3.90) and the above formula that

$$\|L(L^*L+\alpha)^{-1}[K_p]\|_{\mathcal{H}}^2 \leq \frac{1}{4\alpha}\|K_p\|_{\mathcal{H}}^2 = \frac{1}{4\alpha}K(p,p). \tag{3.91}$$

If we insert (3.91) to (3.89), then we obtain (3.88).

Theorem 3.14 means that to obtain good approximate solutions, we must take a sufficiently small α, however, here we have restrictions for them, as we see, when $\mathbf{d}$ moves to $\mathbf{d}'$, by considering $f_{\mathbf{d},,} - f_{\mathbf{d}',,}$ in connection with the relation of the difference $\|\mathbf{d} - \mathbf{d}'\|_{\mathcal{H}}$. This fact is a very reasonable one, because we cannot obtain good solutions from the data containing errors. Here in advance, we wish to know how to take a small parameter α and what is the bound for such α. These problems are very delicate and important practically, and we have many methods.

The basic idea may be given as follows. We examine the corresponding extremal functions for various α tending to zero. By examining the corresponding extremal functions, when it converges to some function numerically and after then when the sequence diverges numerically, this will give the bound for α numerically [163, 164].

For this important problem and the method of L-curve, see [194, 276], for example.

Note that in (3.88), the factor 2 is missing in the result presented in [230]; that is, the estimate (3.88) is improved here [85]. As the example of $\mathscr{H} = H_K(E)$ and $L = I$ shows, we learn that the non trivial equality (3.88) is attained.

3.2.3 Representations of the Solutions of the Tikhonov Functional Equation

We present a natural representation of the Tikhonov extremal function in Theorem 3.14 by using the complete orthogonal systems in the Hilbert spaces $H_K(E)$ and $\mathscr{H}$. Following the natural logic of the proof, however, we have to assume that the operator L is a Hilbert Schmidt operator in the general situation, as we see from the following deduction. Recall that a bounded linear operator T from a Hilbert space to a Hilbert space is said to be a Hilbert Schmidt operator if the eigenvalue of T^*T is summable. Denote by $\mathscr{H}\mathscr{S}(H)$ the set of all Hilbert Schmidt operators from a Hilbert space H into itself. We have the following typical example; see [496] for example:

Example 3.3. Let $(X, \mathscr{M}, \mu)$ and $(Y, \mathscr{N}, \nu)$ be σ-finite measure spaces. Assume that $K : X \times Y \to \mathbb{C}$ is an $\mathscr{M} \otimes \mathscr{N}$-measurable function such that

$$M \equiv \int_{X\times Y} |K(x,y)|^2 \, d\mu(x) \, d\nu(y) < \infty. \tag{3.92}$$

Then the operator

$$Tf(y) \equiv \int_X K(x,y) f(x) \, d\mu(x) \tag{3.93}$$

belongs to $\mathscr{H}\mathscr{S}(H)$ and it satisfies $\|T\|_{\mathscr{H}\mathscr{S}(H)} = \sqrt{M}$.

Let $\{\Phi_\mu\}_{\mu=0}^\infty$ and $\{\Psi_\nu\}_{\nu=0}^\infty$ be any fixed complete orthonormal systems of the Hilbert spaces $\mathscr{H}$ and $H_K(E)$, respectively.

Therefore, from the form (3.85) and the representation (3.87) in Theorem 3.14 of the Tikhonov extremal function $f_{\mathbf{d},\lambda}$, we assume the representation

$$L\tilde{K}_\lambda(\cdot, q) = \sum_{\mu=0}^\infty \sum_{\nu=0}^\infty d_{\mu,\nu} \overline{\Phi_\nu(q)} \Phi_\mu, \tag{3.94}$$

in the sense of the tensor product $\mathscr{H} \otimes \overline{H_K(E)}$. This assumption is a basic condition in this section.

In the expansion of the reproducing kernel $K(p,q)$

$$K(p,q) = \sum_{n=0}^\infty \Psi_n(p) \overline{\Psi_n(q)}, \tag{3.95}$$

since $L[K_q]$ belongs to $\mathscr{H}$, we can set

$$L\Psi_n = \sum_{m=0}^{\infty} D_{n,m}\Phi_m \tag{3.96}$$

with the uniquely determined constants $\{D_{n,m}\}_{m=0}^{\infty} \in \ell^2(\mathbb{N}_0)$ satisfying, for any fixed n. Then we obtain

$$\langle L\tilde{K}_q, LK_p\rangle_{\mathscr{H}} = \sum_{\mu=0}^{\infty}\sum_{\nu=0}^{\infty}\sum_{n=0}^{\infty} d_{\mu,\nu}\overline{D_{n,\mu}}\Psi_n(p)\overline{\Psi_\nu(q)}. \tag{3.97}$$

Indeed, by repeating the Schwarz inequality we obtain the estimate

$$|\langle L\tilde{K}_q, LK_p\rangle_{\mathscr{H}}| \le \sqrt{K(p,p)K(q,q)\sum_{\mu=0}^{\infty}\sum_{\nu=0}^{\infty}|d_{\mu,\nu}|^2\sum_{n=0}^{\infty}\sum_{\mu=0}^{\infty}|D_{n,\mu}|^2}. \tag{3.98}$$

Note that the condition

$$\sum_{n=0}^{\infty}\sum_{\mu=0}^{\infty}|D_{n,\mu}|^2 < \infty \tag{3.99}$$

is equivalent to the operator L being a Hilbert Schmidt operator from $H_K(E)$ into $\mathscr{H}$. By applying L, from (3.87) we obtain

$$L_p[\langle L\tilde{K}_q, LK_p\rangle_{\mathscr{H}}] = \sum_{\mu=0}^{\infty}\sum_{\nu=0}^{\infty}\sum_{n=0}^{\infty} d_{\mu,\nu}\overline{D_{n,\mu}}\sum_{m=0}^{\infty} D_{n,m}\Phi_m \otimes \overline{\Psi_\nu(q)}. \tag{3.100}$$

Here, note that by repeating the Schwarz inequality

$$\sum_{m=0}^{\infty}\left(|\sum_{\mu=0}^{\infty}\sum_{\nu=0}^{\infty}\sum_{n=0}^{\infty} d_{\mu,\nu}\overline{D_{n,\mu}}D_{n,m}\overline{\Psi_\nu(q)}|\right)^2$$
$$\le \left(\sum_{\mu=0}^{\infty}\sum_{\nu=0}^{\infty}|d_{\mu,\nu}|^2\right)\left(\sum_{n=0}^{\infty}\sum_{\mu=0}^{\infty}|D_{n,\mu}|^2\right)^2 K(q,q).$$

Now, by comparing the coefficients of Φ_μ, we obtain for any μ,

$$\lambda\sum_{\nu=0}^{\infty} d_{\mu,\nu}\overline{\Psi_\nu(q)} + \sum_{m=0}^{\infty}\sum_{\nu=0}^{\infty}\sum_{n=0}^{\infty} d_{m,\nu}\overline{D_{n,m}}D_{n,\mu}\overline{\Psi_\nu(q)} = \sum_{n=0}^{\infty} D_{n,\mu}\overline{\Psi_n(q)}. \tag{3.101}$$

By comparing the coefficients $\overline{\Psi_\nu(q)}$, we obtain for any μ, ν,

$$\lambda d_{\mu,\nu} + \sum_{m=0}^{\infty}\sum_{n=0}^{\infty} d_{m,\nu}\overline{D_{n,m}}D_{n,\mu} = D_{\nu,\mu}. \tag{3.102}$$

By setting

$$A_{m,\mu} \equiv \sum_{n=0}^{\infty} \overline{D_{n,m}}D_{n,\mu}$$

we obtain the infinite equations

$$\lambda d_{\mu,\nu} + \sum_{m=0}^{\infty} d_{m,\nu}A_{m,\mu} = D_{\nu,\mu}. \tag{3.103}$$

We assumed their existence and uniqueness of the coefficients $\{d_{\mu,\nu}\}_{\mu,\nu\in\mathbb{Z}_+}$ satisfying

$$\sum_{\mu=0}^{\infty}\sum_{\nu=0}^{\infty} |d_{\mu,\nu}|^2 < \infty. \tag{3.104}$$

Here, note that in any finite case of $\{d_{\mu,\nu}\}_{\mu,\nu=0}^{N}$ we can determine the coefficients $\{d_{\mu,\nu}\}_{\mu,\nu=0}^{N}$ through these equations, because for any $N > 0$, the coefficient matrix, for the unit matrix I_N $\lambda I_N + \{A_{m,\mu}\}_{m,\mu}$ is Hermitian positive definite. However, here, we have to assume that all $\{d_{\mu,\nu}\}_{\mu,\nu=N+1}^{\infty}$ are zero; that is, we approximate the solutions by $\{d_{\mu,\nu}\}_{\mu,\nu=0}^{N}$.

We will see some interesting structure of the equations (3.103).

At first, we can consider the equations, for any fixed ν, by running μ. Then, for $\mu = 0$, we have

$$d_{0,\nu} = \frac{D_{\nu,0} - \sum_{m=1}^{\infty} d_{m,\nu}A_{m,0}}{\lambda + A_{0,0}}. \tag{3.105}$$

This means that $d_{0,\nu}$ is represented by the terms of $\{d_{m,\nu}\}_{m=1}^{\infty}$.

Next, we insert this $d_{0,\nu}$ for the equation for $\mu = 1$. Note that

$$A_{0,1}A_{1,0} = \sum_{n=0}^{\infty}\overline{D_{n,0}}D_{n,1}\sum_{n=0}^{\infty}\overline{D_{n,1}}D_{n,0} \le \sum_{n=0}^{\infty}|D_{n,0}|^2\sum_{n=0}^{\infty}|D_{n,1}|^2 = A_{0,0}A_{1,1}.$$

That is, in particular,

$$(\lambda + A_{0,0})(\lambda + A_{1,1}) - A_{0,1}A_{1,0} > 0, \neq 0.$$

Therefore, we can obtain

$$d_{1,\nu} = \frac{(\lambda + A_{0,0})(D_{\nu,1} - \sum_{m=2}^{\infty} d_{m,\nu}A_{m,1}) + A_{0,1}(-D_{\nu,o} + \sum_{m=2}^{\infty} d_{m,\nu}A_{m,0})}{(\lambda + A_{0,0})(\lambda + A_{1,1}) - A_{0,1}A_{1,0}}. \tag{3.106}$$

That is, $d_{1,\nu}$ is represented by the terms of $\{d_{m,\nu}\}_{m=2}^{\infty}$. Using these procedures, we will be able to represent $d_{\mu,\nu}$ by $\{d_{m,\nu}\}$ for any large m terms of $\{d_{\mu,\nu}\}$. However, these procedures will be very involved. Anyhow and setting

$$\mathbf{d} = \sum_{\mu=0}^{\infty} b_\mu \Phi_\mu, \tag{3.107}$$

we obtain the representation of the Tikhonov extremal function:

Theorem 3.15. *Assume that the operator L is a Hilbert Schmidt operator and system* (3.103) *has the solutions* $\{d_{\mu,\nu}\}_{\mu,\nu\in\mathbb{Z}_+}$ *satisfying*

$$\sum_{\mu=0}^{\infty}\sum_{\nu=0}^{\infty} |d_{\mu,\nu}|^2 < \infty. \tag{3.108}$$

Then, for any $\mathbf{d} = \sum_{\mu=0}^{\infty} b_\mu \Phi_\mu \in \mathscr{H}$, *the Tikhonov extremal function* $f_{\mathbf{d},\lambda}$ *in the sense of Theorem 3.14 is given by*

$$f_{\mathbf{d},\lambda}(p) = \sum_{\mu=0}^{\infty}\sum_{\nu=0}^{\infty} b_\mu \overline{d_{\nu,\mu}} \Psi_\nu(p) \tag{3.109}$$

and the fundamental estimates hold:

$$\begin{aligned} |f_{\mathbf{d},\lambda}(p)| &\le \left(\sum_{\mu=0}^{\infty} |b_\mu|^2\right)^{1/2} \left(\sum_{\mu=0}^{\infty}\left|\sum_{\nu=0}^{\infty} \overline{d_{\nu,\mu}} \Psi_\nu(p)\right|^2\right)^{1/2} \\ &\le \left(\sum_{\mu=0}^{\infty} |b_\mu|^2\right)^{1/2} \left(\sum_{\mu=0}^{\infty}\sum_{\nu=0}^{\infty} |d_{\nu,\mu}|^2\right)^{1/2} K(p,p)^{1/2} \end{aligned}$$

and

$$\|f_{\mathbf{d},\lambda}\|_{H_K(E)}^2 = \sum_{\nu=0}^{\infty}\left|\sum_{\mu=0}^{\infty} b_\mu \overline{d_{\nu,\mu}}\right|^2. \tag{3.110}$$

Here, note that $d_{\nu,\mu}$ *are depending on* λ.

In particular, if $H_K(E)$ is finite dimensional, the infinite equations (3.103) *are solved explicitely and so, then it is enough to check the convergence* (3.104).

Of course, if both spaces $H_K(E)$ and $\mathscr{H}$ are finite dimensional, then we obtain the complete representation for the Tikhonov extremal function $f_{\mathbf{d},\lambda}(p)$.

Meanwhile, when we apply the above complete orthonormal systems to the Moore Penrose generalized problem, we see that we can, in general, obtain no valuable information, because we cannot analyze the structure of the important reproducing kernel Hilbert space $H_k(E)$ in Theorem 3.4.

Many concrete Hilbert Schmidt operators may be considered from self-adjoint compact operators related to ordinary and partial differential equations; and in those cases, the transform constants $D_{m,n}$ are diagonal and the equations (3.103) may be solved easily. See [217] for many examples.

For concrete applications, see [416] for a finite-dimensional case and the matrices theory.

3.2.4 Applications of Tikhonov Regularization: Approximations by Sobolev Spaces

We will give many concrete applications of the above Tikhonov regularization theory. To this end we make broad use of the theory of reproducing kernels to various problems in the subsequent sections. Indeed, doing this is the one main new topics of the book, as we stated in the introduction. However, in this subsection, we state typical applications as prototypes. We use Sobolev Hilbert spaces.

Approximations Using the First Order Sobolev Hilbert Space

For the first order Sobolev Hilbert space $W^{2,1}(\mathbb{R})$, we consider the two bounded linear operators

$$L_1 : W^{2,1}(\mathbb{R}) \mapsto L_1 f \equiv f \in L^2(\mathbb{R})$$

and

$$L^2 : W^{2,1}(\mathbb{R}) \mapsto L^2 f \equiv f' \in L^2(\mathbb{R}).$$

Then, $K_{1,1}(\cdot,\cdot;\lambda)$ and $K_{1,2}(\cdot,\cdot;\lambda)$, the associated reproducing kernels for the RKHSs with norms

$$\|f\|_{H_{K_{1,1}(\cdot,\cdot;\lambda)}(\mathbb{R})} \equiv \sqrt{\lambda\|f\|^2_{W^{2,1}(\mathbb{R})} + \|f\|^2_{L^2(\mathbb{R})}} \tag{3.111}$$

and

$$\|f\|_{H_{K_{1,2}(\cdot,\cdot;\lambda)}(\mathbb{R})} \equiv \sqrt{\lambda\|f\|^2_{W^{2,1}(\mathbb{R})} + \|f'\|^2_{L^2(\mathbb{R})}}, \tag{3.112}$$

are given by

$$K_{1,1}(x,y;\lambda) = \frac{1}{2\pi}\int_{-\infty}^{\infty} \frac{\exp(i\xi(x-y))}{\lambda\xi^2 + (\lambda+1)}d\xi = \frac{1}{2\sqrt{\lambda(\lambda+1)}}\exp\left(-\sqrt{\frac{\lambda+1}{\lambda}}|x-y|\right)$$

and

$$K_{1,2}(x,y;\lambda) = \frac{1}{2\pi}\int_{-\infty}^{\infty} \frac{\exp(i\xi(x-y))}{(\lambda+1)\xi^2 + \lambda}\, d\xi = \frac{1}{2\sqrt{\lambda(\lambda+1)}}\exp\left(-\sqrt{\frac{\lambda}{\lambda+1}}|x-y|\right),$$

respectively; see Theorem 1.7. Hence, for any $g \in L^2(\mathbb{R})$, the best approximate functions $f^*_{1,1}(x;\lambda,g)$ and $f^*_{1,2}(x;\lambda,g)$ in the sense of

$$\inf_{f\in W^{2,1}(\mathbb{R})} \left\{\lambda\|f\|^2_{W^{2,1}(\mathbb{R})} + \|f-g\|^2_{L^2(\mathbb{R})}\right\}$$
$$= \lambda\|f^*_{1,1}(\cdot;\lambda,g)\|^2_{W^{2,1}(\mathbb{R})} + \|f^*_{1,1}(\cdot;\lambda,g) - g\|^2_{L^2(\mathbb{R})}$$

and

$$\inf_{f\in W^{2,1}(\mathbb{R})} \left\{\lambda\|f\|^2_{W^{2,1}(\mathbb{R})} + \|f'-g\|^2_{L^2(\mathbb{R})}\right\}$$
$$= \lambda\|f^*_{1,2}(\cdot;\lambda,g)\|^2_{W^{2,1}(\mathbb{R})} + \|f^{*\prime}_{1,2}(\cdot;\lambda,g) - g\|^2_{L^2(\mathbb{R})}$$

are given by

$$f^*_{1,1}(x;\lambda,g) = \frac{1}{2\sqrt{\lambda(\lambda+1)}}\int_{\mathbb{R}} g(\xi)\exp\left(-\sqrt{\frac{\lambda+1}{\lambda}}|\xi - x|\right)d\xi \tag{3.113}$$

and

$$f^*_{1,2}(x;\lambda,g) = \frac{1}{2\sqrt{\lambda(\lambda+1)}}\int_{\mathbb{R}} g(\xi)\frac{\partial}{\partial\xi}\exp\left(-\sqrt{\frac{\lambda}{\lambda+1}}|\xi - x|\right)d\xi \tag{3.114}$$

for $x \in \mathbb{R}$, respectively. Note that $f^*_{1,2}(\cdot;\lambda,g)$ can be considered as an approximate and generalized solution of the differential equation

$$y' = g(x) \quad \text{on} \quad \mathbb{R} \tag{3.115}$$

in the first order Sobolev Hilbert space $W^{2,1}(\mathbb{R})$.

Approximations Using the Second Order Sobolev Hilbert Space

For the second order Sobolev Hilbert space $W^{2,2}(\mathbb{R})$, whose norm is given by

$$\|f\|_{W^{2,2}(\mathbb{R})} \equiv \sqrt{\|f\|^2_{L^2(\mathbb{R})} + \|f'\|^2_{L^2(\mathbb{R})} + \|f''\|^2_{L^2(\mathbb{R})}},$$

we consider the three bounded linear operators into $L^2(\mathbb{R})$ defined by

$$L_1 : f \mapsto f,\ L^2 : f \mapsto f',\ L_3 : f \mapsto f''. \tag{3.116}$$

Then, the reproducing kernels $K_{2,1}(x, y; \lambda), K_{2,2}(x, y; \lambda)$ and $K_{2,3}(x, y; \lambda)$ for the Hilbert spaces with the norms

$$\sqrt{\lambda\|f\|^2_{W^{2,2}(\mathbb{R})} + \|f\|^2_{L^2(\mathbb{R})}},\ \sqrt{\lambda\|f\|^2_{W^{2,2}(\mathbb{R})} + \|f'\|^2_{L^2(\mathbb{R})}}$$

and

$$\sqrt{\lambda\|f\|^2_{W^{2,2}(\mathbb{R})} + \|f''\|^2_{L^2(\mathbb{R})}}$$

are given respectively by

$$K_{2,1}(x, y; \lambda) = \frac{1}{2\pi}\int_{\mathbb{R}} \frac{\exp(i\xi(x-y))}{\lambda\xi^4 + \lambda\xi^2 + (\lambda+1)}d\xi,$$

$$K_{2,2}(x, y; \lambda) = \frac{1}{2\pi}\int_{\mathbb{R}} \frac{\exp(i\xi(x-y))}{\lambda\xi^4 + (\lambda+1)\xi^2 + \lambda}d\xi,$$

$$K_{2,3}(x, y; \lambda) = \frac{1}{2\pi}\int_{\mathbb{R}} \frac{\exp(i\xi(x-y))}{(\lambda+1)\xi^4 + \lambda\xi^2 + \lambda}d\xi.$$

This can be verified in a similar way as in Theorem 1.12.

We now consider the following approximation problems:

$$\inf\{\lambda\|f\|_{W^{2,2}(\mathbb{R})} + \|f^{(i-1)} - g\|_{L^2(\mathbb{R})}\}$$

for $i = 1, 2, 3$, where $g \in L^2(\mathbb{R})$ and $f^{(0)} = f$. Then the corresponding best approximate functions $f^*_{2,i}(x; \lambda, g)$ are given by

$$f^*_{2,1}(x; \lambda, g) = \frac{1}{2\pi}\int_{\mathbb{R}} g(\xi)\left\{\int_{\mathbb{R}} \frac{\exp(-i\eta(\xi-x))}{\lambda\eta^4 + \lambda\eta^2 + (\lambda+1)}d\eta\right\} d\xi,$$

$$f^*_{2,2}(x; \lambda, g) = \frac{1}{2\pi}\int_{\mathbb{R}} g(\xi)\left\{\int_{\mathbb{R}} \frac{-i\eta\cdot\exp(-i\eta(\xi-x))}{\lambda\eta^4 + (\lambda+1)\eta^2 + \lambda}d\eta\right\} d\xi,$$

$$f_{2,3}^{*}(x;\lambda,g)=\frac{1}{2\pi}\int_{\mathbb{R}}g(\xi)\left\{\int_{\mathbb{R}}\frac{-\eta^{2}\cdot\exp(-i\eta(\xi-x))}{(\lambda+1)\eta^{4}+\lambda\eta^{2}+\lambda}d\eta\right\}\,d\xi,$$

for $i=1,2,3$ respectively.

As a special case, we can consider how elements in $W^{1,2}(\mathbb{R})$ and $W^{2,2}(\mathbb{R})$ approach the function $g=\chi_{[-1,1]}$ as $\lambda\downarrow 0$. We can also consider how elements of $W^{1,2}(\mathbb{R})$ and $W^{2,2}(\mathbb{R})$ approach solutions of the differential equation $y'=\chi_{[-1,1]}$ as $\lambda\downarrow 0$.

Best Approximation with Point Data

Let $H_K(E)$ be a Hilbert space admitting a reproducing kernel $K(p,q)$ on E. Let $\{\xi_j\}_{j=1}^n$ be a finite set of points of E. Then we can apply Theorem 3.4 as follows: For any $\lambda>0$, the approximation $f_\lambda^*(p;\{\xi_j\}_{j=1}^n,\{A_j\}_{j=1}^n)\in H_K(E)$ in the sense of

$$\inf_{f\in H_K(E)}\left\{\lambda\|f\|_{H_K(E)}^2+\sum_{j=1}^{n}|A_j-f(\xi_j)|^2\right\}$$

$$=\lambda\|f_\lambda^*(\cdot;\{\xi_j\}_{j=1}^n,\{A_j\}_{j=1}^n)\|_{H_K(E)}^2+\sum_{j=1}^{n}|A_j-f_\lambda^*(\xi_j;\{\xi_k\}_{k=1}^n,\{A_k\}_{k=1}^n)|^2$$

exists uniquely and it is given by

$$f_\lambda^*(p;\{\xi_j\}_{j=1}^n,\{A_j\}_{j=1}^n)=\sum_{j=1}^{n}A_jK_\lambda(p,\xi_j;\{\xi_k\}_{k=1}^n). \qquad (3.117)$$

We have followed [31] to write this section and we will be able to see their excellent computer graphics when we consider the practical meanings of the approximations.

3.3 General Fractional Functions

3.3.1 *What Is a Fractional Function?*

In Sect. 3.3, we will consider a general fractional function. What is a fractional function

$$\frac{g}{f} \qquad (3.118)$$

for some very general functions g and f on a set E? The point that counts is that f can vanish on some points of E.

In order to consider such fractional function (3.118), we will consider the related equation

$$f_1(p)f(p) = g(p) \quad \text{on} \quad E \tag{3.119}$$

for some functions f_1 and g on the set E. If the solution f_1 of (3.119) on the set E exists, then the solution f_1 will give the meaning of the fractional function (3.118). So the problem may be transformed to the very general and popular equation (3.119).

3.3.2 An Approach by Using RKHSs

In this starting point, the function f is initially given. So, for analyzing the equation (3.119), we must introduce a suitable function space containing the function f_1 and then we find the induced function space containing the product $f_1 \cdot f$. Then, we will consider the solution of the equation (3.119). Here, on this line, we will show that we can discuss the above problem in a very general setting. Indeed, this will be performed for an arbitrary function f on the set E that is non-identically zero on the set E using the theory of reproducing kernels.

At first, we observe that for an arbitrary function $f(p)$, there exist many reproducing kernel Hilbert spaces containing the function $f(p)$; the simplest reproducing kernel is given by $f \otimes \overline{f}$ on $E \times E$. In general, a reproducing kernel Hilbert space $H_K(E)$ on E admitting a reproducing kernel K on $E \times E$ is characterized by the property that any point evaluation $f(p)$ is a bounded linear operator on $H_K(E)$ for any point $p \in E$. Therefore, we will consider such a reproducing kernel Hilbert space $H_{K_1}(E)$ admitting a reproducing kernel K_1 containing the functions $f_1(p)$.

Then we note the very interesting fact that the product $f_1 \cdot f$ determines a reproducing kernel Hilbert space that is induced by the reproducing kernel Hilbert space $H_{K_1}(E)$ and by a second reproducing kernel Hilbert space, say $H_K(E)$, containing the function f. In fact, the space in question is a reproducing kernel Hilbert space $H_{K_1K}(E)$ that is determined by the product $K_1K = K_1 \cdot K$ and, furthermore, we obtain the inequality

$$\|f_1f\|_{H_{K_1 \cdot K}(E)} \leq \|f_1\|_{H_{K_1}(E)}\|f\|_{H_K(E)}. \tag{3.120}$$

See Corollary 2.4. This important inequality (3.120) means that for the linear operator $\varphi_f(f_1)$ on $H_{K_1}(E)$(for a fixed function f), defined by

$$\varphi_f(f_1)(p) \equiv f_1(p)f(p), \tag{3.121}$$

we obtain the inequality

$$\|\varphi_f(f_1)\|_{H_{K_1K}(E)} \leq \|f_1\|_{H_{K_1}(E)}\|f\|_{H_K(E)}. \tag{3.122}$$

This means that the mapping φ_f is a bounded linear operator from $H_{K_1}(E)$ into $H_{K_1K}(E)$.

Now we can consider the operator equation (3.119) in this framework. We will mathematically analyze this situation in a natural way and will develop a consequent theory, however, the operator problem will be sensitive with respect to the functions f.

For some reasonable solutions for the operator equation (3.119), we will be able to introduce *approximate fractional functions* and *generalized fractional functions* in correspondence with the usual fractional function.

Now we shall consider the equation (3.119). At first, we fix a reproducing kernel Hilbert space $H_K(E)$ containing the function f. Next we consider a reproducing kernel Hilbert space $H_{K_1}(E)$ containing the *solutions* f_1. Then, the products f_1f belong to the natural reproducing kernel Hilbert space $H_{K_1K}(E)$ admitting the reproducing kernel $K_1(p,q)K(p,q)$ and we obtain the following inequality:

$$\|f_1f\|_{H_{K_1K}(E)} \leq \|f_1\|_{H_{K_1}(E)}\|f\|_{H_K(E)}. \tag{3.123}$$

That is, for fixed $f \in H_K(E)$, the linear operator $\varphi_f : H_{K_1}(E) \to H_{K_1K}(E)$, given by

$$Lf_1(p) \equiv \varphi_f(f_1)(p) = f_1(p)f(p), \tag{3.124}$$

is bounded on $H_{K_1}(E)$. So, we can consider the Tikhonov functional for any $g \in H_{K_1K}(E)$:

$$\inf_{f_1 \in H_{K_1}(E)} \left\{\lambda\|f_1\|^2_{H_{K_1}(E)} + \|g - \varphi(f_1)\|^2_{H_{K_1K}(E)}\right\}.$$

The extremal function $f_{1,\lambda}$ exists uniquely and we have, if (3.119) has the Moore Penrose generalized inverse $f_1(p)$,

$$\lim_{\lambda \downarrow 0} f_{1,\lambda}(p) = f_1(p) \quad \text{on} \quad E$$

uniformly where $K_1(p,p)$ is bounded. Furthermore, its convergence is also in the sense of the norm of $H_{K_1}(E)$. Sometimes, we can take $\lambda = 0$ and in this case we can represent the Moore Penrose generalized solution in some direct form.

Therefore, at first, we can introduce the *approximate fractional function* by the extremal function $f_{1,\lambda}$ above whose existence is always ensured in the above situation. In case there exists the Moore Penrose generalized inverse, we will call it a *generalized fractional function*. We can examine the above properties using the theory of reproducing kernels by the given propositions in sections of best approximation problems, spectral analysis and the Tikhonov regularization, the Tikhonov regularization and reproducing kernels, and the Moore Penrose inverses using reproducing kernels. The theory will give general numerical solutions, however, for some concrete analytically results, we will need the realization of the reproducing kernel Hilbert spaces that appeared in the above theory. In order to see the corresponding details, we can consider all this in a Fourier analysis context.

It is worth mentioning that when we want to realize the reproducing kernel Hilbert space $H_{K_1 K}(E)$ and when we can look at a reproducing kernel Hilbert space $H_{\mathbf{K}}(E)$ admitting a reproducing kernel $\mathbf{K}$ satisfying

$$K_1(p,q)K(p,q) \ll \mathbf{K}(p,q) \quad (p,q) \in E^2$$

(that is, $\mathbf{K}(p,q) - K_1(p,q)K(p,q)$ is a positive definite quadratic form function on $E \times E$) and its structure is simple, from the properties $H_{K_1 K}(E) \subset H_{\mathbf{K}}(E)$ and $\|g\|_{H_{K_1 K}(E)} \geq \|g\|_{H_{\mathbf{K}}(E)}, \quad g \in H_{K_1 K}(E)$, we see that in the above theory we can use the space $H_{\mathbf{K}}(E)$ instead of the space $H_{K_1 K}(E)$.

The materials were taken from [88]; examine the reference for details.

See [50] for more about applications of the Tikhonov regularization. We followed [72, 73] here.

Chapter 4
Real Inversion Formulas of the Laplace Transform

As stated in the preface, one of our strong motivations for writing this book is given by the historical success of the numerical and real inversion formulas of the Laplace transform which is a famous typical ill-posed and very difficult problem. In this chapter, we will see their mathematical theory and formulas, as a clear evidence of the definite power of the theory of reproducing kernels when combined with the Tikhonov regularization.

4.1 Real Inversion Formulas of the Laplace Transform

4.1.1 Problem and Orientation

We will consider the real inversion formulas of the Laplace transform

$$\mathscr{L}f(p) = F(p) = \int_0^\infty e^{-pt} f(t)\, dt, \quad p > 0 \tag{4.1}$$

on certain function spaces that are applicable in numerical analysis. This integral transform is fundamental in mathematical science and engineering. The inversion of the Laplace transform is, in general, given by a complex form; however, we are still interested in its real inversion, which is the problem to find the original function $f(t)$ from a given image function $F(p), p \geq 0$; this is required in various practical problems. The real inversion is unstable in usual settings, thus the real inversion is ill-posed in the sense of Hadamard, and numerical real inversion methods have not been established [108, 241]. In other words, the image functions of the Laplace transform are analytic on a half complex plane, and the real inversion will be complicated. One is led to think that its real inversion is essentially involved, because we need to grasp *analyticity* from the real and discrete data.

S. Saitoh, Y. Sawano, *Theory of Reproducing Kernels and Applications*,
Developments in Mathematics 44, DOI 10.1007/978-981-10-0530-5_4

In Sect. 4.1, we give our new approach to the numerical real inversion of the Laplace transform based on the compactness of the Laplace transform on some good reproducing kernel Hilbert spaces. We use Tikhonov regularization in combination with the theory of reproducing kernels above; however, in order to obtain good numerical results, we will also need some powerful numerical algorithm and computer system, in which we refer to the details later. We will give here mathematical background for the practical applications.

4.1.2 Known Real Inversion Formulas of the Laplace Transform

In order to know the situation on the real inversion formula of the Laplace transform, we first recall well-known inversion formulas.

The most popular formulas are

$$\lim_{n\to\infty}\frac{(-1)^n}{n!}\left(\frac{n}{t}\right)^{n+1} f^{(n)}\left(\frac{n}{t}\right) = F(t) \tag{4.2}$$

(see Post [361, Section 16] and Widder [482, p. 61 Corollary]), and

$$\lim_{n\to\infty}\prod_{k=1}^{n}\left(1+\frac{t}{k}\frac{d}{dt}\right)\left[\frac{n}{t}f\left(\frac{n}{t}\right)\right] = F(t), \tag{4.3}$$

(see Widder [482]).

The analytical real inversion formula for the Laplace transform after the long steps in the spirit and logic in Chap. 2 was obtained in [388]:

$$f(p) = \int_0^\infty e^{-pt}F(t)dt \tag{4.4}$$

for $p > 0$, where $F : (0,\infty) \to \mathbb{C}$ is a measurable function satisfying

$$\int_0^\infty |F(t)|^2 dt < \infty. \tag{4.5}$$

For the polynomial of degree $2N + 2$,

$$P_N(\xi) = \sum_{0\le\nu\le n\le N}\frac{(-1)^{\nu+1}(2n)!\xi^{n+\nu}}{(n+1)!(n-\nu)!(n+\nu)!\nu!}$$
$$\times\left\{\frac{(2n+1)\xi^2}{n+\nu+1} - \left(\frac{2n+1}{n+\nu+1}+3n+1\right)\xi + n(n+\nu+1)\right\},$$

and we set

$$F_N(t) \equiv \int_0^\infty f(p)e^{-pt}P_N(pt)dp \quad (t > 0). \tag{4.6}$$

Then we have

$$\lim_{N\to\infty} \int_0^\infty |F(t) - F_N(t)|^2 dt = 0. \tag{4.7}$$

Furthermore, the estimate of the truncation error of F_N was also given in [19, Theorem 1] and [401].

We also recall the very complicated formula [365, p. 221]. For the Laplace transform

$$\int_0^b e^{-pt}F(t)dt = f(p), \tag{4.8}$$

we have

$$F(t) = \frac{2tb^{-1}}{\pi}\frac{d}{du}\int_0^u \frac{G(v)}{(u-v)^{\frac{1}{2}}}dv\bigg|_{u=t^2b^{-2}};$$

$$G(v) = \frac{2}{\pi\sqrt{v}}\int_{(0,\infty)^3} \cos(y\cosh^{-1}v^{-1})\cosh\pi y \frac{\cos(zy)}{\sqrt{\cosh z}} f(p)J_0\left(p\frac{b}{\sqrt{\cosh z}}\right)dydzdp.$$

Here J_0 denotes the Bessel function of order 0 :

$$J_0(t) = \sum_{m=0}^\infty \frac{(-1)^m}{m!\Gamma(m+1)}\left(\frac{t}{2}\right)^{2m} \quad (t \in \mathbb{R}).$$

Unfortunately, in this complicated formula (4.8), the characteristic properties of both the functions F and f which make the inversion formula hold are not given.

Peng and Chung [357, Theorem 2.5] gave the following formula:

$$f(t) = \lim_{\sigma\to\infty}\bar{f}_\sigma(t), \quad \bar{f}_\sigma(t) = \sum_{j=1}^\infty \frac{(-1)^{j-1}}{(j-1)!}e^{j\sigma}\frac{\sigma}{t}F\left(\frac{j\sigma}{t}\right), \tag{4.9}$$

while Tuan and Duc [467] gave

$$f(t) = \lim_{n\to\infty} n\sum_{j=1}^\infty \frac{(-1)^{j-1}}{(j-1)!}e^{njt}F(nj). \tag{4.10}$$

Because of the limits in (4.9) and (4.10), (4.9) and (4.10) are not likely to be useful for practical computation.

Furthermore, see the related articles [71, 227, 388, 398, 400]. See also the great references [221, 222]. The problem may be related to analytic extension problems, see [268, Section 4] and [269].

4.1.3 Compactness of the Modified Laplace Transform

In this subsection we give many function spaces on which the modified Laplace transform is compact; then we can apply the theory of Sect. 4.1.3 to singular value decomposition methods for the real inversion formula of the Laplace transform.

For this purpose, we consider the following weighted norm inequalities: For some positive continuous functions u and w on $(0, \infty)$,

$$\int_0^\infty |Lf(p)|^2 u(p)\, dp \lesssim \int_0^\infty |f'(t)|^2 w(t)\, dt, \tag{4.11}$$

where L (the modified Laplace tranform) is given by

$$Lf(p) \equiv p\mathscr{L}f(p) = p\int_0^\infty e^{-tp} f(t)\, dt \tag{4.12}$$

and f is an absolutely continuous function with $f(0) = 0$. These inequalities imply the boundedness of the linear operators from weighted Sobolev spaces into weighted L^2 spaces on the positive real line by the modified Laplace transform. Furthermore, we will actually show their compactness for these bounded linear operators.

Example 4.1. We assume that $F \in C^1[0, \infty)$ satisfies the properties (P): Let $0 < \alpha, \beta < k - \dfrac{1}{2}$ and

$$F'(t) = o(e^{\alpha t}),\ F(t) = o(e^{\beta t}) \quad (t \to \infty). \tag{4.13}$$

Then the function

$$G(t) = \{F(t) - F(0) - tF'(0)\}e^{-kt} \tag{4.14}$$

belongs to $H_K(\mathbb{R}^+)$. Then

$$\mathscr{L}G(p) = f(p+k) - \frac{F(0)}{p+k} - \frac{F'(0)}{(p+k)^2}. \tag{4.15}$$

We remark that the origin of (P) is in [300, p. 10].

Throughout Sect. 4.1.3 we use the following notation : AC$[0, \infty)$ denotes the set of all real-valued absolutely continuous functions on $[0, \infty)$. $\mathbb{R}^+$ denotes $(0, \infty)$;

see (1.32). By the term "weight" we mean a continuous function $\mathbb{R}^+ \to \mathbb{R}^+$. Given a weight w, we define

$$L^2(w) \equiv \{f : \mathbb{R}^+ \to \mathbb{R} : \|f\|_{L^2(u)} < \infty\},$$
$$H_K(w) \equiv \{f \in \mathrm{AC}[0,\infty) : f(0) = 0,\ \|f\|_{H_K(w)} < \infty\},$$

where the norms are given by

$$\|f\|_{L^2(u)} \equiv \left(\int_0^\infty |f(p)|^2\, w(p)\, dp\right)^{\frac{1}{2}}, \quad \|f\|_{H_K(w)} \equiv \left(\int_0^\infty |f'(t)|^2\, w(t)\, dt\right)^{\frac{1}{2}}$$

respectively, so that our problem is rephrased as

$$\|Lf\|_{L^2(u)} \lesssim \|f\|_{H_K(w)} \tag{4.16}$$

and $H_K(w)$-$L^2(u)$ compactness of L.

We establish the following main theorem:

Theorem 4.1. *Assume that for a given pair w and u of weights,*

$$M \equiv \iint_{\mathbb{R}^+ \times \mathbb{R}^+} \exp(-2tp) w(t)^{-1} u(p)\, dt\, dp < \infty. \tag{4.17}$$

Then L is a compact operator from $H_K(w)$ to $L^2(u)$ with norm less than or equal to $\sqrt{M}$.

Corollary 4.1. *Suppose u and w^{-1} belong to $L^1(\mathbb{R}^+)$, namely,*

$$\int_{\mathbb{R}^+} u(p)\, dp < \infty \tag{4.18}$$

and

$$\int_{\mathbb{R}^+} w(t)^{-1}\, dt < \infty. \tag{4.19}$$

Then L is a compact operator from $H_K(w)$ to $L^2(u)$.

We prove Theorem 4.1 via several steps.

Let us begin with quantitative information about $f \in H_K(w)$, which will yield that $H_K(w)$ is complete.

Proposition 4.1. *Let $f \in H_K(w)$. Then*

$$|f(t)| \leq \left(\int_0^t w(s)^{-1}\, ds\right)^{\frac{1}{2}} \|f\|_{H_K(w)} \tag{4.20}$$

for all $t > 0$.

Proposition 4.1 can be used when we show that $H_K(w)$ is complete under assumption (4.17). Indeed, if we assume (4.17), by Fubini's theorem we have

$$\int_0^\infty \exp(-2tp)w(t)^{-1}\,dt < \infty \tag{4.21}$$

for dp-almost every $t > 0$. However, by monotonicity with respect to t, (4.21) is the case for all $t > 0$. Therefore

$$\int_0^t w(s)^{-1}\,ds < e^{2tp}\int_0^\infty e^{-2sp}w(s)^{-1}\,ds < \infty, \tag{4.22}$$

showing the local integrable property of w^{-1}. With (4.22), it is easy to see that $H_K(w)$ is complete.

Proof (of Proposition 4.1). Since f is absolutely continuous with $f(0) = 0$, we have

$$f(t) = \int_0^t f'(s)\,ds \quad (t \geq 0). \tag{4.23}$$

Using the Hölder inequality, we obtain

$$|f(t)| \leq \left(\int_0^t w(s)^{-1}\,ds\right)^{\frac{1}{2}}\left(\int_0^t |f'(s)|^2\,w(s)\,ds\right)^{\frac{1}{2}} \leq \left(\int_0^t w(s)^{-1}\,ds\right)^{\frac{1}{2}}\|f\|_{H_K(w)},$$

which proves the assertion.

Next, let us obtain an approximation procedure. Pick a smooth function ζ that equals 1 near a small neighborhood of origin and vanishes outside $[2, \infty)$. Define

$$I_R f(t) \equiv \int_0^t \zeta\left(\frac{s}{R}\right)f'(s)\,ds. \tag{4.24}$$

It is not so hard to see the following.

Proposition 4.2. *The family of operators* $\{I_R\}_{R>0}$ *given by* (4.24) *converges strongly to the identity operator in the topology of* $H_K(w)$ *as* $R \to \infty$. *Namely,* $\lim_{R\to\infty} I_R f = f$ *for all* $f \in H_K(w)$.

From Proposition 4.2 it follows that the set of all functions in $H_K(w)$ whose derivatives are compactly supported in $[0, \infty)$ is dense.

Corollary 4.2. *Let* $p > 0$. *Assume* (4.17). *Then* $e^{-tp}f(t) \in L^1(\mathbb{R}_t^+)$ *and* $Lf(p) = \mathscr{L}[f'](p)$ *for all* $p > 0$.

Proof. Let $p > 0$ and $f \in H_K(w)$ be fixed. Assume for the time being that there exists T such that the derivative $f'(t)$ vanishes for all $t > T$, that is, f is constant outside a bounded set. An integration by parts gives us that

$$
\begin{aligned}
Lf(p) &= p\int_0^\infty e^{-tp}f(t)\,dt \\
&= \lim_{R\to\infty} p\int_0^R e^{-tp}f(t)\,dt \\
&= \lim_{R\to\infty} \int_0^R (-e^{-tp})'f(t)\,dt \\
&= \lim_{R\to\infty} \left(\left[-e^{-tp}f(t)\right]_{t=0}^{t=R} + \int_0^R e^{-tp}f'(t)\,dt \right).
\end{aligned}
$$

Assuming (4.17) and hence (4.21), we conclude with the help of the Lebesgue convergence theorem that the limit of the far right-hand side converges to $\mathscr{L}[f'](p)$. Meanwhile to verify that the first term vanishes, we use the assumption that f is constant outside a bounded set. The corollary is therefore proved under the additional assumption that f is constant outside a bounded set.

Now let us pass to the general case. Let $p > 0$. By decomposition $f' = (f')_+ - (f')_-$, we may assume that f is increasing and hence positive. If we apply the truncation procedure above, then we have $I_R f \uparrow f$ as $R \to \infty$. Therefore, we are in the position of using the monotone convergence theorem to see that

$$
Lf(p) = \lim_{R\to\infty} L[I_R f](p). \tag{4.25}
$$

As we have shown above, we have

$$
L[I_R f](p) = \mathscr{L}[(I_R f)'](p) = \int_0^\infty \zeta\left(\frac{t}{R}\right) f'(t)e^{-tp}\,dt. \tag{4.26}
$$

By the Hölder inequality, we have

$$
\left|\int_0^\infty f'(t)e^{-tp}\,dt\right|^2 \leq \int_0^\infty |f'(t)|^2 w(t)\,dt \cdot \int_0^\infty e^{-2pt}w(t)^{-1}\,dt < \infty \tag{4.27}
$$

for all $p > 0$ by (4.21), which implies $t \in \mathbb{R}^+ \to f'(t)e^{-tp} \in \mathbb{R}$ is integrable. As a consequence, by the Lebesgue convergence theorem we obtain

$$
\lim_{R\to\infty} \int_0^\infty \zeta\left(\frac{t}{R}\right) f'(t)e^{-tp}\,dt = \int_0^\infty f'(t)e^{-tp}\,dt = \mathscr{L}[f'](p). \tag{4.28}
$$

Putting our observations together, we obtain that $e^{-tp}f(t) \in L^1(\mathbb{R}_t^+)$ for all $p > 0$ and $Lf(p) = \mathscr{L}[f'](p)$. This is the desired result.

Proposition 4.3. *Under the assumption* (4.17), *L is bounded from $H_K(w)$ to $L^2(u)$ with operator norm less than or equal to $\sqrt{M}$ in* (4.17).

Proof. By Corollary 4.2, we have

$$Lf(p) = \mathscr{L}[f'](p) = \int_0^\infty e^{-tp} f'(t)\, dt \tag{4.29}$$

for every $f \in H_K(w)$. Therefore, by the Hölder inequality, we obtain

$$\begin{aligned} |Lf(p)|^2 &\le \left(\int_0^\infty \exp(-2tp) w(t)^{-1}\, dt \right) \left(\int_0^\infty |f'(t)|^2 w(t)\, dt \right) \\ &= \left(\int_0^\infty \exp(-2tp) w(t)^{-1}\, dt \right) \cdot \|f\|^2_{H_K(w)}. \end{aligned} \tag{4.30}$$

Inserting (4.30) into the quantity $\|Lf\|_{L^2(u)}$, we are led to

$$\begin{aligned} \int_0^\infty |Lf(p)|^2 u(p)\, dp &\le \left(\iint_{\mathbb{R}^+ \times \mathbb{R}^+} \exp(-2tp) w(t)^{-1} u(p)\, dt\, dp \right) \cdot \|f\|^2_{H_K(w)} \\ &= M \|f\|^2_{H_K(w)}, \end{aligned}$$

proving the assertion. Here we used the Fubini theorem.

Now let us refer back to the proof of Theorem 4.1.

Proof. It is a trivial that an operator L_R, given by $L_R f(p) \equiv p \int_{R^{-1}}^{R} e^{-tp} f(t)\, dt,\ R > 0$ is compact. Therefore, we have only to see that L_R converges to L as $R \to \infty$ in the operator topology.

An integration by parts gives us that

$$L_R f(p) = e^{-R^{-1}p} f(R^{-1}) - e^{-Rp} f(R) + \int_{R^{-1}}^{R} e^{-tp} f'(t)\, dt. \tag{4.31}$$

As we have seen,

$$|f(R^{-1})| \le \left(\int_0^{R^{-1}} w(s)^{-1}\, ds \right)^{\frac{1}{2}} \|f\|_{H_K(w)}. \tag{4.32}$$

Therefore,

$$\begin{aligned} &\int_0^\infty \exp(-2R^{-1}p) |f(R^{-1})|^2 u(p)\, dp \\ &\le \left(\int_0^\infty \int_0^{R^{-1}} \exp(-2R^{-1}p) w(s)^{-1} u(p)\, ds\, dp \right) \cdot \|f\|^2_{H_K(w)} \\ &\le \left(\iint_{\mathbb{R}^+ \times \mathbb{R}^+} \chi_{[0,R^{-1}]}(s) \exp(-2R^{-1}p) w(s)^{-1} u(p)\, ds\, dp \right) \cdot \|f\|^2_{H_K(w)}. \end{aligned}$$

Since we are assuming $M < \infty$ and

$$\chi_{[0,R^{-1}]}(s)\exp(-2R^{-1}p)w(s)^{-1}u(p) \le e^{-2sp}w(s)^{-1}u(p), \tag{4.33}$$

we are in the position to apply the Lebesgue convergence theorem to see

$$\lim_{R\to\infty}\iint_{\mathbb{R}^+\times\mathbb{R}^+}\chi_{[0,R^{-1}]}(s)\exp(-2R^{-1}p)w(s)^{-1}u(p)\,ds\,dp = 0. \tag{4.34}$$

Therefore, the treatment of the first term in (4.31) is valid. As for the second term, going through a similar argument using Proposition 4.1, we obtain

$$\begin{aligned}
&\int_0^\infty \exp(-2Rp)|f(R)|^2u(p)\,dp\\
&\le \left(\int_0^\infty\int_0^R \exp(-2Rp)w(s)^{-1}u(p)\,ds\,dp\right)\cdot\|f\|^2_{H_K(w)}\\
&\le \left(\iint_{\mathbb{R}^+\times\mathbb{R}^+}\chi_{[0,R]}(s)\exp(-2Rp)w(s)^{-1}u(p)\,ds\,dp\right)\cdot\|f\|^2_{H_K(w)}.
\end{aligned}$$

The integrand is again majorized by a function $e^{-2sp}w(s)^{-1}u(p)$ which is integrable over $\mathbb{R}^+\times\mathbb{R}^+$ and hence, by the Lebesgue convergence theorem, we have

$$\lim_{R\to\infty}\iint_{\mathbb{R}^+\times\mathbb{R}^+}\chi_{[0,R]}(s)\exp(-2Rp)w(s)^{-1}u(p)\,ds\,dp = 0. \tag{4.35}$$

Looking back on the proof of Proposition 4.2 and using the Lebesgue convergence theorem, we see that the operator norm

$$f \mapsto \int_0^\infty e^{-tp}f'(t)\,dt - \int_{R^{-1}}^R e^{-tp}f'(t)\,dt = \int_{\mathbb{R}^+\setminus[R^{-1},R]} e^{-tp}f'(t)\,dt \tag{4.36}$$

tends to 0 as $R\to\infty$. Combining these three observations together, we obtain the desired result.

4.1.4 Examples of the Weights

Let us show several typical examples. These examples give the corresponding real inversion formulas of the Laplace transform. In particular, we need to put a serious condition on the weight u. With the condition, u assures the corresponding real inversion formula of the Laplace transform, because it gives the condition of the images of the Laplace transform. Meanwhile the weight w will not cause any serious problem, since we try to approximate the original function f by the weighted Sobolev space. Another reason is that the space is sufficiently large as the approximating function space.

Example 4.2. If $u(p) \equiv 1$ and $w(t) = \frac{e^t}{t}$, then (4.17) is satisfied with $M = \frac{1}{2}$. However, u and w fail (4.18) and (4.19) in Corollary 4.1. The reproducing kernel of $H_K(w)$ is given by

$$K(s,t) = \int_0^{s\wedge t} \xi\, e^{-\xi}\, d\xi = 1 - \left(s \wedge t + 1\right) e^{-s\wedge t} \quad (s,t > 0). \tag{4.37}$$

This is an example that we took up in [160]. In that framework we can treat only an image function F of the Laplace transform which satisfies $pF(p)$ is square integrable on $\mathbb{R}^+$. That means we cannot treat the images of any polynomial in this framework because $p\mathscr{L}[t^n](p) = n!/p^n$ is not square integrable on $\mathbb{R}^+$.

The next example is interesting in that $H_K(w)$ contains any polynomial without constant terms.

Example 4.3. We define two weights u and w by

$$u(p) = \exp\left(-p^2 - \frac{1}{p^2}\right),\ w(t) = \exp(-\sqrt{t}) \quad (p,t > 0). \tag{4.38}$$

Then this couple satisfies (4.17) but fails (4.19) in Corollary 4.1. In this case the reproducing kernel of $H_K(w)$ is given by

$$K(s,t) = -2\left(\sqrt{s\wedge t} + 1\right)\exp\left(-\sqrt{s\wedge t}\right) + 2, \quad (s,t > 0). \tag{4.39}$$

Proof. By an elementary arithmetical mean, we have $2a + b \geq 3\sqrt[3]{a^2 b}$ for all $a, b > 0$. With this estimate, we obtain

$$\begin{aligned} M &= \iint_{\mathbb{R}^+\times\mathbb{R}^+} \exp\left(-2tp + \sqrt{t} - p^2 - \frac{1}{p^2}\right) dp\, dt \\ &\leq \iint_{\mathbb{R}^+\times\mathbb{R}^+} \exp\left(-3t^{\frac{2}{3}} + \sqrt{t} - p^2\right) dp\, dt < \infty. \end{aligned}$$

Therefore (4.17) is proved. The reproducing kernel is obtained by the formula

$$K(s,t) = \int_0^{s\wedge t} \frac{d\xi}{e\sqrt{\xi}} = 2\int_0^{\sqrt{s\wedge t}} \frac{\xi}{e^\xi}\, d\xi = -2\left(\sqrt{s\wedge t} + 1\right)\exp\left(-\sqrt{s\wedge t}\right) + 2$$

for $s,t > 0$. Here we used the change of variables for the second equality.

Example 4.4. We write $s \wedge t = \min(s,t)$ for $s,t \in \mathbb{R}$.

1. We can even arrange that $w \equiv 1$ by adjusting u. We can verify with ease that $u(p) = p\,e^{-p}$ and $w(t) \equiv 1$ do satisfy (4.17). However, w fails (4.19) in Corollary 4.1. The reproducing kernel of $H_K(w)$ is given by

$$K(s,t) = \int_0^{s\wedge t} d\xi = s \wedge t. \tag{4.40}$$

In this simple example, we still have $e^{at} \in H_K(w)$ with $a < 0$ as well as $\cos \omega t \in H_K(w)$ with $\omega \in \mathbb{R}$.
2. The following is just a generalization of Example 4.4: Set $u(p) = p^{k+2}\,e^{-p}$ and $w(t) = (t+1)^{-k}$ for a fixed $k \in \mathbb{N}$. Then this couple again satisfies (4.17) but w fails (4.19) in Corollary 4.1. The reproducing kernel of $H_K(w)$ is given by

$$K(s,t) = \int_0^{s\wedge t} (\xi+1)^k\,d\xi = \frac{\left(1+s\wedge t\right)^{k+1} - 1}{k+1}, \qquad (s,t>0). \tag{4.41}$$

Note that for the Laplace transforms with all polynomials of order $k-2$, we can apply this example.
3. In connection with the Fourier transform, it would be more convenient to deal with $u(p) = p^{k+2}\,e^{-p^2}$ and $w(t) = (t+1)^{-k}$ for $k \in \mathbb{N}$. Then this couple again satisfies (4.17) but fails (4.19) in Corollary 4.1. The reproducing kernel of $H_K(w)$ is given by (4.41).
4. The functions $u(p) = p^{-\frac{1}{2}}$ and $w(t) = t^{-2}e^t$ do satisfy (4.17) as well as the integrable assumption in Corollary 4.1. But it fails (4.18). The reproducing kernel is given by

$$K(s,t) = \int_0^{s\wedge t} \xi^2\,e^{-\xi}\,d\xi = -\Big\{(s\wedge t)^2 + 2s\wedge t + 2\Big\}e^{-s\wedge t} + 2.$$

5. Let $k \in \mathbb{N}$. For the couple $u(p) = \dfrac{p^k}{p^{2k+3}+1}$ and $w(t) = (t+1)^{-k}$, we have (4.18) and (4.17) but fails (4.19) in Corollary 4.1. Here the reproducing kernel is given by (4.41).

Example 4.5.

1. The functions $u(p) = \exp(\sqrt{p})$ and $w(t) = \exp(t^2 + t^{-2})$ do satisfy (4.17). However, they fail (4.18) in Corollary 4.1.
2. The functions $u(p) = \sinh(\sqrt{p})$ and $w(t) = \exp(t^2 + t^{-2})$ do satisfy (4.17). However, they fail (4.18) in Corollary 4.1.

Example 4.6.

1. Let $u(p) = \dfrac{p}{p^2+1}$ and $w(t) = \log(2+t)$. Then this couple satisfies all requirements (4.17) but fails (4.18) and (4.19) in Corollary 4.1.

2. Let $u(p) = \dfrac{p}{p^2+1}$ and $w(t) = \log(2+\log(2+t))$. Then this couple satisfies all requirements (4.17) but fails (4.18) and (4.19) in Corollary 4.1.
3. For the couple $u(p) = \dfrac{\sqrt{p}}{p^2+1}$ and $w(t) = \log(2+\log(2+t))$, we have (4.18) and (4.17) but fails (4.19) in Corollary 4.1.

4.1.5 Real Inversion of the Laplace Transform Through Singular Values Decomposition

If Theorem 4.1 is valid, then we see that L admits the singular value decomposition. That is, by considering the spectral decomposition of a compact self-adjoint operator L^*L, we can find a real sequence $\{\lambda_n\}_{n=1}^{\infty}$ which shrinks monotonically to 0 and sequences of vectors $\{V_n\}_{n=1}^{\infty} \subset H_K(w)$, $\{U_n\}_{n=1}^{\infty} \subset L^2(u)$ such that

$$LV_n = \lambda_n U_n,\ L^* U_n = \lambda_n V_n$$

for all $n \in \mathbb{N}$. Since the Laplace transform is invertible, so is L. Therefore, $\{V_n\}_{n=1}^{\infty}$ and $\{U_n\}_{n=1}^{\infty}$ form complete orthonormal systems in $H_K(w)$ and $L(H_K(w))$ respectively.

Let F be a function such that $pF(p) \in L(H_K(w))$. Then, keeping in mind that

$$L\mathscr{L}^{-1}F(p) = pF(p),$$

we have

$$\begin{aligned} L\mathscr{L}^{-1}F(p) &= \sum_{n=1}^{\infty} \left(\int_0^{\infty} qF(q)U_n(q)u(q)\,dq \right) U_n \\ &= L\left[\sum_{n=1}^{\infty} \frac{1}{\lambda_n} \left(\int_0^{\infty} qF(q)U_n(q)u(q)\,dq \right) V_n \right] \end{aligned}$$

at least formally. Therefore, we can say that

$$\mathscr{L}^{-1}F(t) = \sum_{n=1}^{\infty} \frac{1}{\lambda_n} \left(\int_0^{\infty} qF(q)U_n(q)u(q)\,dq \right) V_n(t) \tag{4.42}$$

gives us a real inversion formula of the Laplace transform. There are many results in the real inversion formula of the Laplace transform. By this method, we can see good numerical experiments of the Laplace transform in [160].

Meanwhile, we refer to [267, Sections 14.4 and 14.5] for more mathematical related details.

4.1.6 Real Inversion of the Laplace Transform Using Tikhonov Regularization

The Tikhonov regularization scheme with a regularization parameter α transforms the problem into having to find the minimizer of

$$\alpha\|f\|^2_{H_K(w)} + \|Lf(p) - pF(p)\|^2_{L^2(u)}, \tag{4.43}$$

where f runs over all elements in $H_K(w)$ and $pF(p) \in L^2(u)$. In terms of the singular system $\{(\lambda_n, U_n, V_n)\}_{n=1}^{\infty}$, the solution of this problem is expressed by

$$\mathscr{L}_\alpha^{-1}F(t) = \sum_{n=1}^{\infty} \frac{\lambda_n}{\lambda_n^2 + \alpha} \left(\int_0^\infty qF(q)U_n(q)u(q)\,dq \right) V_n(t). \tag{4.44}$$

These methods, which seem elementary, have turned out to be powerful as our numerical computation, as is shown in [160] for some special cases. We will present another representation by the Tikhonov regularized solution, as is developed in this section. Let $K(s,t)$ be the reproducing kernel of $H_K(w)$ and set $K_t(s) = K(s,t)$. For the regularization parameter α and any fixed number $t \geq 0$, the following equation of the second kind:

$$(\alpha I + LL^*)H_{\alpha,t} = LK_t \tag{4.45}$$

has a unique solution $H_{\alpha,t} \in L^2(u)$, where I is the identity on $L^2(u)$. Using the solution $H_{\alpha,t}$ we have the regularized solution as

$$\mathscr{L}_\alpha^{-1}F(t) = \langle G, H_{\alpha,t}\rangle_{L^2(u)} = \int_0^\infty pF(p)H_{\alpha,t}(p)u(p)\,dp. \tag{4.46}$$

Once we obtain the expression of the reproducing kernel K, we can write the adjoint operator of L in terms of integral kernels using

$$L^*g(t) = \langle L^*g, K_t\rangle_{H_K(w)} = \langle g, LK_t\rangle_{L^2(u)}. \tag{4.47}$$

We will use the simple reproducing kernel Hilbert space (RKHS) $H_K[0,\infty)$ consisting of absolutely continuous functions F on the positive real line $\mathbb{R}^+$ with finite norms

$$\left\{ \int_0^\infty |F'(t)|^2 \frac{e^t}{t} dt \right\}^{1/2} \tag{4.48}$$

and satisfying $F(0) = 0$. This Hilbert space admits the reproducing kernel $K(t,t')$

$$K(t,t') = \int_0^{\min(t,t')} \xi \exp(-\xi) d\xi. \tag{4.49}$$

According to (1.66), following [300, Section 7], we calculate

$$K(t,t') = -\min(t,t')e^{-\min(t,t')} - e^{-\min(t,t')} + 1. \tag{4.50}$$

Consequently, we obtain

$$\mathscr{L}K_{t'}(p) = \frac{1}{p(p+1)^2} - e^{-t'p}e^{-t'}\left(\frac{t'}{p(p+1)} + \frac{1}{p(p+1)^2}\right),$$

which in turn yields

$$\int_0^\infty e^{-qt'}\mathscr{L}K_{t'}(p)dt' = \frac{1}{pq(p+q+1)^2}. \tag{4.51}$$

Note that we can represent the adjoint operator by using the reproducing relation $L^*g(t) = \langle L^*g, K_t\rangle_{H_K(\mathbb{R}+)}$ as follows:

$$L^*g(t) = \langle g, LK_t\rangle_{L^2(\mathbb{R}+)} = \int_0^\infty g(\xi)\frac{(1-(\xi+1)^2e^{-t(\xi+1)})(t\xi+t+1)}{(\xi+1)^4}d\xi.$$

Therefore, by setting

$$H_\alpha(\xi,t) \equiv \mathscr{L}[K_\alpha)_t](\xi)\cdot\xi, \tag{4.52}$$

which is needed, we obtain the Fredholm integral equation of the second type

$$\alpha H_\alpha(\xi,t) + \int_0^\infty \frac{H_\alpha(p,t)}{(p+\xi+1)^2}dp = \left(t + \frac{1}{\xi+1}\right) + \frac{1}{(\xi+1)^2} - \frac{e^{-t\xi}e^{-t}}{\xi+1}. \tag{4.53}$$

Numerical experiments using Mathematica for the integral equation (4.53) were given in a positive way. In [299, 300] the sinc method proposed by M. Sugihara and his group was applied. They were able to obtain some numerical results for the real inversion formula. However, by solving (4.53), H. Fujiwara derived quite a reasonable inversion formula and he expanded good algorithms for numerical and real inversion formulas of the Laplace transform. H. Fujiwara made the software and we use it with his kind guide. In particular, H. Fujiwara solved the integral equation (4.53) with 6000 points discretization with **600 digits precision** based on the concept of **infinite precision** which is in turn based on **multiple-precision arithmetic** [158, 162, 164, 225, 226]. Then, the regularization parameters were $\alpha = 10^{-100}, 10^{-400}$ surprisingly. For the integral equation, he used the **DE formula** by H. Takahashi and M. Mori, using double exponential transforms [452]. H. Fujiwara was successful in deriving numerically the inversion for the Laplace transform of the distribution delta which was proposed by V. V. Kryzhniy as a difficult case. This fact will mean that the above results are valid for general functions approximated by the functions of the reproducing kernel Hilbert space $H_K(\mathbb{R}^+)$.

The typical examples are as follows and he gave many numerical computer graphics for showing the effectiveness of the above formulas [162]:

Example 4.7. We consider the function

$$f(t) = \begin{cases} 2t, & \text{if} \quad 0 \le t < 1; \\ 3 - t, & \text{if} \quad 1 \le t < 2; \\ 0, & \text{if} \quad t \ge 2, \end{cases} \tag{4.54}$$

whose Laplace transform is

$$F(p) = \frac{1 - 2e^{-p} + e^{-3p}}{p^2} \qquad (p > 0). \tag{4.55}$$

Fujiwara took up (4.54) as an example in [160, Section 4], where he proposed a truncation in order to calculate the inverse Laplace transform.

Example 4.8. The next example is the case when f is a characteristic function of

$$\left[\frac{1}{2}, 1\right] \cup \left[\frac{3}{2}, \frac{5}{2}\right] \cup \left[3, \frac{7}{2}\right], \tag{4.56}$$

that is a discontinuous function and is a difficult case. However, H. Fujiwara was successful in reconstructing this very precisely [161, Example 3].

Here and below, we let δ_a be the Dirac delta massed at the point $x = a \in \mathbb{R}$.

Example 4.9. For the case

$$F(p) = \exp(-p) \quad (p > 0), \tag{4.57}$$

which is the Laplace transform of the Dirac delta δ_1 in the distribution sense. Even for the Dirac delta function (not a usual function), H. Fujiwara was successful in "catching" the delta function by computers [161, Example 4].

Example 4.10. Let $a > 0$ and $r \in (0, 1)$ be fixed. Let

$$f \equiv \sum_{n=0}^{\infty} \frac{r^n(1-r)}{(a-r)^{n+1}} \delta_n.$$

A problem on the waiting time distribution in a queue

$$\mathscr{L}f(p) = \frac{1-r}{a - r(1 - e^{-p})} \qquad (p > 0). \tag{4.58}$$

See [229, Remark 6.2] for a generalization of (4.58). H. Fujiwara was successful in obtaining good numerical results [162, p. 207, Example 6], where he presented examples of numerical results.

Example 4.11. Let

$$f \equiv \frac{1}{2a^2(a+1)}\delta_0 - \sum_{n=1}^{\infty} \frac{1}{a(a+1)}\delta_{2n}.$$

For a problem on the circuit:

$$\mathcal{L}f(p) = \frac{1}{a(a+1)}\left(\frac{1}{2a} - \frac{e^{-2p}}{1-e^{-2p}}\right), \tag{4.59}$$

H. Fujiwara was again successful in obtaining good numerical results [162, p. 198 Example 1]. He invoked Theorem 1.8 and Example 1.3 with $w(t) = (1+t)^{-2}$ for $t \geq 0$ [162, Section 6.2].

Similarly, he gave many numerical experiments in effective ways and he even made a software for the numerical Laplace transform.

At this moment, H. Fujiwara used 6000 size discrete data. He gave the solutions with $\alpha = 10^{-400}$ and **600 digits precision**. See [162, 164, 406, 415, 419] for basic references.

Furthermore, as a complete version of Fujiwara, see [165]. They solved more numerical problems and gave many numerical and computer graphics.

By a suitable transform, our inversion formula in the previous section is applicable for more general functions as follows:

Therefore, if we know $F(0)$ and $F'(0)$, then from

$$g(p) = \mathcal{L}G(p), \tag{4.60}$$

we obtain $G(t)$ and so, from the identity

$$F(t) = G(t)e^{kt} + F(0) + tF'(0) \tag{4.61}$$

we have the inverse $F(t)$ from the data $f(p), F(0)$ and $F'(0)$ through the above procedures.

4.1.7 Optimal Real Inversion Formulas Using a Finite Number of Data

H. Fujiwara gave the solutions with $\alpha = 10^{-400}$ and **600 digits precision** for linear simultaneous equations of 6000 size by discretization – he used 6000 discrete data.

This fact greately influenced the authors over many years and we have considered more simpler algorithms for the awkward computer powers with great precision. As results, we obtain the discretization method, a kind of collocation methods or a pseudo-spectral method. The basic ideas are to use a finite number of data in order to escape any complicated discretization procedures, and from the finite data, we wish to obtain the optimal inversions.

We shall formulate a real inversion formula for the Laplace transform

$$\mathscr{L}F(p) = f(p) = \int_0^\infty e^{-pt} F(t)dt, \quad p > 0, \tag{4.62}$$

for functions F of some natural function spaces.

We will consider the simple reproducing kernel Hilbert space $H_K(\mathbb{R}^+)$ consisting of absolutely continuous functions F on the positive real line $\mathbb{R}^+$ with finite norms

$$\|f\|_{H_K(\mathbb{R}+)} \equiv \sqrt{\int_0^\infty |F'(t)|^2 \frac{\exp(t)}{t} dt} \tag{4.63}$$

and satisfying $F(0) = 0$. This Hilbert space admits the reproducing kernel $K(t,t')$ defined by

$$\begin{aligned} K(t,t') &\equiv \int_0^{\min(t,t')} \xi \exp(-\xi) d\xi \\ &= 1 - \min(t,t') \exp(-\min(t,t')) - \exp(-\min(t,t')). \end{aligned}$$

See (1.66). Then, the linear point evaluation operator on $H_K(\mathbb{R}^+)$, for any fixed point p

$$LF(p) = p\mathscr{L}F(p) \tag{4.64}$$

is bounded, as we see from (4.30). Now we will consider the bounded linear mapping $\mathbf{L}$, for any fixed different points $\{p_j\}_{j=1}^n$ of $\mathbb{R}^+$,

$$\mathbf{L} : H_K(\mathbb{R}^+) \ni F \mapsto (LF(p_1), LF(p_2), \ldots, LF(p_n)) \in \mathbb{R}^n. \tag{4.65}$$

We take the standard orthonormal system $\{\mathbf{e}_j\}_{j=1}^n$ in the space $\mathbb{R}^n$. Then we see that

$$LF(p_j) = \langle \mathbf{L}F, \mathbf{e}_j \rangle_{\mathbb{R}^n} = \langle F, \mathbf{L}^* \mathbf{e}_j \rangle_{H_K(\mathbb{R}_+)} \quad j = 1, 2, \ldots, n. \tag{4.66}$$

Thus, we are now precisely in the situation of Chap. 4. Therefore, the image space forms the Hilbert space H_A admitting the reproducing kernel

$$a_{j,j'} = \langle \mathbf{L}^* \mathbf{e}_{j'}, \mathbf{L}^* \mathbf{e}_j \rangle_{H_K(\mathbb{R})} \quad j, j' = 1, 2, \ldots, n. \tag{4.67}$$

Here, $a_{j,j'}$ is calculated as follows: First, observe that

$$\mathbf{L}^*\mathbf{e}_j(t) = \langle \mathbf{L}^*\mathbf{e}_j, K_t\rangle_{H_K(\mathbb{R})} \quad j = 1, 2, \ldots, n$$

by the reproducing property. It follows from the definition of the identity that we have

$$\mathbf{L}^*\mathbf{e}_j(t) = \langle \mathbf{e}_j, \mathbf{L}K_t\rangle_{\mathbb{R}^n} = \frac{1}{(p_j+1)^2} - \exp(-t(p_j+1))\left(\frac{t}{(p_j+1)} + \frac{1}{(p_j+1)^2}\right)$$

for $j = 1, 2, \ldots, n$ and

$$LK_{t'}(p) = \frac{1}{p(p+1)^2} - \exp(-t'p - t')\left(\frac{t'}{p(p+1)} + \frac{1}{p(p+1)^2}\right). \tag{4.68}$$

Hence, from (4.67) and (4.68), we derive that

$$a_{j,j'} = \frac{1}{(p_j + p_{j'} + 1)^2} \quad j, j' = 1, 2, \ldots, n. \tag{4.69}$$

In particular, note that the matrix $A = \{a_{j,j'}\}_{j,j'=1}^n$ is positive definite.

Thus, assembling all the just presented facts, especially Theorem 2.33, we obtain the real inversion formula.

Theorem 4.2. *For any given n values* $\mathbf{d} = \{d_j\}_{j=1}^n$, *among the Laplace transforms taking the values*

$$f(p_j) = d_j \quad j = 1, 2, \ldots, n, \tag{4.70}$$

and among their inverses, the uniquely determined function $F_{\mathbf{d}}^*(t)$ *with the minimum norm* $\|F\|_{H_K(\mathbb{R}+)}$ *is uniquely determined and it is represented by*

$$F_{\mathbf{d}}^* = \sum_{j,j'=1}^{n} d_j \tilde{a}_{j,j'} \mathbf{L}^*\mathbf{e}_{j'}. \tag{4.71}$$

This theorem was taken from [82].

The inverse problems in heat conduction and in the real inversion of the Laplace transform are famous as essentially difficult ones, however, by mathematical methods derived; with H. Fujiwara's powerful computer systems, we can solve them practically and numerically. In the case of heat conduction, our formulas for 10^{-22} as α in the Tikhonov regularization solution for using the Sobolev spaces and $1/300$ for h for using the sinc functions were calculated. For the case of the Laplace transform, H. Fujiwara calculates the formula for 10^{-400} as α with 6000 discrete points. Such values will show how it is difficult to solve the inverse problems from the standpoint of the Tikhonov regularization. These results will show some good prospects how we can grasp analytic functions using computers and how our mathematical theory may be applied to wide practical problems. However, we still have difficult problems, for example, to obtain the global data from some local data for analytic functions.

From the inverse analysis point of view, there has been several proposals employing stabilization methods such as the Tikhonov regularization. On the other hand, singular value decomposition is applicable not only for reconstruction of solutions, but also for Hilbert scales and noise reduction of measurement data. Though the singular value decomposition has various applications, its concrete treatments are hard both mathematically [305] and numerically [160, 162, 164].

The Laplace transform is not compact on the usual Lebesgue or Sobolev spaces, and it has continuous spectrum. In the setting, a certain truncation is required for numerical real inversions [241, p. 8, Example 2]. One of the key ideas has to use the reproducing kernel Hilbert spaces, in which we have a concrete representation for the adjoint operator and this enables us to realize an effective numerical real inversion. The proposed approach using numerical singular value decomposition is straightforward, so it is applicable to many inverse problems.

The central problem in the method is now to solve effectively the functional equation as in (4.53), and there we need practically some discretization of the equation to solve it using computers. The practical treatments depend on the concrete equations, case-by-case. When a complete orthonormal system of $\mathscr{H}$ is simply constructed, we can discrete it using the Fourier coefficients. However, the last one stated is a very simple general algorithm and its essential part is to calculate the inversion of positive definite Hermitian matrices. In the near future, because of the great power of computers, we are sure to solve effectively and sufficiently many linear problems by these methods.

For some more recent general discretization principles with many concrete examples, see [92, 93]. There, Fujiwara showed his powerful numerical calculations for the inversion of positive definite Hermitian matrices for already some very large sizes.

We refer to [126] for a vast amount of details on Laplace transform.

Chapter 5
Applications to Ordinary Differential Equations

We will give concrete representation formulas of the approximate solutions for linear ordinary differential equations following the Tikhonov regularization and using a finite number of data.

For other applications of reproducing kernels, see a series of papers; for example, [128, 177–181, 232, 233, 283, 284]. In these papers reproducing kernels are applied to nonlinear differential equations using many concrete reproducing kernels with numerical experiments given as well. See also the recent article [477].

5.1 Use Tikhonov Regularization

We will start with the simplest prototype case.

5.1.1 Ordinary Linear Differential Equations with Constant Coefficients

In this subsection we consider the typical elementary differential operator

$$Ly = y'' + \alpha y' + \beta y \tag{5.1}$$

on the whole real line $\mathbb{R}$. For more formal generalizations, results and formulas are similar in the cases of higher order ordinary differential equations.

In order to consider a generalized solution of $Ly = g$, for $g \in L^2(\mathbb{R})$, we consider the Sobolev space $H_S^2(\mathbb{R})$ on the whole real line $\mathbb{R}$ with finite norms

$$\|f\|_{H_S^2(\mathbb{R})} \equiv \left\{ \int_{-\infty}^{\infty} (f''(x)^2 + 2f'(x)^2 + f(x)^2) dx \right\}^{1/2} < \infty \tag{5.2}$$

S. Saitoh, Y. Sawano, *Theory of Reproducing Kernels and Applications*, Developments in Mathematics 44, DOI 10.1007/978-981-10-0530-5_5

admitting the reproducing kernel

$$K(x, y) = \frac{1}{2\pi}\int_{-\infty}^{\infty} \frac{\exp(i\xi(x-y))}{(1+\xi^2)^2} d\xi = \frac{1}{4} e^{-|x-y|}(1+|x-y|), \tag{5.3}$$

as we saw in Theorem 1.11. From (5.1), we will consider the existence of second order derived functions. First we consider the best approximation problem; for any given $g \in L^2(\mathbb{R})$ and for any $\lambda > 0$,

$$\inf\left\{\lambda \|f\|^2_{H^2_S(\mathbb{R})} + \|Lf - g\|^2_{L^2(\mathbb{R})} : f \in H^2_S(\mathbb{R})\right\}. \tag{5.4}$$

Then for the RKHS $H_{K_\lambda}(\mathbb{R})$ consisting of all the members of $H^2_S(\mathbb{R})$ with the norm

$$\|f\|_{H_{K_\lambda}(\mathbb{R})} \equiv \sqrt{\lambda \|f\|^2_{H^2_S(\mathbb{R})} + \|Lf\|^2_{L^2(\mathbb{R})}}, \tag{5.5}$$

the reproducing kernel $K_\lambda(x, y)$ can be calculated directly using Fourier's integrals as follows:

$$K_\lambda(x, y) = \frac{1}{2\pi}\int_{-\infty}^{\infty} \frac{\exp(i(x-y))\, d\xi}{\lambda(\xi^2+1)^2 + |-\xi^2 + i\alpha\xi + \beta|^2}. \tag{5.6}$$

We thus obtain the member of $H^2_S(\mathbb{R})$ with the minimum norm which attains the infimum (5.4) as follows:

$$f^*_{\lambda,g}(x) = \frac{1}{2\pi}\int_{\mathbb{R}}\left\{g(\xi)\cdot \int_{\mathbb{R}} \frac{(-\eta^2 - i\alpha\eta + \beta)\exp(-i\eta(\xi - x))}{\lambda(\eta^2+1)^2 + |-\eta^2 + i\alpha\eta + \beta|^2} d\eta\right\} d\xi. \tag{5.7}$$

For $g \in L^2(\mathbb{R})$, if there exists a solution $\hat{f}_g$ of the equation

$$Ly(x) = g(x) \quad \text{on} \quad \mathbb{R}, \tag{5.8}$$

then we have the representation using Fourier's integral

$$\hat{f}_g(x) = \frac{1}{2\pi}\int_{\mathbb{R}}\left\{g(\xi)\int_{\mathbb{R}} \frac{\exp(-i\eta(\xi - x))}{-\eta^2 + i\alpha\eta + \beta} d\eta\right\} d\xi. \tag{5.9}$$

Then

$$\lim_{\lambda\downarrow 0} f^*_{\lambda,g}(x) = \hat{f}_g(x), \tag{5.10}$$

uniformly in view of Remark 1.3.

We shall give a general approximation formula:

Theorem 5.1 ([294, Theorem 1]). *Let* $g \in L^2(\mathbb{R})$*,* $\lambda > 0$ *and* $s > \dfrac{1}{2}$*. Let* $H^s(\mathbb{R})$ *have the structure of a reproducing kernel Hilbert space as in Theorem* 1.13*. Define*

$$Q_{\lambda,s}(x) \equiv \frac{1}{\pi}\int_0^\infty \frac{(-p^2+\beta)\cos(px)+\alpha\, p\, \sin(px)}{\lambda(p^2+1)^s+p^4+(\alpha^2-2b)p^2+\beta^2}\, dp \quad (x \in \mathbb{R}) \tag{5.11}$$

and

$$F^*_{\lambda,s,g} \equiv g * Q_{\lambda,s}. \tag{5.12}$$

Then, $F^*_{\lambda,s,g}$ *minimizes*

$$\min_{F \in H^s(\mathbb{R})} \left(\lambda\, \|F\|^2_{H^s(\mathbb{R})} + \|g - L(D)F\|^2_{L^2(\mathbb{R})}\right) \tag{5.13}$$

and the minimizer is unique.

Proof. As we have seen in the last section, the solution of the minimizing problem (5.13) is given by

$$F^*_{\lambda,s,g}(x) = \langle g, L\,(\lambda + L^*L)^{-1}[K_x]\rangle_{L^2(\mathbb{R})}. \tag{5.14}$$

Let us begin by calculating the adjoint $L^* : L^2(\mathbb{R}) \to H^s(\mathbb{R})$. To this end, we let $g \in L^2(\mathbb{R})$ and $f \in H^s(\mathbb{R})$. Then we have

$$\begin{aligned}\langle L^* g, f\rangle_{H^s(\mathbb{R})} &= \langle g, Lf\rangle_{L^2(\mathbb{R})} \\ &= \int_{\mathbb{R}} \mathscr{F}g(\xi)\overline{(-\xi^2+\alpha\, i\,\xi+\beta)\mathscr{F}f(\xi)}\, d\xi = \langle M(D)g, f\rangle_{H^s(\mathbb{R}^n)},\end{aligned}$$

where

$$M(\xi) = (-\xi^2 - \alpha\, i\,\xi + \beta)(1+|\xi|^2)^{-\frac{s}{2}}. \tag{5.15}$$

As a consequence, if we set

$$N(\xi) \equiv \frac{(-\xi^2+\alpha\, i\,\xi+\beta)}{\lambda + (-\xi^2-\alpha\, i\,\xi+\beta)(-\xi^2+\alpha\, i\,\xi+\beta)(1+|\xi|^2)^{-s}} \quad (x \in \mathbb{R}), \tag{5.16}$$

then we have

$$L\left(\lambda + L^*L\right)^{-1} g = N(D)g. \tag{5.17}$$

In view of this, we have

$$\begin{aligned}
&L[(\lambda + L^*L)^{-1} K_x](y) \\
&= \frac{1}{\sqrt{2\pi}} \mathscr{F}^{-1}[N(\xi)e^{-ix\xi}(1+|\xi|^2)^{-\frac{s}{2}}](y) \\
&= \frac{1}{2\pi}\int_{\mathbb{R}} \frac{(-\xi^2 + \alpha\, i\,\xi + \beta)e^{i(y-x)\xi}}{\lambda(1+|\xi|^2)^s + (-\xi^2 - \alpha\, i\,\xi + \beta)(-\xi^2 + \alpha\, i\,\xi + \beta)}\, d\xi \\
&= \frac{1}{\pi}\int_0^{\infty} \frac{(-\xi^2 + \beta)\cos((y-x)\xi) - \alpha\, p \sin((y-x)\xi)}{\lambda(1+|\xi|^2)^s + (-\xi^2 - \alpha\, i\,\xi + \beta)(-\xi^2 + \alpha\, i\,\xi + \beta)}\, d\xi \\
&= Q_{\lambda,s}(x-y).
\end{aligned}$$

Hence $F^*_{\lambda,s,g}$ minimizes (5.13). The uniqueness of this minimizer is clear.

5.1.2 Variable Coefficients Case

Let $\mathbb{R}^+ = \{x \in \mathbb{R} : x > 0\}$ as usual. We will now briefly describe how to consider the previous proposed method for inhomogeneous ordinary differential equations for any $g \in L^2(\mathbb{R}^+)$ and for real valued $L^2(\mathbb{R}^+)$(or bounded integrable) functions a, b, c :

$$Ly(x) \equiv L(D)y(x) = a(x)y'' + b(x)y' + c(x)y = g(x) \quad \text{on} \quad \mathbb{R}^+. \tag{5.18}$$

As a general algorithm, and in the same sense as in the previous setting, we consider the discrete point data case. In view of this, we will take into consideration the corresponding problem of determining

$$\inf_{f \in H_K(\mathbb{R}^+)} \left\{ \lambda \|f\|^2_{H_K(\mathbb{R}^+)} + \sum_{j=1}^{N} |Lf(x_j) - d_j|^2 \right\} \tag{5.19}$$

for some observable elements which may be taken as $d_j = g(x_j)$.

For the just presented general coefficients a, b and c, we are only requiring that $y \mapsto L(D)y$ must be a bounded linear operator on a certain reproducing kernel Hilbert space with kernel K.

To use Theorems 3.13 and 3.14, we need the information regarding $L[K_L(\cdot, p; \lambda)]$. This is provided by (3.85). In (3.85), we operate L and we have

$$\lambda L[\widetilde{K}(\cdot, q; \lambda)] + L\langle L\widetilde{K}_q, LK_\cdot\rangle_{\mathscr{H}} = LK_q. \tag{5.20}$$

Then we see that the required N values $LK_L(\cdot, q; \lambda)$ are determined from the regular N linear equations (5.20) for the N points $\{x_j\}_j$. We state the precise procedure. Here it is also interesting to remark that when we can take $\lambda = 0$ in the numerical sense, we can include, of course, $\lambda = 0$ in those arguments.

We refer to [143] for some general property of the operator L.

5.1.3 Finite Interval Cases

For a finite interval case, $I = (a, b)$, we propose a new method for solving the discrete differential equation (5.40).

Now we will consider the following minimum problem: By taking N points $\{x_j\}_{j=1}^N$ on $I = (a, b)$, where $x_1 = a, x_j < x_{j+1}$ $(j = 1, 2, \ldots, N-1), x_N = b$, for boundary data $A, B \in \mathbb{R}$, and for any fixed $\alpha_1 > 0$ and $\alpha_N > 0$, we look for the solution to

$$\inf_{F \in H^2(\mathbb{R})} \left\{ \lambda \|F\|^2_{H^2(\mathbb{R})} + \alpha_1 |F(a) - A|^2 + \alpha_N |F(b) - B|^2 + \sum_{j=2}^{N-1} |LF(x_j) - g(x_j)|^2 \right\}. \tag{5.21}$$

For the points $\{x_j\}_{j=2}^{N-1}$, we will be able to obtain good approximation solutions

$$Ly(x_j) = a(x_j)y''(x_j) + b(x_j)y'(x_j) + c(x_j)y(x_j) = g(x_j) \quad j = 2, 3, \ldots, N-1 \tag{5.22}$$

subject to the boundary conditions

$$y(a) = A \quad \text{and} \quad y(b) = B. \tag{5.23}$$

Then we can construct the related reproducing kernel by direct linear equations as stated.

In (5.21), by taking fixed different parameters α_1 and α_N, we can adjust the speeds of convergence of the extremal functions $F^*_{g,\lambda,\alpha_1,\alpha_N}(x)$ in (5.21) in a way so that

$$\lim_{\lambda \downarrow 0} F^*_{g,\lambda,\alpha_1,\alpha_N}(a) = A \tag{5.24}$$

and

$$\lim_{\lambda \downarrow 0} F^*_{g,\lambda,\alpha_1,\alpha_N}(b) = B. \tag{5.25}$$

If $\alpha_1 \gg \alpha_N$, then the convergence in (5.24) is faster than in (5.25). See [31] for realizing some example of this phenomenon.

5.1.4 Exact Algorithm

We state now the exact algorithm for the problem (5.21), in conjugation with (5.20) by solving the linear equations.

We set, for $K_\lambda(p,q) \equiv \widetilde{K}(p,q;\lambda)$

$$X_\lambda(p,t') = L_p(D)K_\lambda(p,t'), \quad k(p,t') = L_p(D)K(p,t') \tag{5.26}$$

and

$$\kappa(p,q) = L_q(D)L_p(D)K(p,q). \tag{5.27}$$

However, at the points a and b, we do not operate $L(D)$ in the above formulas. Instead, we understand them as the identity operator at the points a and b ; that is,

$$X_\lambda(a,t') = K_\lambda(a,t'),\ X_\lambda(b,t') = K_\lambda(b,t'), \tag{5.28}$$

$$X_\lambda(a,a) = K_\lambda(a,a),\ X_\lambda(b,b) = K_\lambda(b,b), \tag{5.29}$$

$$k(a,t') = K(a,t'),\ k(b,t') = K(b,t'), \tag{5.30}$$

$$\kappa(a,a) = K(a,a),\ \kappa(b,b) = K(b,b), \tag{5.31}$$

and, letting $g(x) \equiv L(D)[k(a,\cdot)](x)$(see (5.18) above),

$$\kappa(a,x_j) = g(x_j), \quad j = 2,3,\ldots,N-1. \tag{5.32}$$

As the solution to the regular linear equations, we have

$$\begin{aligned} k(x_j,t') &= \lambda X_\lambda(x_j,t') + \alpha_1 X_\lambda(a,t')\kappa(x_j,a) \\ &\quad + \alpha_N X_\lambda(b,t')\kappa(x_j,b) + \sum_{j'=2}^{N-1} X_\lambda(x_{j'},t')\kappa(x_j,x_{j'}), \end{aligned} \tag{5.33}$$

for $j = 1,2,\ldots,N$ and we determine $X_\lambda(x_j,t')$. Then we obtain the approximation solution of (5.21) satisfying the boundary conditions for $g(x_j) = d_j$ $(j = 2,3,\ldots,N-1)$,

$$F_{\mathbf{d},\lambda,\alpha_1,\alpha_N}(x) = A\alpha_1 X_\lambda(a,x) + B\alpha_N X_\lambda(b,x) + \sum_{j=2}^{N-1} d_j X_\lambda(x_j,x). \tag{5.34}$$

5.1.5 One Point Boundary Condition Case

The method presented in Sect. 5.1.4 is also adaptable to other boundary conditions. Namely, we may consider boundary conditions based on the same point of the finite interval $I = (a, b)$. Let

$$a = x_1 < x_2 < \cdots < x_{N-1} < b = x_N.$$

For instance, instead of the boundary conditions used in Sect. 5.1.3, we may consider the boundary conditions

$$y(a+0) = A \quad \text{and} \quad y'(a+0) = B, \tag{5.35}$$

so that we are interested in

$$y(a+0) = A, \quad y'(a+0) = B, \quad Ly(x) = g(x) \quad (x \in (a,b)).$$

As a consequence, in the present case we will be facing the extremal problem

$$\inf_{F \in H^2(a,b)} \left\{ \lambda \|F\|^2_{H^2(a,b)} + \alpha |F(a) - A|^2 + \beta |F'(a) - B|^2 + \sum_{j=2}^{N-1} |LF(x_j) - g(x_j)|^2 \right\}.$$

Then, in comparison with the method of Sect. 5.1.4, instead of the point b, the coefficients $k(p,t')$ and $\kappa(p,q)$ should be replaced (for $j = N$) as follows:

$$X_\lambda(b,t') = \left[\frac{\partial (K_\lambda)_{t'}}{\partial x}(x) \right](a) = \left[\frac{\partial}{\partial x} K_\lambda(x,t') \right](a), \tag{5.36}$$

$$k(b,t') = \left[\frac{\partial K_{t'}}{\partial x}(x) \right](a), \quad \kappa(b,b) = \left[\frac{\partial}{\partial t'} \left(\frac{\partial K_{t'}}{\partial t}(t) \right) \right](t,t') \Big|_{t=t'=a}, \tag{5.37}$$

and so on. Then, from (5.33) we arrive at the approximate solution:

$$F_{\mathbf{d},\lambda,\alpha_1,\alpha_N}(x) = A\alpha_1 X_\lambda(a,x) + B\alpha_N \left[\frac{\partial (X_\lambda)_x}{\partial \xi}(\xi) \right](a) + \sum_{j=2}^{N-1} d_j X_\lambda(x_j,x). \tag{5.38}$$

See [86, Section 8] for details.

5.2 Discrete Ordinary Linear Differential Equations

Since we can obtain practically a finite number of data, we will formulate the representations of some optimal solutions using a finite number of data for typical cases. Of course, we are considering discretization for many analytical problems.

We will give concrete results for typical problems in the following: The argument presented here will show that our results will be valid in a general situation. However, to state the results and methods in a simpler manner, we consider a prototype; e.g., the differential operator

$$Ly = y'' + \alpha y' + \beta y. \tag{5.39}$$

In the first subsection, we will consider (5.39) on the whole real line $\mathbb{R}$, and assume that α and β are constants. Later on we will consider situations other than the real line, and will also consider variable coefficients.

We wish to construct some solution of

$$Ly = g \tag{5.40}$$

satisfying (5.39) on a finite number of points. For this purpose, for the differential operator L, we will consider it as a bounded linear operator from some reproducing kernel Hilbert space into $\mathbb{R}^N$. That is,

$$\mathbf{L}(y)(x_1, x_2, \ldots, x_N) \equiv (Ly(x_1), Ly(x_2), \ldots, Ly(x_N)) \in \mathbb{R}^N, \tag{5.41}$$

for different points x_j on $\mathbb{R}$. For our purpose, we need concrete reproducing kernel spaces and so, we will use two typical reproducing kernels:

$$K_s(x, y) = \frac{1}{2\pi} \int_{\mathbb{R}} (1 + |\xi|^2)^{-s} e^{i(x-y)\cdot\xi} \, d\xi \tag{5.42}$$

and

$$K_h(z, \overline{u}) = \frac{1}{2\pi} \int_{\mathbb{R}} \chi_{(-h,h)}(t) e^{-iz\cdot t} \overline{e^{-iu\cdot t}} dt = \frac{1}{\pi(z - \overline{u})} \sin \frac{\pi}{h}(z - \overline{u}), \tag{5.43}$$

for $z, u \in \mathbb{C}$ introduced in detail in Chap. 1; see Sect. 1.2.

5.2.1 *Use a Finite Number of Data*

For different N points $\{x_j\}_{j=1}^N$, we will consider the bounded linear operators from the RKHS $H_{K_h}(\mathbb{R})$ into $\mathbb{R}$, which is an evaluation mapping

$$H_{K_h}(\mathbb{R}) \ni F \mapsto LF(x_j) \in \mathbb{R}, \quad (j = 1, 2, \ldots, N) \tag{5.44}$$

and we will write them in the vector form

$$\mathbf{L} : H_{K_h}(\mathbb{R}) \ni F \mapsto \begin{pmatrix} LF(x_1) \\ LF(x_2) \\ \vdots \\ LF(x_N) \end{pmatrix} \in \mathbb{R}^N. \tag{5.45}$$

Given a point $z \in \mathbb{R}^N$, our problem is to find a function $F \in H_{K_h}(\mathbb{R})$ such that $\mathbf{L}F = z \in \mathbb{R}^N$. Generally speaking, this type of problem admits infinitely many solutions. However, as we have seen, with some minimality condition, we are able to obtain a solution mathematically. This is performed as follows:

We take a standard orthonormal system $\{\mathbf{e}_j\}_{j=1}^N$ in the space $\mathbb{R}^N$. Then, from (5.45), we obtain that

$$LF(x_j) = \langle \mathbf{L}F, \mathbf{e}_j \rangle_{\mathbb{R}^N} = \langle F, \mathbf{L}^* \mathbf{e}_j \rangle_{H_{K_h}(\mathbb{R})}. \tag{5.46}$$

Keeping in mind that $H_{K_h}(\mathbb{R})$ is given by (5.43) and that $x_1, x_2, \ldots, x_N$ are distinct points, in the statements of Sect. 2.5.2 we can say that we are considering the case when

$$E = \{x_1, x_2, \ldots, x_N\}, \mathscr{H} = H_{K_h}(\mathbb{R}), \mathbf{h}(x_j) = L^* \mathbf{e}_j$$

with $L : H_{K_h}(\mathbb{R}) \to \mathscr{F}(E)$ in (2.228) given by (5.46).

Therefore, in view of Theorem 2.36, the image space of the linear mapping $\mathbf{L}$ forms the Hilbert space $H_A(E)$ admitting the reproducing kernel

$$A := \{\langle \mathbf{L}^* \mathbf{e}_{j'}, \mathbf{L}^* \mathbf{e}_j \rangle_{H_{K_h}(\mathbb{R})}\}_{j,j'=1,2,\ldots,N} := \{a_{j,j'}\}_{j,j'=1,2,\ldots,N}. \tag{5.47}$$

The definition (5.47) is based on the fact that we can and do identify functions on $E \times E$ with $N \times N$ matrices. In the statements of functions on $E \times E$, (5.47) corresponds to

$$K_A(x_j, x_{j'}) = \langle \mathbf{L}^* \mathbf{e}_{j'}, \mathbf{L}^* \mathbf{e}_j \rangle_{H_{K_h}(E)} \quad (j, j' = 1, 2, \ldots, N).$$

If the matrix $\{a_{j,j'}\}_{j,j'=1,2,\ldots,N}$ is strictly positive definite, then from Example 1.1, the norm in $H_A(E)$ is defined by

$$\|\mathbf{L}F\|_{H_A(E)}^2 = {}^t(\mathbf{L}F)\widetilde{A}(\mathbf{L}F) = (LF(x_1), LF(x_2), \ldots, LF(x_N))\widetilde{A} \begin{pmatrix} LF(x_1) \\ LF(x_2) \\ \vdots \\ LF(x_N)) \end{pmatrix}$$

where

$$\widetilde{A} = \overline{A^{-1}} = \{\widetilde{a_{j,j'}}\}_{j,j'=1,2,\dots,N}. \tag{5.48}$$

The precise form of $a_{j,j'}$ is given as follows:

Lemma 5.1. *Let $X \equiv \pi/h$ and $b \equiv |x_j - x_{j'}|$, with $j, j' = 1, 2, \dots, N$ fixed. The elements $a_{j,j'}$ introduced in (5.47) are explicitly given by*

$$a_{j,j'} = \frac{4X(-6 + b^2X^2)\cos(bX)}{\pi b^4} + \frac{(24 - 12b^2X^2 + b^4X^4)\sin(bX)}{\pi b^5} \tag{5.49}$$

$$+ \frac{(\alpha^2 - 2\beta)(2 - b^2X^2)\sin(bX)}{\pi b^3} - \frac{2(\alpha^2 - 2\beta)X\cos(bX)}{\pi b^2} + \frac{\beta^2}{\pi b}\sin(bX).$$

Proof. Recall first that K_h is given by (5.43). In view of the computation of $a_{j,j'}$, we start by understanding that

$$\mathbf{L}^*\mathbf{e}_j(x) = \langle \mathbf{e}_j, \mathbf{L}(K_h)_x \rangle_{\mathbb{R}^N} = \frac{1}{2\pi}\int_{-\frac{\pi}{h}}^{\frac{\pi}{h}} (-\eta^2 - i\alpha\eta + \beta)\exp(-i\eta(x - x_j))d\eta.$$

Hence

$$a_{j,j'} = \frac{1}{2\pi}\int_{-\frac{\pi}{h}}^{\frac{\pi}{h}} |-\eta^2 - i\alpha\eta + \beta|^2 e^{-i\eta(x_j - x_{j'})}d\eta. \tag{5.50}$$

In particular, we conclude from (5.50) that the matrix $A = \{a_{j,j'}\}_{j,j'=1}^N$ is positive definite.

Furthermore, in view of the symmetry, we have

$$\begin{aligned} a_{j,j'} &= \frac{1}{2\pi}\int_{-\frac{\pi}{h}}^{\frac{\pi}{h}} (\eta^4 - 2\beta\eta^2 + \beta^2 + \alpha^2\eta^2)\exp(-i\eta(x_j - x_{j'}))d\eta \\ &= \frac{1}{2\pi}\int_{-\frac{\pi}{h}}^{\frac{\pi}{h}} (\eta^4 + (\alpha^2 - 2\beta)\eta^2 + \beta^2)\exp(-i\eta|x_j - x_{j'}|)d\eta \\ &= \frac{1}{\pi}\int_0^{\pi/h} (\eta^4 + (\alpha^2 - 2\beta)\eta^2 + \beta^2)\cos(|x_j - x_{j'}|\eta)d\eta. \end{aligned}$$

If we calculate the integrals on the right-hand side, we obtain (5.49).

Therefore, for the linear transform (5.45), we have the inequality

$$\|\mathbf{L}F\|_{H_A} \leq \|F\|_{H_{K_h}(E)} \tag{5.51}$$

by (2.235) and there exists a uniquely determined $F^*_{\mathbf{d}}$ for which the equality holds; that is, for the minimum members, the linear mapping corresponds isometrically onto $\mathbb{R}^N$. Hence, we denote the minimum member by $F^*_{\mathbf{d}}$ satisfying (5.45). Then, from Theorem 2.36, which is the general theory in such a situation, we obtain the desired inversion formula:

$$\mathbf{L}F \in H_A(E) \mapsto F^*_{\mathbf{d}} \in H_{K_h}(E).$$

All the just derived elements are now assembled in the following desired result:

Theorem 5.2. *For any given N values* $\mathbf{d} = \{d_j\}_{j=1}^N$, *among the* $H_{K_h}(\mathbb{R})$ *functions F taking the values*

$$LF(x_j) = d_j, \quad j = 1, 2, \ldots, N, \tag{5.52}$$

the function $F^*_{\mathbf{d}}$ *with the minimum norm* $\|F\|_{H_{K_h}(\mathbb{R})}$ *is uniquely determined and it is represented as follows:*

$$F^*_{\mathbf{d}} = \sum_{j,j'=1}^{N} d_j \widetilde{a_{j,j'}} \mathbf{L}^* \mathbf{e}_{j'}, \tag{5.53}$$

where the $\widetilde{a_{j,j'}}$ *'s are determined by* (5.48) *and* (5.49).

Proof. This is a special case of Theorem 2.33.

Assume that the equality in (5.51) holds. As we see from the isometric relation between the data (5.45) and the inverse with the minimum norm $F^*_{\mathbf{d}}$, the observation data and the corresponding inverse are rigid, and our inversion formula is then given in the sense of a well-posed problem.

In the Sobolev space $H_{K_s}(\mathbb{R})$ of order s ($s \geq 3$), where K_s is given by (5.42), we have the corresponding formulas:

$$\mathbf{L}^*\mathbf{e}_j(x) = \langle \mathbf{L}^*\mathbf{e}_j, (K_s)_x \rangle_{H_{K_s}(\mathbb{R})} = \frac{1}{2\pi} \int_{\mathbb{R}} \frac{(\beta - i\alpha\eta - \eta^2)\exp(-i\eta(x - x_j))}{(1+\eta^2)^s} d\eta,$$

which follows from $\langle \mathbf{L}^*\mathbf{e}_j, (K_s)_x \rangle_{H_{K_s}(\mathbb{R})} = \langle \mathbf{e}_j, \mathbf{L}(K_s)_x \rangle_{\mathbb{R}^N}$ Hence, we have in this case

$$a_{j,j'} = \frac{1}{2\pi} \int_{\mathbb{R}} \frac{|-\eta^2 - i\alpha\eta + \beta|^2 e^{i\eta(x_j - x_{j'})}}{(1+\eta^2)^s} d\eta. \tag{5.54}$$

Therefore, the calculation is completely analogous to the Paley Wiener case above. For the sake of convenience we present an algorithm of calculating $a_{j,j'}$ when s is an integer. Note that

$$
\begin{aligned}
a_{j,j'} &= \frac{1}{2\pi}\int_{\mathbb{R}} \frac{\{|\eta^2-\beta|^2+\alpha^2\eta^2\}e^{i\eta(x_j-x_{j'})}}{(1+\eta^2)^s}d\eta \\
&= \frac{1}{2\pi}\int_{\mathbb{R}} \frac{\{(\eta^2+1)^2+(\alpha^2-2\beta-2)(\eta^2+1)-\alpha^2+(\beta+1)^2\}e^{i\eta(x_j-x_{j'})}}{(1+\eta^2)^s}d\eta \\
&= I_{s-2}(x_j-x_{j'})+(\alpha^2-2\beta-2)I_{s-1}(x_j-x_{j'})-(\alpha^2-(\beta+1)^2)I_s(x_j-x_{j'}),
\end{aligned}
$$

where

$$
I_s(t)=\frac{1}{2\pi}\int_{\mathbb{R}}\frac{\exp(i\eta t)}{(1+\eta^2)^s}\,d\eta=\frac{1}{2\pi}\int_{\mathbb{R}}\frac{\exp(i\eta|t|)}{(1+\eta^2)^s}\,d\eta \quad (t\in\mathbb{R}). \tag{5.55}
$$

Here, due to the iterative construction, note that we are considering I_s elements not only for $s\geq 3$ but also for $s\geq 1$. In view of this, considering (5.55), observe that

$$
I_1(t)=i\mathrm{Res}\left(\frac{\exp(i\eta|t|)}{1+\eta^2},\eta=i\right)=\frac{1}{2}e^{-|t|}. \tag{5.56}
$$

Also, if $s\geq 2$, we differentiate $I_s(t)$ to obtain

$$
\begin{aligned}
I_s'(t) &= \frac{1}{2\pi}\int_{\mathbb{R}}\frac{i\eta\exp(i\eta t)}{(1+\eta^2)^s}\,d\eta \\
&= -\frac{i}{4\pi(s-1)}\int_{\mathbb{R}}\left(\frac{1}{(1+\eta^2)^{s-1}}\right)'\exp(i\eta t)\,d\eta \\
&= \frac{t}{4\pi(s-1)}\int_{\mathbb{R}}\frac{1}{(1+\eta^2)^{s-1}}\exp(i\eta t)\,d\eta \\
&= \frac{t}{2(s-1)}I_{s-1}(t) \quad (t\in\mathbb{R}).
\end{aligned}
$$

If $t=0$, then we have

$$
I_s(0)=\frac{1}{2\pi}\int_{\mathbb{R}}\frac{d\eta}{(1+\eta^2)^s}=\frac{1}{\pi}\int_0^{\frac{\pi}{2}}\cos^{2s-2}\theta\,d\theta=\frac{1}{2}\cdot\frac{(2s-1)!!}{(2s-2)!!} \quad (t\in\mathbb{R}). \tag{5.57}
$$

Thus, we have just obtained that

$$
I_s(0)=\frac{1}{2}\frac{(2s-1)!!}{(2s-2)!!},\quad I_{s+1}'(t)=\frac{t}{2s}I_s(t),\quad I_1(t)=\frac{1}{2}e^{-|t|} \quad (t\in\mathbb{R}). \tag{5.58}
$$

5.2.2 Discrete Inverse Source Problems

For different N points $\{x_j\}_{j=1}^N$, we consider the bounded linear operators from the RKHS $H_{K_h}(\mathbb{R})$ into $\mathbb{R}$:

$$H_{K_h}(\mathbb{R}) \ni g \mapsto y(x_j) \in \mathbb{R} \quad (j = 1, 2, \ldots, N) \tag{5.59}$$

by the integral representation (5.12) and with $y(x) \equiv f_{\lambda,g}^*(x)$ by way of (5.7). It is understood that it is an approximate solution of the equation (5.40) and we consider its inversion. We will write them in the form

$$\mathbf{M} : H_{K_h}(\mathbb{R}) \ni g \mapsto (y(x_1), y(x_2), \ldots, y(x_N)) \in \mathbb{R}^N. \tag{5.60}$$

It is important to realize that

$$y(x_j) = \langle \mathbf{M}g, \mathbf{e}_j \rangle_{\mathbb{R}^N} = \langle g, \mathbf{M}^* \mathbf{e}_j \rangle_{H_{K_h}(\mathbb{R})}. \tag{5.61}$$

Therefore, the image space of the linear mapping $\mathbf{M}$ forms the Hilbert space H_B admitting the reproducing kernel

$$B := \langle \mathbf{M}^* \mathbf{e}_{j'}, \mathbf{M}^* \mathbf{e}_j \rangle_{H_{K_h}(\mathbb{R})} := \{b_{j,j'}\}_{j,j'=1,2,\ldots,N}. \tag{5.62}$$

Theorem 5.3. *The matrix $B = \{b_{j,j'}\}_{j,j'=1}^N$ above is positive definite, and its entries are given by*

$$b_{j,j'} = \frac{1}{2\pi} \int_{-\frac{\pi}{h}}^{\frac{\pi}{h}} \frac{|-\eta^2 - i\alpha\eta + \beta|^2 e^{-i\eta(x_j - x_{j'})}}{(\lambda(\eta^2+1)^2 + |-\eta^2 + i\alpha\eta + \beta|^2)^2} d\eta. \tag{5.63}$$

Proof. Using the structure of $H_{K_h}(\mathbb{R})$ and the definition of $\mathbf{M}^*$, we observe that

$$\mathbf{M}^* \mathbf{e}_j(x) = \langle \mathbf{e}_j, \mathbf{M}(K_h)_x \rangle_{\mathbb{R}^N} = \frac{1}{2\pi} \int_{-\frac{\pi}{h}}^{\frac{\pi}{h}} \frac{(-\eta^2 - i\alpha\eta + \beta) e^{-i\eta(x - x_j)}}{\lambda(\eta^2+1)^2 + |-\eta^2 + i\alpha\eta + \beta|^2} d\eta.$$

Thus, the formula (5.63) follows, which by itself yields the conclusion that the matrix $B = \{b_{j,j'}\}_{j,j'=1}^N$ is positive definite.

Therefore, in an analogous way as in Sect. 5.2.1, we are in the position of obtaining the following conclusion directly:

Theorem 5.4. *For any given N values $\mathbf{d} = \{d_j\}_{j=1}^N$, among the $H_{K_h}(\mathbb{R})$-functions g taking the values*

$$Lf_{\lambda,g}^*(x_j) = g(x_j), \quad f_{\lambda,g}^*(x_j) = d_j, \quad (j = 1, 2, \ldots, N), \tag{5.64}$$

the function $g_{\mathbf{d}}^*(x)$ with the minimum norm $\|g\|_{H_{K_h}(\mathbb{R})}$ is uniquely determined and it is represented as follows:

$$g_{\mathbf{d}}^* = \sum_{j,j'=1}^{N} d_j \widetilde{b_{j,j'}} \mathbf{M}^* \mathbf{e}_{j'}, \tag{5.65}$$

where the $b_{j,j'}$ are provided by (5.63).

In the Sobolev space of order s with $s \geq 1$, by using

$$\mathbf{M}^* \mathbf{e}_j(x) = \langle \mathbf{M}^* \mathbf{e}_j, (K_s)_x \rangle_{H_{K_s}(\mathbb{R})},$$

we obtain the following corresponding formulas:

$$\begin{aligned} \mathbf{M}^* \mathbf{e}_j(x) &= \langle \mathbf{e}_j, \mathbf{M}(K_s)_x \rangle_{\mathbb{R}^N} \\ &= \frac{1}{2\pi} \int_{\mathbb{R}} \frac{(-\eta^2 - i\alpha\eta + \beta) e^{-i\eta(x-x_j)}}{(1+\eta^2)^s (\lambda(\eta^2+1)^2 + |-\eta^2 + i\alpha\eta + \beta|^2)} d\eta. \end{aligned}$$

Hence, we have for this case

$$b_{j,j'} = \frac{1}{2\pi} \int_{\mathbb{R}} \frac{|-\eta^2 - i\alpha\eta + \beta|^2 e^{i\eta(x_j - x_{j'})}}{(1+\eta^2)^s (\lambda(\eta^2+1)^2 + |-\eta^2 + i\alpha\eta + \beta|^2)^2} d\eta, \tag{5.66}$$

which can be used in (5.65).

We followed [86] in this section.

Chapter 6
Applications to Partial Differential Equations

In this chapter, we gather concrete results on typical linear partial differential equations using Tikhonov regularization and a finite number of data.

6.1 Poisson's Equation

We start with the fundamental Poisson equation. In Sect. 6.1 we consider the Poisson equation on $\mathbb{R}^n$:

$$\Delta u = g.$$

Here we suppose $g \in L^2(\mathbb{R}^n)$.

6.1.1 Construction of the Solutions

Recall that $H^s(\mathbb{R}^n)$ is defined in Theorem 1.13 by the norm

$$\|f\|_{H^s(\mathbb{R}^n)} = \left(\int_{\mathbb{R}^n} (1 + |\xi|^2)^s |\mathscr{F}f(\xi)|^2 \, d\xi \right)^{\frac{1}{2}}. \tag{6.1}$$

Using Tikhonov regularization, we obtain the following:

Theorem 6.1 ([293, Theorem 1]). *Assume that $s \geq 2$ if $n \leq 3$ and $s > \frac{n}{2}$ if $n \geq 4$. Set*

$$Q_{\lambda,s}(x) \equiv \frac{1}{(2\pi)^n} \int_{\mathbb{R}} \frac{-|p|^2 \exp(ip \cdot x) \, dp}{\lambda(|p|^2 + 1)^s + |p|^4} \quad (x \in \mathbb{R}^n). \tag{6.2}$$

S. Saitoh, Y. Sawano, *Theory of Reproducing Kernels and Applications*, Developments in Mathematics 44, DOI 10.1007/978-981-10-0530-5_6

Then the following minimizing problem

$$\min_{F \in H^s(\mathbb{R}^n)} \left\{ \lambda \, \|F\|^2_{H^s(\mathbb{R}^n)} + \|g - \Delta F\|^2_{L^2(\mathbb{R}^n)} \right\} \tag{6.3}$$

admits a unique solution $F^*_{\lambda,s,g} = g * Q_{\lambda,s}$. *Furthermore, if* $F \in H^s(\mathbb{R}^n)$, *then we have*

$$\lim_{\lambda \downarrow 0} F^*_{\lambda,s,\Delta F} = F \tag{6.4}$$

in the topology of $H^s(\mathbb{R}^n)$ *and therefore in the uniform topology.*

Proof. This is a generality of the Tikhonov regularization except for the representation of the convolution kernel $Q_{\lambda,s}$. We will prove (6.2).

In this minimizing problem (6.3) we regard Δ as a continuous operator from $H^s(\mathbb{R}^n)$ to $L^2(\mathbb{R}^n)$. The minimizer of this problem is given by

$$F^*_{\lambda,s,g}(x) = \Delta(\lambda + \Delta^* \Delta)^{-1}[K_x]. \tag{6.5}$$

Hence we are concerned with the expression of the adjoint $\Delta^* : L^2(\mathbb{R}^n) \to H^s(\mathbb{R}^n)$; look for the operator satisfying

$$\langle \Delta f, g \rangle_{L^2(\mathbb{R}^n)} = \langle f, \Delta^* g \rangle_{L^2(\mathbb{R}^n)}$$

for all $f \in H^s(\mathbb{R}^n)$ and $g \in L^2(\mathbb{R}^n)$. Let $g \in L^2(\mathbb{R}^n)$.

If there exists such an operator Δ^*, then we have

$$\int_{\mathbb{R}^n} \Delta f(x) \cdot \overline{g(x)} \, dx = \int_{\mathbb{R}^n} (1 + |\xi|^2)^s \mathscr{F} f(\xi) \cdot \overline{\mathscr{F}[\Delta^* g](\xi)} \, d\xi$$

for all $f \in H^s(\mathbb{R}^n)$. If we define $F \in L^2(\mathbb{R}^n)$ by

$$\mathscr{F} F(\xi) \equiv (1 + |\xi|^2)^s \mathscr{F} f(\xi) \quad (\xi \in \mathbb{R}^n), \tag{6.6}$$

then we have

$$\Delta f = -\mathscr{F}^{-1}[|\xi|^2 (1 + |\xi|^2)^{-s} \mathscr{F} F]. \tag{6.7}$$

Hence it follows that the adjoint Δ^* is given by

$$\Delta^* g = -\mathscr{F}^{-1}[|\xi|^2 (1 + |\xi|^2)^{-s} \mathscr{F} g] \quad (g \in L^2(\mathbb{R}^n)). \tag{6.8}$$

As a consequence, we obtain

$$\Delta(\lambda + \Delta^* \Delta)^{-1} g = \mathscr{F}^{-1}[-|\xi|^2 (\lambda + |\xi|^4 (1 + |\xi|^2)^{-s})^{-1} \mathscr{F} g]. \tag{6.9}$$

Using the reproducing kernel by (6.1), we have

$$
\begin{aligned}
&\Delta K_x(y)\\
&= \left[\frac{1}{(2\pi)^{\frac{n}{2}}}\mathscr{F}^{-1}\left[\frac{-|\xi|^2}{(\lambda+|\xi|^4(1+|\xi|^2)^{-s})(1+|\xi|^2)^s}\right](\cdot - x)\right](y)\\
&= \mathscr{F}^{-1}\left[-\frac{|\xi|^2}{(2\pi)^{\frac{n}{2}}(\lambda(1+|\xi|^2)^s+|\xi|^4)}e^{-ix\cdot\xi}\right](y)\\
&= -(2\pi)^{-n}\int_{\mathbb{R}^n}\frac{|\xi|^2}{\lambda(1+|\xi|^2)^s+|\xi|^4}e^{i(y-x)\cdot\xi}\,d\xi.
\end{aligned}
$$

As a consequence, we obtain

$$
F^*_{\lambda,s,g}(x) = -\frac{1}{(2\pi)^n}\int_{\mathbb{R}^n} g(y)\left(\int_{\mathbb{R}^n}\frac{|\xi|^2\exp(i(x-y)\cdot\xi)}{\lambda(1+|\xi|^2)^s+|\xi|^4}\,d\xi\right)dy. \tag{6.10}
$$

This proves (6.2). The proof of (6.4) is similar to the other cases.

We have the following estimate of $F^*_{\lambda,s,g}$:

Theorem 6.2. *Let s satisfy $s > \frac{n}{2}$ and $s \geq 2$ and let $\lambda > 0$. Then the function $F^*_{\lambda,s,g}$, given by* (6.5)*, satisfies*

$$
\|F^*_{\lambda,s,g}\|^2_{H^s(\mathbb{R}^n)} \leq \frac{1}{4\lambda}\int_{\mathbb{R}^n}|g(x)|^2\,dx. \tag{6.11}
$$

Proof. We have

$$
\mathscr{F}F^*_{\lambda,s,g}(\xi) = \mathscr{F}g(\xi)\cdot\frac{-|p|^2\exp(ip\cdot\xi)}{\lambda(|p|^2+1)^s+|p|^4} \tag{6.12}
$$

from (6.10). Therefore, it follows that

$$
\begin{aligned}
\|F^*_{\lambda,s,g}\|^2_{H^s(\mathbb{R}^n)} &= \int_{\mathbb{R}^n}|\mathscr{F}g(p)|^2\left(\frac{|p|^2(|p|^2+1)^s}{\lambda(|p|^2+1)^s+|p|^4}\right)^2 dp\\
&= \frac{1}{\lambda}\int_{\mathbb{R}^n}|\mathscr{F}g(p)|^2\left(\frac{\lambda^{\frac{1}{2}}|p|^2(|p|^2+1)^{\frac{s}{2}}}{\lambda(|p|^2+1)^s+|p|^4}\right)^2 dp\\
&\leq \frac{1}{4\lambda}\int_{\mathbb{R}^n}|\mathscr{F}g(p)|^2\,dp = \frac{1}{4\lambda}\int_{\mathbb{R}^n}|g(x)|^2\,dx,
\end{aligned}
$$

proving (6.11).

In general, we have:

Theorem 6.3. *Let s satisfy $s > \dfrac{n}{2}$ and $s \geq 2$ and let $\lambda > 0$. Then the error $F^*_{\lambda,s,\Delta F} - F$ can be written as*

$$F^*_{\lambda,s,\Delta F} - F = -\mathscr{F}^{-1}\left(\frac{\lambda(1+|\xi|^2)^{\frac{s}{2}}}{\lambda(1+|\xi|^2)^s + |\xi|^4}\mathscr{F}F\right) \tag{6.13}$$

for any $F \in H^s(\mathbb{R}^n)$.

Proof. This is just a matter of simple arithmetic. Indeed, by (6.10) we have

$$\begin{aligned} F^*_{\lambda,s,\Delta F}(x) &= \frac{1}{(2\pi)^n}\int_{\mathbb{R}^n} F(y)\left(\int_{\mathbb{R}^n} \frac{|\xi|^4 e^{i(x-y)\cdot\xi}\,d\xi}{\lambda(1+|\xi|^2)^s + |\xi|^4}\right) dy \\ &= \mathscr{F}^{-1}\left(\frac{|\xi|^4\mathscr{F}F}{\lambda(1+|\xi|^2)^s + |\xi|^4}\right)(x) \end{aligned}$$

from (6.10). Hence we obtain (6.13).

Now we are oriented forwards the error estimate. In terms of the Sobolev norm $\|\cdot\|_{H^s(\mathbb{R}^n)}$, we obtain:

Theorem 6.4. *Let s satisfy $s > \dfrac{n}{2}$ and $s \geq 2$ and let $\lambda > 0$. Let $\delta > 0$ and assume that $g, g_\delta \in L^2(\mathbb{R}^n)$ satisfy*

$$\|g - g_\delta\|_{L^2(\mathbb{R}^n)} \leq \delta. \tag{6.14}$$

Then

$$\|F^*_{\lambda,s,g_\delta} - F^*_{\lambda,s,g}\|_{H^s(\mathbb{R}^n)} \leq \frac{\delta}{2\sqrt{\lambda}}. \tag{6.15}$$

Proof. First, we have

$$\mathscr{F}F^*_{\lambda,s,g_\delta}(p) = -\frac{\mathscr{F}g_\delta(p)|p|^2}{\lambda(|p|^2+1)^s + |p|^4}, \quad \mathscr{F}F^*_{\lambda,s,g}(p) = -\frac{\mathscr{F}g(p)|p|^2}{\lambda(|p|^2+1)^s + |p|^4}.$$

Hence we obtain

$$\begin{aligned} (|p|^2+1)^s|\mathscr{F}F^*_{\lambda,s,g_\delta}(p) - \mathscr{F}F^*_{\lambda,s,g}(p)|^2 &\leq \frac{|\mathscr{F}g_\delta(p) - \mathscr{F}g(p)|^2(|p|^2+1)^s}{4\lambda(|p|^2+1)^s} \\ &= \frac{|\mathscr{F}g_\delta(p) - \mathscr{F}g(p)|^2}{4\lambda}. \end{aligned}$$

Integrating this over $\mathbb{R}^n$ and using the Plancherel theorem as well as (6.14), we obtain the desired result (6.15).

Let us denote by $W^{r+n+1,1}(\mathbb{R}^n)$ the set of all integrable functions f such that weak derivatives exist up to order $r+n+1$ and all partial derivatives belong to $L^1(\mathbb{R}^n)$. We set

$$\|f\|_{W^{r+n+1,1}(\mathbb{R}^n)} \equiv \sum_{|\alpha|\le r+n+1} \|\partial^\alpha f\|_{L^1(\mathbb{R}^n)}.$$

For the speed of convergence in (6.10), we obtain:

Theorem 6.5. *Let $s=2$ when $n\le 3$ and $s>n/2$ when $n\ge 4$. Fix $F\in H^s(\mathbb{R}^n)$ and set $g=\Delta F$. Define*

$$\eta_1(\lambda,r) \equiv \int_{B(\sqrt[4]{\lambda})} (|p|^2+1)^r |\mathscr{F}F(p)|^2\,dp$$

$$\eta_2(\lambda,r) \equiv \int_{\mathbb{R}^n\setminus B(\sqrt[2s-4]{\lambda})} (|p|^2+1)^r |\mathscr{F}F(p)|^2 dp$$

for $\lambda<1$ and $r\in[0,s]$.

1. For all $0<\lambda<1$,

$$\|F^*_{\lambda,s,g}-F\|_{H^r(\mathbb{R}^n)} \lesssim \lambda\|F\|_{H^r(\mathbb{R}^n)} + \sqrt{\eta_1(\lambda,r)}. \tag{6.16}$$

Moreover, if $F\in W^{r+n+1,1}(\mathbb{R}^n)\cap H^s(\mathbb{R}^n)$, then

$$\|F^*_{\lambda,s,g}-F\|_{H^r(\mathbb{R}^n)} \lesssim \lambda\|F\|_{H^s(\mathbb{R}^n)} + \lambda^{n/8}\|F\|_{W^{r+n+1,1}(\mathbb{R}^n)}. \tag{6.17}$$

2. For all $0<\lambda<1$,

$$\|F^*_{\lambda,s,g}-F\|_{H^r(\mathbb{R}^n)} \lesssim \lambda\|F\|_{H^s(\mathbb{R}^n)} + \sqrt{\eta_1(\lambda,r)} + \sqrt{\eta_2(\lambda,r)}. \tag{6.18}$$

Furthermore, when $s>2$,

$$\|F^*_{\lambda,s,g}-F\|_{H^r(\mathbb{R}^n)} \lesssim (\lambda+\lambda^{\frac{s-r}{2(s-2)}})\|F\|_{H^s(\mathbb{R}^n)} + \sqrt{\eta_1(\lambda,r)}. \tag{6.19}$$

Moreover, if we assume in addition that $F\in W^{r+n+1,1}(\mathbb{R}^n)\cap H^s(\mathbb{R}^n)$, then we have

$$\|F^*_{\lambda,s,g}-F\|_{H^r(\mathbb{R}^n)} \lesssim (\lambda+\lambda^{\frac{s-r}{2(s-2)}})\|F\|_{H^s(\mathbb{R}^n)} + \lambda^{n/8}\|F\|_{W^{r+n+1,1}(\mathbb{R}^n)}. \tag{6.20}$$

Proof. We have

$$\mathscr{F}F(p) = -\mathscr{F}g(p)|p|^{-2},\ \mathscr{F}F^*_{\lambda,s,g}(p)$$

$$= \frac{-\mathscr{F}g(p)|p|^2}{\lambda^2(|p|^2+1)^s + |p|^4} = \frac{\mathscr{F}F(p)}{\lambda^2(|p|^2+1)^s|p|^{-4}+1}.$$

Hence

$$\mathscr{F}F^*_{\lambda,s,g}(p) - \mathscr{F}F(p) = \frac{-\mathscr{F}F(p)\lambda^2(|p|^2+1)^s|p|^{-4}}{\lambda^2(|p|^2+1)^s|p|^{-4}+1}. \tag{6.21}$$

We distinguish two cases; $s = 2$ and $s > 2$.

1. For $s = 2$, the function $f(t) = (t+1)^s t^{-2}$ is decreasing on $(0, +\infty)$. Therefore, for any p with $|p| > \sqrt[4]{\lambda}$, we have

$$\lambda^2(|p|^2+1)^s|p|^{-4} = \lambda^2 f(|p|^2) \le \lambda^2 f(\lambda^{1/2}) = \lambda(\lambda^{1/2}+1)^s \lesssim \lambda.$$

Therefore, from (6.21) we have

$$(|p|^2+1)^{r/2}|\mathscr{F}F^*_{\lambda,s,g}(p) - \mathscr{F}F(p)| \lesssim \lambda|\mathscr{F}F(p)|(|p|^2+1)^{r/2} \tag{6.22}$$

for $|p| > \sqrt[4]{\lambda}$. It follows that

$$\int_{\mathbb{R}\backslash B(\sqrt[4]{\lambda})} (|p|^2+1)^r|\mathscr{F}F^*_{\lambda,s,g}(p) - \mathscr{F}F(p)|^2 dp$$

$$\lesssim \lambda^2 \int_{\mathbb{R}\backslash B(\sqrt[4]{\lambda})} (|p|^2+1)^r|\mathscr{F}F(p)|^2 dp \lesssim \lambda^2\|F\|^2_{H^r}. \tag{6.23}$$

Meanwhile, from (6.21) we have

$$\int_{B(\sqrt[4]{\lambda})} (|p|^2+1)^r|\mathscr{F}F^*_{\lambda,s,g}(p) - \mathscr{F}F(p)|^2 dp$$

$$\le \int_{B(\sqrt[4]{\lambda})} (|p|^2+1)^r|\mathscr{F}F(p)|^2 dp \equiv \eta_1(\lambda, r). \tag{6.24}$$

Combining (6.23) and (6.24) gives

$$\int_{\mathbb{R}} (|p|^2+1)^r|\mathscr{F}F^*_{\lambda,s,g}(p) - \mathscr{F}F(p)|^2 dp \lesssim \lambda^2\|F\|^2_{H^r} + \eta_1(\lambda, r). \tag{6.25}$$

If $F \in W^{r+n+1,1}(\mathbb{R}^n) \cap H^s(\mathbb{R}^n)$, then $(|p|^2+1)^{r/2}\mathscr{F}F \in L^\infty(\mathbb{R}) \cap L^2(\mathbb{R})$. Consequently, we obtain

$$\eta_1(\lambda, r) \le \sup_{p\in\mathbb{R}}(|p|^2+1)^r|\mathscr{F}F(p)|^2 \int_{|p|<\sqrt[4]{\lambda}} dp. \tag{6.26}$$

Hence we have

$$\eta_1(\lambda, r) \lesssim \|F\|_{W^{r+n+1,1}(\mathbb{R})}{}^2 \lambda^{n/4}. \tag{6.27}$$

From (6.25) and (6.26), we deduce (6.16) and (6.17).

2. For $s > 2$, the function $f(x) = (x+1)^s x^{-2}$ never attains a local maximum in $(0,\infty)$. Hence, for $\sqrt[4]{\lambda} \le |p| \le \sqrt[2s-4]{\lambda^{-1}}$,

$$\lambda^2(|p|^2+1)^s|p|^{-4} \le \lambda^2 \max\{f(\lambda^{1/2}), f(\sqrt[s-2]{\lambda^{-1}})\}. \tag{6.28}$$

We therefore have

$$\lambda^2(|p|^2+1)^s|p|^{-4} \lesssim \lambda, \tag{6.29}$$

whenever $\sqrt[4]{\lambda} \le |p| \le \sqrt[2s-4]{\lambda^{-1}}$. From (6.21) and (6.29), we have

$$\begin{aligned}
&\int_{B(\sqrt[2s-4]{\lambda^{-1}})\backslash B(\sqrt[4]{\lambda})} (|p|^2+1)^r|\mathscr{F}F^*_{\lambda,s,g}(p) - \mathscr{F}F(p)|^2 dp \\
&\lesssim \lambda^2 \int_{B(\sqrt[2s-4]{\lambda^{-1}})\backslash B(\sqrt[4]{\lambda})} (|p|^2+1)^r|\mathscr{F}F(p)|^2 dp \\
&\lesssim \lambda^2\|F\|^2_{H^r(\mathbb{R}^n)}.
\end{aligned}$$

Going through the same argument as above, we have (6.18) by criteria of this estimate.

Now, for every $r \in (0, s)$, we have

$$\begin{aligned}
\eta_2(\lambda, r) &\le \frac{1}{(1+\sqrt[s-2]{\lambda^{-1}})^{s-r}} \int_{\mathbb{R}^n\backslash B(\sqrt[2s-4]{\lambda})} (|p|^2+1)^s|\mathscr{F}F(p)|^2 dp \\
&\le C\lambda^{\frac{s-r}{s-2}}\|F\|_{H^s(\mathbb{R})}.
\end{aligned}$$

As a result we have (6.19). If we have, in addition, that $F \in W^{r+n+1,1}(\mathbb{R}^n)$, then in the same way, we obtain (6.20).

We modified [293, Theorem 4] somehow to formulate Theorem 6.5.

The solution (6.10) yields a practical formula for the Poisson equation. Experimental results using computers are also given in [293, Section 7]. There, we will see that in order to overcome the difficulty in the equation, we need a lot of work;

that is, we must take λ very small and we must calculate the integral (6.10) in the numerical sense. Computers help us with this hard work – calculating the integral; they calculate the integral with a very small λ.

See [208, Theorem 1] for an approach similar to Theorem 6.5 in spirit.

See [293, Section 7] for more details.

6.1.2 Source Inversions in the Poisson Equation

We will consider the Poisson equation

$$\Delta u = -\rho(\mathbf{r}) \quad \text{in} \quad r < a \tag{6.30}$$

for a real-valued source function ρ whose support is contained in the interior of a sphere $r < a$; r denotes the distance $r = |\mathbf{r}|$ from the origin. We wish to look for the source function ρ satisfying (6.30) from some information outside of the support of ρ. This problem has a practical application to determining the source function ρ from the potential u. Indeed, this inverse problem was proposed by Laplace about 200 years ago and many mathematicians have worked on this problem. Mathematical solutions for this problem in the framework of $L^2(d\mathbf{r})$ spaces based on the general theory of Chap. 2 and in miscellaneous situations were given in [388].

The Poisson equation (6.30) has physical interpretations, namely the correspondence

$$\rho \longleftrightarrow u \tag{6.31}$$

can be understood as

$$\begin{aligned} \text{mass distribution} &\longleftrightarrow \text{gravitational potential} \\ \text{current} &\longleftrightarrow \text{magnetic field} \\ \text{charge} &\longleftrightarrow \text{electrostatic field} \\ \text{heat source} &\longleftrightarrow \text{temperature,} \end{aligned}$$

where we consider only steady state situations. Hence we can apply our inverse formula to various physical situations.

In particular, note that from the whole data outside of support for the density ρ, we cannot determine the density fully and its part is, in general, determined. In particular, note that we can determine the harmonic part of the source, even if we use all the data outside of the support of the source. See [51, 388] for the detailed structure. Meanwhile, we used the whole data of a sphere $\{|\mathbf{r}| = a'\}$ $(a' > a)$ outside of the support of the density. So, following the recent general idea in Chap. 3, we will give new inversion formulas using a finite number of observation data outside of the support of the density.

For the sake of our method, for the outside information of the potential u, we will be able to use various physical data; we, however, use the potential values $u(P)$ and some of its derivatives at some points Q, as typical physical data. We will see, in this situation the method, and we can consider a very general situation for the problem.

We will, for example, consider our problems for the 3-dimensional case and the 2-dimensional case clearly (for other interesting situations, we can consider the cases similarly, see also [388]).

We assume that the support of ρ is contained in the sphere $r < a$. So, for $\mathbf{r} \in \mathbb{R}^3$ and $|\mathbf{r}| = r$, we have the integral representation of the solution u of Poisson's equation (6.30)

$$u(\mathbf{r}'; \rho) = \frac{1}{4\pi} \int_{r<a} \frac{1}{|\mathbf{r}' - \mathbf{r}|} \rho(r) d\mathbf{r} \tag{6.32}$$

in $\mathbb{R}^3$, where the source function ρ satisfies

$$\int_{r<a} \rho(\mathbf{r})^2 d\mathbf{r} < \infty. \tag{6.33}$$

For the two-dimensional potential, we have

$$u(\mathbf{r}'; \rho) = \frac{1}{2\pi} \int_{r<a} \left(\log \frac{1}{|\mathbf{r}' - \mathbf{r}|} \right) \rho(\mathbf{r}) d\mathbf{r} < \infty \tag{6.34}$$

for a source function ρ satisfying

$$\int_{r<a} \rho(\mathbf{r})^2 d\mathbf{r} < \infty \tag{6.35}$$

whose support is contained in the disk $r < a$.

Following the general principle, when we consider inversions for these singular integral equations, we must calculate the related singular integrals of the reproducing kernels as we see also from the following arguments; also their calculations are very involved, and so we will use the regularized integral representation in Lemma 5.1 based on Tikhonov regularization and the theory of reproducing kernels.

Good numerical experiments were given for the Poisson equation. Therefore, instead of the exact solutions with involved singular integral representations for the Poisson equation, we will apply the approximate solutions in Lemma 5.1 with regular integral representations; that is, we consider the bounded linear operator L from an RKHS $H_K(\mathbb{R}^3) \ni \rho \mapsto F^*_{\lambda,s,\rho} \equiv u(\cdot; \lambda, s, \rho)$, which is given in the theorem.

Recall that the Paley Wiener space $H_{K_h}(\mathbb{R})$ was defined by (1.6). Meanwhile, in Lemma 5.1 when we apply Paley Wiener spaces, we obtain the following theorem: Of course, the function spaces are strongly restricted, however, in the idea of Tikhonov regularization, we consider approximations for solutions and in the case of the Paley Wiener spaces, representations of the formulas will be much simpler than in the case of the Sobolev spaces as we will see. So, we refer to the case of Paley Wiener spaces obtained by Lemma 5.1 in the same way as in the Sobolev spaces.

Theorem 6.6. *Let $n \geq 1$ and define*

$$Q_{\lambda,h}(\eta) \equiv -\frac{1}{(2\pi)^n}\int_{(-\frac{\pi}{h},\frac{\pi}{h})^n} \frac{|p|^2 \exp(-ip\eta)}{\lambda + |p|^4} dp \quad (\eta \in \mathbb{R}). \tag{6.36}$$

For any given function $g \in L^2(\mathbb{R}^n)$ and for any $\lambda > 0$, the best approximate function $F^*_{\lambda,h}$ in the sense of

$$\inf_{F \in H_{K_h}(\mathbb{R}^n)} \left\{ \lambda \|F\|^2_{H_{K_h}(\mathbb{R}^n)} + \|g - \Delta F\|^2_{L^2(\mathbb{R}^n)} \right\} = \lambda \|F^*_{\lambda,h}\|^2_{H_{K_h}(\mathbb{R}^n)} + \|g - \Delta F^*_{\lambda,h}\|^2_{L^2(\mathbb{R}^n)}$$

exists uniquely and $F^*_{\lambda,h}$ is represented by $F^*_{\lambda,h} = g * Q_{\lambda,h}$. If, for $F \in H_{K_h}(\mathbb{R}^n)$, we consider the solution u_F to $\Delta u_F(x) = F(x)$ and we take $g = \Delta u_F$, then

$$F^*_{\lambda,h} \to u_F, \tag{6.37}$$

as $\lambda \downarrow 0$ uniformly in view of Remark 1.3.

Recall that the Paley Wiener space $H_{K_h}(\mathbb{R})$ was defined by (1.6). Note that the difference between cases in the Sobolev spaces and Paley Wiener spaces is that *the factor $\lambda(|p|^2 + 1)^s$ is changed just into λ, and put $\chi_h(p)$ in the case of Paley Wiener spaces.* Given these changes in the formulas obtained in the Sobolev spaces, parallel formulas for the Paley Wiener spaces are valid as well.

Direct Representations of the Sources Using Finite Data

For the sake of regularization, we discuss our problems in a unified way and in a very general situation. As one typical problem, for example, we have the boundary condition that the normal derivative of the potential u be zero on the boundary, and as the physical observation data, we can consider the values of the potential on the boundary. Therefore, we can realize these conditions in the followings. For a unit vector ν, we consider the directional derivative at a point Q as follows:

$$\frac{\partial u}{\partial \nu}(Q) \equiv \nu \cdot \nabla u(Q). \tag{6.38}$$

For some points $\{P_j\}_{j=1}^N$ and $\{Q_j\}_{j=1}^M$ on the space $\mathbb{R}^n$ that may be points on the support of ρ, we will consider the bounded linear operators L

$$H_K(\mathbb{R}^n) \ni g = \rho \mapsto F^*_{\lambda,s,g}(P) = u(P; \lambda, s, \rho).$$

We take M unit vectors b_j for $j = 1, 2, \ldots, M$ and we consider M derivatives at Q_j as follows:

$$\frac{\partial u}{\partial \nu}(Q_j). \tag{6.39}$$

We will introduce the bounded linear operator $\mathbf{L} : H_K(\mathbb{R}^n) \to \mathbb{R}^{N+M}$ for ρ to

$$\left(u(P_1;\lambda,s,\rho), u(P_2;\lambda,s,\rho), \ldots, u(P_N;\lambda,s,\rho), \frac{\partial u}{\partial\nu}(Q_1), \frac{\partial u}{\partial\nu}(Q_2), \ldots, \frac{\partial u}{\partial\nu}(Q_M)\right).$$

We will choose a standard orthonormal system $\{\mathbf{e}_j\}_{j=1}^{N+M}$ in the space $\mathbb{R}^{N+M}$. Then we see that

$$u(P_j;\lambda,s,\rho) = \langle \mathbf{L}\rho, \mathbf{e}_j\rangle_{\mathbb{R}^{N+M}} = \langle \rho, \mathbf{L}^*\mathbf{e}_j\rangle_{H_K(\mathbb{R}^n)} \tag{6.40}$$

and

$$\frac{\partial u}{\partial\nu}(Q_j) = \langle \mathbf{L}\rho, \mathbf{e}_{N+j}\rangle_{\mathbb{R}^{N+M}} = \langle \rho, \mathbf{L}^*\mathbf{e}_{N+j}\rangle_{H_K(\mathbb{R}^n)}. \tag{6.41}$$

Therefore the image space of a_j forms the Hilbert space H_A admitting the reproducing kernel

$$a_{j,j'} \equiv \langle \mathbf{L}^*\mathbf{e}_{j'}, \mathbf{L}^*\mathbf{e}_j\rangle_{H_K(\mathbb{R}^n)}. \tag{6.42}$$

Here note that $a_{j,j'}$ is calculated using $\langle \mathbf{L}^*\mathbf{e}_j, K_x\rangle_{H_K} = \langle \mathbf{e}_j, \mathbf{L}[K_x]\rangle_{\mathbb{R}^{N+M}}$ as follows: for $1 \le j, j' \le N$ or M,

$$\mathbf{L}^*\mathbf{e}_j(x) = \langle \mathbf{L}^*\mathbf{e}_j, K_x\rangle_{H_K} = \frac{1}{(2\pi)^n}\int_{\mathbb{R}^n} \frac{-|\xi|^2\exp(-i\xi\cdot(x-P_j))d\xi}{(\lambda(|\xi|^2+1)^s+|\xi|^4)(1+|\xi|^2)^s}$$

and using $\langle \mathbf{L}^*\mathbf{e}_{N+j}, K_x\rangle_{H_K} = \langle \mathbf{e}_{N+j}, \mathbf{L}K_x\rangle_{\mathbb{R}^{N+M}}$, we obtain

$$\mathbf{L}^*\mathbf{e}_{N+j}(x) = \langle \mathbf{L}^*\mathbf{e}_{N+j}, K_x\rangle_{H_K} = \frac{1}{(2\pi)^n}\int_{\mathbb{R}^n} \frac{-|\xi|^2 i(\xi\cdot b_j)\exp(-i\xi\cdot(x-Q_j))d\xi}{(\lambda(|\xi|^2+1)^s+|\xi|^4)(1+|\xi|^2)^s}.$$

Hence,

$$a_{j,j'} = \frac{1}{(2\pi)^n}\int_{\mathbb{R}^n} \frac{|\xi|^4\exp(i\xi\cdot(P_j-P_{j'}))d\xi}{(1+|\xi|^2)^s(\lambda(|\xi|^2+1)^s+|\xi|^4)^2},$$

$$a_{j,N+j'} = \frac{1}{(2\pi)^n}\int_{\mathbb{R}^n} \frac{|\xi|^4(i\xi\cdot b_{j'})\exp(i\xi\cdot(Q_{j'}-P_j))d\xi}{(1+|\xi|^2)^s(\lambda(|\xi|^2+1)^s+|\xi|^4)^2},$$

$$a_{N+j,j'} = \frac{1}{(2\pi)^n}\int_{\mathbb{R}^n} \frac{|\xi|^4(-i\xi\cdot b_j)\exp(i\xi\cdot(P_{j'}-Q_j))d\xi}{(1+|\xi|^2)^s(\lambda(|\xi|^2+1)^s+|\xi|^4)^2},$$

$$a_{N+j,N+j'} = \frac{1}{(2\pi)^n}\int_{\mathbb{R}^n} \frac{|\xi|^4(\xi\cdot b_j)(\xi\cdot b_{j'})\exp(i\xi\cdot(Q_j-Q_{j'}))d\xi}{(1+|\xi|^2)^s(\lambda(|\xi|^2+1)^s+|\xi|^4)^2}.$$

In particular, note that the matrix $A = \{a_{j,j'}\}_{j,j'=1}^{N+M}$ is positive definite. We thus obtain the inversion formula.

Theorem 6.7. *Let $P_1, P_2, \ldots, P_N, Q_1, Q_2, \ldots, Q_M \in \mathbb{R}^N$. For any given $N + M$ complex values $\mathbf{d} = \{d(P_j)\}_{j=1}^{N+M}$, where it is understood that $P_{N+j} \equiv Q_j$ for $j = 1, 2, \ldots, M$, among the regular solutions of the Poisson integral taking the values*

$$u(P_j; \lambda, s, \rho) = d(P_j) \quad j = 1, 2, \ldots, N \tag{6.43}$$

and

$$\frac{\partial u}{\partial \nu}(Q_j) = d(Q_j) \quad j = 1, 2, \ldots, M, \tag{6.44}$$

and among their inverses, the uniquely determined function $\rho^*(x; \lambda, s, \mathbf{d})$ with the minimum norm $\|\rho\|_{H_K}$ is uniquely determined and represented as follows:

$$\rho^*(x; \mathbf{d}) = \sum_{j,j'=1}^{N+M} d(P_j)\tilde{a}_{j,j'}(\mathbf{L}^*\mathbf{e}_{j'})(x), \tag{6.45}$$

where $\tilde{A} = \overline{A^{-1}} = \{\tilde{a}_{j,j'}\}_{j,j'=1}^{N+M}$.

As before, we calculate that $\mathbf{L}^*\mathbf{e}_j(x) = \langle \mathbf{L}^*\mathbf{e}_j, (K_h)_x\rangle_{H_{K_h}(\mathbb{R})} = \langle \mathbf{e}_j, \mathbf{L}(K_h)_x\rangle_{\mathbb{R}^{N+M}}$. For the Paley Wiener spaces, we have the following corresponding results: for $1 \leq j, j' \leq N$ or M

$$\mathbf{L}^*\mathbf{e}_j(x) = \frac{-1}{(2\pi)^n}\int_{(-\frac{\pi}{h},\frac{\pi}{h})^n} \frac{|\xi|^2\exp(-i\xi\cdot(x-P_j))}{\lambda(|\xi|^2+1)^s+|\xi|^4}d\xi$$

and

$$\begin{aligned}
\mathbf{L}^*\mathbf{e}_{N+j}(x) &= \langle \mathbf{L}^*\mathbf{e}_{N+j}, (K_h)_x\rangle_{H_{K_h}(\mathbb{R})} \\
&= \langle \mathbf{e}_{N+j}, \mathbf{L}[(K_h)_x]\rangle_{\mathbb{R}^{N+M}} \\
&= \frac{-i}{(2\pi)^n}\int_{(-\frac{\pi}{h},\frac{\pi}{h})^n} \frac{|\xi|^2\exp(-i\xi\cdot(x-Q_j))}{\lambda(|\xi|^2+1)^s+|\xi|^4}\xi\cdot b_j d\xi.
\end{aligned}$$

Hence for $j, j' = 1, 2, \ldots, N$,

$$a_{j,j'} = \frac{1}{(2\pi)^n}\int_{(-\frac{\pi}{h},\frac{\pi}{h})^n} \frac{|\xi|^4 e^{i\xi\cdot(P_j-P_{j'})}}{(\lambda(|\xi|^2+1)^s+|\xi|^4)^2}d\xi,$$

$$a_{j,N+j'} = \frac{1}{(2\pi)^n} \int_{(-\frac{\pi}{h},\frac{\pi}{h})^n} \frac{|\xi|^4 (i\xi \cdot b_{j'}) e^{i\xi\cdot(Q_{j'}-P_j)}}{(\lambda(|\xi|^2+1)^s + |\xi|^4)^2} d\xi,$$

$$a_{N+j,j'} = \frac{1}{(2\pi)^n} \int_{(-\frac{\pi}{h},\frac{\pi}{h})^n} \frac{(-i\xi \cdot b_j) \cdot |\xi|^4 e^{i\xi\cdot(P_{j'}-Q_j)}}{(\lambda(|\xi|^2+1)^s + |\xi|^4)^2} d\xi$$

and

$$a_{N+j,N+j'} = \frac{1}{(2\pi)^n} \int_{(-\frac{\pi}{h},\frac{\pi}{h})^n} \frac{|\xi|^4 (\xi \cdot b_j)(\xi \cdot b_{j'}) \exp(i\xi \cdot (Q_j - Q_{j'}))}{(\lambda(|\xi|^2+1)^s + |\xi|^4)^2} d\xi.$$

6.1.3 Constructing Solutions Satisfying Boundary Conditions

Since the Poisson equation is fundamental, we will refer to the construction of the solutions of the Poisson equation satisfying boundary conditions. Our method is valid in a similar way in a general situation.

We consider the problem by the Paley Wiener spaces. Recall again that the Paley Wiener space $H_{K_h}(\mathbb{R})$ was defined by (1.6). We introduce the bounded linear operator, $\mathbf{M} : H_{K_h}(\mathbb{R}) \to \mathbb{R}^{N+M}$ for F to

$$\left((\nabla F)(P_1), \ldots, (\nabla F)(P_N), \frac{\partial(\nabla F)}{\partial \nu}(Q_1), \ldots, \frac{\partial(\nabla F)}{\partial \nu}(Q_M)\right).$$

Then we see that

$$\nabla F(P_j) = \langle \mathbf{M}F, \mathbf{e}_j \rangle_{\mathbb{R}^{N+M}} = \langle F, \mathbf{M}^* \mathbf{e}_j \rangle_{H_{K_h}(\mathbb{R})} \tag{6.46}$$

and

$$\frac{\partial(\nabla F)}{\partial \nu}(Q_j) = \langle \mathbf{M}F, \mathbf{e}_{N+j} \rangle_{\mathbb{R}^{N+M}} = \langle F, \mathbf{M}^* \mathbf{e}_{N+j} \rangle_{H_{K_h}(\mathbb{R})}. \tag{6.47}$$

Therefore the image space forms the Hilbert space H_B admitting the reproducing kernel. Let us denote

$$b_{j,j'} := \langle \mathbf{M}^* \mathbf{e}_{j'}, \mathbf{M}^* \mathbf{e}_j \rangle_{H_{K_h}(\mathbb{R})}. \tag{6.48}$$

Here note that $b_{j,j'}$, given by (6.48), is calculated using $\langle \mathbf{M}^* \mathbf{e}_j, (K_{K_h})_x \rangle_{H_{K_h}(\mathbb{R})} = \langle \mathbf{e}_j, \mathbf{M}K_x \rangle_{\mathbb{R}^{N+M}}$ as follows: for $1 \le j, j' \le N$ or M,

$$\mathbf{M}^* \mathbf{e}_j(x) = \langle \mathbf{M}^* \mathbf{e}_j, (K_{K_h})_x \rangle_{H_{K_h}(\mathbb{R})} = \frac{-1}{(2\pi)^n} \int_{(-\frac{\pi}{h},\frac{\pi}{h})^n} \frac{|\xi|^2}{e^{i\xi\cdot(x-P_j)}} d\xi$$

and

$$\mathbf{M}^*\mathbf{e}_{N+j}(x) = \langle \mathbf{M}^*\mathbf{e}_{N+j}, (K_h)_x\rangle_{H_{K_h}(\mathbb{R})} = \frac{-i}{(2\pi)^n}\int_{(-\frac{\pi}{h},\frac{\pi}{h})^n} |\xi|^2 \frac{\xi\cdot b_j}{\exp(i\xi\cdot(x-Q_j))}d\xi.$$

Hence, for $j, j' = 1, 2, \ldots, N$,

$$b_{j,j'} = \frac{1}{(2\pi)^n}\int_{(-\frac{\pi}{h},\frac{\pi}{h})^n} |\xi|^4 \exp(i\xi\cdot(P_j - P_{j'}))d\xi,$$

$$b_{j,N+j'} = \frac{i}{(2\pi)^n}\int_{(-\frac{\pi}{h},\frac{\pi}{h})^n} |\xi|^4(\xi\cdot b_{j'}) \exp(i\xi\cdot(Q_{j'} - P_j))d\xi,$$

$$b_{N+j,j'} = \frac{-i}{(2\pi)^n}\int_{(-\frac{\pi}{h},\frac{\pi}{h})^n} |\xi|^4(\xi\cdot b_j) \exp(i\xi\cdot(P_{j'} - Q_j))d\xi,$$

$$b_{N+j,N+j'} = \frac{1}{(2\pi)^n}\int_{(-\frac{\pi}{h},\frac{\pi}{h})^n} |\xi|^4(\xi\cdot b_j)(\xi\cdot b_{j'}) \exp(i\xi\cdot(Q_j - Q_{j'}))d\xi.$$

Note also that the matrix $B = \{b_{j,j'}\}_{j,j'=1}^{N+M}$ is positive definite. We thus obtain the inversion formula.

Theorem 6.8. *Let $P_1, P_2, \ldots, P_N, Q_1, Q_2, \ldots, Q_M \in \mathbb{R}^N$. For any $N+M$ values $\mathbf{d} = \{d_j\}_{j=1}^{N+M} = \{d(P_j)\}_{j=1}^{N+M}$, where it is understood that $d(P_{N+j}) \equiv d(Q_j)$, among the solutions of the Poisson integral taking the values*

$$(\nabla F)(P_j) = d_j \quad j = 1, 2, \ldots, N \tag{6.49}$$

and

$$\frac{\partial(\nabla F)}{\partial \nu}(Q_j) = d_{j+N} \quad j = 1, 2, \ldots, M, \tag{6.50}$$

the solution with the minimum norm $\|F\|_{H_{K_h}(\mathbb{R}^N)}$ is uniquely determined and it is represented as follows:

$$F^*(x;\lambda, h, \mathbf{d}) = \sum_{j,j'=1}^{N+M} d_j\tilde{b}_{j,j'}(\mathbf{M}^*\mathbf{e}_{j'})(x) \quad (x\in\mathbb{R}^N), \tag{6.51}$$

where $\tilde{B} = \{\tilde{b}_{j,j'}\}_{j,j'=1}^{n} = \overline{B^{-1}}$.

See [228, Foreword and References] and, for general references on inverse source problems, see, in particular, [51, 105].

6.2 Laplace's Equation

6.2.1 Numerical Dirichlet Problems

Here we work on a bounded smooth domain D.

We recall the Dirichlet principle that the harmonic function $u(x)$ satisfying the boundary condition

$$u(x) = g(x) \quad \text{on} \quad \partial D \tag{6.52}$$

is the extremal function minimizing the Dirichlet integral on D among a class of functions on D satisfying the boundary condition (6.52). Here, the famous historical fact is the existence problem of the extremal function. So we are considering the Dirichlet problem. Look for a function $u \in H^s(\mathbb{R}^n)$ such that

$$\begin{cases} \Delta u|D = 0 & \text{in } D, \\ u|\partial D = g & \text{on } \partial D. \end{cases}$$

Here s satisfies $s \geq 2$ and $s > n/2$.

Now we would like to clarify the Dirichlet principle by modifying and simplifying it from the viewpoint of numerical analysis as follows: In order to use our theory, as a function space we use the Sobolev Hilbert space $H^s(\mathbb{R}^n)$ and we consider the extremal problem

$$\inf_{F \in H^s(\mathbb{R}^n)} \int_D |\Delta F(x)|^2 dx, \tag{6.53}$$

satisfying the boundary condition (6.52). Intuitively, we can see that the harmonic function $u(x)$ satisfying the boundary condition (6.52) will be the extremal function of (6.53). We wish to discuss clearly and simply the existence of the extremal functions, and furthermore we wish to obtain some good representation of the extremal functions, when they exist. For this purpose, we consider the problem as follows:

To make the matter fall under the scope of our framework, we let

$$L : u \in H^s(\mathbb{R}^n) \mapsto (-\Delta u|D, u|\partial D) \in L^2(\mathbb{R}^n) \times L^2(\partial D).$$

Note that $L^2(\mathbb{R}^n) \times L^2(\partial D)$ bears the structure of a Hilbert space. This Hilbert space leads us to the extremal problem, for fixed $\lambda > 0$ and for any $g \in L^2(\partial D)$

$$\inf_{F \in H^s(\mathbb{R}^n)} \left\{ \lambda \|F\|^2_{H^s(\mathbb{R}^n)} + \|\Delta F\|^2_{L^2(D)} + \|F - g\|^2_{L^2(\partial D)} \right\}. \tag{6.54}$$

In order to simplify the problem, we consider the following.

1. As a function space, we use the Sobolev Hilbert space on the whole space; in this case the space admits the simple reproducing kernel. In the extremal problem, for the flexibility of the Sobolev space, we will be able to use the Sobolev space. In this case the reproducing kernel which is used essentially in our method is extremely simple.
2. For the integral of the ΔF, we consider it on the whole space, not on the domain D. Then the extremal problem will become very simple. Furthermore, in this setting we can consider the inner Dirichlet problem on D and the outer Dirichlet problem on $\overline{D}^c$ at the same time. This will mean that it permits the domain to be simply connected.

For (6.54), after we simplify the matter, we see that the reproducing kernel K_λ of the Hilbert space $H_{K_\lambda}(\mathbb{R})$ with norm square

$$\lambda\|F\|^2_{H^s(\mathbb{R})} + \|\Delta F\|^2_{L^2(\mathbb{R})} \tag{6.55}$$

is

$$K_\lambda(x,y) = \frac{1}{(2\pi)^n}\int_{\mathbb{R}^n} \frac{\exp(ip\cdot(x-y))}{\lambda(|p|^2+1)^s + |p|^4}dp. \tag{6.56}$$

Boundary Conditions

Our strategy is first to represent the extremal function $F^*_{s,\lambda,g}(x)$ in (6.54) in an explicit form and next to consider the limit of this extremal function as $\lambda \downarrow 0$. These procedures are given precisely as follows:

We look for the reproducing kernel $K_{s,\lambda,\Delta}$ for the Hilbert space $H_{K_{s,\lambda,\Delta}}(\mathbb{R})$ with the norm square, for $F \in H^s(\mathbb{R})$,

$$\lambda\|F\|^2_{H^s(\mathbb{R}^n)} + \|\Delta F\|^2_{L^2(D)} + \|F\|^2_{L^2(\partial D)}. \tag{6.57}$$

The reproducing kernel $K_{s,\lambda,\Delta}$ is determined by the functional equation

$$K_\lambda(x,y) = (\lambda I + L^*L)[(K_{s,\lambda,\Delta})_y](x), \tag{6.58}$$

where $L : H_{K_\lambda}(\mathbb{R}) \to L^2(\partial D)$ is the bounded linear operator from

$$H_{K_\lambda}(\mathbb{R}) \quad \text{into} \quad L^2(\partial D). \tag{6.59}$$

In this case, the functional equation is the Fredholm integral equation of the second kind containing the reproducing kernel K_λ. Then the extremal function is represented by g directly as follows:

$$F^*_{s,\lambda,g}(x) = \langle g, L[(K_{s,\lambda,\Delta})_x]\rangle_{L^2(\partial D)}. \tag{6.60}$$

6.2.2 Algorithm by the Iteration Method

In order to consider the boundary value problem in (6.52), we will consider it in (6.54) as follows: For any fixed points $\{x_j\}_{j=1}^N$ on the boundary ∂D and for any given values $\{A_j\}_{j=1}^N$, we consider the extremal problem, for any fixed $\{\lambda_j\}_{j=1}^N(\lambda_j > 0)$

$$\inf_{F \in H^s(\mathbb{R}^n)} \left\{ \lambda \|F\|^2_{H^s(\mathbb{R}^n)} + \|\Delta F\|^2_{L^2(\mathbb{R}^n)} + \sum_{j=1}^N \lambda_j |F(x_j) - A_j|^2 \right\}; \tag{6.61}$$

that is, we approximate the integral in (6.54) by summation. This translation will be reasonable in the sense that

$$\|F - g\|^2_{L^2(\partial D)} \tag{6.62}$$

is replaced by

$$\sum_{j=1}^N \lambda_j |F(x_j) - A_j|^2. \tag{6.63}$$

Then the reproducing kernel K_{λ_j} of the Hilbert space $H_{K_{\lambda_j}}(\mathbb{R})$ with norm square

$$\lambda \|F\|^2_{H^s(\mathbb{R}^n)} + \|\Delta F\|^2_{L^2(\mathbb{R}^n)} + \sum_{j=1}^N \lambda_j |F(x_j)|^2 \tag{6.64}$$

is given in terms of K_λ as follows:

$$K_{\lambda_j}(x, y) = K_\lambda(x, y) - \sum_{j,j'=1}^N \lambda_j K_\lambda(x, x_j) A_{j,j'} K_\lambda(x_{j'}, y), \tag{6.65}$$

where $\{A_{j,j'}\}_{j,j'=1}^N$ is the inverse of the positive matrix $\left\{\delta_{j,j'} + \lambda_j K_\lambda(x_{j'}, x_j)\right\}_{j,j'=1}^N$. See also [388, pp. 81–83] for details. For this direct representation of the reproducing kernel K_{λ_j} which we need, for many points $\{x_j\}_{j=1}^N$ of the boundary ∂D, the size of the matrix $\{A_{j,j'}\}_{j,j'=1}^N$ is large and so this direct representation will not be effective in some case.

In order to overcome this difficulty, we propose a new approach. We start with one point x_1. The reproducing kernel $K_\lambda^{(1)}$ of the Hilbert space with the norm square

$$\lambda \|F\|^2_{H^s(\mathbb{R}^n)} + \|\Delta F\|^2_{L^2(\mathbb{R}^n)} + \sum_{j=1}^1 \lambda_j |F(x_j)|^2$$

$$= \lambda \|F\|^2_{H^s(\mathbb{R}^n)} + \|\Delta F\|^2_{L^2(\mathbb{R}^n)} + \lambda_1 |F(x_1)|^2 \tag{6.66}$$

is given by

$$K_\lambda^{(1)}(x,y) = K_\lambda(x,y) - \frac{\lambda_1 K_\lambda(x,x_1)K_\lambda(x_1,y)}{1+\lambda_1 K_\lambda(x_1,x_1)} \tag{6.67}$$

[388, p. 81], or as we see directly from Theorem 2.11. For two points x_1, x_2, the reproducing kernel $K_\lambda^{(2)}$ of the Hilbert space with the norm square

$$\lambda \|F\|^2_{H^s(\mathbb{R}^n)} + \|\Delta F\|^2_{L^2(\mathbb{R}^n)} + \sum_{j=1}^{2} \lambda_j |F(x_j)|^2 \tag{6.68}$$

is

$$K_\lambda^{(2)}(x,y) = K_\lambda^{(1)}(x,y) - \frac{\lambda_2 K_\lambda^{(1)}(x,x_2)K_\lambda^{(1)}(x_2,y)}{1+\lambda_2 K_\lambda^{(1)}(x_2,x_2)} \quad (x,y \in \partial D) \tag{6.69}$$

using the reproducing kernel $K_\lambda^{(1)}$. In this way, we can proceed many steps further using the similar calculation. For this procedure, to write the computer program is very easy. When we have the N-th kernel $K_\lambda^{(N)}$, it is the reproducing kernel of the Hilbert space $H_{K_\lambda^{(N)}}(\mathbb{R})$ with the norm square (6.64) and it coincides with K_{λ_j}. The extremal function in the minimum problem in (6.61) is

$$F^*_{\lambda,s,x_j,A_j}(x) = \sum_{j=1}^{N} A_j \lambda_j K_\lambda^{(N)}(x,x_j). \tag{6.70}$$

By taking a small λ, we will be able to obtain the approximate solution of the problem:

$$\Delta u \sim 0, \tag{6.71}$$

and

$$u(x_j) \sim A_j \quad j = 1,2,\ldots,N. \tag{6.72}$$

Letting λ tend to zero, we obtain mathematically the solution u of the problem:

$$\Delta u = 0 \tag{6.73}$$

and

$$u(x_j) = A_j \quad j = 1,2,\ldots,N, \tag{6.74}$$

for any finite points $\{x_j\}_{j=1}^N$ and for any values $\{A_j\}_{j=1}^N$.

We considered the extremal problem in (6.61) by considering a general weight $\{\lambda_j\}_{j=1}^N$. This means that for a larger λ_{j0}, the speed of the convergence

$$u(x_{j0}) \to A_{j0} \tag{6.75}$$

is higher. Once the power of our computers is greater, then our algorithm seems to be more effective. Then, by our algorithm we will be able to solve many concrete problems using computers.

Note that analytical theory depends seriously on the domain D, for example, recall the Poisson integral formula; however, here our method does essentially not depend on the domain D.

We followed [297] in this section.

6.2.3 The Cauchy Problem

We propose a new algorithm for constructing approximate solutions for the Laplace equation

$$\Delta u = 0 \tag{6.76}$$

in a C^1-domain $D \subset \mathbb{R}^n$ satisfying the boundary conditions

$$u = f \quad \text{on} \quad \Gamma \tag{6.77}$$

and

$$\frac{\partial u}{\partial \nu} = g \quad \text{on} \quad \Gamma \tag{6.78}$$

where $\dfrac{\partial}{\partial \nu}$ denotes the outer normal derivative $\Gamma \subset \partial D$ as before. For functions f and g, they must satisfy strong conditions, in order that equations (6.76), (6.77) and (6.78) are satisfied. However, in our method, for any $L^2(\Gamma)$ functions f, g, we can consider the best approximation u for the equations (6.76), (6.77) and (6.78).

For our problem (6.76), (6.77), and (6.78), we will consider the extremal problem: for a fixed parameter $\lambda > 0$ and for any $f, g \in L^2(\Gamma)$,

$$\inf_{F \in H^s(\mathbb{R}^n)} \left\{ \lambda \|F\|_{H^s(\mathbb{R}^n)}^2 + \|\Delta F\|_{L^2(\mathbb{R}^n)}^2 + \|F - f\|_{L^2(\Gamma)}^2 + \left\| \frac{\partial F}{\partial \nu} - g \right\|_{L^2(\Gamma)}^2 \right\}. \tag{6.79}$$

In order to simplify the problem, we take the following strategy:

1. As a function space, we use $H^s(\mathbb{R}^n)$; in this case the space admits the reproducing kernel, which is extremely simple;

2. for the integral of ΔF, we will consider it in the whole space, not in the domain D. Then the extremal problem will become very simple.

After we simply the matter in the way described above, we see that the reproducing kernel K_λ of the Hilbert space $H_{K_\lambda}(\mathbb{R})$ with the norm square

$$\lambda \|F\|^2_{H^s(\mathbb{R}^n)} + \|\Delta F\|^2_{L^2(\mathbb{R}^n)} \tag{6.80}$$

is given by (6.56).

6.2.4 Use a Finite Number of Data for the Cauchy Problem

Our strategy is first to represent the extremal function $F^*_{s,\lambda,g}$ in (6.79) in an explicit form and then to consider the limit of this extremal function as $\lambda \downarrow 0$. We can formulate these procedures and discuss them theoretically, using a Fredholm integral equation of the second kind. For this we propose a new numerical approach for the present problem.

We can find an approach in [275, 473, 494] similar to our current approach in spirit.

In order to consider the boundary value problems, we will consider it for (6.79) as follows:

For any set of fixed points $\{x_j\}_{j=1}^N$ on the boundary Γ and for any set of given values $\{A_j\}_{j=1}^N$ and $\{B_j\}_{j=1}^N$, we consider the extremal problem, for any fixed sets of weights $\{\lambda_j\}_{j=1}^N \subset (0,\infty)$ and $\{\mu_j\}_{j=1}^N \subset (0,\infty)$ we want to minimize

$$\left\{ \lambda \|F\|^2_{H^s(\mathbb{R}^n)} + \|\Delta F\|^2_{L^2(\mathbb{R}^n)} + \sum_{j=1}^N \lambda_j |F(x_j) - A_j|^2 + \sum_{j=1}^N \mu_j \left| \frac{\partial F}{\partial \nu}(x_j) - B_j \right|^2 \right\},$$

that is we approximate the integral in (6.79) by weighted sums. Then we wish to look for the reproducing kernel K_{λ_j,μ_j} of the Hilbert space $H_{K_{\lambda_j,\mu_j}}(\mathbb{R})$ with the norm square

$$\lambda \|F\|^2_{H^s(\mathbb{R})} + \|\Delta F\|^2_{L^2(\mathbb{R})} + \sum_{j=1}^N \lambda_j |F(x_j)|^2 + \sum_{j=1}^N \mu_j \left| \frac{\partial F}{\partial \nu}(x_j) \right|^2. \tag{6.81}$$

We can represent the reproducing kernel K_{λ_j,μ_j} in terms of K_λ directly; however, for many points $\{x_j\}_{j=1}^N$ on the boundary Γ, the size of a matrix is so large that this direct representation will not be effective in some case.

In order to overcome this difficulty, we will propose an approach as in the previous case: We use Theorem 2.4. The reproducing kernel $K^{(1,0)}$ of the Hilbert space with the norm square

$$\lambda\|F\|^2_{H^s(\mathbb{R})} + \|\Delta F\|^2_{L^2(\mathbb{R})} + \sum_{j=1}^{1} \lambda_j |F(x_j)|^2 \tag{6.82}$$

is

$$K^{(1,0)}(x,y) = K_\lambda(x,y) - \frac{\lambda_1 K_\lambda(x,x_1) K_\lambda(x_1,y)}{1+\lambda_1 K_\lambda(x_1,x_1)}. \tag{6.83}$$

Write

$$\partial_{\nu,x}\partial_{\nu,y} K^{(1,0)}(x_1,x_1) = \frac{\partial^2}{\partial t_1 \partial t_2} K^{(1,0)}(x_1 + t_1\nu, x_2 + t_2\nu)\bigg|_{t_1=t_2=0}.$$

Next, the reproducing kernel $K^{(1,1)}$ of the Hilbert space with the norm square

$$\lambda\|F\|^2_{H^s(\mathbb{R}^n)} + \|\Delta F\|^2_{L^2(\mathbb{R}^n)} + \sum_{j=1}^{1} \lambda_j |F(x_j)|^2 + \sum_{j=1}^{1} \mu_j \left|\frac{\partial F}{\partial \nu}(x_j)\right|^2 \tag{6.84}$$

is

$$K^{(1,1)}(x,y) = K^{(1,0)}(x,y) - \frac{\mu_1 \partial_{\nu,x} K^{(1,0)}(x,x_1) \partial_{\nu,y} K^{(1,0)}(x_1,y)}{1+\mu_1 \partial_{\nu,x}\partial_{\nu,y} K^{(1,0)}(x_1,x_1)}. \tag{6.85}$$

For two points x_1, x_2, the reproducing kernel $K_\lambda^{(2,1)}$ of the Hilbert space with the norm square

$$\lambda\|F\|^2_{H^s(\mathbb{R}^n)} + \|\Delta F\|^2_{L^2(\mathbb{R}^n)} + \sum_{j=1}^{2} \lambda_j |F(x_j)|^2 + \sum_{j=1}^{1} \mu_j \left|\frac{\partial F}{\partial \nu}(x_j)\right|^2 \tag{6.86}$$

is

$$K^{(2,1)}(x,y) = K^{(1,1)}(x,y) - \frac{\lambda_2 K^{(1,1)}(x,x_2) K^{(1,1)}(x_2,y)}{1+\lambda_2 K^{(1,1)}(x_2,x_2)} \tag{6.87}$$

in terms of the reproducing kernel $K^{(1,1)}$. In this way, we can iterate this procedure further. When we have the N-th kernel $K^{(N,N)}$, it is the reproducing kernel of the Hilbert space with the norm square (6.79) and the extremal function in the minimum problem in (6.79) is

$$F^*_{\lambda,s,\{x_j\}_{j=1}^N,\{\lambda_j\}_{j=1}^N,\{\mu_j\}_{j=1}^N,\{A_j\}_{j=1}^N,\{B_j\}_{j=1}^N} \tag{6.88}$$

$$= \sum_{j=1}^{N} A_j \lambda_j (K^{(N,N)})_{x_j} + \sum_{j=1}^{N} B_j \mu_j \partial_\nu (K^{(N,N)})_{x_j}.$$

6.2.5 Error Estimates

In the representation (6.88), when the data A_j and B_j contain noises, we have the error estimate:

Theorem 6.9. *In* (6.88), *we obtain the estimate*

$$|F^*_{\lambda,s,\{x_j\}_{j=1}^N,\{\lambda_j\}_{j=1}^N,\{\mu_j\}_{j=1}^N,\{A_j\}_{j=1}^N,\{B_j\}_{j=1}^N}(x)|$$

$$\leq \frac{1}{\sqrt{\lambda}} \sqrt{\frac{\Gamma(s-n/2)}{(4\pi)^{n/2}\Gamma(s)}} \sum_{j=1}^{N} (\lambda_j |A_j|^2 + \mu_j |B_j|^2).$$

We omit the proof of Theorem 6.9.

For numerics for the Cauchy problem of the Laplace equation, there exist many papers; we refer to some of them and the readers can consult also the references therein [24, p. 435], [51, 58, 76, 104, 125, 195, 218, 228, 257, 282, 344, 348, 370, 481]. Moreover as a substantial source book on inverse problems, we refer to Isakov [228] for example.

For the extremal problem (6.61), Yan Mo and Tao Qian [308] applied the support vector machine method and compared it with other typical numerical methods. The finite difference method (FDM) [434] and the finite element method (FEM) [224] require defining a mesh (domain discretization) where the functions are approximated locally.

The construction of a mesh in two or more dimensions is a non trivial problem. A major disadvantage of those methods, however, is their mesh-dependent characteristics which normally require an enormous computational effort and induce numerical instability when a large number of grids or elements are required.

Another approach for solving partial differential equations is to use artificial neural networks (ANNs); see [273, 274, 277]. The approach using ANNs to solve partial differential equations relies on the functional approximation capability of feedforward neural networks and results in constructing a solution written in a differentiable and closed analytic form. This form employs the feedforward neural network as the basic approximation element, whose parameters (weights and bases) are adjusted to minimize an appropriate error function. The solution in terms of artificial neural networks has several attractive features. One is that the solution is infinitely differentiable and has a closed analytic form which can be easily used in any subsequent calculation. The other is that neural networks possess a smaller number of parameters compared to other solution technique, see [273, 274].

However, ANNs techiniques suffer from their theoretical weakness. For example, back-propagation may not converge to an optimal global solution.

SVM (support vector machine), developed by Vapnik and his coworkers in 1995 [471], is based on statistical learning theory which seeks to minimize an upper bound of the generalization error consisting of the sum of the training error and a confidence interval. This principle is different from the commonly used empirical risk minimization (ERM) principle which only minimizes the training error. Based on this, SVMs usually achieve higher generalization performance than ANNs which implement ERM principle. As consequence, SVMs can be used wherever ANNs can, and usually achieve better results. Another key characteristic of SVM is that training SVM is equivalent to solving a linearly constrained quadratic programming problem so that the solution of SVM is unique and global, unlike ANNs' training which requires nonlinear optimization with the possibility of getting stuck in local minima.

For the extremal problem (6.61), the introduced algorithms here are for the powerful computers as in H. Fujiwara's infinite precision method, but the application of the SVMs to the problem may be applied with the usual level of computer power and furthermore dealt with errorness data.

6.3 Heat Equation

6.3.1 Applications to the Inversion to 1*-Dimensional Heat Conduction*

At first, we refer to the original results in [409] where Tikhonov regularlization was applied to the ill-posed 1-dimensional heat conduction problem.

We consider the heat equation on $\mathbb{R} \times (0, \infty)$:

$$\partial_t u(x,t) = \partial_{xx} u(x,t) \quad (x,t) \in \mathbb{R} \times (0,\infty) \tag{6.89}$$

subject to $u(\cdot, 0) = F \in L^2(\mathbb{R})$.

Then define the Weierstrass (Gauss) transform

$$u_F(x,t) \equiv \frac{1}{\sqrt{4\pi t}} \int_{\mathbb{R}} F(y) \exp\left(-\frac{(x-y)^2}{4t}\right) dy. \tag{6.90}$$

Then we recall the following facts from partial differential equations:

1. u_F is a solution to the heat equation

$$\partial_t u_F - \partial_{xx} u_F = 0 \tag{6.91}$$

on $(x,t) \in \mathbb{R} \times (0,\infty)$.

2. u_F satisfies the initial condition in the following sense :

$$\lim_{t\downarrow 0} \|u_F(\cdot,t) - F\|_{L^2(\mathbb{R})} = 0. \tag{6.92}$$

Following the idea of image identification using the theory of reproducing kernels, in order to determine the image space, we formed the positive definite quadratic function (2.261) and we obtained the inversion formulas in Theorem 2.48.

For applying the Tikhonov regularization method for (6.89). we consider the integral operator $L_t : H_S(\mathbb{R}) \to L^2(\mathbb{R})$ defined by

$$L_t f(x) = u_F(x,t) = \frac{1}{\sqrt{4\pi t}} \int_{\mathbb{R}} f(y) \exp\left(-\frac{(y-x)^2}{4t}\right) dy, \tag{6.93}$$

for $t > 0$. Here $H_S = H_S(\mathbb{R})$ is the first-order Sobolev Hilbert space on the whole real line with norm defined by

$$\|f\|_{H_S(\mathbb{R})} = \sqrt{\int_{\mathbb{R}} (f'(x)^2 + f(x)^2) dx} \tag{6.94}$$

admitting the reproducing kernel

$$K(x,y) = \frac{1}{2\pi} \int_{-\infty}^{\infty} \frac{\exp(i\xi(x-y))}{(1+\xi^2)} d\xi = \frac{1}{2} e^{-|x-y|}. \tag{6.95}$$

See Theorem 1.3 for details.

We now consider the following best approximation problem; that is, the Tikhonov functional: For any $g \in L^2(\mathbb{R})$ and $\lambda > 0$,

$$\inf_{f\in H_S(\mathbb{R})} \left\{ \lambda \|f\|^2_{H_S(\mathbb{R})} + \|L_t f - g\|^2_{L^2(\mathbb{R})} \right\}. \tag{6.96}$$

Then for the RKHS $H_{K_\lambda}(\mathbb{R})$ consisting of all the members of $H_S(\mathbb{R})$ with the norm

$$\|f\|_{H_{K_\lambda}(\mathbb{R})} = \sqrt{\lambda \|f\|^2_{H_S(\mathbb{R})} + \|L_t f\|^2_{L^2(\mathbb{R})}}, \tag{6.97}$$

the reproducing kernel $K_\lambda(x,y)$ can be calculated directly using the Fourier integrals as follows:

$$K_\lambda(x,y;t) = \frac{1}{2\pi} \int_{\mathbb{R}} \frac{\exp(i\xi(x-y))d\xi}{\lambda(1+\xi^2) + e^{-2\xi^2 t}} \quad (t > 0). \tag{6.98}$$

Hence the unique member of $H_S(\mathbb{R})$ with the minimum $H_S(\mathbb{R})$ – the function $f^*_{\lambda,g}$ which attains the infimum is given by

$$f^*_{\lambda,g}(x) = \frac{1}{2\pi}\int_{-\infty}^{\infty}\left\{g(\xi)\int_{\mathbb{R}}\frac{e^{-ip(\xi-x)}\,dp}{\lambda(p^2+1)e^{p^2t}+e^{-p^2t}}\right\}d\xi \quad (x\in\mathbb{R}).$$

See [298, Theorem 4] for the proof. The proof is similar to the previous cases. For $f \in H_S(\mathbb{R})$ and for $g = L_t f$, we have the formula

$$\lim_{\lambda\downarrow 0} f^*_{\lambda,g}(x) = f(x) \tag{6.99}$$

uniformly on $\mathbb{R}$ from Remark 1.3.

The inversion formula for heat conduction is a typical and ill-posed problem. For example, recall the classical real inversion of the Gaussian convolution formula [212, p. 182]. Define

$$e^{-D^2} \equiv \sum_{j=0}^{\infty}\frac{1}{j!}(-D^2)^j.$$

For a bounded and continuous function $f(x)$ and for $t = 1$,

$$e^{-D^2}[L_1 f](x) = f(x) \quad \text{pointwise} \quad \text{on} \quad \mathbb{R}. \tag{6.100}$$

T. Matsuura performed numerical experiments for the formula and he was successful in realizing an effectiveness of the formulas [296, 298]. From this fact, various inversion formulas from the viewpoint of the numerical and practical inversion formulas were given, following the above method: In this part in this book, we will see various inversion formulas for which numerical experiments were given. Their algorithms for realizing approximate solutions for computer graphics were also given.

6.3.2 Use Sobolev Spaces

We will consider the n-dimensional Gaussian convolution (the Weierstrass transform)

$$u_F(x,t) = L_t F(x) = \frac{1}{(4\pi t)^{n/2}}\int_{\mathbb{R}^n} F(\xi)\exp\left(-\frac{|\xi-x|^2}{4t}\right)d\xi \tag{6.101}$$

for functions F of the Sobolev Hilbert space $H^s(\mathbb{R}^n)$ of order s on the whole real space $\mathbb{R}^n$ $(n \geq 1, s > n/2)$. This integral transform which represents the solution $u(x,t)$ of the heat equation

$$\begin{cases} u_t = u_{xx} & \text{in} \quad \mathbb{R}^n \times \{t > 0\}, \\ u(\cdot,0) = F & \text{on} \quad \mathbb{R}^n \end{cases} \tag{6.102}$$

is fundamental and has many applications to the mathematical sciences.

Our formulations and results are stated as follows:

Theorem 6.10 ([298, Theorem 4] and [409, Theorem 1.1]). *Fix $t > 0$. Let also $\lambda > 0$ and let s be such that $s \geq 2$ when $n \leq 3$ and $s > \frac{n}{2}$ when $n \geq 4$. Define*

$$Q_{\lambda,s}(\eta) \equiv \frac{1}{(2\pi)^n} \int_{\mathbb{R}^n} \frac{\exp(-ip \cdot \eta)\, dp}{\lambda(|p|^2 + 1)^s \exp(|p|^2 t) + \exp(-|p|^2 t)} \quad (\eta \in \mathbb{R}^n). \tag{6.103}$$

1. *For any function $g \in L^2(\mathbb{R}^n)$ and for any $\lambda > 0$, the best approximate function $F^*_{\lambda,s,g}$ in the sense that*

$$\inf_{F \in H^s(\mathbb{R}^n)} \left\{ \lambda \|F\|^2_{H^s(\mathbb{R}^n)} + \|g - u_F(\cdot, t)\|^2_{L^2(\mathbb{R}^n)} \right\}$$

$$= \lambda \|F^*_{\lambda,s,g}\|^2_{H^s(\mathbb{R}^n)} + \|g - u_{F^*_{\lambda,s,g}}(\cdot, t)\|^2_{L^2(\mathbb{R}^n)}$$

exists uniquely and $F^*_{\lambda,s,g}$ is represented by

$$F^*_{\lambda,s,g} = g * Q_{\lambda,s}. \tag{6.104}$$

2. *Let $F \in H^s(\mathbb{R}^n)$ and $g = u_F(\cdot, t)$. Then: as $\lambda \downarrow 0$*

$$F^*_{\lambda,s,g} \to F, \tag{6.105}$$

uniformly.

The proof is similar to those for similar situation and we omit the proof.

Recall that we considered (6.104) via (6.103). Meanwhile, for any $\lambda > 0$ and any $t > 0$, we define a linear mapping

$$M_{\lambda,s,t} : L^2(\mathbb{R}^n) \to H^s(\mathbb{R}^n) \tag{6.106}$$

by $M_{\lambda,s,t} g \equiv F^*_{\lambda,s,g}$. Now we consider the composite operators $L_t M_{\lambda,s,t}$ and $M_{\lambda,s,t} L_t$. Using Fourier integrals it can be shown that for $F \in H^s(\mathbb{R}^n)$,

$$M_{\lambda,s,t} L_t F(x) = \frac{1}{(2\pi)^n} \int_{\mathbb{R}^n} \left\{ F(\xi) \cdot \int_{\mathbb{R}^n} \frac{\exp(-ip \cdot (\xi - x))dp}{\lambda(|p|^2 + 1)^s \exp(2|p|^2 t) + 1} \right\} d\xi \tag{6.107}$$

and for $g \in L^2(\mathbb{R}^n)$,

$$L_t M_{\lambda,s,t} g(x) = \frac{1}{(2\pi)^n} \int_{\mathbb{R}^n} \left\{ g(\xi) \cdot \int_{\mathbb{R}^n} \frac{\exp(-ip \cdot (\xi - x))dp}{\lambda(|p|^2 + 1)^s \exp(2|p|^2 t) + 1} \right\} d\xi. \tag{6.108}$$

Setting

$$\Delta_{\lambda,s}(x-\xi) = \frac{1}{(2\pi)^n}\int_{\mathbb{R}^n} \frac{\exp(-ip\cdot(\xi-x))}{\lambda(|p|^2+1)^s\exp(2|p|^2t)+1}dp \tag{6.109}$$

in (6.107) and (6.3.2), we have

$$M_{\lambda,s,t}L_tF = F * \Delta_{\lambda,s}, \quad (F\in H_S(\mathbb{R}^n)) \tag{6.110}$$

and

$$L_tM_{\lambda,s,t}g = g * \Delta_{\lambda,s}, \quad (g\in L^2(\mathbb{R}^n)). \tag{6.111}$$

Then we obtain

$$\lim_{\lambda\downarrow 0}\Delta_{\lambda,s}(\cdot-\xi) = \delta_\xi \tag{6.112}$$

from the general theory; see (3.76) and (3.77) in Theorem 3.11. More precisely, we have

$$\lim_{\lambda\downarrow 0} M_{\lambda,s,t}L_t = I \tag{6.113}$$

in the strong operator topology of $H^s(\mathbb{R})$ and

$$\lim_{\lambda\downarrow 0} L_tM_{\lambda,s,t} = I \tag{6.114}$$

in the strong operator topology of $L^2(\mathbb{R})$. Direct consequences of (6.113) and (6.114) are

$$\lim_{\lambda\downarrow 0} M_{\lambda,s,t}L_tF = F$$

uniformly on $\mathbb{R}^n$ for any $F\in H^s(\mathbb{R}^n)$ and

$$\lim_{\lambda\downarrow 0} L_tM_{\lambda,s,t}g = g$$

in $L^2(\mathbb{R}^n)$ for any $g\in L^2(\mathbb{R}^n)$, respectively.

6.3.3 Use of Paley Wiener Spaces (Sinc Methods)

We place ourselves in $\mathbb{R}$ here. When we use Paley Wiener reproducing kernels, we have more effective formulas than using the Sobolev spaces for computers. Recall that Paley Wiener space $H_{K_h}(\mathbb{R})$ was defined by (1.6).

Our formulations and results are stated as follows:

Theorem 6.11. *Let $t, h, \lambda > 0$. Write*

$$Q_{t,\lambda,h}(\zeta) \equiv \frac{1}{2\pi}\int_{-\frac{\pi}{h}}^{\frac{\pi}{h}} \frac{\exp(-ip\cdot\zeta)dp}{\lambda\exp(|p|^{2t}) + \exp(-|p|^2 t)} \tag{6.115}$$

for $\zeta \in \mathbb{R}$.

1. *For any function $g \in L^2(\mathbb{R})$ and for any $\lambda > 0$, the best approximate function exists uniquely in the following sense:*

$$\lambda\|F\|^2_{H_{K_h}(\mathbb{R})} + \|g - u_F(\cdot,t)\|^2_{L^2(\mathbb{R})}$$
$$\geq \lambda\|F^*_{t,\lambda,h,g}\|^2_{H_{K_h}(\mathbb{R})} + \|g - u_{F^*_{t,\lambda,h,g}}(\cdot,t)\|^2_{L^2(\mathbb{R})}$$

for all $F \in H_{K_h}(\mathbb{R})$. The function $F^*_{t,\lambda,h,g}$ is represented by

$$F^*_{t,\lambda,h,g} = g * Q_{t,\lambda,h}. \tag{6.116}$$

2. *Let $F \in H_{K_h}(\mathbb{R})$. If we take $g = u_F(\cdot,t)$, then we have, as $\lambda \downarrow 0$*

$$F^*_{t,\lambda,h,g} \to F, \tag{6.117}$$

uniformly.

We again omit the proof due to similarity.

Here we note the interesting fact that for the Sobolev space case, for $\lambda = 0$ the corresponding representation (6.104) does not exist, meanwhile for the Paley Wiener space $W\left(\frac{\pi}{h}\right)$ case of (6.116), for $\lambda = 0$ the representation (6.116) is still valid; that is, the result is valid for even $\lambda = 0$. Hence, we can consider the results for $\lambda = 0$ in the spirit of the Tikhonov regularization in which we are interested in a small λ or λ tending to zero. That is, when we use the Paley Wiener space $W\left(\frac{\pi}{h}\right)$, we need not consider the Tikhonov regularization. Then, since

$$L_t F^*_{t,0,h,g}(x) = \langle g, (K_h)_x\rangle_{L^2(\mathbb{R})}, \tag{6.118}$$

and since the output is the orthogonal projection of g onto the Paley Wiener space $W\left(\frac{\pi}{h}\right)$, we can estimate the difference of the output of our inverse $F^*_{t,0,h,g}$ and g clearly as

$$\|L_t F^*_{t,0,h,g} - g\|_{L^2(\mathbb{R})} \tag{6.119}$$

which is the distance from g onto the Paley Wiener space $W\left(\frac{\pi}{h}\right)$. Of course, $F^*_{t,0,h,g}$ is the Moore Penrose generalized inverse of the operator equation, for any $g \in L^2(\mathbb{R})$ and $F \in W\left(\frac{\pi}{h}\right)$,

$$L_t F = g. \tag{6.120}$$

For the Paley Wiener space $W\left(\frac{\pi}{h}\right)$, we need not use the Tikhonov regularization and we can look for the Moore Penrose generalized inverse $F^*_{t,0,h,g}$ using the theory of reproducing kernels. However, we had better calculate the extremal functions $F^*_{t,\lambda,h,g}$ in the Tikhonov regularization and to set $\lambda = 0$, because the structure of the Moore Penrose generalized inverses is involved.

Meanwhile, we are interested in the convergence property of the extremal functions, when h tends to zero for a general g.

However, in (6.118) we can establish

$$\lim_{h\downarrow 0} L_t F^*_{t,0,h,g} = g \tag{6.121}$$

in $L^2(\mathbb{R})$. Note that (6.121) is a consequence of (3.77).

We are interested in numerical experiments for both cases, i.e., the Sobolev space H_S and the Paley Wiener space $W\left(\frac{\pi}{h}\right)$ for their convergences as $\lambda \downarrow 0$ and $h \downarrow 0$.

The real inversion formula (6.116) will give a practical formula for Gaussian convolution. Experimental results using computers are also given. When we use the Sobolev spaces as the reproducing kernels, we will see that in order to overcome the high "ill-posedness" in the real inversion and in order to grasp "analyticity" of the image of (6.101), we must work seriously; that is, we must take a very small λ and we must barely calculate the corresponding integral in the numerical sense. Computers help us with this hard work calculating the integral for a very small λ.

However, when we usee Paley Wiener spaces for $\lambda = 0$ and for h tending to 0, we can obtain a very simple and improved inversion formula. See the details [296, 298] for numerical experiments.

Now let us consider minimizing (6.124). We summarize the results and proofs in a brief manner as follows:

Theorem 6.12. *Let $h, \lambda, t > 0$ and define*

$$K_h(x,y) \equiv \frac{1}{2\pi}\int_{\mathbb{R}} \chi_{\left(-\frac{\pi}{h},\frac{\pi}{h}\right)}(v)\exp(-i(x-y)\cdot v)\,dv \quad (x,y\in\mathbb{R}) \tag{6.122}$$

$$Q_{t,\lambda,h}(x) \equiv \frac{1}{2\pi}\int_{\mathbb{R}} \frac{\chi_{(-\pi,\pi)}(hp)\exp(ip\cdot x)}{\lambda\exp(|p|^2 t)+\exp(-|p|^2 t)}\,dp \quad (x\in\mathbb{R}). \tag{6.123}$$

Then $F^*_{t,\lambda,h,g} \equiv Q_{t,\lambda,h} * g$ attains the minimum of

$$\lambda\,\|F\|^2_{H_{K_h}(\mathbb{R})} + \|g - u_F(\cdot,t)\|^2_{L^2(\mathbb{R})},\ F\in H_{K_h}(\mathbb{R}). \tag{6.124}$$

Proof. Let us calculate the adjoint of L^*. Note that the adjoint is given by $m(D)$, where

$$m(\xi) = \chi_{(-\pi,\pi)}(h\xi)\exp(-t\,|\xi|^2). \tag{6.125}$$

As a result, we have $L^*(\lambda + L^*L)^{-1} = M(D)$, where

$$M(\xi) = \frac{\chi_{(-\pi,\pi)}(h\xi)\exp(-t\,|\xi|^2)}{\lambda + \chi_{(-\pi,\pi)}(h\xi)\exp(-2t\,|\xi|^2)} = \frac{\chi_{(-\pi,\pi)}(h\xi)}{\lambda\,\exp(t\,|\xi|^2) + \exp(-t\,|\xi|^2)}. \tag{6.126}$$

Hence it follows that

$$\begin{aligned} L^*(\lambda + L^*L)^{-1}[(K_h)_x] &= \frac{1}{\sqrt{2\pi}} L^*(\lambda + L^*L)^{-1}\left(\mathscr{F}^{-1}\left[[M_{-x}\chi_{(-\pi,\pi)}](h\cdot)\right]\right) \\ &= \frac{1}{\sqrt{2\pi}} L^*(\lambda + L^*L)^{-1}\left(\mathscr{F}^{-1}\left[[M_{-x}\chi_{(-\pi,\pi)}](h\cdot)\right]\right) \\ &= \frac{1}{2\pi}\int_{\mathbb{R}} \frac{\chi_{(-\pi,\pi)}(h\xi)\exp(-i(x-\cdot)\xi)}{\lambda\,\exp(t\,|\xi|^2) + \exp(-t\,|\xi|^2)}\,d\xi. \end{aligned}$$

This is the desired result.

For any $\lambda > 0$ and any $t > 0$, we define a linear mapping

$$M_{t,\lambda,h} : L^2(\mathbb{R}) \to H_{K_h}(\mathbb{R}) \tag{6.127}$$

by $M_{t,\lambda,h}g \equiv F^*_{t,\lambda,h,g}$. Note that

$$\lambda\,\|F\|^2_{H_{K_h}(\mathbb{R})} + \|g - u_F(\cdot,t)\|^2_{L^2(\mathbb{R})} = \lambda\,\|F\|^2_{H_{K_h}(\mathbb{R})} + \|g - L_tF\|^2_{L^2(\mathbb{R})}$$

for $F \in H_{K_h}(\mathbb{R})$.

Corollary 6.1. *Let $h > 0$ and $t > 0$ be fixed.*

1. *Let $F \in H_{K_h}(\mathbb{R})$. Then one has*

$$\lim_{\lambda \downarrow 0} M_{t,\lambda,h}L_tF = F \tag{6.128}$$

in $H_{K_h}(\mathbb{R})$.

2. *Let $g \in L^2(\mathbb{R})$. Then one has*

$$\lim_{\lambda \downarrow 0} L_tM_{t,\lambda,h}g = g \tag{6.129}$$

in $L^2(\mathbb{R})$.

Proof. The results (6.128) and (6.129) follow from (3.76) and (3.77), respectively.

Now we consider the composite operators $L_tM_{t,\lambda,h}$ and $M_{t,\lambda,h}L_t$. Let $F \in H_{K_h}(\mathbb{R})$ and $g \in L^2(\mathbb{R})$. Using the Fourier integrals as we did in Theorem 6.11, we have

Proposition 6.1. *Keep to the same setting as above. Then we have*

$$M_{t,\lambda,h}L_tF(x) = \frac{1}{2\pi}\int_{\mathbb{R}}\left\{F(\xi)\int_{-\frac{\pi}{h}}^{\frac{\pi}{h}}\frac{\exp(-ip\cdot(\xi-x))}{\lambda\exp(2|p|^2t)+1}dp\right\}d\xi \tag{6.130}$$

and

$$L_tM_{t,\lambda,h}g(x) = \frac{1}{2\pi}\int_{\mathbb{R}}\left\{g(\xi)\int_{(-\frac{\pi}{h},\frac{\pi}{h})}\frac{\exp(-ip\cdot(\xi-x))}{\lambda\exp(2|p|^2t)+1}dp\right\}d\xi. \tag{6.131}$$

Proof. We concentrate on (6.130), the proof of (6.131) being similar. This is immediate from

$$M_{t,\lambda,h} = m_0(D), \quad L_t(D) = m_1(D) \tag{6.132}$$

where

$$m_0(\xi) = \mathscr{F}^{-1}\left[\frac{\chi_{(-\frac{\pi}{h},\frac{\pi}{h})}}{\lambda\ \exp(2t|\cdot|^2)+1}\right](\xi),\ m_1(\xi) = \exp(-t\,|\xi|^2). \tag{6.133}$$

This is the desired result.

Let $F \in H_{K_h}(\mathbb{R})$ and $g \in L^2(\mathbb{R})$. Setting

$$\Delta_{t,\lambda,h}(x) \equiv \frac{1}{2\pi}\int_{-\frac{\pi}{h}}^{\frac{\pi}{h}}\frac{\exp(ip\cdot x)}{\lambda\exp(2|p|^2t)+1}dp, \tag{6.134}$$

from (6.130) and (6.131), we obtain, for $x \in \mathbb{R}$,

$$M_{t,\lambda,h}L_tF(x) = \int_{\mathbb{R}}F(\xi)\Delta_{t,\lambda,h}(x-\xi)\,d\xi, \tag{6.135}$$

$$L_tM_{t,\lambda,h}g(x) = \int_{\mathbb{R}}g(\xi)\Delta_{t,\lambda,h}(x-\xi)\,d\xi. \tag{6.136}$$

As a special case when $\lambda = 0$, we have $\lim_{h\downarrow 0}\Delta_{t,0,h}(\cdot-\xi) = \delta_\xi$ in the following sense:

Proposition 6.2. *Let $F, g \in L^2(\mathbb{R})$. Then*

$$\lim_{h\downarrow 0}M_{t,0,h}L_tF = F \tag{6.137}$$

and

$$\lim_{h\downarrow 0}L_tM_{t,0,h}g = g \tag{6.138}$$

in $L^2(\mathbb{R})$.

Proof. In view of expressions (6.135) and (6.136), we have only to prove (6.137). Observe that

$$\Delta_{t,0,h}(x) \equiv \frac{1}{2\pi}\int_{-\frac{\pi}{h}}^{\frac{\pi}{h}} \exp(ip\cdot x)\,dp \quad (x \in \mathbb{R})$$

and hence

$$\mathscr{F}\Delta_{t,0,h}(\xi) \equiv \frac{1}{\sqrt{2\pi}}\chi_{(-\frac{\pi}{h},\frac{\pi}{h})}(\xi) \quad (\xi \in \mathbb{R}).$$

Therefore, by the Plancherel theorem, we have (6.137).

6.3.4 Source Inversions of Heat Conduction Using a Finite Number of Data

For different points $\{(x_j, t_j)\}_{j=1}^n$ such as the points of the time and space, we consider the bounded linear operators from the RKHS $H_{K_h}(\mathbb{R})$ in (1.6) into $\mathbb{R}$ in the Weierstrass transform (6.101):

$$H_{K_h}(\mathbb{R}) \ni F \mapsto u_F(x_j, t_j) = \frac{1}{\sqrt{4\pi t_j}}\int_{\mathbb{R}} F(\xi)\exp\left(-\frac{|\xi - x_j|^2}{4t_j}\right)d\xi \qquad (6.139)$$

for $j = 1, 2, \ldots, n$, and we write them in the vector form:

$$\mathbf{L} : H_{K_h}(\mathbb{R}) \ni F \mapsto (u_F(x_1, t_1), u_F(x_2, t_2), \ldots, u_F(x_n, t_n)) \in \mathbb{R}^n. \qquad (6.140)$$

We take the standard orthonormal system $\{\mathbf{e}_j\}_{j=1}^n$ in the space $\mathbb{R}^n$. Then we see

$$u_F(x_j, t_j) = \langle \mathbf{L}F, \mathbf{e}_j\rangle_{\mathbb{R}} = \langle F, \mathbf{L}^*\mathbf{e}_j\rangle_{H_{K_h}(\mathbb{R})} \quad j = 1, 2, \ldots, n. \qquad (6.141)$$

Therefore, the image space of the linear mapping $\mathbf{L}$ forms the Hilbert space H_A admitting the reproducing kernel

$$a_{j,j'} \equiv \langle \mathbf{L}^*\mathbf{e}_{j'}, \mathbf{L}^*\mathbf{e}_j\rangle_{H_{K_h}(\mathbb{R})} \quad j, j' = 1, 2, \ldots, n. \qquad (6.142)$$

Here note that $a_{j,j'}$ are calculated as follows:

$$\mathbf{L}^*\mathbf{e}_j(x) = \langle \mathbf{L}^*\mathbf{e}_j, (K_h)_x\rangle_{H_{K_h}(\mathbb{R})} = \langle \mathbf{e}_j, \mathbf{L}(K_h)_x\rangle_{\mathbb{R}} = u_{(K_h)_x}(x_j, t_j).$$

Recall that u_F is given by (6.90):

$$u_F(x, t) = \frac{1}{\sqrt{4\pi t}}\int_{\mathbb{R}} F(y)\exp\left(-\frac{(x-y)^2}{4t}\right)dy.$$

In particular,

$$u_{(K_h)_x}(x_j, t_j) = \frac{1}{\sqrt{4\pi t_j}} \int_{\mathbb{R}} K_h(y, x) \exp\left(-\frac{(x_j - y)^2}{4t_j}\right) dy. \tag{6.143}$$

If we insert (1.6) to (6.143), then we obtain

$$\begin{aligned}\mathbf{L}^*\mathbf{e}_j(x) &= \frac{1}{\sqrt{4\pi t_j}} \int_{\mathbb{R}} \left(\int_{-\frac{\pi}{h}}^{\frac{\pi}{h}} \exp\left(i(y-x)t\right) \frac{dt}{2\pi} \right) \exp\left(-\frac{(x_j - y)^2}{4t_j}\right) dy \\ &= \frac{1}{h} \int_0^1 \exp\left(-\frac{t_j \pi^2}{h^2}\xi^2\right) \cos\left(\frac{\pi}{h}(x_j - x)\xi\right) d\xi.\end{aligned}$$

For later consideration, we need the following lemma:

Lemma 6.1. *Let $h > 0$. Then*

$$\lim_{L\to\infty} \int_{\mathbb{R}^2} \frac{h\varphi(\xi,\eta)}{\pi(\xi-\eta)} \sin\left(\frac{\pi L}{h}(\xi-\eta)\right) d\xi\, d\eta = \pi \int_{\mathbb{R}} \varphi(\xi,\xi)\, d\xi$$

for all $\varphi \in \mathscr{S}(\mathbb{R} \times \mathbb{R})$.

Proof. It suffices to show that

$$\lim_{L\to\infty} \int_{\mathbb{R}^2} \frac{h\varphi(\xi,\eta)}{\pi\eta} \sin\left(\frac{\pi}{h}L\eta\right) d\xi\, d\eta = \pi \int_{\mathbb{R}} \varphi(\xi,0)\, d\xi.$$

We know that

$$\lim_{L\to\infty} \frac{h}{\pi\eta} \sin\left(\frac{\pi}{h}L\eta\right) = \pi\delta_0$$

in the topology of $\mathscr{S}'(\mathbb{R})$ and hence there exists a constant $N \in \mathbb{N} \cap [2, \infty)$ such that

$$\left| \int_{\mathbb{R}} \frac{h}{\pi\eta} \sin\left(\frac{\pi}{h}L\eta\right) \tau(\eta)\, d\eta \right| \le N \sum_{|\alpha|\le N} \sup_{x\in\mathbb{R}} (1+|x|)^N |\partial^\alpha \tau(x)|$$

for all $\tau \in \mathscr{S}$. Consequently,

$$\begin{aligned}\left| \int_{\mathbb{R}} \frac{h\varphi(\xi,\eta)}{\pi\eta} \sin\left(\frac{\pi}{h}L\eta\right) \tau(\eta)\, d\eta \right| &\le N \sum_{|\alpha|\le N} \sup_{x\in\mathbb{R}} (1+|x|)^N |\partial^\alpha \varphi(\xi,x)| \\ &\le \frac{N}{(1+|\xi|)^N} \sum_{|\alpha|\le N} \sup_{x,\xi'\in\mathbb{R}} (1+|x|)^N (1+|\xi'|)^N |\partial^\alpha \varphi(\xi',x)|.\end{aligned} \tag{6.144}$$

With (6.144), we can use the dominated convergence theorem to conclude that

$$\begin{aligned}\lim_{L\to\infty}\int_{\mathbb{R}^2}\frac{h\varphi(\xi,\eta)}{\pi\eta}\sin\left(\frac{\pi}{h}L\eta\right)d\xi\,d\eta &= \lim_{L\to\infty}\int_{\mathbb{R}}\left[\int_{\mathbb{R}}\frac{h\varphi(\xi,\eta)}{\pi\eta}\sin\left(\frac{\pi}{h}L\eta\right)d\eta\right]d\xi\\ &= \int_{\mathbb{R}}\lim_{L\to\infty}\left[\int_{\mathbb{R}}\frac{h\varphi(\xi,\eta)}{\pi\eta}\sin\left(\frac{\pi}{h}L\eta\right)d\eta\right]d\xi\\ &= \pi\int_{\mathbb{R}}\varphi(\xi,0)\,d\xi.\end{aligned}$$

Thus we obtain the desired result.

By applying the variable changing $y=\frac{h}{\pi}s+x$, the Plancherel formula and the Fourier transforms of the Sinc function and the Gaussian function, and

$$\mathscr{F}\left[\exp\left(-\frac{|\cdot|^2}{4a}\right)\right](\xi)=\sqrt{2a\pi}\exp(-a|\xi|^2), \tag{6.145}$$

and by direct calculation, we have

$$\begin{aligned}&\frac{1}{\sqrt{4\pi t_j}}\int_{\mathbb{R}}\frac{1}{\pi(y-x)}\sin\left(\frac{\pi}{h}(y-x)\right)\exp\left(-\frac{(y-x_j)^2}{4t_j}\right)dy\\ &=\frac{1}{\sqrt{4\pi t_j}}\int_{\mathbb{R}}\frac{\mathrm{sinc}(s)}{h}\exp\left(-\frac{1}{4t_j}\left(\frac{h}{\pi}s+x-x_j\right)^2\right)\frac{h}{\pi}ds\\ &=\frac{1}{\pi\sqrt{4\pi t_j}}\int_{\mathbb{R}}\mathrm{sinc}(s)\exp\left(-\frac{h^2}{4t_j\pi^2}\left(s-\frac{\pi}{h}(x_j-x)\right)^2\right)ds\\ &=\frac{1}{\pi\sqrt{4\pi t_j}}\int_{\mathbb{R}}\mathrm{sinc}(s)\exp\left(-\frac{h^2}{4t_j\pi^2}\left(s-\frac{\pi}{h}(x_j-x)\right)^2\right)ds.\end{aligned}$$

Recall that

$$\mathrm{sinc}(x)\equiv\begin{cases}1 & x=0,\\ \dfrac{\sin x}{x} & x\in\mathbb{R}\setminus\{0\},\end{cases}$$

is called the *Sinc function*. Then, we have

$$\mathscr{F}[\mathrm{sinc}](\xi)=\frac{\sqrt{2\pi}}{2}\chi_{(-1,1)}(\xi),\quad (\xi\in\mathbb{R}).$$

Therefore, we have

$$\frac{1}{\sqrt{4\pi t_j}}\int_{\mathbb{R}}\frac{1}{\pi(y-x)}\sin\left(\frac{\pi}{h}(y-x)\right)\exp\left(-\frac{(y-x_j)^2}{4t_j}\right)dy$$

$$=\frac{1}{\pi\sqrt{4\pi t_j}}\int_{-1}^{1}\frac{\sqrt{2\pi}}{2}\exp\left(-i\frac{\pi}{h}(x_j-x)\xi\right)\sqrt{\frac{2t_j\pi}{h^2}}\exp\left(-\frac{t_j\pi^2}{h^2}\xi^2\right)d\xi$$

$$=\frac{1}{h}\int_{-1}^{1}\exp\left(-\frac{t_j\pi^2}{h^2}\xi^2\right)\exp\left(-i\frac{\pi}{h}(x_j-x)\xi\right)d\xi.$$

By Euler's formula, we have

$$\frac{1}{\sqrt{4\pi t_j}}\int_{\mathbb{R}}\frac{1}{\pi(y-x)}\sin\left(\frac{\pi}{h}(y-x)\right)\exp\left(-\frac{(y-x_j)^2}{4t_j}\right)dy$$

$$=\frac{1}{h}\int_{-1}^{1}\exp\left(-\frac{t_j\pi^2}{h^2}\xi^2\right)\cos\left(\frac{\pi}{h}(x_j-x)\xi\right)d\xi$$

$$=\frac{\sqrt{2\pi}}{h}\int_{0}^{1}\exp\left(-\frac{t_j\pi^2}{h^2}\xi^2\right)\cos\left(\frac{\pi}{h}(x_j-x)\xi\right)d\xi.$$

Choose $\{\psi_M\}_{M=1}^{\infty}\subset C_c^{\infty}(\mathbb{R})$ so that for every $M=1,2,\ldots,$

$$0\le\psi_M\le\psi_{M+1}\le\chi_{(-1,1)},\quad \lim_{M\to\infty}\psi_M=\chi_{(-1,1)}.$$

If we use the monotone convergence theorem, then we obtain

$$a_{j,j'}=\frac{1}{4h^2}\int_{\mathbb{R}}\left|\int_{-1}^{1}\exp\left(-\frac{t_j\pi^2}{h^2}\xi^2-i\frac{\pi}{h}(x_j-x)\xi\right)d\xi\right|^2dx$$

$$=\lim_{M\to\infty}\lim_{L\to\infty}\frac{1}{4h^2}\int_{-L}^{L}\left|\int_{\mathbb{R}}\psi_M(\xi)\exp\left(-\frac{t_j\pi^2}{h^2}\xi^2-i\frac{\pi}{h}(x_j-x)\xi\right)d\xi\right|^2dx.$$

We write

$$\Phi(\xi,\eta)\equiv\exp\left(-\frac{\pi^2}{h^2}(t_j\xi^2+t_{j'}\eta^2)-i\frac{\pi}{h}(x_j\xi-x_{j'}\eta)\right).$$

If we change the order of integrations, then we obtain

$$
\begin{aligned}
&a_{j,j'}\\
&= \lim_{M\to\infty}\lim_{L\to\infty}\frac{1}{4h^2}\int_{\mathbb{R}^2}\Phi(\xi,\eta)\psi_M(\xi)\psi_M(\eta)\left(\int_{-L}^{L}\exp\left(i\frac{\pi}{h}x(\xi-\eta)\right)dx\right)d\xi d\eta\\
&= \lim_{M\to\infty}\lim_{L\to\infty}\frac{1}{2h^2}\int_{\mathbb{R}^2}\Phi(\xi,\eta)\psi_M(\xi)\psi_M(\eta)\frac{h}{\pi(\xi-\eta)}\sin\left(\frac{\pi}{h}L(\xi-\eta)\right)d\xi d\eta.
\end{aligned}
$$

By Lemma 6.1, we have

$$
a_{j,j'} = \lim_{M\to\infty}\frac{\pi}{2h^2}\int_{\mathbb{R}}\Phi(\xi,\xi)\psi_M(\xi)\psi_M(\xi)\,d\xi = \frac{\pi}{2h^2}\int_{\mathbb{R}}\Phi(\xi,\xi)\,d\xi.
$$

Consequently, inserting the definition of Φ, we have

$$
\begin{aligned}
a_{j,j'} &= \frac{\pi}{2h^2}\int_{-1}^{1}\exp\left(-\frac{\pi^2}{h^2}(t_j+t_{j'})\xi^2 - i\frac{\pi}{h}(x_j-x_{j'})\xi\right)d\xi\\
&= \frac{\pi}{h^2}\int_{0}^{1}\exp\left(-\frac{\pi^2}{h^2}(t_j+t_{j'})\xi^2\right)\cos\left(\frac{\pi}{h}(x_j-x_{j'})\xi\right)d\xi.
\end{aligned}
\tag{6.146}
$$

In particular, from (6.146) we have

Proposition 6.3. *The matrix* $A=\{a_{j,j'}\}_{j,j'=1,2,\ldots,n}$ *is strictly positive definite.*

Proof. The matrix $A=\{a_{j,j'}\}_{j,j'=1,2,\ldots,n}$ is positive definite from (6.146). We need to prove $C_1=C_2=\cdots=C_n=0$ if a finite set $\{C_1,C_2,\ldots,C_n\}\subset\mathbb{C}$ satisfies

$$
\sum_{j,j'=1}^{n} C_j a_{j,j'} C_{j'} = 0.
$$

Since $a_{j,j'} = \langle \mathbf{L}^*\mathbf{e}_j, \mathbf{L}^*\mathbf{e}_{j'}\rangle$, we have

$$
\sum_{j=1}^{n} C_j\mathbf{L}^*\mathbf{e}_j \equiv 0, \tag{6.147}
$$

or equivalently,

$$
\sum_{j=1}^{n} C_j\frac{1}{(4\pi t_j)^{1/2}}\exp\left(-\frac{(\xi-x_j)^2}{4t_j}\right) \equiv 0. \tag{6.148}
$$

For a nonzero number C_j, we can assume that all the t_j 's are identical, which we let just t. Indeed, if, for example, $t_1 > \max\{t_2,t_3,\ldots,t_n\}$, then we have

$$\sum_{j=1}^{n} \frac{C_j}{(4\pi t_j)^{1/2}} \exp\left(\frac{(\xi - x_1)^2}{4t_1} - \frac{(\xi - x_j)^2}{4t_j}\right) \equiv 0.$$

Letting $t \to \infty$, we obtain $C_1 = 0$. Assuming $t_1 = t_2 = \cdots = t_n = t$, we conclude that

$$\sum_{j=1}^{n} C_j \exp\left(-\frac{(\xi - x_j)^2}{4t}\right) \equiv 0, \tag{6.149}$$

and therefore,

$$\sum_{j=1}^{n} C_j \exp\left(\frac{x_j \xi}{2t} - \frac{x_j^2}{4t}\right) \equiv 0 \quad (\xi \in \mathbb{R}). \tag{6.150}$$

We see that for nonzero C_j, all x_j must be the same; (6.150) implies that the matrix $A = \{a_{j,j'}\}_{j,j'=1}^{n}$ is strictly positive definite.

We thus obtain the real inversion formula by way of Theorem 2.33.

Theorem 6.13 ([78, Theorem 3.3]). *For any given n values $\mathbf{d} = \{d(x_j, t_j)\}_{j=1}^{n}$, among the Weierstrass transforms satisfying*

$$u_F(x_j, t_j) = d(x_j, t_j) \quad j = 1, 2, \ldots, n, \tag{6.151}$$

and among their inverses, the uniquely determined function $F^*_{\mathbf{d}}(x)$ with the minimum norm $\|F\|_{H_{K_h}(\mathbb{R})}$ is represented uniquely as follows:

$$F^*_{\mathbf{d}} = \sum_{j,j'=1}^{n} d(x_j, t_j)\tilde{a}_{j,j'} \mathbf{L}^* \mathbf{e}_{j'}. \tag{6.152}$$

In the Sobolev space with $m = n = 1$, the reproducing kernel is

$$K(x, y) = \frac{1}{2} \exp(-|y - x|) \quad (x, y \in \mathbb{R})$$

according to Theorem 1.3. Then we have the corresponding formulas:

Lemma 6.2. *We have*

$$\begin{aligned} \mathbf{L}^* \mathbf{e}_j(x) &= \frac{1}{\sqrt{2\pi}} \int_{\mathbb{R}} \frac{\exp(-t_j \xi^2)}{\xi^2 + 1} \exp(-i(x - x_j)\xi) d\xi \\ &= \frac{1}{\sqrt{2\pi}} \int_{\mathbb{R}} \frac{\exp(-t_j \xi^2)}{\xi^2 + 1} \cos\left((x - x_j)\xi\right) d\xi \end{aligned}$$

for $x \in \mathbb{R}$ and $j = 1, 2, \ldots, n$.

Proof. First we observe that

$$\mathbf{L}^*\mathbf{e}_j(x) = \langle \mathbf{L}^*\mathbf{e}_j, (K_h)_x\rangle_{H_{K_h}(\mathbb{R})} = \langle \mathbf{e}_j, \mathbf{L}(K_h)_x\rangle_{\mathbb{R}^n}.$$

Now we calculate the following

$$\frac{1}{\sqrt{4\pi t_j}}\int_{\mathbb{R}}\frac{1}{2}\exp\left(-|y-x|-\frac{(y-x_j)^2}{4t_j}\right)dy$$

$$=\frac{1}{4\sqrt{\pi t_j}}\int_{\mathbb{R}}\exp\left(-|y|-\frac{(y+x-x_j)^2}{4t_j}\right)dy$$

$$=\frac{1}{4\sqrt{\pi t_j}}\int_{\mathbb{R}}\left(\frac{\sqrt{2\pi}}{\pi}\frac{1}{\xi^2+1}\right)(2\sqrt{\pi t_j}\exp\{-i(x-x_j)\xi-t_j\xi^2\})\,d\xi$$

$$=\frac{1}{\sqrt{2\pi}}\int_{\mathbb{R}}\frac{\exp(-t_j\xi^2)}{\xi^2+1}\exp(-i(x-x_j)\xi)d\xi.$$

We set

$$\mathrm{I}\equiv\int_{\mathbb{R}}\left(\int_{\mathbb{R}}\frac{\exp((x_j-x)\xi i-t_j\xi^2)}{\xi^2+1}d\xi\right)\left(\int_{\mathbb{R}}\frac{\exp((x-x_{j'})\eta i-t_{j'}\eta^2)}{\eta^2+1}d\eta\right)dx,$$

and

$$\mathrm{J}_L$$
$$\equiv\int_{-L}^{L}\left(\int_{\mathbb{R}}\frac{\xi\exp((x-x_j)\xi i-t_j\xi^2)}{\xi^2+1}d\xi\right)\left(\int_{\mathbb{R}}\frac{\eta\exp((x-x_{j'})\eta i-t_{j'}\eta^2)}{\eta^2+1}d\eta\right)dx.$$

Observe that $a_{j,j'}=\lim_{L\to\infty}\dfrac{\mathrm{I}+\mathrm{J}_L}{2\pi}$. Then we have

$$\mathrm{I}=\int_{\mathbb{R}}\int_{\mathbb{R}}\frac{\exp((x_j\xi-x_{j'}\eta)i-t_j\xi^2-t_{j'}\eta^2)}{(\xi^2+1)(\eta^2+1)}\cdot\left(\int_{\mathbb{R}}\exp(-ix(\xi-\eta))\,dx\right)d\xi d\eta$$

$$=2\pi\int_{\mathbb{R}}\int_{\mathbb{R}}\frac{\exp(-t_j\xi^2-t_{j'}\eta^2)}{(\xi^2+1)(\eta^2+1)}\exp(i(x_j\xi-x_{j'}\eta))\delta(\xi-\eta)d\xi d\eta$$

$$=2\pi\int_{\mathbb{R}}\frac{\exp(-(t_j+t_{j'})\xi^2)}{(\xi^2+1)^2}e^{i(x_j-x_{j'})\xi}d\xi$$

$$=2\pi\int_{\mathbb{R}}\frac{\exp(-(t_j+t_{j'})\xi^2)}{(\xi^2+1)^2}\cos((x_j-x_{j'})\xi)d\xi.$$

As for J_L, we have

$$
\begin{aligned}
&\mathrm{J}_L \\
&= \int_{-L}^{L} \left(\int_{\mathbb{R}} \frac{\xi \exp((x_j - x)\xi i - t_j \xi^2)}{\xi^2 + 1} d\xi \right) \left(\int_{\mathbb{R}} \frac{\eta \exp((x - x_{j'})\eta i - t_{j'} \eta^2)}{\eta^2 + 1} d\eta \right) dx \\
&= \int_{\mathbb{R}} \int_{\mathbb{R}} \frac{\xi \eta \exp((x_j \xi - x_{j'} \eta) i - (t_j \xi^2 + t_{j'} \eta^2))}{(\xi^2 + 1)(\eta^2 + 1)} \left(\int_{-L}^{L} \exp(-ix(\xi - \eta))\, dx \right) d\xi d\eta.
\end{aligned}
$$

Recall $t_1, t_2, \ldots, t_n > 0$. By Lemma 6.1 we have

$$
\begin{aligned}
\lim_{L \to \infty} \mathrm{J}_L &= \int_{\mathbb{R}} \int_{\mathbb{R}} \frac{\xi \eta e^{-(t_j \xi^2 + t_{j'} \eta^2)}}{(\xi^2 + 1)(\eta^2 + 1)} \exp(i(x_j \xi - x_{j'} \eta)) \delta(\xi - \eta) d\xi d\eta \\
&= \int_{\mathbb{R}} \frac{\xi^2 \exp(-(t_j + t_{j'})\xi^2)}{(\xi^2 + 1)^2} \exp(i(x_j - x_{j'})\xi) d\xi \\
&= \int_{\mathbb{R}} \frac{\xi^2 \exp(-(t_j + t_{j'})\xi^2)}{(\xi^2 + 1)^2} \cos((x_j - x_{j'})\xi) d\xi.
\end{aligned}
$$

Therefore it follows that

$$
a_{j,j'} = \int_{\mathbb{R}} \frac{\exp(-(t_j + t_{j'})\xi^2) \cos((x_j - x_{j'})\xi)}{\xi^2 + 1} d\xi \quad (j, j' = 1, 2, \ldots, n), \tag{6.153}
$$

as was to be shown.

The source in this subsection is taken from [78].

6.3.5 *Natural Outputs and Global Inputs of Linear Systems*

We now consider a new concept for a finite number of input data in some general linear system, i.e., we would like to consider some natural global output and some corresponding global natural input satisfying the input and output relation. We will be able to formulate such a problem by considering the generalized inverse by means of the Tikhonov regularization. We will consider concrete examples for heat conduction problems and the wave equation.

A General Concept and Introduction

At first, we introduce a simple general concept for a general linear system. In order to state our concept simply, we shall state it with a very simple example of heat conduction (6.139).

Now, how will a natural solution of the heat equation (6.139) be for a finite number of initial data $\{F(x_j)\}_{j=1}^{n}$? The first natural answer will be given by considering some good function taking the data $\{F(x_j)\}_{j=1}^{n}$ and using the integral representation (6.101). However, with this idea, what kind of interpolation will be suitable satisfying the given data $F(x_j)$? For this general problem, we would like to give some interpolation, and so the corresponding natural heat conduction. Indeed a rough idea for this and the construction of this subsection are as follows: We consider a natural inversion of the integral transform (6.101) and we will consider the solutions $u(x, t)$ satisfying the conditions $u(x_j, t_j) = F(x_j)$. With the solution $u(x, t)$ with some minimum norm among the solutions $u(x, t)$ satisfying (6.101), we can determine the initial heat distribution $F(x)$ globally. Note that our global interpolation $F(x)$ depends on the system (6.101) and so, we will call our interpolation *one with sensibility to the system.* We formulate this method for some general setting in the next section and we state the concrete formulations in the below two sections for the above heat conduction and the wave equation case.

General Formulations

In order to formulate a general theory for the above idea, we will need a general inversion for a general linear system. We again use Theorem 2.33. Let $L : H_K(\mathbb{R}^n) \to \mathscr{H}$ be a bounded linear mapping from $H_K(\mathbb{R}^n)$ to a Hilbert space $\mathscr{H}$. Now we will find $\mathbf{d} \in \mathscr{H}$ such that for any finite data $\{f_{\mathbf{d},\alpha}(p_j)\}_{j=1}^{N}$, given by (3.87), to this end, we will consider the linear mapping

$$\mathbf{d} \in \mathscr{H} \mapsto \{\langle \mathbf{d}, L[K_L(\cdot, p_j; \alpha)]\rangle_{\mathscr{H}}\}_{j=1}^{N} \in \mathbb{C}^N.$$

Here $K_L(p, q; \alpha) \equiv K_\alpha(p, q)$ is a function appearing in Theorem 3.13. Then, according to Theorem 2.33, the image space forms the reproducing kernel Hilbert space admitting the kernel

$$k(p_j, p_{j'}) \equiv a_{j,j'} \equiv \big\langle L[K_L(\cdot, p_{j'}; \alpha)], L[K_L(\cdot, p_j; \alpha)]\big\rangle_{\mathscr{H}}, \tag{6.154}$$

and we can derive the inverse with the minimum norm in $\mathscr{H}$. Precisely, the minimum norm inverse $\mathbf{d}^*$ for any given $f_j = \langle \mathbf{d}, K_L(\cdot, p_j; \alpha)\rangle_{\mathscr{H}}$ for $j = 1, 2, \ldots, N$ is

$$\mathbf{d}^* = \sum_{j,j'=1}^{N} f_j \tilde{a}_{j,j'} L[K_L(\cdot, p_{j'}; \alpha)], \tag{6.155}$$

if the system $\{L[K_L(\cdot, p_j; \alpha)]\}_{j=1}^{N}$ is linearly independent in $\mathscr{H}$ and for $\tilde{A} = \overline{A^{-1}} = \{\tilde{a}_{j,j'}\}_{j,j'=1}^{N}$ for $A = \{a_{j,j'}\}_{j,j'=1}^{N}$ and

$$\|\mathbf{d}^*\|_{\mathscr{H}} = \sqrt{\sum_{j,j'=1}^{N} \tilde{a}_{j,j'} f_j \overline{f_{j'}}}. \tag{6.156}$$

Then we obtain the desired interpolation of the initial input function using *(6.155)* with $\mathbf{d}^*$. We thus obtain the general theorem as a corollary of Theorem 2.33:

Theorem 6.14. *We assume that the system* $\{L[K_L(\cdot, p_j; \alpha)]\}_{j=1}^N$ *is linearly independent in* $\mathscr{H}$. *Then for any input data* $f_j = F(p_j)$, *we give a vector* $\mathbf{d}^*$ *by* (6.155). *Then by using* (6.155) *with this* $\mathbf{d}^*$, *we obtain the global initial function which is determined by the linear system L.*

We will call the interpolation of an input function determined by Theorem 6.14 *a sensible interpolation of the input function* determined by the linear system L. Since we used Tikhonov regularization for considering the generalized inverses that are effective numerically and practically, the sensible interpolation depends on the Tikhonov parameter α. The sensible interpolation makes a sense intuitively for a small parameter α ; however it will depend on how large the data error is.

We now wish to consider a formula in (2.252); for a finite number of initial heat values $F(x_j)$, $j = 1, 2, \ldots, N$, we need to determine g such that $F(x_j) = F^*_{t,\lambda,h,g}(x_j)$, $j = 1, 2, \ldots, N$, in (6.104).

For that purpose, from a finite number of data $F^*_{t,\lambda,h,g}(x_j)$, $j = 1, 2, \ldots, N$, we wish to determine the optimal heat distribution $g(\xi)$ in the sense of (2.252).

Therefore, we consider the bounded linear operator from $L^2(\mathbb{R}^n)$ into $\mathbb{R}^N$:

$$L^2(\mathbb{R}^n) \ni g \mapsto \{F^*_{t,\lambda,h,g}(x_j)\} \in \mathbb{R}^N; \quad j = 1, 2, \ldots, N. \tag{6.157}$$

Let us define

$$Q_{t,\lambda,h}(x) \equiv \frac{1}{(2\pi)^n} \int_{(-\frac{\pi}{h}, \frac{\pi}{h})^n} \frac{\exp(-i\eta \cdot x)}{\lambda e^{|\eta|^2 t} + e^{-|\eta|^2 t}} d\eta.$$

The image space of the linear mapping forms the Hilbert space H_B admitting the reproducing kernel

$$\begin{aligned} b_{j,j'} &\equiv \int_{\mathbb{R}^n} Q_{t,\lambda,h}(\xi - x_{j'}) Q_{t,\lambda,h}(\xi - x_j) d\xi \\ &= \frac{1}{(2\pi)^n} \int_{(-\frac{\pi}{h}, \frac{\pi}{h})^n} \frac{\exp(-ip \cdot (x_j - x_{j'}))dp}{(\lambda \exp(|p|^{2t}) + e^{-|p|^2 t})^2}. \end{aligned}$$

If the matrix $B \equiv \{b_{j,j'}\}_{j,j'=1}^N$ is positive definite, then the corresponding norm in H_B is determined by

$$\|\mathbf{F}\|_{H_B}^2 = \mathbf{F}^* \tilde{B} \mathbf{F}, \tag{6.158}$$

where $\tilde{B} = \overline{B^{-1}} = \{\tilde{b}_{j,j'}\}_{j,j'=1}^N$ and for $\mathbf{F} \equiv \{F^*_{t,\lambda,h,g}(x_j)\}$.

We thus obtain the desired result from Theorem 6.14:

Theorem 6.15. *For any given N values* $\mathbf{F} = \{F^*_{t,\lambda,h,g}(x_j)\}_{j=1}^N \in \mathbb{R}^N$, *among the* $L^2(\mathbb{R}^n)$ *functions g satisfying* (6.157) *for* $j = 1, 2, \ldots, N$, *the function* $g^*_{\mathbf{F}}(x)$ *with the minimum norm* $\|g\|_{L^2(\mathbb{R}^n)}$ *is uniquely determined and it is represented as follows:*

$$g^*_{\mathbf{F}} = \sum_{j,j'=1}^{N} F^*_{t,\lambda,h,g}(x_j)\tilde{b}_{j,j'} Q_{t,\lambda,h}(\cdot - x_{j'}). \tag{6.159}$$

Furthermore, we have the global initial heat distribution

$$F^*_{t,\lambda,h,g^*_{\mathbf{F}}}(x) = \sum_{j,j'=1}^{N} \frac{F^*_{t,\lambda,h,g}(x_j)\tilde{b}_{j,j'}}{(2\pi)^n} \int_{(-\frac{\pi}{h},\frac{\pi}{h})^n} \frac{\exp(-ip\cdot(x-x_{j'}))}{(\lambda\exp(|p|^2t)+\exp(-|p|^2t))^2}dp. \tag{6.160}$$

Theorem 6.15 will have a delicate meaning and we can interpret it as follows: For any initial data N values $\mathbf{F} = \{F(x_j)\}_{j=1}^N$, we will consider the function $g^*_{\mathbf{F}}(x)$ given by (6.159) as $F^*_{t,\lambda,h,g}(x_j) = F(x_j)$ for $j = 1, 2, \ldots, N$. Then the extremal function $g^*_{\mathbf{F}}(x)$ is determined in the sense of Theorem 6.15. The corresponding function $F^*_{t,\lambda,h,g^*_{\mathbf{F}}}(x)$, determined by $g^*_{\mathbf{F}}(x)$ and the integral (6.101), of course satisfies $F^*_{t,\lambda,h,g^*_{\mathbf{F}}}(x_j) = F(x_j) \quad j = 1, 2, \ldots, N$. Then the heat distribution $u(\cdot, t)$ $(t > 0)$ with the initial heat distribution $F^*_{t,\lambda,h,g^*_{\mathbf{F}}}$ is given by the formula

$$u_{F^*_{t,\lambda,h,g^*_{\mathbf{F}}}}(x,t) = \frac{1}{\sqrt{4\pi t}}\int_{\mathbb{R}^n} \exp\left(-\frac{|x-y|^2}{4t}\right) F^*_{t,\lambda,h,g^*_{\mathbf{F}}}(y)\,dy. \tag{6.161}$$

Then we will be able to consider this heat distribution as a natural heat distribution with the initial heat distribution of N values $\mathbf{F} = \{F(x_j)\}_{j=1}^N$.

We will introduce an interesting phenomenon that we considered to be the integral transform (6.101) in the space $L^2(\mathbb{R}^n)$. The Paley Wiener reproducing kernel Hilbert spaces $H_{K_h}(\mathbb{R}^n)$ are very small subspaces of $L^2(\mathbb{R}^n)$. See (1.6) for the definition of K_h. When we consider the above problem in the RKHS $H_{K_h}(\mathbb{R})$, curiously we obtain the same results. This means, in particular, that the sensibility interpolation functions $g^*_{\mathbf{F}}$ in Theorem 6.15 belong to the Paley Wiener spaces $H_{K_h}(\mathbb{R})$ automatically consisting of entire functions of exponential type. However, this smoothness property may be viewed also from the explicit representations.

Indeed, we consider the bounded linear operators from the RKHS $H_{K_h}(\mathbb{R})$ into $\mathbb{R}$:

$$H_{K_h} \ni g \mapsto Mg(x_j) \equiv F^*_{t,\lambda,h,g}(x_j); \quad j = 1, 2, \ldots, N. \tag{6.162}$$

We set

$$\mathbf{M} : H_{K_h}(\mathbb{R}) \ni g \mapsto (Mg(x_1), Mg(x_2), \ldots, Mg(x_N)) \in \mathbb{R}^N. \tag{6.163}$$

Then we see that by taking an orthonormal system $\mathbf{e}_j$ in $\mathbb{R}^N$, we have

$$Mg(x_j) = \langle \mathbf{M}g, \mathbf{e}_j \rangle_{\mathbb{R}^N} = \langle g, \mathbf{M}^* \mathbf{e}_j \rangle_{H_{K_h}(\mathbb{R})}. \tag{6.164}$$

The image space of the linear mapping $\mathbf{M}$ forms the Hilbert space admitting the reproducing kernel

$$\langle \mathbf{M}^* \mathbf{e}_{j'}, \mathbf{M}^* \mathbf{e}_j \rangle_{H_{K_h}(\mathbb{R})}. \tag{6.165}$$

However, this matrix coincides with B. Indeed, by the reproducing property, we have

$$\mathbf{M}^* \mathbf{e}_j(x) = \langle \mathbf{M}^* \mathbf{e}_j, (K_h)_x \rangle_{H_{K_h}} = \langle \mathbf{e}_j, \mathbf{M}(K_h)_x \rangle_{\mathbb{R}^N} = M[(K_h)_x](x_j).$$

If we use the expression of M, we obtain

$$\mathbf{M}^* \mathbf{e}_j(x) = \frac{1}{(2\pi)^n} \int_{(-\frac{\pi}{h}, \frac{\pi}{h})^n} \frac{\exp(-i\eta \cdot (x - x_j))}{\lambda e^{|\eta|^2 t} + e^{-|\eta|^2 t}} d\eta = Q_{t,\lambda,h}(x - x_j).$$

Furthermore, of course,

$$\langle \mathbf{M}^* \mathbf{e}_{j'}, \mathbf{M}^* \mathbf{e}_j \rangle_{H_{K_h}(\mathbb{R})} = \frac{1}{(2\pi)^n} \int_{(-\frac{\pi}{h}, \frac{\pi}{h})^n} \frac{\exp(-i\eta \cdot (x_j - x_{j'}))d\eta}{(\lambda \exp(|\eta|^2 t) + \exp(-|\eta|^2 t))^2} = b_{j,j'}.$$

Note that the Sobolev spaces $H^s(\mathbb{R}^n)$ belong to the intermediate spaces of the Hilbert spaces $L^2(\mathbb{R}^n)$ and $H_{K_h}(\mathbb{R})$, however, the results obtained are different. Indeed, in the above operator $\mathbf{M}$, the corresponding results are as follows:

$$\mathbf{M}^* \mathbf{e}_j(x) = \frac{1}{(2\pi)^n} \int_{(-\frac{\pi}{h}, \frac{\pi}{h})^n} \frac{e^{-i\eta \cdot (x - x_j)}}{(1 + |\eta|^2)^s (\lambda e^{|\eta|^2 t} + e^{-|\eta|^2 t})} d\eta \tag{6.166}$$

and

$$\langle \mathbf{M}^* \mathbf{e}_{j'}, \mathbf{M}^* \mathbf{e}_j \rangle_{H^s(\mathbb{R}^n)} = \frac{1}{(2\pi)^n} \int_{(-\frac{\pi}{h}, \frac{\pi}{h})^n} \frac{\exp(-i\eta \cdot (x_j - x_{j'}))d\eta}{(1 + |\eta|^2)^{2s} (\lambda \exp(|\eta|^2 t) + \exp(-|\eta|^2 t))^2}.$$

We relied upon [84] for the materials of this subsection.
See [337] for the case of $\partial_t - (-\Delta)^\alpha$, where $0 < \alpha < 1$.

6.4 Wave Equation

6.4.1 Inverse Problems in the 1-Dimensional Wave Equation

In the one-dimensional case, we now examine the details.

Let $c > 0$ be a constant. We shall consider the wave equation

$$\frac{\partial^2 u}{\partial t^2}(x,t) = c^2 \frac{\partial^2 u}{\partial x^2}(x,t). \tag{6.167}$$

We write

$$U(x,t) \equiv \chi_{(-ct,ct)}(x) \tag{6.168}$$

for $t > 0$ and $x \in \mathbb{R}$. The solutions u satisfying the boundary conditions

$$\begin{cases} u(0,t) = F_1(t) & t \geq 0, \\ u(x,0) = 0 & x \in \mathbb{R}, \end{cases} \tag{6.169}$$

$$\begin{cases} u(x,0) = 0 & x \in \mathbb{R}, \\ u_t(x,0) = F_2(x) & x \in \mathbb{R}, \end{cases} \tag{6.170}$$

$$\begin{cases} u(x,0) = F_3(x) & x \in \mathbb{R}, \\ u_t(x,0) = 0 & x \in \mathbb{R}, \end{cases} \tag{6.171}$$

are represented as follows:

$$u_{F_1}(x,t) = \frac{\partial}{\partial t}\left(\int_0^t F_1(\xi)U(x,t-\xi)d\xi\right), \tag{6.172}$$

$$u_{F_2}(x,t) = \frac{F_2(x+ct) - F_2(x-ct)}{2c}, \tag{6.173}$$

$$u_{F_3}(x,t) = \frac{F_3(x+ct) + F_3(x-ct)}{2}, \tag{6.174}$$

respectively. Of course, the solution to (6.169) is given by

$$u(x,t) = G(x+ct) - G(x-ct),$$

where $G(ct) - G(-ct) = F_1(t)$ for $t \geq 0$. This implies that we have freedom in choosing the even part of G. However, here we are particularly concerned with the function u_{F_1} given by (6.172).

We will use the first-order Sobolev Hilbert space $H_S(\mathbb{R})$ and $H_{S,e}(\mathbb{R})$, both of which were defined in Example 1.2:

$$\|F\|_{H_S(\mathbb{R})} \equiv \sqrt{\int_{\mathbb{R}} (F(x)^2 + F'(x)^2)dx}.$$

As we see from (1.54) and (1.55), any member $F \in H_{S,e}(\mathbb{R})$ is represented uniquely by a function $\mathbf{F}$ in the form

$$F(x) = \int_{\mathbb{R}+} \frac{\cos(x\eta)}{1+\eta^2}\mathbf{F}(\eta)d\eta = \frac{1}{2}\int_{\mathbb{R}} \frac{e^{ix\eta}}{1+\eta^2}\mathbf{F}(\eta)d\eta \tag{6.175}$$

satisfying

$$\|F\|_{H_{S,e}(\mathbb{R})} \equiv \sqrt{\int_{\mathbb{R}+} \frac{|\mathbf{F}(\eta)|^2}{1+\eta^2}d\eta} < \infty. \tag{6.176}$$

In the typical situations (6.170) and (6.173), we have the results:

Theorem 6.16 ([295, Theorem 1]). *Let $t > 0$ be fixed. Define u_F by* (6.172). *For $\lambda > 0$, define*

$$P_{\lambda,t}(\zeta) = -\frac{ci}{2\pi}\int_{\mathbb{R}} \frac{\sin(ct\eta)e^{-i\eta\zeta}d\eta}{\lambda c^2(1+\eta^2)+\sin^2(ct\eta)} \quad (\zeta \in \mathbb{R}). \tag{6.177}$$

1. *For any function $g \in L^2(\mathbb{R})$ and for any $\lambda > 0$, the best approximate function $F^*_{g,\lambda,t}$ in the sense of*

$$\inf_{F\in H_S(\mathbb{R})} \left\{\lambda\|F\|^2_{H_S(\mathbb{R})} + \|g - \partial_x u_F(\cdot,t)\|^2_{L^2(\mathbb{R})}\right\}$$
$$= \lambda\|F^*_{g,\lambda,t}\|^2_{H_S(\mathbb{R})} + \|g - \partial_x u_{F^*_{g,\lambda,t}}(\cdot,t)\|^2_{L^2(\mathbb{R})}$$

exists uniquely and $F^*_{g,\lambda,t}$ is represented by

$$F^*_{g,\lambda,t} = g * P_{\lambda,t}. \tag{6.178}$$

2. *Let $F \in H_S(\mathbb{R})$ and $g \equiv \partial_x u_F(\cdot,t)$. Then, as $\lambda \downarrow 0$,*

$$F^*_{g,\lambda,t} \to F \tag{6.179}$$

uniformly.

Remark that

$$\eta \mapsto \frac{d^m}{d\eta^m}\left(\frac{\sin(ct\eta)e^{-i\eta\zeta}}{\lambda c^2(1+\eta^2)+\sin^2(ct\eta)}\right)$$

is an integrable function and hence g is integrable.

The proof is omitted due to similarity.

For the equation (6.172), we have the following conclusion:

Theorem 6.17 ([295, Theorem 2]). *For $\lambda > 0$, we write*

$$Q_{\lambda,0}(t,\xi) \equiv \int_0^\infty \frac{\cos(ct\eta)\cos(\xi\eta)d\eta}{\lambda(1+\eta^2)+\pi/(2c)} \quad (t,\xi \in \mathbb{R}). \tag{6.180}$$

1. *For any function $g \in L^2(\mathbb{R}^+)$, for any $\lambda > 0$ and for fixed $x = 0$, the best approximate function $F^*_{g,\lambda,0}$ in the sense of*

$$\inf_{F\in H_{S,e}(\mathbb{R})} \left\{\lambda\|F\|^2_{H_{S,e}(\mathbb{R})} + \|g - \partial_t u_F(0,\cdot)\|^2_{L^2(\mathbb{R}+)}\right\}$$

$$= \lambda\|F^*_{g,\lambda,0}\|^2_{H_{S,e}(\mathbb{R})} + \|g - \partial_t u_{F^*_{g,\lambda,0}}(0,\cdot)\|^2_{L^2(\mathbb{R}+)}$$

exists uniquely and $F^*_{g,\lambda,0}$ is represented by

$$F^*_{g,\lambda,0}(\xi) = \int_0^\infty g(t)Q_{\lambda,0}(t,\xi)dt \quad (\xi \in \mathbb{R}). \tag{6.181}$$

2. *If, for $F \in H_{S,e}(\mathbb{R})$ we consider the output $u_F(\cdot,t)$ and we take $g = \partial_t u_F(0,\cdot)$, then we have: as $\lambda \downarrow 0$ in the sense of pointwise convergence*

$$F^*_{g,\lambda,0} \to F \tag{6.182}$$

uniformly.

Proof. The uniform convergence of (6.182) is a consequence of Remark 1.3 as usual.

The mapping L_x

$$L_x : F \in H_S(\mathbb{R}^+) \mapsto \partial_t u_F(x,t) \in L^2(\mathbb{R}^+) \tag{6.183}$$

is a bounded linear operator from $H_S(\mathbb{R}^+)$ into $L^2(\mathbb{R}^+)$ for any fixed x. Then we see that

$$K_\lambda(x',x'';L_0) = \int_{\mathbb{R}+} \frac{\cos(x'\eta)\cos(x''\eta)}{\lambda(1+\eta^2)+\pi/(2c)}d\eta \tag{6.184}$$

satisfies the desired functional equation in our situation; that is, it is the reproducing kernel for the Hilbert space with the norm square

$$\lambda\|F\|^2_{H_{S,e}(\mathbb{R})} + \|\partial_t u_F(0,\cdot)\|^2_{L^2(\mathbb{R}+)}. \tag{6.185}$$

In particular, we thus obtain (6.181). We omit the further detail.

Let us prove (6.182). We insert F defined by (6.175) into (6.173) and we have $u_F(x,t)$. Then we set $\partial_t u_F(0,t) = g(t)$ in (6.181) and we directly obtain

$$F^*_{g,\lambda,0}(\xi) = \frac{\pi}{2c}\int_{\mathbb{R}+} \frac{\cos(\xi\eta)}{(\lambda(1+\eta^2)+\pi/(2c))(1+\eta^2)}\mathbf{F}(\eta)d\eta. \tag{6.186}$$

From (6.186), we thus obtain the desired result (6.182).

The motivations and results in Theorems 6.16 and 6.17 are clear; that is, we are establishing the inversion formulas: from the observation $\partial_x u_F(x,t)$ for any fixed $t > 0$, we determine the initial velocity F, and from the observation $\partial_t u_F(0,t)$ for fixed $x = 0$, we determine the initial velocity F ; indeed, we can only determine the even part of F with respect to the origin, respectively. See the details [295, pp. 146–157].

The inversion formula (6.178) will give a practical formula. For any $\lambda > 0$ and any $t > 0$, we define a linear mapping

$$M_{\lambda,t} : L^2(\mathbb{R}) \to H_S(\mathbb{R}) \tag{6.187}$$

by $M_{\lambda,t}g \equiv F^*_{g,\lambda,t}$. Now we consider the composite operators $L_t M_{\lambda,t}$ and $M_{\lambda,t}L_t$. Using the Fourier integrals we can show for $F \in H_S(\mathbb{R})$, using the representation;

$$M_{\lambda,t}L_t F(x) = \frac{1}{2\pi}\int_{\mathbb{R}}\left\{F(\xi)\cdot\int_{\mathbb{R}}\frac{\sin^2(ct\eta)\exp(-i\eta(\xi-x))d\eta}{\lambda c^2(1+\eta^2)+\sin^2(ct\eta)}\right\}d\xi, \tag{6.188}$$

and for $g \in L^2(\mathbb{R})$,

$$L_t M_{\lambda,t}g(x) = \frac{1}{2\pi}\int_{\mathbb{R}}\left\{g(\xi)\cdot\int_{\mathbb{R}}\frac{\sin^2(ct\eta)\exp(-i\eta(\xi-x))d\eta}{\lambda c^2(1+\eta^2)+\sin^2(ct\eta)}\right\}d\xi. \tag{6.189}$$

Setting

$$\Delta_{\lambda,t}(\xi - x) \equiv \frac{1}{2\pi}\int_{\mathbb{R}}\frac{\sin^2(ct\eta)\exp(-i\eta(\xi-x))d\eta}{\lambda c^2(1+\eta^2)+\sin^2(ct\eta)} \tag{6.190}$$

in (6.188) and (6.189), we have

$$M_{\lambda,t}L_t F = F * \Delta_{\lambda,t}, \quad (F \in H_S(\mathbb{R})) \tag{6.191}$$

and

$$L_t M_{\lambda,t} g = g * \Delta_{\lambda,t}, \quad (g \in L^2(\mathbb{R})). \tag{6.192}$$

Then we obtain

$$\lim_{\lambda \downarrow 0} \Delta_{\lambda,t}(\xi - x) = \delta(\xi - x) \tag{6.193}$$

in the following sense:

Proposition 6.4.

1. *In the strong topology of* $H_S(\mathbb{R})$, *we have*

$$\lim_{\lambda \downarrow 0} M_{\lambda,t} L_t = I \tag{6.194}$$

In particular, (6.194) *holds in the uniform topology.*

2. *In the strong topology of* $L^2(\mathbb{R})$, *we have* $\lim_{\lambda \downarrow 0} L_t M_{\lambda,t} = I$.

Proof. Again this is a generality; see (3.76) and (3.77). Or you may calculate this directly from the related definitions.

The inversion formula (6.181) will also give a practical formula.

For any $\lambda > 0$ and for fixed $x = 0$, we will define the linear mapping

$$M_{\lambda,0} : g \in L^2(\mathbb{R}^+) \mapsto H_{S,e}(\mathbb{R}) \tag{6.195}$$

by $M_{\lambda,0} g \equiv F^*_{g,\lambda,0}$. We consider the composite operators $L_0 M_{\lambda,0}$ and $M_{\lambda,0} L_0$. Then for $F \in H_{S,e}(\mathbb{R})$

$$M_{\lambda,0} L_0 F(\xi) = \frac{\pi}{2c} \int_{\mathbb{R}} \frac{\mathbf{F}(\eta) \cos(\xi \eta)}{(\lambda(1+\eta^2) + \pi/(2c))(1+\eta^2)} d\eta, \tag{6.196}$$

and for $g \in L^2(\mathbb{R}^+)$

$$L_0 M_{\lambda,0} g(t') = \int_{\mathbb{R}+} \left\{ g(t) \cdot \int_{\mathbb{R}+} \frac{\cos(ct\eta)\cos(ct'\eta) d\eta}{\lambda(1+\eta^2) + \pi/(2c)} \right\} dt. \tag{6.197}$$

Setting

$$\Delta_{\lambda,0}(t, t') \equiv \int_{\mathbb{R}+} \frac{\cos(ct\eta)\cos(ct'\eta) d\eta}{\lambda(1+\eta^2) + \pi/(2c)}, \tag{6.198}$$

we have

$$L_0 M_{\lambda,0} g(t') = \int_{\mathbb{R}+} g(t) \Delta_{\lambda,0}(t, t') dt, \quad (g \in L^2(\mathbb{R})). \tag{6.199}$$

Then

$$\lim_{\lambda \downarrow 0} \Delta_{\lambda,0}(\cdot, t') = \delta_{t'} \tag{6.200}$$

for all $t' \in \mathbb{R}$ and

$$\lim_{\lambda \downarrow 0} L_0 M_{\lambda,0} = I, \tag{6.201}$$

from (3.76) and (3.77). By (6.196), we see that for any $F \in H_{S,e}(\mathbb{R})$

$$\lim_{\lambda \downarrow 0} M_{\lambda,0} L_0 F(x) = F(x) \tag{6.202}$$

uniformly on $\mathbb{R}$.

Example 6.1. Let Erfc denote the *complementary error function* given by

$$\mathrm{Erfc}(x) \equiv \frac{2}{\sqrt{\pi}} \int_x^\infty e^{-t^2}\, dt \quad (x \in \mathbb{R}).$$

1. For $g(t) \equiv \chi_{[0,1]}(t)$, we have,

$$F^*_{g,\lambda,0}(x) = \sum_{j=0}^{1} \frac{\pi \cdot \mathrm{sgn}(c + (-1)^j x)}{2(2\lambda c + \pi)} \left\{ 1 - \exp\left(-|c + (-1)^j x| \sqrt{1 + \frac{\pi}{2\lambda c}} \right) \right\}$$

in (6.181).

2. For $g(t) \equiv \exp(-t^2) \quad (t \geq 0)$, we have, in (6.181)

$$\begin{aligned} F^*_{g,\lambda,0}(x) = {} & \frac{\pi^{3/2}}{8\lambda} \exp\left(\frac{c\pi}{8\lambda} + \frac{1}{4} c^2 \right) \left(1 + \frac{\pi}{2\lambda c} \right)^{-1/2} \\ & \times \sum_{k=0}^{1} \left\{ \exp\left((-1)^k x \sqrt{1 + \frac{\pi}{2\lambda c}} \right) \mathrm{Erfc}\left(\frac{(-1)^k}{2} \sqrt{1 + \frac{\pi}{2\lambda} x} \right) \right\}. \end{aligned} \tag{6.203}$$

We can derive the above results as follows, directly: Let us set

$$u_F(x,t) = \frac{1}{2c} \int_{\mathbb{R}} F(\xi) \chi_{(x-ct, x+ct)}(\xi)\, d\xi = F * \Theta_{ct}(x), \tag{6.204}$$

where $\Theta_{ct}(x) = \frac{1}{2c} \chi_{(-ct,ct)}(x)$. Then we have:

Theorem 6.18. *Define*

$$P_{\lambda,t}(x) \equiv -\frac{c\,i}{2\pi} \int_{\mathbb{R}} \frac{\sin(c\,t\,\eta) \exp(i\,\eta\,x)}{\lambda\, c^2\,(1 + \eta^2) + \sin^2(c\,t\,\eta)}\, d\eta \quad (x \in \mathbb{R}). \tag{6.205}$$

1. $P_{\lambda,t} \in L^1(\mathbb{R})$.
2. Let $g \in L^2(\mathbb{R})$. *Then* $F^*_{g,\lambda,t} \equiv g * P_{\lambda,t}$ *is the unique minimizer of the problem*

$$\min_{F \in H^1(\mathbb{R})} \left(\lambda \, \|F\|^2_{H^1(\mathbb{R})} + \|g - \partial_x u_F(\cdot, t)\|^2_{L^2(\mathbb{R})} \right). \tag{6.206}$$

Proof.

1. This is seen directly; just use

$$\left| \left(\frac{d}{d\eta} \right)^m \frac{\sin(c\,t\,\eta)\exp(i\,\eta\,x)}{\lambda\,c^2\,(1+\eta^2) + \sin^2(c\,t\,\eta)} \right| \le C_{m,t,c,\lambda,x}(1+\eta^2)^{-m},$$

where $C_{m,t,c,\lambda,x}$ does not depend on η.
2. We have to calculate

$$L(\lambda + L^* L)^{-1} K_x, \quad L F = \partial_x u_F(\cdot, t) \tag{6.207}$$

for this purpose. Note that

$$u_F(x,t) = \frac{1}{2c} \int_{x-c\,t}^{x+c\,t} F(\xi)\, d\xi. \tag{6.208}$$

Hence we obtain

$$LF(x) = \partial_x u_F(x,t) = \frac{F(x+c\,t) - F(x-c\,t)}{2c} = M_1(D)F(x), \tag{6.209}$$

where

$$M_1(\xi) = \frac{\exp(i\,c\,t\,\xi) - \exp(-i\,c\,t\,\xi)}{2c}. \tag{6.210}$$

Let us calculate the adjoint operator.

$$\begin{aligned} \langle L^* g, F \rangle_{H^1(\mathbb{R})} &= \langle g, LF \rangle_{L^2(\mathbb{R})} \\ &= \int_{\mathbb{R}} g(x) \overline{\left(\frac{F(x+c\,t) - F(x-c\,t)}{2c} \right)} dx \\ &= \int_{\mathbb{R}} \left(\frac{g(x-c\,t) - g(x+c\,t)}{2c} \right) \overline{F(x)}\, dx \\ &= \int_{\mathbb{R}} \mathscr{F} g(\xi) \frac{\exp(-i\,c\,t\,\xi) - \exp(i\,c\,t\,\xi)}{2c} \mathscr{F} F(\xi)\, d\xi. \end{aligned}$$

Thus, the adjoint of L is given by

$$L^*g(x) = M_2(D)g(x),\ M_2(\xi) = \frac{\exp(-i\,c\,t\,\xi) - \exp(i\,c\,t\,\xi)}{2c\,(1 + |\xi|^2)}. \tag{6.211}$$

Therefore it follows that

$$(\lambda + L^*L)F = (\lambda + M_1(D)M_2(D))F. \tag{6.212}$$

As a result, we obtain

$$L(\lambda + L^*L)F = M_3(D)F, \tag{6.213}$$

where

$$\begin{aligned} M_3(\xi) &\equiv M_1(\xi)\left(\lambda + \frac{2 - \exp(-2i\,c\,t\,\xi) - \exp(2i\,c\,t\,\xi)}{4c^2(1 + |\xi|^2)}\right)^{-1} \\ &= \frac{c\,i\,\sin(c\,t\,\xi)(1 + |\xi|^2)}{\lambda\,c^2(1 + |\xi|^2) + \sin^2(c\,t\,\xi)} \quad (\xi \in \mathbb{R}^n). \end{aligned}$$

Now that

$$K(x, y) = \frac{1}{\sqrt{2\pi}}\mathscr{F}^{-1}\left(\frac{1}{1 + |\xi|^2}\right)(x - y), \tag{6.214}$$

it follows that

$$\begin{aligned} L(\lambda + L^*L)^{-1}[K_x](y) &= \frac{1}{\sqrt{2\pi}}\mathscr{F}^{-1}\left[\frac{c\,i\,\sin(c\,t\,\xi)\exp(-i\,x\,\xi)}{\lambda\,c^2(1 + |\xi|^2) + \sin^2(c\,t\,\xi)}\right](y) \\ &= \frac{c\,i}{2\pi}\int \frac{\sin(c\,t\,\xi)\exp(i\,(y - x)\,\xi)\,d\xi}{\lambda\,c^2(1 + |\xi|^2) + \sin^2(c\,t\,\xi)}. \end{aligned}$$

Hence it follows that

$$\begin{aligned} F^*_{g,\lambda,t}(x) &= \int g(y)\overline{L(\lambda + L^*L)^{-1}[K_x](y)}\,dy \\ &= -\frac{c\,i}{2\pi}\int g(y)\left(\int \frac{\sin(c\,t\,\xi)\exp(\,(x - y)\,\xi)}{\lambda\,c^2(1 + |\xi|^2) + \sin^2(c\,t\,\xi)}\,d\xi\right)dy \\ &= g * P_{\lambda,t}(x). \end{aligned}$$

This is the desired result.

Let us denote by $H^1_e(\mathbb{R})$ the reproducing kernel Hilbert space of all the even continuous functions in $H^1(\mathbb{R})$.

Theorem 6.19. *Let $\lambda > 0$. Define*

$$Q_\lambda(t,\xi) \equiv 2c \int_{\mathbb{R}} \frac{\cos(c\,t\,\eta)\exp(i\,\eta\,t)}{2c\lambda\,(1+\eta^2)+\pi}\,d\eta \quad (t \in \mathbb{R}). \tag{6.215}$$

Then we obtain the results:

1. $Q_\lambda(\cdot,\xi) \in L^2(0,\infty)$ *for all* $\xi \in \mathbb{R}$.
2. *Let* $g \in L^2(\mathbb{R})$. *Then* $F^*_{g,\lambda,t}(\xi) = \displaystyle\int_0^\infty g(t)Q_\lambda(t,\xi)\,dt$ *is the unique minimizer of the problem*

$$\min_{F \in H^1_e(\mathbb{R})} \left(\lambda\,\|F\|^2_{H^1_e(\mathbb{R})} + \|g - \partial_t u_F(\cdot,t)\|^2_{L^2(\mathbb{R}+)}\right). \tag{6.216}$$

The proof is again analogous.

6.4.2 Inverse Problems Using a Finite Number of Data

Recall that we considered $F^*_{g,\lambda,t}(\xi) = \displaystyle\int_0^\infty g(t)Q_\lambda(t,\xi)\,dt$ in Theorem 6.19. We will consider the bounded linear operator from $L^2(\mathbb{R})$ into $\mathbb{R}^N$:

$$L^2(\mathbb{R}) \ni g \mapsto \{F^*_{g,\lambda,t}(x_j)\} \in \mathbb{R}^N; \quad j = 1,2,\ldots,N. \tag{6.217}$$

The image space of the linear mapping forms the Hilbert space H_W admitting the reproducing kernel

$$w_{j,j'} \equiv \int_{\mathbb{R}} P_{\lambda,t}(\xi - x_{j'})P_{\lambda,t}(\xi - x_j)\,d\xi = \frac{c^2}{\pi}\int_{\mathbb{R}+} \frac{\sin^2(ct\eta)\cos[\eta(x_j - x_{j'})]\,d\eta}{(\lambda c^2(1+\eta^2)+\sin^2(ct\eta))^2}. \tag{6.218}$$

If the matrix $W \equiv \{w_{j,j'}\}_{j,j'=1}^N$ is positive definite, then the corresponding norm in H_W is determined by

$$\|\mathbf{F}\|^2_{H_W} = \mathbf{F}^*\tilde{W}\mathbf{F} = (F^*_{g,\lambda,t}(x_1), F^*_{g,\lambda,t}(x_2),\ldots,F^*_{g,\lambda,t}(x_n))\tilde{W}\begin{pmatrix} F^*_{g,\lambda,t}(x_1) \\ F^*_{g,\lambda,t}(x_2) \\ \vdots \\ F^*_{g,\lambda,t}(x_n)) \end{pmatrix}, \tag{6.219}$$

for $\mathbf{F} \equiv \{F^*_{g,\lambda,t}(x_j)\}$.

We thus obtain the desired result:

Theorem 6.20. *For any N values* $\mathbf{F} = \{F_j\}_{j=1}^N \in \mathbb{R}^N$, *among the* $L^2(\mathbb{R})$*-functions* g, *the function* $g^*_{\mathbf{F}}(x)$ *with the minimum norm* $\|g\|_{L^2(\mathbb{R})}$ *is uniquely determined and we can represent it as follows:*

$$g^*_{\mathbf{F}}(x) = \sum_{j,j'=1}^{N} F_j \tilde{w}_{j,j'} P_{\lambda,t}(x - x_{j'}). \tag{6.220}$$

The corresponding initial wave speed is

$$F^*_{g^*_{\mathbf{F}},\lambda,t}(x) = \sum_{j,j'=1}^{N} F_j \tilde{w}_{j,j'} \frac{c^2}{\pi} \int_{\mathbb{R}^+} \frac{\sin^2(ct\eta)\cos[\eta(x - x_{j'})]d\eta}{(\lambda c^2(1+\eta^2) + \sin^2(ct\eta))^2}. \tag{6.221}$$

When we consider the Paley Wiener spaces $H_{K_h}(\mathbb{R})$, defined by (1.6), then the results are different. Indeed, we consider te real numbers for $f \in H_{K_h}(\mathbb{R})$:

$$H_{K_h}(\mathbb{R}) \ni g \mapsto Og(x_j) \equiv F^*_{g,\lambda,t}(x_j); \quad j = 1, 2, \ldots, N. \tag{6.222}$$

Define a mapping $\mathbf{O}$ by

$$\mathbf{O} : H_{K_h}(\mathbb{R}) \ni g \mapsto (Og(x_1), Og(x_2), \ldots, Og(x_N)) \in \mathbb{R}^N. \tag{6.223}$$

Then, we see, by taking an orthonormal system $\mathbf{e}_j$ in $\mathbb{R}^N$, that

$$Og(x_j) = \langle \mathbf{O}g, \mathbf{e}_j \rangle_{\mathbb{R}^N} = \langle g, \mathbf{O}^*\mathbf{e}_j \rangle_{H_{K_h}(\mathbb{R})}. \tag{6.224}$$

The image space of the linear mapping $\mathbf{O}$ forms the Hilbert space admitting the reproducing kernel

$$K(j, j') \equiv \langle \mathbf{O}^*\mathbf{e}_{j'}, \mathbf{O}^*\mathbf{e}_j \rangle_{H_{K_h}(\mathbb{R})}. \tag{6.225}$$

We calculate

$$\mathbf{O}^*\mathbf{e}_j(x) = \langle \mathbf{O}^*\mathbf{e}_j, (K_h)_x \rangle_{H_{K_h}(\mathbb{R})} = \langle \mathbf{e}_j, \mathbf{O}[(K_h)_x] \rangle_{\mathbb{R}^N} = O[(K_h)_x](x_j),$$

but we need to insert the different kernel

$$\mathbf{O}^*\mathbf{e}_j(x) = \frac{-ci}{2\pi} \int_{(-\frac{\pi}{h},\frac{\pi}{h})} \frac{\sin(ct\eta)e^{-i\eta(\xi - x)}}{\lambda c^2(1+\eta^2) + \sin^2(ct\eta)} d\eta.$$

Furthermore, we obtain

$$\langle \mathbf{O}^*\mathbf{e}_{j'}, \mathbf{O}^*\mathbf{e}_j\rangle_{H_{K_h}(\mathbb{R})} = \frac{c^2}{\pi}\int_0^{\pi/h} \frac{\sin^2(ct\eta)\cos(\eta(x_j - x_{j'}))}{(\lambda c^2(1+\eta^2) + \sin^2(ct\eta))^2} d\eta. \tag{6.226}$$

The source is taken from [292].

6.5 General Inhomogeneous PDEs on Whole Spaces

6.5.1 Approximate Solutions by Tikhonov Regularization

We first consider very simple approximate solutions for the general inhomogeneous partial differential equation

$$L(D)u = g \quad \text{on} \quad \mathbb{R}^n, \tag{6.227}$$

in the class of the functions of the s-th order Sobolev Hilbert space $H^s(\mathbb{R}^n)$ on the whole real space $\mathbb{R}^n (n \geq 1, s \geq m \geq 1, s > n/2)$, and for any complex-valued $L^2(\mathbb{R}^n)$-function g. Here $L(D)$ denotes a nontrivial general linear partial differential operator with complex constant coefficients on $\mathbb{R}^n$ of order m. That is, we are given a linear partial differential operator

$$L(D) = \sum_{|\alpha| \leq m} a_\alpha \left(\frac{\partial}{\partial x}\right)^\alpha, \tag{6.228}$$

with the a_α's being complex numbers and there is a multi-index α_0 of length m such that $a_\alpha \neq 0$. Many constant coefficient partial differential equations fall under the scope of the main results for inhomogeneous linear partial differential equations with complex constant coefficients of all types on the whole space $\mathbb{R}^n$.

For simplicity, we write

$$\mathbf{L}(\xi) \equiv e^{-ix\cdot\xi} L(D) e^{ix\cdot\xi} \tag{6.229}$$

using a complex polynomial L.

Theorem 6.21. *Let $n \geq 1, s \geq m \geq 1$ and $s > n/2$.*

1. *For any complex-valued function $g \in L^2$ and for any $\lambda > 0$,*

$$\inf_{F \in H^s(\mathbb{R}^n)} \left\{\lambda \|F\|^2_{H^s(\mathbb{R}^n)} + \|g - L(D)F\|^2_{L^2(\mathbb{R}^n)}\right\} \tag{6.230}$$

is attained by a unique function $F^*_{\lambda,s,g}$.

2. *Let us write*

$$Q_{\lambda,s}(\eta) \equiv \frac{1}{(2\pi)^n}\int_{\mathbb{R}^n} \frac{\overline{\mathbf{L}(p)}\exp(-ip\cdot\eta)}{\lambda(|p|^2+1)^s+|\mathbf{L}(p)|^2}dp \quad (\eta \in \mathbb{R}^n). \tag{6.231}$$

Then the extremal function $F^*_{\lambda,s,g}$ *is represented by*

$$F^*_{\lambda,s,g} = g * Q_{\lambda,s}. \tag{6.232}$$

3. *If* g *is expressed as* $g = L(D)F$*, for a function* $F \in H^s(\mathbb{R}^n)$*, then as* $\lambda \downarrow 0$ *we have*

$$F^*_{\lambda,s,g} \to F, \tag{6.233}$$

uniformly.

Let $c > 0$ be a fixed number. The examples we envisage are the following:

1. The $\overline{\partial}$-operator:

$$\frac{\partial}{\partial \overline{z}} = \frac{1}{2}\left(\frac{\partial}{\partial x} - i\frac{\partial}{\partial y}\right) \quad (x,y) \in \mathbb{R}\times\mathbb{R}. \tag{6.234}$$

2. The heat operator:

$$\partial_t u - c^2\Delta_x u \quad (x,t) \in \mathbb{R}^{n-1}\times(0,\infty). \tag{6.235}$$

When we consider the heat operator, we freeze $t > 0$.

3. The wave operator:

$$\partial_t^2 - c^2\Delta_x \quad (x,t) \in \mathbb{R}^{n-1}\times\mathbb{R}. \tag{6.236}$$

Although there are no global solutions in (6.235), the main result is still applicable.

From concrete examples, we learn to compute the representations (6.232) and we come to know the approximate solutions (6.232) converge to the analytical solutions of (6.227) as in (6.233).

Proof. First, since $s \geq m$, we have the inequality

$$\|L(D)F\|^2_{L^2(\mathbb{R}^n)} \leq M\|F\|^2_{H^s(\mathbb{R}^n)}, \tag{6.237}$$

for some positive constant M depending on L. That is, the operator $L(D)$ is a bounded linear operator from $H^s(\mathbb{R}^n)$ into $L^2(\mathbb{R}^n)$.

Before we come to the proof of Theorem 6.21, let us check the fundamental functional equation directly (3.52). Recall that

$$K_s(x,y) = \frac{1}{(2\pi)^n}\int_{\mathbb{R}^n}(1+|\xi|^2)^{-s}\exp(i(x-y)\cdot\xi)\,d\xi. \tag{6.238}$$

Thus if we set

$$K_\lambda(x,y;L(D)) = \frac{1}{(2\pi)^n}\int_{\mathbb{R}^n}\frac{\exp(ip\cdot(x-y))}{\lambda(|p|^2+1)^s+|\mathbf{L}(p)|^2}dp, \tag{6.239}$$

then we have

$$\begin{aligned}
&\lambda K_\lambda(x,y;L(D)) + \langle L(D)[K_\lambda(\cdot,y;L(D))], L(D)[(K_s)_x]\rangle_{L^2(\mathbb{R}^n)} \\
&= \lambda K_\lambda(x,y;L(D)) + \langle \mathscr{F}[L(D)K_\lambda(\cdot,y;L(D))], \mathscr{F}[L(D)[(K_s)_x]]\rangle_{L^2(\mathbb{R}^n)} \\
&= \frac{1}{(2\pi)^n}\int_{\mathbb{R}^n}\frac{\lambda\exp(ip\cdot(x-y))}{\lambda(|p|^2+1)^s+|\mathbf{L}(p)|^2}dp \\
&\quad + \frac{1}{(2\pi)^n}\int_{\mathbb{R}^n}\frac{(|p|^2+1)^{-s}|\mathbf{L}(p)|^2\exp(ip\cdot(x-y))}{(\lambda(|p|^2+1)^s+|\mathbf{L}(p)|^2)}dp \\
&= K_s(x,y).
\end{aligned}$$

Consequently, $K_\lambda(\cdot,\cdot;L(D))$ satisfies the functional equation (3.52). That is, it is the reproducing kernel for the Hilbert space with the norm square

$$\lambda\|F\|^2_{H^s(\mathbb{R}^n)} + \|L(D)F\|^2_{L^2(\mathbb{R}^n)}. \tag{6.240}$$

In particular, Proposition 4.1 yields

$$\begin{aligned}
F^*_{\lambda,s,g}(x) &= \langle g, L(D)[K_\lambda(\cdot,x;L(D))]\rangle_{L^2(\mathbb{R}^n)} \\
&= \frac{1}{(2\pi)^{\frac{n}{2}}}\int_{\mathbb{R}^n}\mathscr{F}g(p)\frac{\overline{\mathbf{L}(p)}\exp(ip\cdot x)}{\lambda(|p|^2+1)^s+|\mathbf{L}(p)|^2}dp \\
&= \mathscr{F}^{-1}\left[\mathscr{F}g(p)\frac{\overline{\mathbf{L}(p)}}{\lambda(|p|^2+1)^s+|\mathbf{L}(p)|^2}\right](x).
\end{aligned}$$

Thus it follows that

$$F^*_{\lambda,s,g}(x) = \frac{1}{(2\pi)^{\frac{n}{2}}}\int_{\mathbb{R}^n}\mathscr{F}g(p)\frac{\overline{\mathbf{L}(p)}\exp(ip\cdot x)}{\lambda(|p|^2+1)^s+|\mathbf{L}(p)|^2}\,dp. \tag{6.241}$$

Dealing carefully with the convolution, we are led to

$$\begin{aligned} F^*_{\lambda,s,g}(x) &= \frac{1}{(2\pi)^{\frac{n}{2}}}\, g * \mathscr{F}^{-1}\left[\frac{\overline{\mathbf{L}(p)}}{\lambda(|p|^2+1)^s+|\mathbf{L}(p)|^2}\right](x)\\ &= \frac{1}{(2\pi)^n}\int_{\mathbb{R}^n} g(\xi)\left(\int_{\mathbb{R}^n}\frac{\overline{\mathbf{L}(p)}\exp(ip\cdot(x-\xi))}{\lambda(|p|^2+1)^s+|\mathbf{L}(p)|^2}\,dp\right)d\xi\\ &= \frac{1}{(2\pi)^n}\int_{\mathbb{R}^n} g(\xi)\left(\int_{\mathbb{R}^n}\frac{\overline{\mathbf{L}(p)}\exp(-ip\cdot(\xi-x))}{\lambda(|p|^2+1)^s+|\mathbf{L}(p)|^2}\,dp\right)d\xi\\ &= \int_{\mathbb{R}^n} g(\xi)Q_{\lambda,s}(\xi-x)\,d\xi. \end{aligned}$$

Hence we obtain (6.232).

In order to prove the result (6.233), we shall make full use of the formula (6.241) with $g = L(D)F$ for some $F \in H^s(\mathbb{R}^n)$. Then, we have

$$F^*_{\lambda,s,L(D)F}(x) - F(x) = -\frac{1}{(2\pi)^{\frac{n}{2}}}\int_{\mathbb{R}^n}\frac{\lambda(|p|^2+1)^s\mathscr{F}F(p)\exp(ip\cdot x)}{\lambda(|p|^2+1)^s+|\mathbf{L}(p)|^2}dp. \tag{6.242}$$

If we take the Fourier transform, then we obtain

$$\mathscr{F}F^*_{\lambda,s,L(D)F}(p) - \mathscr{F}F(p) = \frac{-\lambda(|p|^2+1)^s\mathscr{F}F(p)}{\lambda(|p|^2+1)^s+|\mathbf{L}(p)|^2}. \tag{6.243}$$

Thus it follows that

$$\|F^*_{\lambda,s,L(D)F} - F\|_{H^s(\mathbb{R}^n)} = \sqrt{\int_{\mathbb{R}^n}\frac{\lambda^2(|p|^2+1)^{3s}|\mathscr{F}F(p)|^2}{(\lambda(|p|^2+1)^s+|\mathbf{L}(p)|^2)^2}\,dp}. \tag{6.244}$$

Noticing that $(|p|^2+1)^s|\mathscr{F}F(p)|^2$ is integrable as a function of p, we conclude that

$$\lim_{\lambda\downarrow 0}\|F^*_{\lambda,s,L(D)F} - F\|_{H^s(\mathbb{R}^n)} = 0 \tag{6.245}$$

due to the Lebesgue convergence theorem. This implies $F^*_{\lambda,s,L(D)F}$ converges to F as $\lambda \downarrow 0$ uniformly in view of Remark 1.3.

6.5.2 *Convergence Properties of the Approximate Solutions*

For any $\lambda > 0$, we will define a linear mapping

$$M_{\lambda,s} : L^2(\mathbb{R}^n) \to H^s(\mathbb{R}^n), \quad M_{\lambda,s}(g) \equiv F^*_{\lambda,s,g} \tag{6.246}$$

and investigate its behavior as $\lambda \downarrow 0$. We hope to show that $M_{\lambda,s}$ plays the role of the approximation inverse of $L(D)$.

To this end, we consider the composite operators $L(D)M_{\lambda,s}$ and $M_{\lambda,s}L(D)$. In view of (6.241), for $F \in H^s(\mathbb{R}^n)$ we have

$$M_{\lambda,s}L(D)F(x) = \frac{1}{(2\pi)^n}\int_{\mathbb{R}^n}\left\{F(\xi)\int_{\mathbb{R}^n}\frac{|\mathbf{L}(p)|^2\exp(-ip\cdot(\xi-x))dp}{\lambda(|p|^2+1)^s+|\mathbf{L}(p)|^2}\right\}d\xi \tag{6.247}$$

and for $g \in L^2$ we have

$$L(D)M_{\lambda,s}g(x) = \frac{1}{(2\pi)^n}\int_{\mathbb{R}^n}\left\{g(\xi)\int_{\mathbb{R}^n}\frac{|\mathbf{L}(p)|^2\exp(-ip\cdot(\xi-x))dp}{\lambda(|p|^2+1)^s+|\mathbf{L}(p)|^2}\right\}d\xi. \tag{6.248}$$

Setting

$$\Delta_{\lambda,s}(x-\xi) \equiv \frac{1}{(2\pi)^n}\int_{\mathbb{R}^n}\frac{|\mathbf{L}(p)|^2\exp(-ip\cdot(\xi-x))}{\lambda(|p|^2+1)^s+|\mathbf{L}(p)|^2}dp \tag{6.249}$$

in (6.247) and (6.248), we have

$$M_{\lambda,s}L(D)F(x) = \int_{\mathbb{R}^n}F(\xi)\Delta_{\lambda,s}(x-\xi)d\xi \quad (F \in H^s(\mathbb{R}^n)) \tag{6.250}$$

and

$$L(D)M_{\lambda,s}g(x) = \int_{\mathbb{R}^n}g(\xi)\Delta_{\lambda,s}(x-\xi)d\xi \quad (g \in L^2). \tag{6.251}$$

Then, by the Fourier transform, we have

$$\lim_{\lambda\downarrow 0}\Delta_{\lambda,s}(x-\xi) = \delta(x-\xi) \tag{6.252}$$

in (6.250) and (6.251). More precisely

$$\lim_{\lambda\downarrow 0}M_{\lambda,s}L(D) = \mathrm{id}_{H^s(\mathbb{R}^n)} \tag{6.253}$$

in the strong topology of $H^s(\mathbb{R}^n)$ and

$$\lim_{\lambda\downarrow 0}L(D)M_{\lambda,s} = \mathrm{id}_{L^2(\mathbb{R}^n)} \tag{6.254}$$

in the strong topology of $L^2(\mathbb{R}^n)$.

Use of Paley Wiener Spaces

In Theorem 6.21, we used the Sobolev spaces as approximate function spaces in the Tikhonov regularization, and they may be considered as the natural ones from the viewpoint of the solutions of the differential equations. However, as approximate function spaces we can use the Paley Wiener spaces, given by (1.6). Then, the corresponding formulas may be simpler, and so for some numerical studies we will expect to be more effective, as was the case in the heat conduction problem. For the Paley Wiener spaces, we can obtain the following results as in Theorem 6.21.

Theorem 6.22. *Let $n \geq 1$ be an integer and $\lambda, h > 0$. Let us write*

$$Q_{\lambda,h}(\eta) \equiv \frac{1}{(2\pi)^n} \int_{(-\pi/h,\pi/n)^n} \frac{\overline{\mathbf{L}(p)} \exp(-ip \cdot \eta)}{\lambda + |\mathbf{L}(p)|^2} dp \quad (\eta \in \mathbb{R}^n). \tag{6.255}$$

1. *For any complex-valued function $g \in L^2(\mathbb{R}^n)$ and for any $\lambda > 0$,*

$$\inf_{F \in H_{K_h}(\mathbb{R}^n)} \left\{ \lambda \|F\|^2_{H_{K_h}(\mathbb{R}^n)} + \|g - L(D)F\|^2_{L^2(\mathbb{R}^n)} \right\} \tag{6.256}$$

   is attained by a unique function $F^*_{\lambda,h}$.
2. *The extremal function $F^*_{\lambda,h}$ is represented by $F^*_{\lambda,h} = g * Q_{\lambda,h}$.*
3. *Let $F \in H_{K_h}(\mathbb{R}^n)$ and $g \equiv L(D)F$. Then, as $\lambda \downarrow 0$,*

$$F^*_{\lambda,h} \to F, \tag{6.257}$$

   uniformly.

Note that the resulting differences between the cases of the Sobolev and the Paley Wiener spaces $H_{K_h}(\mathbb{R})$ is that the factor $\lambda(|p|^2 + 1)^s$ is changed to just λ, and that it is used $\chi_h(p)$ in the case of the Paley Wiener spaces. See (1.6). Taking into account these changes and the formulas obtained in case of the Sobolev spaces, we see that the corresponding formulas are valid for the Paley Wiener spaces framework.

For applications and connections to the Dunkel transform, Fock spaces, Sobolev spaces and Sturm Liouville hypergroups, see [176, 435, 441, 442].

6.5.3 *General Discrete Partial Differential Equations and Inverse Formulas*

We will first consider the typical Helmholtz equation

$$\Delta u + k^2 u = -\rho(\mathbf{r}) \quad \text{on} \quad \mathbb{R}^3 \tag{6.258}$$

for a complex-valued $L^2(d\mathbf{r})$ source function ρ whose support is contained in a sphere $r < a$; r denotes the distance $r = |\mathbf{r}|$ from the origin. This equation is fundamental in mathematical physics, in particular, mathematical acoustics and electromagnetics. See, for example, [51, 107, 311, 365].

We saw the characterization and natural representation of the wave $u(\mathbf{r})$ outside of the support of ρ. As an application, we expressed ρ^* simply in terms of u outside of the support of ρ, which has a minimum $L^2(d\mathbf{r})$ norm among the set of source functions ρ satisfying (6.258). These representations provide a useful means for determining the source ρ^* from the field u. Secondly, in terms of the radiation pattern, that is,

$$g\left(\frac{\mathbf{r}'}{r'}\right) = \frac{1}{4\pi}\int_{r<a} \exp\left(-i\frac{k}{r'}(\mathbf{r}', \mathbf{r})\right)\rho(\mathbf{r})d\mathbf{r}, \tag{6.259}$$

we saw the solution of the inverse problem representing the visible component of ρ. See also [51, pp. 22–26] for example for this problem.

In particular, note that from the whole data of the outside of support of the density ρ, we cannot determine the density fully and its part is, in general, determined. We refer to [388] for the detailed structure. Meanwhile, we used the whole data of a sphere $\{|\mathbf{r}| = a'\}$ $(a' > a)$ of the outside of the support of the density. So, we shall give inversion formulas using a finite number of observation data of the outside of the support of the density.

We assume that the support of ρ is contained in the sphere $r < a$. Since the fundamental solution of (6.258) is

$$\frac{\exp(ik|\mathbf{r} - \mathbf{r}'|)}{4\pi|\mathbf{r} - \mathbf{r}'|}, \tag{6.260}$$

we shall examine the integral representation of the solution u

$$u(\mathbf{r}'; \rho) = \frac{1}{4\pi}\int_{r<a} \frac{\exp(ik|\mathbf{r}' - \mathbf{r}|)}{|\mathbf{r}' - \mathbf{r}|}\rho(\mathbf{r})d\mathbf{r} \tag{6.261}$$

of Helmholtz's equation (6.258) satisfying the Sommerfeld radiation condition

$$\lim_{r\to\infty} r\left(\frac{\partial u}{\partial r} - iku\right) = 0, \tag{6.262}$$

for a source function ρ satisfying

$$\int_{r<a} |\rho(\mathbf{r})|^2 d\mathbf{r} < \infty. \tag{6.263}$$

Next, we will consider the case of using radiation patterns. The behavior at infinity of the solution $u(\mathbf{r})$ of (6.258) is

$$u(\mathbf{r};\rho) = g\left(\frac{\mathbf{r}}{r}\right)\frac{\exp(ik\mathbf{r})}{4\pi r}(1+O(r^{-1})), \tag{6.264}$$

where the radiation pattern g is given by (6.259) (see, for example, [107, p. 20]). We will give the solution of the inverse problem representing the source function ρ in terms of a finite number of the data radiation pattern g of the outside of the support of the density.

When we consider the inversion, we must calculate the related singular integrals of the reproducing kernels; their calculations are very involved, and so we apply the regularized integral representations in Theorem 6.21 based on Tikhonov regularization and the theory of reproducing kernels. Then we can deal with the general situation in Theorem 6.21 containing the Helmholtz equation.

Use of Paley Wiener Spaces

Here we work $H_{K_h}(\mathbb{R})$, the Paley Wiener reproducing kernel Hilbert space described in Theorem 6.21. Recall that any element in $H_{K_h}(\mathbb{R})$ is smooth. For different points $\{x_j\}_{j=1}^N$, we shall consider the complex numbers given by

$$H_{K_h}(\mathbb{R}) \ni F \mapsto LF(x_j) \equiv L(D)F(x_j); \quad j = 1, 2, \ldots, N$$

for $f \in H_{K_h}(\mathbb{R})$. Needless to say, $H_{K_h}(\mathbb{R})$ being agreeable, the mapping $F \mapsto L(D)F$ is continuous. We define

$$\mathbf{L} : H_{K_h}(\mathbb{R}) \ni F \mapsto (LF(x_1), LF(x_2), \ldots, LF(x_N)) \in \mathbb{C}^N. \tag{6.265}$$

We consider a standard orthonormal system $\{\mathbf{e}_j\}_{j=1}^N$ in the space $\mathbb{C}^N$. Then we see that

$$LF(x_j) = \langle \mathbf{L}F, \mathbf{e}_j\rangle_{\mathbb{C}^N} = \langle F, \mathbf{L}^*\mathbf{e}_j\rangle_{H_{K_h}(\mathbb{R})}. \tag{6.266}$$

Let us set

$$A \equiv \{a_{j,j'}\}_{j,j'=1}^N \equiv \{\langle \mathbf{L}^*\mathbf{e}_{j'}, \mathbf{L}^*\mathbf{e}_j\rangle_{H_{K_h}(\mathbb{R})}\}_{j,j'=1}^N. \tag{6.267}$$

Here, note that $a_{j,j'}$ are calculated as follows:

$$\mathbf{L}^*\mathbf{e}_j(x) = \langle \mathbf{e}_j, \mathbf{L}[(K_h)_x]\rangle_{\mathbb{C}^N} = \mathbf{L}[(K_h)_x](x_j) = \frac{1}{(2\pi)^n}\int_{(-\frac{\pi}{h},\frac{\pi}{h})^n} \frac{\mathbf{L}(-\eta)d\eta}{\exp(i\eta\cdot(x-x_j))}.$$

Hence it follows that

$$a_{j,j'} = \frac{1}{(2\pi)^n} \int_{(-\pi/h,\pi/n)^n} |\mathbf{L}(-\boldsymbol{\eta})|^2 \exp(-i\eta \cdot (x_j - x_{j'}))d\eta. \tag{6.268}$$

In particular, we conclude from (6.268) that the matrix $A = \{a_{j,j'}\}_{j,j'=1}^N$ is strictly positive definite.

Denote by H_A the image of $\mathbf{L}$. Since the matrix $\{a_{j,j'}\}_{j,j'=1}^N$ is invertible, then the norm in H_A is

$$\|\mathbf{L}F\|_{H_A}^2 = (\mathbf{L}F)^* A^* (\mathbf{L}F), \tag{6.269}$$

where $A^* = \overline{A^{-1}} = \{\tilde{a}_{j,j'}\}_{j,j'=1}^N$.

Therefore, we obtain the desired result.

Theorem 6.23. *For any N values* $\mathbf{d} = \{d_j\}_{j=1}^N \in \mathbb{C}^N$, *among the* $H_{K_h}(\mathbb{R}^n)$*-functions F satisfying*

$$LF(x_j) = d_j, \quad (j = 1, 2, \ldots, N), \tag{6.270}$$

the function $F_{\mathbf{d}}^*$ *with the minimum norm* $\|F\|_{H_{K_h}(\mathbb{R}^n)}$ *is uniquely determined and it is represented as follows:*

$$F_{\mathbf{d}}^* = \sum_{j,j'=1}^{N} d_j \tilde{a}_{j,j'} \cdot \mathbf{L}^* \mathbf{e}_{j'}. \tag{6.271}$$

Proof. The function given by (6.271) belongs to the range of $\mathbf{L}^*$ and agrees with g at x_j, $j = 1, 2, \ldots, N$. So, the result is an immediate consequence of Proposition 4.3.

Since the numerical calculation of inverses for large size matrices is possible to accomplish for exact data, we will be able to directly apply Theorem 6.23.

As we see from the isometric relation between the data and the inverse with the minimum norm $F_{\mathbf{d}}^*(x)$, the observation data and the corresponding inverse are rigid, and our inversion formula is given in the form of a well-posed problem.

Remark 6.1. In the Sobolev space of $s \geq m$ and $s > \frac{n}{2}$, one has the corresponding formulas:

$$\mathbf{L}^*\mathbf{e}_j(x) = \langle \mathbf{L}^*\mathbf{e}_j, (K_s)_x \rangle_{H_{K_s}(\mathbb{R})} = \langle \mathbf{e}_j, \mathbf{L}(K_s)_x \rangle_{\mathbb{C}^N} = \int_{\mathbb{R}^n} \frac{\overline{\mathbf{L}(\eta)} \exp(-i\eta \cdot (x - x_j))}{(2\pi)^n (1 + |\eta|^2)^s} d\eta$$

and

$$a_{j,j'} = \frac{1}{(2\pi)^n} \int_{\mathbb{R}^n} \frac{|\mathbf{L}(\eta)|^2 \exp(i\eta \cdot (x_j - x_{j'}))}{(1 + |\eta|^2)^s} d\eta. \tag{6.272}$$

The source in this section is taken from [85].

6.5.4 PDEs and Inverse Problems: A General Approach

We will be able to apply our theory to various inverse problems to look for the whole data from the local data of the domain or from some boundary data. Here, we will refer to these problems with a prototype example in order to show this basic idea.

If $F_1 = 0$ and λ is very close to zero, then the problem may be transformed into one that we wish to construct for the solution u of the differential equation

$$Lu = 0 \quad \text{on} \quad G \tag{6.273}$$

satisfying

$$Bu = F_2 \quad \text{on} \quad D. \tag{6.274}$$

Our general theory gives a practical construction method for this inverse problem; i.e., from the observation F_2 on the part D, we construct u on the whole domain G satisfying the equation $Lu = 0$.

We recall the Sobolev embedding theorem [42, pp. 18–19]. In order to use the results in the framework of Hilbert spaces, we assume $p = q = 2$ there.

Let $W_2^\ell(G)$ ($\ell = 0, 1, 2, \ldots$) be the Sobolev Hilbert space on G, where $G \subset \mathbb{R}^n$ is a bounded domain with a one piecewise-smooth continuously differentiable boundary $\Gamma \equiv \partial G$. We assume that

$$k \geq \ell - \frac{n}{2}. \tag{6.275}$$

Let $m = 0, 1, 2, \ldots$ be such that

$$m > n - 2(\ell - k). \tag{6.276}$$

Let $D = D^m \subset G \cup \Gamma$ be any C^ℓ-manifold of dimension m. Then for any $u \in W_2^\ell(G)$, the derivative $\partial_\alpha u \in L^2(D)(x \in D)$, where $\|\alpha\| \leq k$, and there exists $M > 0$ such that

$$\|\partial_\alpha u\|_{L^2(D)} \leq M\|u\|_{W_2^\ell(G)}, \quad (u \in W_2^\ell(G)). \tag{6.277}$$

Of course

$$\|u|G\|_{W_2^\ell(G)} \leq \|u\|_{W_2^\ell(\mathbb{R}^n)} \quad (u \in W_2^\ell(\mathbb{R}^n)), \tag{6.278}$$

and we can construct the reproducing kernel for the space $W_2^\ell(\mathbb{R}^n)$ by using the Fourier integral with $2\ell > n$. Then for any linear differential operator L with variable coefficients on G satisfying

$$\|Lu\|_{L^2(G)} \lesssim \|u\|_{W_2^\ell(G)} \tag{6.279}$$

and for any linear (boundary) operator B with variable coefficients on D satisfying

$$\|Bu\|_{L^2(D)} \lesssim \|u\|_{W_2^\ell(G)}, \tag{6.280}$$

we can discuss the best approximation: For any $F_1 \in L^2(G)$, for any $F_2 \in L^2(D)$ and for any $\lambda > 0$,

$$\inf_{u \in W_2^\ell(\mathbf{R}^n)} \left\{ \lambda \|u\|^2_{W_2^\ell(\mathbb{R}^n)} + \|F_1 - L(u|G)\|^2_{L^2(G)} + \|F_2 - B(u|G)\|^2_{L^2(D)} \right\}. \tag{6.281}$$

For some more recent general discretization principle with many concrete examples, see [92, 93], containing numerical experiments and numerical viewpoints.

Chapter 7
Applications to Integral Equations

7.1 Singular Integral Equations

Singular integral equations are presently encountered in a wide range of mathematical models, for instance in acoustics, fluid dynamics, elasticity and fracture mechanics. Together with these models, a variety of methods and applications for these integral equations has been developed. For fracture mechanics and crack, see [101, 102, 144] for example. Singular integral equations are derived from physical equations using the Fourier transform, the Plemeli formula, and distribution theory.

7.1.1 Mathematical Formulation and an Approach via the Paley Wiener Spaces

One of the typical singular integral equations is the Carleman equation over a real interval $(-1, 1)$; for any $L^2(-1, 1)$ function g and for real or complex valued-bounded measurable functions a, b, look for y satisfying

$$Ly(t) \equiv a(t)y(t) + \frac{b(t)}{\pi i}\text{p.v.}\int_{-1}^{+1} \frac{y(\zeta)}{\zeta - t}\,\mathrm{d}\zeta = g(t) \quad \text{on} \quad -1 < t < 1 \tag{7.1}$$

in the class of functions of the Paley Wiener space $W_h(\mathbb{R})$ for $h > 0$. See (1.6) for the definition of Paley Wiener spaces $W_h(\mathbb{R})$. Recall that

$$K_h(x, x') = \frac{1}{h}\operatorname{sinc}\frac{\pi}{h}(x - x') \quad (x, x' \in \mathbb{R}) \tag{7.2}$$

is the reproducing kernel of $W_h(\mathbb{R})$. We denote by p.v. *Cauchy's principal value* of the integral. According to [312], the operator L, satisfying a condition

S. Saitoh, Y. Sawano, *Theory of Reproducing Kernels and Applications*,
Developments in Mathematics 44, DOI 10.1007/978-981-10-0530-5_7

$a(t)^2 - b(t)^2 \neq 0$ for $-1 < t < 1$, is called a *regular type* operator. It is well known that an explicit solution of equation (7.1) for a regular type operator exists. See also [142, 306]. When $a(t)^2 - b(t)^2 = 0$, solutions exist if and only if g satisfies a special condition. The analysis of this case is important for the kinked crack problem. Accordingly, we will see a new method which gives simple and natural approximate solutions for linear singular integral equations including the case where the condition of a regular type operator is violated. We can deal with a general linear singular integral equation, however, for simplicity, we state the results for this most typical case.

Indeed, here we introduce a new approach for some general linear singular integral equations with bounded linear integral operators transforming the integral equations into Fredholm integral equations of the second type with sufficiently smooth coefficients through two theories of the Tikhonov regularization and reproducing kernels.

We use Paley Wiener spaces for approximate function spaces and we give approximate solutions for the equation (7.1). We can use Sobolev spaces by a simple modification, and from the viewpoint of flexibility in the numerical sense, the Sobolev spaces will be more suitable. We will firstly give a concrete analytic representation of the approximate solutions for the Carleman integral equation (7.1) for the case of the whole space with constant coefficients.

Let $a(t) = a$ and $b(t) = b$. We consider the extremal problem

$$\inf\left\{\lambda\|f\|^2_{W_h(\mathbb{R})} + \|Lf - g\|^2_{L^2(-1,1)} : f \in W_h(\mathbb{R})\right\}. \tag{7.3}$$

Note that L is a bounded linear operator from $W_h(\mathbb{R})$ into $L^2(-1,1)$, as we see from the Schwarz inequality and boundedness

$$\int_{-1}^{1}\left|\frac{1}{\pi}\,\mathrm{p.v.}\int_{-1}^{1}\frac{F(\xi)}{\xi-\eta}\,d\xi\right|^2 \mathrm{d}\eta \leq \int_{-1}^{1}|F(\xi)|^2\,\mathrm{d}x \left(\leq \int_{\mathbb{R}}|F(\xi)|^2\,\mathrm{d}x\right) \tag{7.4}$$

of the finite Hilbert transform; see [134, 306, 443, 444, 446].

We wish to construct the reproducing kernel for the Hilbert space $W_h(\mathbb{R})$ with the norm square

$$\lambda\|f\|^2_{W_h(\mathbb{R})} + \|Lf\|^2_{L^2(-1,1)}. \tag{7.5}$$

We calculate the associated reproducing kernel $K_\lambda(t,t')$ by solving Fredholm integral equation of the second kind according to Theorem 2.10:

$$\frac{1}{\lambda}K_h(t,t') = K_\lambda(t,t') + \frac{1}{\lambda}\langle L(K_\lambda)_{t'}, L(K_h)_t\rangle_{L^2(-1,1)}. \tag{7.6}$$

Then the extremal function f_λ^* of (7.3) is given by

$$f_\lambda^*(t) \equiv \langle g, L(K_\lambda)_t\rangle_{L^2(-1,1)} = \langle L^*g, (K_\lambda)_t\rangle_{W_h(\mathbb{R})} \quad (t \in \mathbb{R}). \tag{7.7}$$

By Corollary 2.1, we have the following:

Lemma 7.1. *Let $\lambda, t' \in \mathbb{R}$ be fixed. Then the mapping*

$$t \in \mathbb{R} \mapsto \langle L(K_\lambda)_{t'}, L(K_h)_t \rangle_{L^2(-1,1)} \in \mathbb{C}$$

belongs to $W_h(\mathbb{R})$.

By applying the operator L with respect to functions of t, we have

$$\frac{1}{\lambda}[L(K_h)_{t'}] = L[(K_\lambda)_{t'}] + \frac{1}{\lambda} L\left[\langle L[(K_\lambda)_{t'}], L(K_h)_\cdot \rangle_{L^2(-1,1)}\right]. \tag{7.8}$$

Therefore the function $L[(K_\lambda)_t]$ is given as the solution of the Fredholm integral equation of the second kind for fixed t.

7.1.2 *The Complex Constant Coefficients Case on the Whole Line*

Unlike (7.1) we place ourselves in $\mathbb{R}$ instead of $(-1, 1)$. Here we consider approximate solutions for Carleman's equation for the case of the whole line with complex constant coefficients a, b :

$$ay(t) + \frac{b}{\pi i} \text{ p.v.} \int_{\mathbb{R}} \frac{y(\zeta)}{\zeta - t} \, d\zeta = g(t) \quad \text{on} \quad -\infty < t < \infty. \tag{7.9}$$

For the Fourier transform $\mathscr{F}$,

$$\mathscr{F}f(\xi) = \frac{1}{\sqrt{2\pi}} \int_{-\infty}^{\infty} f(t) \exp(-i\xi t) \, dt, \tag{7.10}$$

the Hilbert transform $\mathscr{H}$ can be described as follows:

$$\mathscr{H}y(t) = \frac{1}{\pi} \text{p.v.} \int_{-\infty}^{\infty} \frac{y(\zeta)}{\zeta - t} \, d\zeta \quad y \in L^2(\mathbb{R}),$$

or equivalently,

$$\mathscr{H}F = \mathscr{F}^{-1}[-i\text{sgn} \cdot \mathscr{F}F] \quad F \in L^2(\mathbb{R}).$$

We write

$$\tilde{L}y \equiv ay + \frac{b}{i}\mathscr{H}y.$$

Then (7.12) reads as

$$\tilde{L}y = ay + \frac{b}{i}\mathscr{H}y = g. \tag{7.11}$$

In the same way as above, we obtain

$$\begin{aligned}\frac{1}{\lambda}\tilde{L}[(K_h)_{t'}] &= \tilde{L}[(K_\lambda)_{t'}] + \frac{1}{\lambda}\tilde{L}\langle\tilde{L}[(K_\lambda)_{t'}], \tilde{L}[(K_h)_\cdot]\rangle_{L^2(\mathbb{R})} \\ &= \tilde{L}[(K_\lambda)_{t'}] + \frac{1}{\lambda}\tilde{L}\left[\int_{\mathbb{R}^n} \tilde{L}[(K_\lambda)_{t'}](p)\overline{\tilde{L}[(K_h)_\cdot](p)}\,dp\right] \quad (t' \in \mathbb{R}).\end{aligned} \tag{7.12}$$

If we can find the solution $\tilde{L}(K_\lambda)_{t'}$ in (7.12), we can obtain the approximate solution of (7.9). We assume

$$\tilde{L}_t[(K_\lambda)_{t'}](t) = \tilde{L}_t[(K_\lambda)_0](t-t'). \tag{7.13}$$

Note that (7.13) will be justified because the solution (7.27) actually satisfies (7.12). Note that the function $\tilde{L}[(K_h)_{t'}]$ is calculated using the Fourier integral and the formula: Here, recall the fundamental formulas: We let

$$p(\xi) \equiv \chi_{(-\frac{\pi}{h},\frac{\pi}{h})}(\xi) + \frac{1}{2}\chi_{\{-\pi/h,\pi/h\}}(\xi), \tag{7.14}$$

$$q_1(\xi) \equiv i\chi_{(-\infty,0)}(\xi) - i\chi_{[0,\infty)}(\xi) = i\mathrm{sgn}(\xi), \tag{7.15}$$

$$q_2(\xi) \equiv i\chi_{(-\infty,-\pi)}(h\xi) - i\chi_{(\pi,\infty)}(h\xi), \tag{7.16}$$

$$q(\xi) \equiv i\chi_{(0,\pi)}(h\xi) - i\chi_{(-\pi,0)}(h\xi) = q_1(\xi)\chi_{(-\pi,\pi)}(h\xi). \tag{7.17}$$

Lemma 7.2. *The Hilbert transform of*

$$\frac{1}{\pi x}\sin\frac{\pi x}{h} = \frac{1}{h}\mathrm{sinc}\frac{\pi x}{h} = \frac{1}{\sqrt{2\pi}}\mathscr{F}^{-1}\chi_{(-\frac{\pi}{h},\frac{\pi}{h})}(x)$$

is

$$\frac{1}{\pi x}\left(\cos\frac{\pi x}{h} - 1\right) = -\frac{1}{\sqrt{2\pi}}\mathscr{F}q(x). \tag{7.18}$$

Proof. Noting that $q = q_1 - q_2$, we obtain

$$p(\xi) = \sqrt{\frac{2}{\pi}}\mathscr{F}\left[\frac{1}{t}\sin\frac{\pi t}{h}\right](\xi) \tag{7.19}$$

$$q_1(\xi) = \sqrt{\frac{2}{\pi}}\mathscr{F}\left[\mathrm{p.v.}\frac{1}{t}\right](\xi) \tag{7.20}$$

$$q_2(\xi) = \sqrt{\frac{2}{\pi}}\mathscr{F}\left[\text{p.v.}\frac{1}{t}\cos\frac{\pi t}{h}\right](\xi) \tag{7.21}$$

$$q(\xi) = \sqrt{\frac{2}{\pi}}\mathscr{F}\left[\frac{1}{t}\left(1-\cos\frac{\pi t}{h}\right)\right](\xi) \tag{7.22}$$

for all $\xi \in \mathbb{R}$. Since

$$\mathscr{F}\chi_{(-\frac{\pi}{h},\frac{\pi}{h})}(\xi) = \sqrt{\frac{2}{\pi}}\frac{1}{\xi}\sin\frac{\pi}{h}\xi \tag{7.23}$$

for all $\xi \in \mathbb{R}$, we obtain (7.18) from (7.17) and (7.23).

With this preliminary observation, we consider the problem.
Firstly, we rewrite (7.12) as

$$\begin{aligned}\frac{1}{\lambda}\tilde{L}[(K_h)_{t'}](t_0) &= \tilde{L}[(K_\lambda)_{t'}](t_0) + \frac{1}{\lambda}\tilde{L}\left[\int_{\mathbb{R}}\tilde{L}[(K_\lambda)_{t'}](p)\overline{\tilde{L}[(K_h).](p)}\,dp\right](t_0)\\ &= \tilde{L}[(K_\lambda)_{t'}](t_0) + \frac{1}{\lambda}\int_{\mathbb{R}}\tilde{L}[(K_\lambda)_{t'}](p)\tilde{L}[\overline{\tilde{L}[(K_h).](p)}](t_0)\,dp \end{aligned}\tag{7.24}$$

for all $t_0 \in \mathbb{R}$. We observe

$$\frac{1}{\lambda}\tilde{L}[(K_h)_q](p) = \frac{1}{\lambda}\left[\frac{a}{h}\text{sinc}\frac{\pi(p-q)}{h} - \frac{bi}{\pi(p-q)}\left(\cos\frac{\pi}{h}(p-q)-1\right)\right].$$

from (7.2) and (7.18). Moreover, the kernel of the integral on the right-hand side of (7.24) becomes

$$\tilde{L}\left[\overline{\tilde{L}[(K_h).](p)}\right] = \tilde{L}_{t'}\left[\frac{\bar{a}}{h}\text{sinc}\frac{\pi(\cdot-p)}{h} + \frac{\bar{b}i}{\pi(\cdot-p)}\left(1-\cos\frac{\pi(\cdot-p)}{h}\right)\right]$$

for each fixed $t \in \mathbb{R}$ and hence

$$\tilde{L}\left[\overline{\tilde{L}[(K_h).](t')}\right](t) = \frac{|a|^2+|b|^2}{h}\text{sinc}\left(\frac{\pi(t-t')}{h}\right) + i\frac{a\bar{b}+\bar{a}b}{\pi(t-t')}\left(1-\cos\frac{\pi(t-t')}{h}\right).$$

If we take the Fourier transform of both sides of (7.24) with respect to t, then, using the convolution theorem, we have

$$\mathscr{F}K_h(\xi) = \frac{ap(\xi)-biq(\xi)}{\lambda + (|a|^2+|b|^2)p(\xi) - (a\bar{b}+\bar{a}b)iq(\xi)}. \tag{7.25}$$

From (7.14) and (7.17), (7.25) is rewritten as

$$\mathscr{F}\varphi(\xi) = \frac{a-b}{\lambda+|a-b|^2}\chi_{(-\pi,0)}(h\xi) + \frac{a+b}{\lambda+|a+b|^2}\chi_{(0,\pi)}(h\xi)$$

for $\xi \in \mathbb{R}$. We can find the solution of (7.24) to be

$$\tilde{L}[(K_\lambda)_{t'}](t) = \frac{(a-b)(1-\exp(-i\pi(t-t')/h))}{2\pi i(\lambda+|a-b|^2)(t-t')} + \frac{(a+b)(\exp(i\pi(t-t')/h)-1)}{2\pi i(\lambda+|a+b|^2)(t-t')}. \tag{7.26}$$

Thus, the reproducing kernel $K_\lambda(t,t')$ is represented as follows:

$$K_\lambda(t,t') = \frac{1}{2\pi(\lambda+|a+b|^2)} \int_{-\pi/h}^{0} e^{-i\xi(t-t')}\, d\xi + \frac{1}{2\pi(\lambda+|a-b|^2)} \int_{0}^{\pi/h} e^{-i\xi(t-t')}\, d\xi. \tag{7.27}$$

The following is the expression of $f^*_{\lambda,h,g}$ and some information as $\lambda \downarrow 0$:

Theorem 7.1. *In* (7.9), *we represent the best approximate solution* $f^*_{\lambda,h,g}$ *by:*

$$f^*_{\lambda,h,g}(t) = \frac{1}{2\pi} \int_{-\infty}^{\infty} g(\xi) \left[\frac{\overline{a}+\overline{b}}{\lambda+|a+b|^2} \int_{-\pi/h}^{0} e^{i\eta(\xi-t)}\, d\eta + \frac{\overline{a}-\overline{b}}{\lambda+|a-b|^2} \int_{0}^{\pi/h} e^{i\eta(\xi-t)}\, d\eta \right] d\xi. \tag{7.28}$$

If g *is realized as* $g = \tilde{L}f$ *with some* $f \in W_h(\mathbb{R})$ *by* $\tilde{L}$, *then we obtain*

$$\lim_{\lambda \downarrow 0} f^*_{\lambda,h,g} = f \tag{7.29}$$

uniformly.

Proof. The uniform convergence of (7.29) is a consequence of Remark 1.3 as usual. The latter result in Theorem 7.1 may be derived directly using the representations (7.26) and $f^*_{\lambda,h,g}$.

Note that in the regular type case, $a^2 - b^2 \neq 0$, we can take $\lambda = 0$; that is, we do not need the Tikhonov regularization in our problem.

In Theorem 7.1, a direct calculation yields

$$\tilde{L}f^*_{\lambda,h,g}(t) = \frac{1}{2\pi} \int_{-\infty}^{\infty} g(\xi) \left[\frac{|a+b|^2}{\lambda+|a+b|^2} \int_{-\pi/h}^{0} e^{i\eta(\xi-t)}\, d\eta + \frac{|a-b|^2}{\lambda+|a-b|^2} \int_{0}^{\pi/h} e^{i\eta(\xi-t)}\, d\eta \right] d\xi.$$

Therefore, for cases $\lambda = 0$ and $a^2 - b^2 \neq 0$

$$\lim_{h \downarrow 0} \tilde{L}_t f^*_{0,h,g} = g, \tag{7.30}$$

provided $g \in C_c^\infty(\mathbb{R})$.

In particular, when $a = 0$ and $b = i$, that is, in the Hilbert transform case, we obtain for $\lambda = 0$:

Corollary 7.1. *For the extremal problem*

$$\inf_{f \in W_h(\mathbb{R})} \|\mathscr{H} f - g\|_{L^2(\mathbb{R})}, \tag{7.31}$$

the extremal function $f^*_{H,h,g}$ attaining the infimum exists uniquely and it is given by

$$f^*_{H,h,g}(t) = \frac{1}{\pi} \int_{-\infty}^{\infty} \frac{g(\xi)}{\xi - t} \left\{ \cos\left[\frac{\pi}{h}(\xi - t)\right] - 1 \right\} d\xi \quad (t \in \mathbb{R}), \tag{7.32}$$

and in this case

$$\mathscr{H} f^*_{H,h,g}(t) = \langle g, (K_h)_t \rangle_{L^2(\mathbb{R})} \quad (t \in \mathbb{R}); \tag{7.33}$$

that is, $\mathscr{H} f^*_{H,h,g}$ is the orthogonal projection of g onto the Paley Wiener space $W_h(\mathbb{R})$.

Corollary 7.1 also gives an approximate Hilbert transform for any $L^2(\mathbb{R})$-function g by an ordinary integral; that is,

$$\lim_{h \downarrow 0} \frac{1}{\pi} \int_{-\infty}^{\infty} \frac{g(\xi)}{\xi - t} \left\{ 1 - \cos\left[\frac{\pi}{h}(\xi - t)\right] \right\} d\xi = \mathscr{H} g(t) \tag{7.34}$$

in the topology of $L^2(\mathbb{R})$.

In particular, note that for the singular cases $a^2 - b^2 = 0$, the integral equations have solutions only for very special functions g.

For example, if $a = \pm b$, since $\mathscr{H}(\mathscr{H} y) = -y$, from (7.9),

$$a \mathscr{H} y \mp \frac{a}{i} y = \mathscr{H} g, \tag{7.35}$$

and therefore, we see that g must satisfy the relation

$$\mathscr{H}^2 g = a \mathscr{H}^2 y \mp \frac{a}{i} \mathscr{H} y = -\left(ay \pm \frac{a}{i} \mathscr{H} y\right) = \pm g. \tag{7.36}$$

The following two corollaries give the solutions for the singular cases for general L^2-functions g, and coincide with results for very special functions g in [142, p. 270]:

Corollary 7.2. *Let $a = b = 1$ and let $g \in L^2(\mathbb{R})$.*

1. *In the topology of locally uniform we have*

$$\lim_{\lambda\downarrow 0} f^*_{\lambda,h,g}(t) \equiv f^*_{+0,h,g}(t) = \frac{1}{4}\left[if^*_{H,h,g}(t) + \langle g, (K_h)_t\rangle_{L^2(\mathbb{R})}\right]$$

in $L^2(\mathbb{R})$. *Furthermore,*

$$\lim_{h\downarrow 0} \tilde{L}_t f^*_{+0,h,g}(t) = \frac{i}{2} f^*_{H,h,g}(t) + \frac{1}{2}\langle g, (K_h)_t\rangle_{L^2(\mathbb{R})}$$

in $L^2(\mathbb{R})$.

2. *If, in addition,* $\mathscr{H} g = ig$, *then*

$$\lim_{h\downarrow 0} f^*_{+0,h,g} = \frac{1}{2} g$$

and

$$\lim_{h\downarrow 0} \tilde{L}_t f^*_{+0,h,g} = g$$

in $L^2(\mathbb{R})$.

Corollary 7.3. *Let* $a = 1, b = -1$ *and let* $g \in L^2(\mathbb{R})$.

1. *We have*

$$f^*_{+0,h,g}(t) = \frac{1}{4}\left[-if^*_{H,h,g}(t) + \langle g, (K_h)_t\rangle_{L^2(\mathbb{R})}\right]$$

and

$$\lim_{h\downarrow 0} \tilde{L}_t f^*_{+0,h,g}(t) = -\frac{i}{2} f^*_{H,h,g}(t) + \frac{1}{2}\langle g, (K_h)_t\rangle_{L^2(\mathbb{R})}.$$

in $L^2(\mathbb{R})$.

2. *If, in addition,* $\mathscr{H} g = -ig$, *then*

$$\lim_{h\downarrow 0} f^*_{+0,h,g} = \frac{1}{2} g$$

and

$$\lim_{h\downarrow 0} \tilde{L}_t f^*_{+0,h,g} = g$$

in $L^2(\mathbb{R})$.

Therefore, in Theorem 7.1, we represent explicitly approximate solutions including the singular cases. Surprisingly enough, we can obtain the explicit representations of the "solutions" for any L^2-function g.

7.1.3 Construction of Approximate Solutions by Using a CONS Expansion

We consider (7.1) again. In order to do without Fredholm's integral equation and to derive an effective discretization, we consider a new algorithm in the following way: At first, we introduce a complete orthonormal system $\{\varphi_j\}_{j=1}^{\infty}$ for the Hilbert space $L^2(-1,1)$. For example, we can take the normalized Legendre functions

$$\begin{aligned}\varphi_j(t) &= \sqrt{\frac{2j+1}{2}}P_j(t)\\ &= \frac{1}{2^j j!}\sqrt{\frac{2j+1}{2}}\frac{d^j}{dt^j}(t^2-1)^j\\ &= \sqrt{\frac{2j+1}{2}}\sum_{k=0}^{[j/2]}\frac{(-1)^k(2j-2k)!}{2^j k!(j-k)!(j-2k)!}t^{j-2k}\end{aligned}$$

for $j = 1, 2, \ldots$. See (1.169).

Now we go back to (7.1). Let L be a bounded operator from $W_h(\mathbb{R})$ to $L^2(-1,1)$, which is given by

$$Ly(t) \equiv a(t)y(t) + b(t)\mathscr{H}y(t), \tag{7.37}$$

where $a, b \in L^\infty(-1,1)$. For a sufficiently large N and for a fixed finite set $\Lambda = \{\lambda_j\}_{j=1}^N \subset (0,\infty)$, we consider the following extremal problem corresponding to the Tikhonov functional

$$\inf_{f\in W_h(\mathbb{R})}\left\{\lambda\|f\|_{W_h(\mathbb{R})}^2 + \sum_{j=1}^N \lambda_j|\langle Lf-g,\varphi_j\rangle_{L^2(-1,1)}|^2\right\}. \tag{7.38}$$

That is, when we consider the usual Tikhonv regularization, we need to deal with

$$\|Lf-g\|_{L^2(-1,1)}^2, \tag{7.39}$$

but this is replaced by

$$\sum_{j=1}^N \lambda_j|\langle Lf-g,\varphi_j\rangle_{L^2(-1,1)}|^2. \tag{7.40}$$

Then we obtain:

Theorem 7.2. *Let L be an operator given by* (7.37). *For any* $g \in L^2(-1,1)$, *the extremal function* $f_{\lambda,\Lambda}^{(N)}$ *in the extremal problem* (7.38) *is given by*

$$f_{\lambda,\Lambda}^{(N)}(t) = \sum_{j=1}^{N} \lambda_j \langle g, \varphi_j \rangle_{L^2(-1,1)} \langle \varphi_j, L[(K_{\lambda,\Lambda}^{(N)})_t] \rangle_{L^2(-1,1)}. \tag{7.41}$$

Proof. In the Tikhonov regularization problem, we consider the following problem:

$$\sum_{j=1}^{N} \langle Lf, \varphi_j \rangle_{L^2(-1,1)} \varphi_j = \sum_{j=1}^{N} \langle g, \varphi_j \rangle_{L^2(-1,1)} \varphi_j.$$

See (3.87). We replace L with

$$f \mapsto \sum_{j=1}^{N} \langle Lf, \varphi_j \rangle_{L^2(-1,1)} \varphi_j$$

and g with

$$\sum_{j=1}^{N} \langle g, \varphi_j \rangle_{L^2(-1,1)} \varphi_j.$$

If we replace L ang g as above, we obtain (7.41).

We thus are fascinated with the kernel $K_{\lambda,\Lambda}^{(N)}$. Then we give an algorithm constructing the reproducing kernel $K_{\lambda,\Lambda}^{(N)}$ of the Hilbert space $H_{K_{\lambda,\Lambda}^{(N)}}(\mathbb{R})$ with the norm square

$$\lambda \|f\|_{W_h(\mathbb{R})}^2 + \sum_{j=1}^{N} \lambda_j |\langle Lf, \varphi_j \rangle_{L^2(-1,1)}|^2. \tag{7.42}$$

We consider the issue inductively. We set

$$K^{(0)}(t,t') \equiv \frac{1}{\lambda} K_h(t,t') \quad (t,t' \in \mathbb{R}). \tag{7.43}$$

The reproducing kernel $K^{(1)}(t,t')$ of the Hilbert space with norm square

$$\lambda \|f\|_{W_h(\mathbb{R})}^2 + \sum_{j=1}^{1} \lambda_j |\langle Lf, \varphi_j \rangle_{L^2(-1,1)}|^2 = \lambda \|f\|_{W_h(\mathbb{R})}^2 + \lambda_1 |\langle f, L^* \varphi_1 \rangle_{L^2(-1,1)}|^2 \tag{7.44}$$

is given by

$$K^{(1)}(t,t') = K^{(0)}(t,t') - \frac{\lambda_1 \langle \varphi_1, L[(K^{(0)})_t] \rangle_{L^2(-1,1)} \langle L[(K^{(0)})_{t'}], \varphi_1 \rangle_{L^2(-1,1)}}{1 + \lambda_1 \langle L^* \varphi_1, L^* \varphi_1 \bowtie K^{(0)} \rangle_{L^2(-1,1)}} \tag{7.45}$$

for $t, t' \in \mathbb{R}$, where $\bowtie$ is the operation defined by (2.28). For the second step, the reproducing kernel $K^{(2)}$ of the Hilbert space with norm square

$$\lambda \|f\|^2_{W_h(\mathbb{R})} + \sum_{j=1}^{2} \lambda_j |\langle Lf, \varphi_j \rangle_{L^2(-1,1)}|^2 \quad (t, t' \in \mathbb{R}) \tag{7.46}$$

is given by

$$K^{(2)}(t,t') = K^{(1)}(t,t') - \frac{\lambda_2 \langle \varphi_2, L[(K^{(1)})_t] \rangle_{L^2(-1,1)} \langle L[K^{(1)}_{t'}], \varphi_2 \rangle_{L^2(-1,1)}}{1 + \lambda_2 \langle L^* \varphi_2, L^* \varphi_2 \bowtie K^{(1)} \rangle_{L^2(-1,1)}} \tag{7.47}$$

for $(t, t' \in \mathbb{R})$, using the reproducing kernel $K^{(1)}$. In this way, recursively we can obtain the desired representation of $K^{(N)}_{\lambda,\Lambda}$.

We consider a general extremal problem (7.38) by considering a general weight Λ. This means that for a larger λ_{j_0}, the speed of the convergence

$$\langle Lf, \varphi_{j_0} \rangle_{L^2(-1,1)} \to \langle g, \varphi_{j_0} \rangle_{L^2(-1,1)} \tag{7.48}$$

is higher. This technique is very important for practical applications. For examples, see [31].

Error Estimates

In Theorem 7.2, when the data g contain errors or noises, we need the estimation of our solutions $f^{(N)}_{\lambda,\Lambda}(t)$ in terms of g. For this, we can obtain a good estimation in the form:

Theorem 7.3. *In Theorem 7.2, we obtain the estimate*

$$|f^{(N)}_{\lambda,\Lambda}(t)| \leq \frac{\|a\|_{L^\infty(-1,1)} + \|b\|_{L^\infty(-1,1)}}{\lambda h} \|g\|_{L^2(-1,1)} \sqrt{\sum_{j=1}^{N} \lambda_j^2}. \tag{7.49}$$

Proof. By the Schwarz inequality and (7.41), we obtain

$$|f^{(N)}_{\lambda,\Lambda}(t)|^2 \leq \left(\sum_{j=1}^{N} |\langle g, \varphi_j \rangle_{L^2(-1,1)}|^2 \right) \left(\sum_{j=1}^{N} \lambda_j^2 |\langle \varphi_j, [L(K^{(N)}_{\lambda,\Lambda})_t] \rangle_{L^2(-1,1)}|^2 \right).$$

Since $\{\varphi_j\}_{j=1}^N$ is orthonormal, we have

$$|f_{\lambda,\Lambda}^{(N)}(t)|^2 \le \|g\|_{L^2(-1,1)}^2 \left(\sum_{j=1}^N \lambda_j^2 |\langle \varphi_j, [L(K_{\lambda,\Lambda}^{(N)})_t] \rangle_{L^2(-1,1)}|^2 \right).$$

By the Schwarz inequality once again, we have

$$|f_{\lambda,\Lambda}^{(N)}(t)|^2 \le \|g\|_{L^2(\mathbb{R})}^2 \left(\sum_{j=1}^N \lambda_j^2 \|L[(K_{\lambda,\Lambda}^{(N)})_t]\|_{L^2(\mathbb{R})}^2 \right).$$

We estimate the operator norm as follows using (7.4):

$$\begin{aligned}\|L\|_{L^2(\mathbb{R})\mapsto L^2(-1,1)} &\le \|a\|_{L^\infty(-1,1)} + \|b\|_{L^\infty(-1,1)} \|\mathscr{H}\|_{L^2(\mathbb{R})\to L^2(-1,1)} \\ &= \|a\|_{L^\infty(-1,1)} + \|b\|_{L^\infty(-1,1)}.\end{aligned}$$

Finally, we estimate

$$\|(K_{\lambda,\Lambda}^{(N)})_t\|_{H_{K_{\lambda,\Lambda}^{(N)}}(\mathbb{R})} = K_{\lambda,\Lambda}^{(N)}(t,t) \le K^{(0)}(t,t) \le \frac{1}{\lambda^2 h^2}$$

from (7.45), (7.47) and so on. Therefore, we have the desired result (7.49).

7.1.4 General Singular Integral Equations

In this Sect. 7.1.4, we discuss general linear singular integral equations.

At first, we are interested in the following general Cauchy singular integral (as it appears in (4.21)):

$$\text{p.v.} \int_\Gamma \frac{y(\zeta)}{\zeta - t} \, d\zeta, \tag{7.50}$$

where, for example, Γ is an open arc or a simple closed curve satisfying a smoothness condition on $\mathbb{R}^2$ and having the parameter t defined on the curve Γ. Then, when we use Paley Wiener spaces in 2-dimensions we can establish the bounded property for the corresponding linear operator in the form of

$$\left\| \text{p.v.} \int_\Gamma \frac{y(\zeta)}{\zeta - \cdot} \, d\zeta \right\|_{L^2(\Gamma)} \le C \|y\|_{H_{K_h}(\mathbb{R})},$$

for a constant C. That is, we can consider the Cauchy singular integral operator as a bounded linear operator from the Paley Wiener space $H_{K_h}(\mathbb{R})$ into $L^2(\Gamma)$.

Sometimes it is also useful to consider different types of singular integral operators with shift facing of the form

$$y \mapsto \text{ p.v.}\int_{\Gamma} \frac{y(\zeta)}{\zeta - t + s}\, \mathrm{d}\zeta,$$

for a complex or real number s or a point s on Γ. In many such cases, we can similarly formulate bounded linear operators from the Paley Wiener spaces into $L^2(\Gamma)$. In those general singular integral equations, sometimes not only the function y appears but also their many order derivatives; see, for example, [77]. For these cases, there is no problem in obtaining the boundedness property for the corresponding linear operators since Paley Wiener spaces are very good function spaces for such a purpose.

In the above way, we can consider many singular integral equations in our framework. Therefore, we will have the similar representation (7.7) with (7.6). Numerically, we will calculate the kernel K_λ. By its discretization, we can obtain the representation (7.7) by solving the simultaneous linear equations. The detailed algorithm clearly is stated in [85].

For many multidimensional singular integrals, we have similar situations, since their bounded linear operator properties are widely valid.

By the way, for singular integral operators we will be able to calculate numerically and simply by the following regularization.

For example, we will consider in the following way: for a domain D on the x_1, x_2 plane and for an integrable function f on D, given the singular integral

$$\iint_D \frac{1}{|x-y|} f(x)\, \mathrm{d}x_1 \mathrm{d}x_2, \tag{7.51}$$

we calculate the corresponding regularized integral for a very small $\varepsilon > 0$,

$$\iint_D \frac{1}{|x-y| + \varepsilon} f(x)\, \mathrm{d}x_1 \mathrm{d}x_2. \tag{7.52}$$

For this regularized integral, its computational calculation is very simple. For numerical experiments, see [491]. Furthermore, their error estimates are given in a very general framework in [420].

For the case in (7.50), we can consider it as

$$\text{p.v.}\int_{\Gamma} \frac{y(\zeta)\overline{(\zeta - t)}}{|\zeta - t|^2} d\zeta, \tag{7.53}$$

and therefore, we can calculate it by taking a small ε

$$\lim_{\varepsilon \to 0} \int_{\Gamma} \frac{y(\zeta)\overline{(\zeta - t)}}{|\zeta - t|^2 + \varepsilon} d\zeta. \tag{7.54}$$

Here we are considering the parameters in complex-variables.

See, for example, [115, 142, 210, 211, 235, 252, 266, 285, 289, 306, 312].

Meanwhile, several works are known in which some general singular integral equations are transformed into Riemann boundary value problems by using operator theory and factorization theory. Such an approach is analytical, but the consequent reduction procedures are in general very complicated. Additionally, by using such methods it is not always possible to obtain all the properties and representations of the solutions of corresponding singular integral equations. In particular, a large number of associated questions which are formulated again in terms of factorization problems remains open. See [266, 285] for such a theory.

7.1.5 Discrete Singular Integral Equations

The following arguments are valid similarly in a very general setting stated as in the framework of the previous section. However, in order to state results explicitly, we will discuss here the typical and concrete situation considered in the introduction of Chap. 7.1. Note that the representation of approximate solutions (7.1) in the singular integral equation in Theorem 7.1 may not yield any solution of the discrete singular integral equation (7.9) for a given finite number of the data g. Therefore, we are interested in the corresponding *discrete singular integral equation* because in this case (and having in mind a practical point of view) we only need a finite number of values of g as observation data.

Let L be an operator of the form

$$Ly(t) = ay(t) + b\mathscr{H}y(t),$$

where $a, b \in \mathbb{C}$. Recall that any element in $H_{K_h}(\mathbb{R})$ is very smooth. For different points $\{x_j\}_{j=1}^N$, we will consider the bounded linear operators from the RKHS $H_{K_h}(\mathbb{R})$ into $\mathbb{C}$:

$$H_{K_h}(\mathbb{R}) \ni F \mapsto \tilde{L}F(x_j) \in \mathbb{C}; \quad j = 1, 2, \ldots, N. \tag{7.55}$$

We define $\mathbf{L}$ by

$$\mathbf{L} : H_{K_h}(\mathbb{R}) \ni F \mapsto (\tilde{L}F(x_1), \tilde{L}F(x_2), \ldots, \tilde{L}F(x_N)) \in \mathbb{C}^N \tag{7.56}$$

and we will take a standard orthonormal system $\{\mathbf{e}_j\}_{j=1}^N$ in the space $\mathbb{C}^N$. Then, we see

$$\tilde{L}F(x_j) = \langle \mathbf{L}F, \mathbf{e}_j \rangle_{\mathbb{C}^N} = \langle F, \mathbf{L}^*\mathbf{e}_j \rangle_{H_{K_h}(\mathbb{R})}.$$

Let us set

$$A \equiv \{a_{j,j'}\}_{j,j'=1}^N \equiv \{\langle \mathbf{L}^*\mathbf{e}_{j'}, \mathbf{L}^*\mathbf{e}_j \rangle_{H_{K_h}(\mathbb{R})}\}_{j,j'=1}^N. \quad (7.57)$$

The explicit form of $A = \{a_{j,j'}\}_{j,j'=1}^N$ is given in the next theorem.

Theorem 7.4 ([89, Theorem 7.1]). *We have*

$$a_{j,j'} = \frac{|a|^2 + |b|^2}{\pi(x_j - x_{j'})} \sin \frac{\pi(x_j - x_{j'})}{h} + \frac{2i\mathrm{Re}(a\bar{b})}{\pi(x_j - x_{j'})} \left(\cos \frac{\pi(x_j - x_{j'})}{h} - 1 \right) \quad (7.58)$$

for $j, j' = 1, 2, \ldots, N$.

Proof. We start by interpreting the element $\mathbf{L}^*\mathbf{e}_j$, which we can deduce in the following way:

$$\mathbf{L}^*\mathbf{e}_j(x) = \langle \mathbf{L}^*\mathbf{e}_j, (K_h)_x \rangle_{H_{K_h}(\mathbb{R})} = \langle \mathbf{e}_j, \mathbf{L}[(K_h)_x] \rangle_{\mathbb{C}^N} = \tilde{L}[(K_h)_x](x_j) \quad (x \in \mathbb{R}).$$

If we write it out in full, then we obtain

$$\mathbf{L}^*\mathbf{e}_j(x) = \frac{a}{\pi(x_j - x)} \sin \frac{\pi(x_j - x)}{h} + \frac{bi}{\pi i(x_j - x)} \left(\cos \frac{\pi(x_j - x)}{h} - 1 \right).$$

Hence, using the reproducing property of K_h, we immediately obtain

$$\begin{aligned} a_{j,j'} &= \frac{|a|^2}{\pi(x_j - x_{j'})} \sin \frac{\pi(x_j - x_{j'})}{h} + \frac{2i\mathrm{Re}(a\bar{b})}{\pi(x_j - x_{j'})} \left\{ \cos \frac{\pi(x_j - x_{j'})}{h} - 1 \right\} \\ &+ \frac{|b|^2}{\pi^2} \int_{-\infty}^{\infty} \frac{1}{(x - x_{j'})(x - x_j)} \left(\cos \frac{\pi(x - x_{j'})}{h} - 1 \right) \left(\cos \frac{\pi(x - x_j)}{h} - 1 \right) dx. \end{aligned}$$

From (7.18), we conclude that the inversion of the Hilbert transform preserves the inner product in the $L^2(\mathbb{R})$-space and hence we obtain (7.58).

The next result shows that A is positive definite.

Theorem 7.5 ([89, Theorem 7.2]). *The matrix* $A = \{a_{j,j'}\}_{j,j'=1}^N$ *given by* (7.57) *or* (7.58) *is strictly positive definite.*

Proof. Let us start by assuming that for any complex numbers C_j,

$$\sum_{j=1}^N C_j \mathbf{L}^*\mathbf{e}_j(x) \equiv 0; \quad (7.59)$$

that is,

$$a\sum_{j=1}^{N} C_j K_h(x,x_j) + \frac{b}{i}\sum_{j=1}^{N} C_j \mathscr{H}([(K_h)_{x_j}])(x) \equiv 0. \tag{7.60}$$

When we take the L^2 norm square of this function, by using the reproducing property of $K_h(x,x_j)$, the orthogonal property of the Hilbert transform and of the original function and the isometric property of the Hilbert transform, we obtain

$$|a|^2 \sum_{j=1,j'=1}^{N} C_j\overline{C_{j'}}K_h(x_{j'},x_j) + |b|^2 \sum_{j=1,j'=1}^{N} C_j\overline{C_{j'}}K_h(x_{j'},x_j) = 0; \tag{7.61}$$

that is,

$$\sum_{j=1,j'=1}^{N} C_j\overline{C_{j'}}K_h(x_{j'},x_j) = 0. \tag{7.62}$$

From the positive definiteness of the matrix $\{K_h(x_{j'},x_j)\}_{j,j'=1}^{N}$, we have the desired result that all C_j are zero.

Denote by H_A the image space of $\mathbf{L}$. Since the matrix $\{a_{j,j'}\}_{j,j'=1}^{N}$ is invertible, then the norm in H_A is given by

$$\|\mathbf{L}F\|_{H_A}^2 = (\mathbf{L}F)^* A^* (\mathbf{L}F),$$

where $A^* = \overline{A^{-1}} = \{\tilde{a}_{j,j'}\}_{j,j'=1}^{N}$.

Theorem 7.6. *For any given N values $\mathbf{d} = \{d_j\}_{j=1}^{N} \in \mathbb{C}^N$, among the $H_{K_h}(\mathbb{R})$-functions F taking the values*

$$\tilde{L}F(x_j) = d_j \quad (j = 1,2,\ldots,N), \tag{7.63}$$

the function $F_{\mathbf{d}}^*(x)$ with the minimum norm $\|F\|_{H_{K_h}(\mathbb{R})}$ is uniquely determined and is represented as follows:

$$F_{\mathbf{d}}^* = \sum_{j,j'=1}^{N} d_j \tilde{a}_{j,j'} \mathbf{L}^* \mathbf{e}_{j'}. \tag{7.64}$$

Proof. The function given by (7.64) belongs to the range of L^* and agrees with g at x_j, $j = 1,2,\ldots,N$. Then, the result follows as a consequence of the general theory and the calculations above.

7.1.6 Inverse Source Problems by a Finite Number of Data

We shall consider the inversion problem for (7.55) and (7.56): That is, from a finite number of observation data $y(x_j), j = 1, 2, \ldots, N$, we wish to determine the optimal source g in (7.9). For this purpose, we will use the integral representation; that is, for $y(x_j)$ we use the approximate data $f^*_{\lambda,h,g}(x_j)$. Therefore, we consider the bounded linear operators from the RKHS $H_{K_h}(\mathbb{R})$ into $\mathbb{C}$:

$$H_{K_h}(\mathbb{R}) \ni g \mapsto Mg(x_j) \equiv f^*_{\lambda,h,g}(x_j); \quad j = 1, 2, \ldots, N. \tag{7.65}$$

We set

$$\mathbf{M} : H_{K_h}(\mathbb{R}) \ni g \mapsto (Mg(x_1), Mg(x_2), \ldots, Mg(x_N)) \in \mathbb{C}^N. \tag{7.66}$$

Then we see that

$$Mg(x_j) = \langle \mathbf{M}g, \mathbf{e}_j \rangle_{\mathbb{C}^N} = \langle g, \mathbf{M}^*\mathbf{e}_j \rangle_{H_{K_h}(\mathbb{R})}. \tag{7.67}$$

Let us set

$$B \equiv \{b_{j,j'}\}_{j,j'=1}^N \equiv \{(\mathbf{M}^*\mathbf{e}_{j'}, \mathbf{M}^*\mathbf{e}_j)_{H_{K_h}(\mathbb{R})}\}_{j,j'=1}^N. \tag{7.68}$$

Theorem 7.7 ([89, Theorem 8.1]). *Suppose that $x_1, x_2, \ldots, x_N$ are distinct points in $\mathbb{R}$. The element $b_{j,j'}$ given by* (7.68) *has the following explicit expression:*

$$\begin{aligned} b_{j,j'} = |a-b|^2 \frac{\exp(i\pi(x_j - x_{j'})/h) - 1}{2\pi i(\lambda + |a-b|^2)^2(x_j - x_{j'})} \\ + |a+b|^2 \frac{1 - \exp(-i\pi(x_j - x_{j'})/h)}{2\pi i(\lambda + |a+b|^2)^2(x_j - x_{j'})}. \end{aligned} \tag{7.69}$$

Proof. In view of computing the value of $b_{j,j'}$, we will first consider $\mathbf{M}^*\mathbf{e}_j$ which may be calculated as follows:

$$\mathbf{M}^*\mathbf{e}_j(x) = \langle \mathbf{M}^*\mathbf{e}_j, (K_h)_x \rangle_{H_{K_h}(\mathbb{R})} = \langle \mathbf{e}_j, \mathbf{M}[(K_h)_x] \rangle_{\mathbb{C}^N} = M[(K_h)_x](x_j).$$

If we write it out in full, then we obtain

$$\begin{aligned} \mathbf{M}^*\mathbf{e}_j(x) = \frac{\overline{a} - \overline{b}}{\lambda + |a-b|^2} \times \frac{1}{2\pi i(x - x_j)} \cdot \left(\exp\left(\frac{\pi(x - x_j)}{h} i \right) - 1 \right) \\ + \frac{\overline{a} + \overline{b}}{\lambda + |a+b|^2} \times \frac{1}{2\pi i(x - x_j)} \cdot \left(1 - \exp\left(-\frac{\pi(x - x_j)}{h} i \right) \right). \end{aligned}$$

Therefore, having in mind (7.68), we obtain (7.69).

Theorem 7.8 ([89, Theorem 8.2]). *The matrix* $B = \{b_{j,j'}\}_{j,j'=1}^{N}$ *given by* (7.68) *and* (7.69) *is strictly positive definite.*

Proof. We shall assume that for any complex constants C_j,

$$\sum_{j=1}^{N} C_j \mathbf{M}^* \mathbf{e}_j(x) \equiv 0 \quad (x \in \mathbb{R}); \tag{7.70}$$

that is,

$$\sum_{j=1}^{N} \frac{(\overline{a}+\overline{b})C_j}{\lambda + |a+b|^2} \int_{-\frac{\pi}{h}}^{0} e^{i\eta(x-x_j)}\, d\eta + \sum_{j=1}^{N} \frac{(\overline{a}-\overline{b})C_j}{\lambda + |a-b|^2} \int_{0}^{\frac{\pi}{h}} e^{i\eta(x-x_j)}\, d\eta \equiv 0.$$

Hence

$$\int_{-\infty}^{\infty} \left\{ \sum_{j=1}^{N} \frac{(\overline{a}+\overline{b})C_j \chi_{(-\pi,0)}(h\eta)}{\lambda + |a+b|^2} + \sum_{j=1}^{N} \frac{(\overline{a}-\overline{b})C_j \chi_{(0,\pi)}(h\eta)}{\lambda + |a-b|^2} \right\} e^{i\eta x} e^{-i\eta x_j}\, d\eta \equiv 0,$$

and so, by the uniqueness of the Fourier transform, we obtain

$$\left\{ \sum_{j=1}^{N} \frac{(\overline{a}+\overline{b})C_j}{\lambda + |a+b|^2} \chi_{(-\pi,0)}(h\eta) + \sum_{j=1}^{N} \frac{(\overline{a}-\overline{b})C_j}{\lambda + |a-b|^2} \chi_{(0,\pi)}(h\eta) \right\} e^{-i\eta x_j} \equiv 0 \tag{7.71}$$

for all $\eta \in \mathbb{R}$. Therefore we obtain

$$\sum_{j=1}^{N} C_j \chi_{(-\pi,0)}(h\eta) e^{-i\eta x_j} \equiv 0 \quad (\eta \in \mathbb{R}) \tag{7.72}$$

and

$$\sum_{j=1}^{N} C_j \chi_{(0,\pi)}(h\eta) e^{-i\eta x_j} \equiv 0 \quad (\eta \in \mathbb{R}). \tag{7.73}$$

Finally, we conclude $C_1 = C_2 = \cdots = C_N = 0$ from the linear independence of the system of the functions $e^{-i\eta x_j}$ for $j = 1, 2, \ldots, N$.

We have now the desired result similar to Theorem 7.6:

Theorem 7.9. *For any given* N *values* $\mathbf{d} = \{d_j\}_{j=1}^{N} \in \mathbb{C}^N$, *we can determine uniquely the function* $g_{\mathbf{d}}^*$ *with the minimum norm* $\|g\|_{H_{K_h}(\mathbb{R})}$ *among the* $H_{K_h}(\mathbb{R})$*-functions* g *satisfying* $Mg(x_j) = d_j$, $j = 1, 2, \ldots, N$, *and represented as follows:*

$$g_{\mathbf{d}}^* = \sum_{j,j'=1}^{N} d_j \tilde{b}_{j,j'} \mathbf{L}^* \mathbf{e}_{j'}.$$

We followed [89, Section 7] in this section. See [92, Section 16] and [103] for some more recent general discretization principles with many concrete examples.

See [79–81] for more about this type of equations.

7.2 Convolution Integral Equations

7.2.1 Example of Convolution Integral Equations

In Sect. 7.2 we consider convolution integral equations on the real line of the prototype form

$$2\pi F_1(t) + \int_{\mathbb{R}} F_1(\xi) F_2(t-\xi) d\xi + \int_{\mathbb{R}} F_1(\xi) F_3(t-\xi) d\xi = G(t), \quad t \in \mathbb{R}, \tag{7.74}$$

where the right-hand side function G and the sought solution F_1 belong to some very general spaces to be specified later on. Here the general kernel functions F_j are also given so that the integrals in (7.74) exist in a Lebesgue sense for every $t \in \mathbb{R}$. The technique introduced in this work is also valid for more general equations than (7.74), where we would face more than two integrals of that type on the left-hand side of (7.74).

In particular, note that in (7.74) we can deal with functions F_2 and F_3 with different supports and different properties.

As the first step to study (7.74), we consider its Fourier transform and then we face a corresponding algebraic equation which we would like to solve by inversion. Consequently, the central initial questions in this line are:

- what kind of kernel functions may be considered in the method?
- what is the meaning of the consequent fractional equation in the solution of the mentioned algebraic equation?
- and, depending on what kind of functions G are given, what kind of solutions exist?

7.2.2 An Approach Using RKHSs

We will examine the details for the integral equation using Fourier integrals. Anyway, the situation will be very general containing finite and infinite order Sobolev Hilbert spaces and Paley Wiener spaces. See (1.6). As general reproducing kernels represented by the Fourier integral, we define, for any given nonnegative integrable functions ρ_1, ρ_2, ρ_3 on $\mathbb{R}$ (which are not identically zero measurable functions), the positive definite quadratic form functions K_j $(j = 1, 2, 3)$:

$$K_j(x, y) \equiv \int_{\mathbb{R}} \exp(i(x-y) \cdot t)\rho_j(t)\, dt = \langle E_x, E_y \rangle_{L^2(\rho_j)},$$

where $E_x(t) = \exp(ix \cdot t)$. Then we consider the induced integral transforms $L_j : L^2(\mathbb{R}; \rho_j) \to H_{K_j}(\mathbb{R})$ by

$$f_j(x) = L_j F(x) = \int_{\mathbb{R}} F(t)\rho_j(t) \exp(-it \cdot x) dt = \langle F, E_y \rangle_{L^2(\rho_j)} \tag{7.75}$$

for the measurable functions F satisfying

$$\|F\|_{L^2(\rho_j)} = \sqrt{\int_{\mathbb{R}} |F(t)|^2 \rho_j(t) dt} < \infty, \tag{7.76}$$

for $j = 1, 2, 3$. Meanwhile, the function spaces $H_{K_j}(\mathbb{R})$ are composed of complex-valued functions on the whole space $\mathbb{R}$. Therefore, for the reproducing kernel Hilbert spaces $H_{K_j}(\mathbb{R})$ admitting the kernels K_j, we have the isometric identity

$$\|L_j F\|_{H_{K_j}(\mathbb{R})} = \sqrt{\int_{\mathbb{R}} |F(t)|^2 \rho_j(t) dt}, \tag{7.77}$$

for $j = 1, 2, 3$ since $\{E_x\}_{x \in \mathbb{R}}$ spans a dense subspace in $L^2(\rho_j)$, or equivalently, a function in $L^2(\rho_j)$ is zero when it is perpendicular to $\{E_x\}_{x \in \mathbb{R}}$.

Now, we consider the algebraic equation with the linear operator φ_{f_2, f_3} equation from $H_{K_1}(E)$, for fixed $f_j \in H_{K_j}(E)$ $(j = 2, 3)$:

$$\varphi_{f_2, f_3} f_1 \equiv f_1 + f_1 f_2 + f_1 f_3 = g \tag{7.78}$$

for a function g of a function space that is determined naturally as the image space of the operator φ_{f_2, f_3}. Then we obtain the identity

$$\begin{aligned}\varphi_{L_2F_2, L_3F_3}(L_1F_1)(x) = &\int_{\mathbb{R}} F_1(t)\rho_1(t) \exp(-ix \cdot t)\, dt \\ &+ 2\pi \int_{\mathbb{R}} (F_1\rho_1) * ((F_2\rho_2) + (F_3\rho_3))(t) \exp(-ix \cdot t)\, dt\end{aligned}$$

assuming that $F_2\rho_2, F_3\rho_3 \in L^1(\mathbb{R})$. We set

$$\Omega(t; \rho_1, \rho_2, \rho_3) \equiv \rho_1 + 2\pi\rho_1 * (\rho_2 + \rho_3) \tag{7.79}$$

(for the usual convolution $*$). Note that the convolution makes sense since $\rho_2, \rho_3 \in L^1(\mathbb{R})$. Following the operator φ_{f_2,f_2}, we will consider the identity

$$\begin{aligned}
&\mathbf{K}(x,y)\\
&\equiv K_1(x,y) + 2\pi K_1(x,y)K_2(x,y) + 2\pi K_1(x,y)K_3(x,y)\\
&= L_1[1](y-x) + 2\pi L_1[1](y-x)L_2[1](y-x) + 2\pi L_1[1](y-x)L_3[1](y-x)\\
&= \int_{\mathbb{R}} \Omega(t;\rho_1;\rho_2,\rho_3)\exp(i(x-y)\cdot t)\,dt. \qquad (7.80)
\end{aligned}$$

Then, by the structure of the reproducing kernel Hilbert spaces of sum and product, we see that the image g of the operator φ_{f_2,f_3} belongs to the reproducing kernel Hilbert space $H_{\mathbf{K}}(E)$ defined by (7.80) and furthermore, we obtain the inequality

$$\|g\|^2_{H_{\mathbf{K}}(E)} \le \|f_1\|^2_{H_{K_1}(E)}(1 + 4\pi^2\|f_2\|^2_{H_{K_2}(E)} + 4\pi^2\|f_3\|^2_{H_{K_3}(E)}); \qquad (7.81)$$

meanwhile, in the t space, we obtain the convolution inequality

$$\begin{aligned}
&\int_{\mathbb{R}} \frac{|(F_1\rho_1)(t) + 2\pi(F_1\rho_1)*(F_2\rho_2)(t) + 2\pi(F_1\rho_1)*(F_3\rho_3)(t)|^2}{\Omega(t;\rho_1;\rho_2,\rho_3)}\,dt\\
&\le \int_{\mathbb{R}} |F_1(t)|^2\rho_1(t)\,dt\cdot\left(1 + 4\pi^2\int_{\mathbb{R}}|F_2(t)|^2\rho_2(t)\,dt + 4\pi^2\int_{\mathbb{R}}|F_3(t)|^2\rho_3(t)\,dt\right).
\end{aligned}$$

Now, we wish to solve the convolution equation of the type

$$F_1\rho_1 + 4\pi^2(F_1\rho_1)*(F_2\rho_2) + 4\pi^2(F_1\rho_1)*(F_3\rho_3) \equiv \widetilde{G}. \qquad (7.82)$$

The unknown is the function F_1. This is introduced in the first integral equation (7.74), by setting $F_j\rho_j = F_j$ and that was transformed into the algebraic equation (7.78).

The fundamental inequality (7.81) means that the operator φ_{f_2,f_2} is bounded on $H_{K_1}(E)$ into the space $H_{\mathbf{K}}(E)$. Therefore, by the general method, we can construct the approximate solutions.

See [83, Section 3.2] for a version of mapping which simplifies (7.80), where the relation between the Tikhonov regularization and the linear mapping is discussed.

7.2.3 Integral Equations with Mixed Toeplitz Hankel Kernel

We will consider the integral equation with mixed Toeplitz Hankel kernel

$$\lambda\varphi(x) + \int_E \left[k_1(x-y) + k_2(x+y)\right]\varphi(y)dy = f(x) \qquad (7.83)$$

where k_1, k_2 are called the Toeplitz and Hankel kernels, respectively, and the domain E is one of the following different representative possibilities:

(i) a finite interval $E = (0, a)$;
(ii) the half-line $E = (0, +\infty)$;
(iii) the whole real line $E = (-\infty, +\infty)$.

Except for the case $E = (-\infty, +\infty)$, the domain of k_1 and that of k_2 are not identical; for example, in case $E = (0, a)$, the domain of k_1 is $(-a, a)$ and that of k_2 is $(0, 2a)$. Equation (7.83) was posed a long time ago, and it is an interesting subject in the theory of integral equations as it has many applications. Actually, many mathematicians are interested in (7.83), and it is still an open problem in general cases [22, 27, 99, 140, 237, 463, 465].

We set

$$\Omega(t; \rho) \equiv \rho_1 + 2\pi\rho_1 * \rho_2 + \int_{\mathbb{R}} \rho_1(\xi)\rho_3(\xi + t)d\xi, \tag{7.84}$$

for the usual convolution $*$.

Let us consider any given nonnegative integrable functions ρ_1, ρ_2, ρ_3 on $\mathbb{R}$ that are not zero identically. For the purpose of considering very general reproducing kernels represented by the Fourier integral, for $j = 1, 2, 3$, we will define positive definite quadratic form functions K_j by

$$K_j(x, y) \equiv \int_{\mathbb{R}} \exp(i(x - y) \cdot t)\rho_j(t)\, dt \tag{7.85}$$

and we consider the induced integral transforms $L_j : L^2(\rho_j) \to H_{K_j}(\mathbb{R})$ with $j = 1, 2, 3$ given by

$$f_j(x) = L_j F(x) = \int_{\mathbb{R}} F(t)\, \rho_j(t) \exp(it \cdot x)dt, \tag{7.86}$$

for the measurable functions F satisfying

$$\int_{\mathbb{R}} |F(t)|^2 \rho_j(t)dt < \infty, \tag{7.87}$$

respectively. Consequently, for the reproducing kernel Hilbert spaces $H_{K_j}(\mathbb{R})$ admitting the kernels K_j, we have the isometric identities [95, (3.8)]:

$$\|f_j\|_{H_{K_j}(\mathbb{R})} = \sqrt{\int_{\mathbb{R}} |F_j(t)|^2 \rho_j(t)dt}, \tag{7.88}$$

for $j = 1, 2, 3$. In particular, note that for each j, $F_j = 0$, on the outside of the support of ρ_j. Meanwhile, the function spaces $H_{K_j}(\mathbb{R})$ are composed of complex-valued functions on the whole space $\mathbb{R}$.

Now, we consider the algebraic equation, with the non-linear operator φ_{f_2,f_3}, from $H_{K_1}(\mathbb{R})$, for fixed $f_j \in H_{K_j}(\mathbb{R})$ $(j = 2, 3\ )$,

$$\varphi_{f_2,f_3} f_1(x) = f_1(x) + f_1(x)f_2(x) + \overline{f_1(x)}f_3(x) = g(x), \quad \text{on} \quad \mathbb{R} \tag{7.89}$$

for a function g of a function space to be determined as the image space of the operator φ_{f_2,f_3}. We define

$$G_1 * *G_2(t) \equiv \int_{\mathbb{R}} G_1(\xi)G_2(t+\xi)d\xi. \tag{7.90}$$

Then, we obtain the identity in the t space

$$\begin{aligned}&\varphi_{L_2F_2,L_3f_3} L_1F_1(x)\\ &= \frac{1}{(2\pi)^2}\int_{\mathbb{R}} \exp(ix\cdot t)\{(F_1\rho_1) * (2\pi + F_2\rho_2)(t) + (F_1\rho_1) * *(F_3\rho_3)(t)\}\, dt.\end{aligned} \tag{7.91}$$

Note that F_1 are real-valued functions (otherwise, here, we must put the complex conjugate). Following the operator φ_{f_2,f_3}, we will consider the identity

$$\begin{aligned}\mathbf{K}(x,y) &\equiv K_1(x,y) + K_1(x,y)K_2(x,y) + \overline{K_1(x,y)}K_3(x,y)\\ &= \frac{1}{(2\pi)^2}\int_{\mathbb{R}} [M_{x-y}\Omega(\cdot;\rho)(t)]\, dt\\ &= \frac{1}{(2\pi)^2}\int_{\mathbb{R}} \exp(i(x-y)\cdot t)\cdot \Omega(t;\rho)\, dt.\end{aligned}$$

Then, by the structure of the reproducing kernel Hilbert spaces of sum and product, we see that the image g of the nonlinear operator φ_{f_2,f_3} belongs to the reproducing kernel Hilbert space $H_{\mathbf{K}}(\mathbb{R})$ defined by $\mathbf{K}(x,y)$ and, furthermore, we obtain the inequality

$$\|g\|^2_{H_{\mathbf{K}}(\mathbb{R})} \le \|f_1\|^2_{H_{K_1}(\mathbb{R})}(1 + \|f_2\|^2_{H_{K_2}(\mathbb{R})} + \|f_3\|^2_{H_{K_3}(\mathbb{R})}). \tag{7.92}$$

Meanwhile, in the t space, we obtain the convolution inequality

$$\begin{aligned}&\int_{\mathbb{R}} \frac{1}{\Omega(t;\rho)} |(F_1\rho_1) * (2\pi + (F_2\rho_2))(t) + (F_1\rho_1) * [(F_1\rho_1) * *(F_3\rho_3)](t)|^2\, dt\\ &\le \int_{\mathbb{R}} |F_1(t)|^2\rho_1(t)\, dt \cdot \left(2\pi + \int_{\mathbb{R}} |F_2(t)|^2\rho_2(t)\, dt + \int_{\mathbb{R}} |F_3(t)|^2\rho_3(t)\, dt\right).\end{aligned}$$

Now, by letting $**$ be the operation given by (7.90), we wish to solve the convolution equation of the type

$$\psi_{F_2,F_3}(F_1) \equiv (F_1\rho_1) * (2\pi + F_2\rho_2) + (F_1\rho_1) * *(F_3\rho_3) = \widetilde{G}, \tag{7.93}$$

by setting $F_j\rho_j$ in place of F_j, and that was transformed into the algebraic equation (7.89).

The fundamental inequality (7.92) means that the operator φ_{f_2,f_2} is bounded on $H_{K_1}(\mathbb{R})$ into the space $H_{\mathbf{K}}(\mathbb{R})$, however the mapping is nonlinear. We need a bounded linear operator from some reproducing kernel Hilbert space into a Hilbert space. In order to introduce a suitable reproducing kernel Hilbert space, we assume that ρ_1 is positive continuous on the support $[a,b)$ of ρ_1, with $-\infty < a < b \leq \infty$. In particular, we must assume that the support of ρ_1 is an interval $[a,b)$. Then, note that $F_1 \in L^1(a,b')$ for any $b' < b$. We will consider the reproducing kernel Hilbert space $H_{K_{\rho_1}}[a,b)$. In view of that, we start by defining the positive definite quadratic form function:

$$K_{\rho_1}(t,\tau) \equiv \int_a^{\min(t,\tau)} \frac{d\xi}{\rho_1(\xi)} \quad (t,\tau \in [a,b)). \tag{7.94}$$

Then, the reproducing kernel Hilbert space $H_{K_{\rho_1}}[a,b)$ is composed of real-valued functions f on $[a,b)$, absolutely continuous, $f(a) = 0$, with inner product

$$\langle f_1, f_2 \rangle_{H_{K_{\rho_1}}[a,b)} \equiv \int_a^b f_1'(t) f_2'(t) \rho_1(t) dt. \tag{7.95}$$

Here, when we set

$$f_{\rho_1}(t) \equiv \int_a^t F_1(\xi) d\xi \quad (t \in [a,b)), \tag{7.96}$$

we obtain the isometric identity

$$\|f_{\rho_1}\|_{H_{K_{\rho_1}}[a,b)} = \sqrt{\int_{\mathbb{R}} F_1(t)^2 \rho_1(t) dt} = \sqrt{2\pi}\, \|f_1\|_{H_{K_1}[a,b)}, \tag{7.97}$$

in (7.88). When we consider the operator (7.93) from this reproducing kernel Hilbert space $H_{K_{\rho_1}}[a,b)$ through the derivative into the Hilbert space $L^2(\Omega_\rho)$ consisting of the complex-valued functions F with norm square

$$\frac{1}{(2\pi)^2} \int_{\mathbb{R}} |F(t)|^2 \frac{dt}{\Omega(t;\rho)} < \infty, \tag{7.98}$$

we have a bounded linear operator. Here, the integral is considered on the support of $\Omega(t;\rho)$.

We assume that ρ_1 and ρ_1^{-1} are both integrable on $(a, b]$ and so we must assume that $a \neq -\infty$. However, we can consider the integral on the interval $(-\infty, b]$ for $b < \infty$ similarly, by considering the reproducing kernel

$$K_{\rho_1}(t, \tau) \equiv \int_{\max(t,\tau)}^{b} \frac{d\xi}{\rho_1(\xi)} \tag{7.99}$$

and the reproducing kernel Hilbert space consisting of the absolutely continuous and real-valued functions on $(-\infty, b]$ satisfying $f(b) = 0$ and with the inner product

$$\langle f_1, f_2 \rangle_{H_{K_{\rho_1}}[a,b)} = \int_{-\infty}^{b} f_1'(t) f_2'(t) \rho_1(t) dt. \tag{7.100}$$

By the above setting we can derive the natural theory for the integral equation. See [95] for details. However, we will see the very complicated structure of the integral equation.

Indeed, we see some of the difficulties that arise when applying a *classical method* to the integral equation under study, and that were overcome by the above presented method. In view of this, we recall that in the classical approach to solve our integral equation, we wish to solve the related algebraic equation (7.89). In the present case, the equation is not linear, however, by separating all the functions into the real parts and the imaginary parts, we can consider the reduced simultaneous equations. Then, for the existence of the unique solution, we need the coefficient determinant to be nonzero; that is, precisely,

$$3 + |f_2(x)|^2 - |f_3(x)|^2 \neq 0. \tag{7.101}$$

This assumption implies the injectivity of the operator L, but the condition is too strong; cf. the results in [95]. Furthermore, the meaning of the condition for the original integral kernels is unclear.

Furthermore, when we assume (7.101), the property of the solution of the simultaneous equations is unclear and we cannot obtain the solution of our integral equation. This means that the classical approach is not able to solve the existence and representation problems of the solution of the integral equations in the largest number of possibilities.

For the details of this subsection, see [95].

Meanwhile, for some more general discretization principles with many concrete examples, see [92, 93].

See [53] for more about applications of the Toeplitz and Hankel operator.

Chapter 8
Special Topics on Reproducing Kernels

In this chapter, we will gather simply several special hot and important topics from the viewpoint of the general theory of reproducing kernels.

8.1 Norm Inequalities

8.1.1 The Bergman Norm and the Szegö Norm: An Overview

Here we take an overview of an inequality which arose from the theory of reproducing kernels. We content ourselves with stating the results without giving a proof. First, for a domain (open and connected set) O on the complex $z = x + iy$ plane, we define $A^2(O)$ as the set of all square integrable holomorphic functions on O. Note that $A^2(O)$ is a Hilbert space (Bergman space) with the norm given by

$$\|f\|_{A^2(O)} \equiv \sqrt{\iint_O |f(x+iy)|^2 dxdy}.$$

Thanks to the mean value property of holomorphic functions, we see that $A^2(O)$ is a reproducing kernel Hilbert space.

In 1976, the generalized isoperimetric inequality was obtained by applying of the general theory of reproducing kernels in [374]: For a bounded regular region G in the complex $z = x + iy$ plane whose boundary is surrounded by a finite number of analytic Jordan curves and for any analytic functions φ and ψ on $\overline{G} = G \cup \partial G$,

$$\frac{1}{\pi}\iint_G |\varphi(z)\psi(z)|^2 dxdy \le \left(\frac{1}{2\pi}\oint_{\partial G} |\varphi(z)|^2 |dz|\right)\left(\frac{1}{2\pi}\oint_{\partial G} |\psi(z)|^2 |dz|\right). \tag{8.1}$$

S. Saitoh, Y. Sawano, *Theory of Reproducing Kernels and Applications*,
Developments in Mathematics 44, DOI 10.1007/978-981-10-0530-5_8

In order to prove (8.1), we have to use the former historical results of the following great mathematicians:

G. F. B. Riemann (1826–1866); F. Klein (1849–1925); S. Bergman; G. Szegö; Z. Nehari; M. M. Schiffer; P. R. Garabedian; D. A. Hejhal (1972, thesis).

In particular, a profound result of D. A. Hejhal, which establishes the fundamental relationship between the Bergman and the Szegö reproducing kernels of G [202] must be applied. Furthermore, we must use the general theory of reproducing kernels by N. Aronszajn [28] described in 1950. These circumstances still remain today since the paper [374] was published in 1979.

Meanwhile, the main ingredient in the paper [374] was to determine the equality case in the above inequality; some hard analysis was used in the function theory of one complex variable, stated in the above line. A very deep and general proof appeared 26 years later in A. Yamada [485]. He gave a general framework for such equality problems in the tensor products of reproducing kernel Hilbert spaces. We will introduce his theory in Sect. A.1.

See [487] for more details.

Inequality (8.1) is indeed valid for the Hardy $H_2(G)$ space (the Szegö space) consisting of analytic functions on G with nontangential boundary values with the norm. Now, we see an important meaning or application of the inequality (8.1); that is, when we fix any member ψ of $H_2(G)$, the multiplication operator

$$\varphi \longmapsto \varphi(z)\psi(z) \tag{8.2}$$

on $H_2(G)$ to the Bergman space is bounded. Therefore, by the general theory in Sect. 3.3 for general fractional functions, we can consider the generalized fractional functions; that is, for any Bergman function $f(z)$ on the domain G

$$\frac{f(z)}{\psi(z)}, \tag{8.3}$$

at least in the sense of Tikhonov wherein we can consider the best approximation problem for the functions $\psi(z)^{-1}f(z)$ by the functions $H_2(G)$.

This paper was a milestone in the development of the theory of reproducing kernels. Starting with this paper, various applications of the general theory of reproducing kernels were developed. See also [388] for details. It seems that the general theory of reproducing kernels was, in a strict sense, not active in the theory of concrete reproducing kernels until the publication of the paper. Indeed, after its publication, various fundamental norm inequalities containing quadratic norm inequalities in matrices were derived. Furthermore, a general idea for linear transforms was obtained as stated in this book.

8.1.2 Examples of Norm Inequalities via RHKSs

We will provide a typical example of norm inequalities derived from the theory of reproducing kernels.

Example 8.1. Recall that $K(x, y) \equiv \min(x, y)$, $0 \leq x, y < \infty$ is the reproducing kernel for the Hilbert space $H_0 \equiv H_K([0, \infty))$ consisting of all real-valued and absolutely continuous functions $f(x)$ on $[0, \infty)$ such that $f(0) = 0$ and

$$\|f\|_{H_0} \equiv \sqrt{\int_0^\infty |f'(x)|^2 dx} < \infty \tag{8.4}$$

according to Theorem 1.6.

Now we are going to use Theorem 2.16 and Corollary 2.4. Let us set

$$\varphi_N(t) \equiv 1 + t + t^2 + \cdots + t^{N-1} = \begin{cases} \dfrac{t(1-t^N)}{1-t} & t \neq 1, \\ N & t = 1. \end{cases}$$

Then

$$\mathbf{K}_N(x, y) \equiv \sum_{n=1}^{N} K(x, y)^n = \min\{\varphi_N(x), \varphi_N(y)\} \quad (x, y \in [0, \infty)) \tag{8.5}$$

is the reproducing kernel for the Hilbert space $H_{0,N} \equiv H_{\mathbf{K}_N}([0, \infty))$ consisting of all real-valued and absolutely continuous functions f on $[0, \infty)$ such that $f(0) = 0$ and f has finite norms

$$\|f\|_{\mathbf{K}_N([0,\infty))} = \sqrt{\int_0^\infty f'(x)^2 \left[\left(\frac{x(1-x^N)}{1-x}\right)'\right]^{-1} dx} < \infty. \tag{8.6}$$

Hence, for the nonlinear transform of $f \in H_0$,

$$\varphi_N(f)(x) = \sum_{n=1}^{N} f(x)^n = \frac{f(x)(1-f(x)^N)}{1-f(x)} \quad (x \geq 0), \tag{8.7}$$

we have the inequality

$$\|\varphi_N(f)\|^2_{H_{\varphi(\mathbf{K}_N)}} = \int_0^\infty \frac{\left|\left(\dfrac{f(x)(1-f(x)^N)}{1-f(x)}\right)'\right|^2}{\left|\left(\dfrac{x(1-x^N)}{1-x}\right)'\right|} dx \leq \frac{a(1-a^N)}{1-a} \tag{8.8}$$

thanks to Theorem 2.16 and Corollary 2.4, where

$$a = a_f \equiv \int_0^1 f'(x)^2 dx = \|f\|_{H_0}^2. \tag{8.9}$$

In particular, for $f \in H_0$ with $a_f \in (0,1)$ and $|f(x)| < 1$, by letting $N \to \infty$, we have the inequality

$$\int_0^1 \left(\frac{f(x)}{1-f(x)}\right)'^2 (1-x)^2 dx \le \frac{a}{1-a}. \tag{8.10}$$

Here is an application of Theorem 2.16, which yields another inequality.

Proposition 8.1 ([390, (7.3.9)]). *Assume that $\{C_j\}_{j=0}^{\infty}$ is a positive sequence. Then we have*

$$\sum_{j=0}^{\infty}\left[\left(\sum_{\nu=0}^{j} C_\nu^{(1)} C_{j-\nu}^{(2)}\right)^{-1}\left(\sum_{\nu=0}^{j} a_\nu^{(1)} a_{j-\nu}^{(2)}\right)^2\right] \le \left(\sum_{j=0}^{\infty}\frac{|a_j^{(1)}|^2}{C_j^{(1)}}\right)\left(\sum_{j=0}^{\infty}\frac{|a_j^{(2)}|^2}{C_j^{(2)}}\right)$$

for all real sequences $\{a_j^{(1)}\}_{j=0}^{\infty}$ and $\{a_j^{(2)}\}_{j=0}^{\infty}$.

Before we come to the proof, we present an example:

$$\sum_{j=0}^{\infty}\left[\frac{1}{j+1}\left(\sum_{\nu=0}^{j} a_\nu^{(1)} a_{j-\nu}^{(2)}\right)^2\right] \le \left(\sum_{j=0}^{\infty}|a_j^{(1)}|^2\right)\left(\sum_{j=0}^{\infty}|a_j^{(2)}|^2\right).$$

Note that this is a special case of Proposition 8.1 when $C_0^{(1)} = C_1^{(1)} = \cdots = C_j^{(1)} = 1$ and $C_0^{(2)} = C_1^{(2)} = \cdots = C_j^{(2)} = 1$.

Proof. We may assume that the right-hand side is not zero and that the right-hand side is finite. By the monotone convergence theorem, we may assume that

$$\limsup_{j\to\infty} \sqrt[j]{|C_j|} < \infty, \sharp\{j \in \mathbb{N}_0 : a_j = 0\} < \infty.$$

Let us set

$$f_k(z) \equiv \sum_{j=0}^{\infty} a_j^{(k)} z^j, \quad K_k(z,u) \equiv \sum_{j=0}^{\infty} C_j^{(k)} z^j \overline{u}^j$$

for $k = 1, 2$. Then we have

$$f_k \in H_{K_k}(\mathbb{C}), \quad \|f_k\|_{H_{K_k}(\mathbb{C})}^2 = \sum_{j=0}^{\infty}\frac{|a_j^{(k)}|^2}{C_j^{(k)}}$$

by Proposition 1.2 and hence $f_1 \cdot f_2 \in H_{K_1K_2}(\mathbb{C})$. If we write out the inequality in Corollary 2.4 completely,

$$\sum_{j=0}^{\infty}\left[\left(\sum_{\nu=0}^{j} C_{\nu}^{(1)} C_{j-\nu}^{(2)}\right)^{-1}\left(\sum_{\nu=0}^{j} a_{\nu}^{(1)} a_{j-\nu}^{(2)}\right)^{2}\right] = \|f_1 \cdot f_2\|_{H_{K_1K_2}(\mathbb{C})} \leq \|f_1\|_{H_{K_1}(\mathbb{C})}$$

$$\times \|f_2\|_{H_{K_2}(\mathbb{C})} = \left(\sum_{j=0}^{\infty} \frac{|a_j^{(1)}|^2}{C_j^{(1)}}\right)\left(\sum_{j=0}^{\infty} \frac{|a_j^{(2)}|^2}{C_j^{(2)}}\right). \tag{8.11}$$

we obtain the desired estimate.

Proposition 8.1 can be proved directly, whose details are left for interested readers.

Remark 8.1. In Example 8.1, A. Yamada [486] gave a direct proof of (8.10) in general forms and he unified their generalizations with the Opial inequality, obtained from composition functions. See Sect. A.2 for details. Incidentally, in the same year, N. D. V. Nhan, D. T. Duc and Vu Kim Tuan [329] gave another direct proof with some generalizations, independently of A. Yamada. For some general inequalities and their applications in connection with reproducing kernel Hilbert spaces, see [129–132, 323–330, 395, 396].

8.1.3 General Nonlinear Transforms of Reproducing Kernels

The following Theorem 8.1 states the property of the sum of reproducing kernel Hilbert spaces.

Theorem 8.1. *Suppose that J is a countable or finite index set. Suppose also that we have a family $\{K_j\}_{j\in J}$ of positive definite quadratic functions on E. Assume in addition*

$$\sum_{j\in J} K_j(p,p) < \infty \quad (p \in E). \tag{8.12}$$

1. Define K by

$$K \equiv \sum_{j\in J} K_j. \tag{8.13}$$

Then the sum defining K converges pointwise.

2. *If we have $f_j \in H_{K_j}(E)$ for each $j \in J$ and the f_j satisfy*

$$\sum_{j \in J} \|f_j\|^2_{H_{K_j}(E)} < \infty, \tag{8.14}$$

then the sum

$$\Phi(\{f_j\}_{j \in J}) \equiv \sum_{j \in J} f_j \tag{8.15}$$

converges absolutely in the norm topology of $H_K(E)$.

Proof.

1. Since

$$|K_j(p,q)| \leq \sqrt{K_j(p,p)K_j(q,q)} \quad p,q \in E, j \in J,$$

(8.12) guarantees the convergence on the right-hand side of (8.13).

2. Assume that $J_0 \subset J$ is a finite set. Then we have

$$\|\Phi(\{f_j\}_{j \in J_0})\|_{H_K(E)} \leq \sqrt{\sum_{j \in J_0} \|f_j\|^2_{H_{K_j}(E)}} < \infty$$

by Theorem 2.16, which yields the desired result.

Let us consider a special case.

Theorem 8.2. *Let K be a positive definite quadratic form function on E. Suppose that we are given a sequence of functions $\{d_n\}_{n=0}^{\infty}$ on E satisfying*

$$\sum_{n=0}^{\infty} |d_n(p)|^2 K(p,p)^n < \infty \tag{8.16}$$

for all $p \in E$. Define

$$K_d(p,q) \equiv \sum_{n=0}^{\infty} d_n(p)\overline{d_n(q)}K(p,q)^n \quad (p,q \in E). \tag{8.17}$$

If $f \in H_K(E)$ satisfies

$$\sum_{n=0}^{\infty} (\|f\|_{H_K(E)})^{2n} < \infty, \tag{8.18}$$

or equivalently $\|f\|_{H_K(E)} < 1$, *then the sum* $\Phi f \equiv \sum_{n=0}^{\infty} d_n \cdot f^n$ *converges pointwise and absolutely on E in* $H_{K_d}(E)$. *Furthermore,*

$$\|\Phi f\|_{H_{K_d}(E)} \leq \sqrt{\sum_{n=0}^{\infty} \left(\|f\|_{H_K(E)}\right)^{2n}}. \tag{8.19}$$

As before, by (8.16) the right-hand side of (8.17) converges absolutely.

Proof. The proof is obtained by letting

$$K_j \equiv d_j \otimes \overline{d_j} \cdot K^j. \tag{8.20}$$

Indeed, using Corollaries 2.4 and 2.5, we easily see that

$$\|d_j \cdot f^j\|_{H_{K_j}(E)} \leq \|d_j\|_{H_{d_j \otimes \overline{d_j}}(E)} \|f^j\|_{H_{K^j}(E)} = \|f^j\|_{H_{K^j}(E)} \leq (\|f\|_{H_K(E)})^j, \tag{8.21}$$

yielding $\sum_{j=1}^{\infty} \|d_j \cdot f^j\|_{H_{K_j}(E)}^2 < \infty$ and the absolute convergence of Φf. Furthermore, we have $\sum_{j=1}^{\infty} K_j(p,p) < \infty$ by (8.16). Therefore, Theorem 8.1 is applicable.

Theorem 8.2 above is actually an example of a nonlinear transform.

Theorem 8.3 ([89, Theorem 8.3]). *Assume that* $\{d_n\}_{n=1}^{\infty}$ *and* $\{v_n\}_{n=1}^{\infty}$ *are sequences of complex constants, that*

$$\sum_{n=0}^{\infty} K(p,p)^n < \infty \tag{8.22}$$

and that $f \in H_K(E)$ *satisfies*

$$\sum_{n=0}^{\infty} \frac{|d_n|^2}{|v_n|^2} \left(\|f\|_{H_K(E)}\right)^{2n} < \infty. \tag{8.23}$$

1. *Define*

$$\mathbb{K} \equiv \sum_{j=0}^{\infty} |v_j|^2 K^j. \tag{8.24}$$

   Then the sum defining $\mathbb{K}$ *converges absolutely.*
2. *Define*

$$\phi(f) \equiv \sum_{n=0}^{\infty} d_n \cdot f^n. \tag{8.25}$$

The $\phi(f)$ converges absolutely in $H_{\mathbb{K}}(E)$ and

$$\|\phi(f)\|_{H_{\mathbb{K}}(E)} \le \sqrt{\sum_{n=0}^{\infty} \frac{|d_n|^2}{|v_n|^2} (\|f\|_{H_K(E)})^{2n}}. \tag{8.26}$$

Proof. The proof is analogous to that of Theorem 8.2.

We can prove the following in the spirit similar to Theorem 8.2:

Corollary 8.1 ([389, Theorem]). *If d_n is a complex constant for each $n \in \mathbb{Z}_+$ and $f \in H_K(E)$ satisfies*

$$\sum_{n=0}^{\infty} n!|d_n|^2 (\|f\|_{H_K(E)})^{2n} < \infty,$$

then

$$\left\| \sum_{n=0}^{\infty} d_n \cdot f^n \right\|_{H_{\exp K}(E)} \le \sqrt{\sum_{n=0}^{\infty} n!|d_n|^2 (\|f\|_{H_K(E)})^{2n}}, \tag{8.27}$$

where (8.27) *implicitly includes that $\sum_{n=0}^{\infty} d_n \cdot f^n$ converges in $H_{\exp K}(E)$.*

Definition 8.1 (A notation). Let $N \in \mathcal{O}(\Delta(R))$ and expand it into the Taylor series

$$N(z) = \sum_{j=0}^{\infty} a_j z^j. \tag{8.28}$$

Define

$$N^+(z) \equiv \sum_{j=0}^{\infty} |a_j| z^j. \tag{8.29}$$

We remark that (8.29) via (8.28) is from [389, Corollary 2]. It follows for example from the definition that $\sin^+ z \equiv \sinh z$ and $\cos^+ z \equiv \cosh z$.

Example 8.2 ([389, Corollary 1]). Keep to the same notation (8.28) and (8.29) above. By Theorem 2.16 and Corollary 2.4

$$\|N \circ f\|_{H_{N^+ \circ K}(E)} \le \sqrt{N^+(\|f\|^2_{H_K(E)})}, \tag{8.30}$$

provided $N^+(\|f\|^2_{H_K(E)}) < \infty$.

Example 8.3 ([389, p. 230]). By Theorem 2.16 and Corollary 2.4, for example, we have

$$\|\sin f\|_{H_{\exp K}(E)} \leq \sin^{+}(\|f\|_{H_K(E)}) = \sinh(\|f\|_{H_K(E)}) \tag{8.31}$$

and

$$\|\cos f\|_{H_{\exp K}(E)} \leq \cos^{+}(\|f\|_{H_K(E)}) = \cosh(\|f\|_{H_K(E)}), \tag{8.32}$$

if the right-hand sides of (8.31) and (8.32) converge, respectively.

In the theory of nonlinear partial differential equations, we encounter nonlinear transforms related to the KDV equation etc.,

$$u = u(x,t) \mapsto u_t + 6uu_x + u_{xxx} \tag{8.33}$$

and

$$u = u(x,t) \mapsto u_{tt} - u_{xx} + m^2 \sin u, \tag{8.34}$$

where $m > 0$ is a constant. For such nonlinear transforms we aim to show that similar results are valid.

In order simply to state the result, we will assume that I is an open interval on $\mathbb{R}$. Then, for the smoothness of $H_K(I)$, note that if

$$\frac{\partial^{(j+j')}K}{\partial x^j \partial y^{j'}}(x,y) \tag{8.35}$$

are continuously differentiable on $I \times I$, then any member f of $H_K(I), f^{(j)} (j \leq n)$ are also continuously differentiable on I (see Theorem 2.5), and we have

$$f^{(n)} \in H_{K^{n,n}(I)} \tag{8.36}$$

and

$$\|f^{(n)}\|_{K^{n,n}(I)} \leq \|f\|_{H_K(I)}, \tag{8.37}$$

for the reproducing kernel Hilbert space $H_{K^{n,n}}(I)$ admitting the reproducing kernel

$$K^{n,n}(x,y) = \frac{\partial^{2n}K}{\partial x^n \partial y^n}(x,y) \quad \text{on} \quad I \tag{8.38}$$

by Theorem 2.5. Hence, for example, in the nonlinear transform

$$\psi : f \in H_K(I) \mapsto h_1 f'' + h_2 f'^2 + h_3 |f|^2 \in \mathbb{C}^I \tag{8.39}$$

for any complex-valued functions $\{h_j\}_{j=1}^3$ on I, the images $\psi(f)$ belong to the reproducing kernel Hilbert space $H_{\psi^+(K)}(I)$ admitting the reproducing kernel

$$\psi^+K \equiv h_1 \otimes \overline{h_1} K^{2,2} + h_2 \otimes \overline{h_2}(K^{1,1})^2 + h_3 \otimes \overline{h_3}|K|^2,$$

and by Theorems 2.5 and 2.16 and Corollaries 2.4 and 2.5, we obtain the inequality

$$\begin{aligned}
&\|\psi(f)\|^2_{H_{\psi^+K}(I)} \\
&\le \|h_1 \otimes \overline{h_1} f''\|^2_{h_1 \otimes \overline{h_1} K^{2,2}(I)} + \|h_2 f'^2\|^2_{h_2 \otimes \overline{h_2}(K^{1,1})^2(I)} + \|h_3 |f|^2\|^2_{h_3 \otimes \overline{h_3}|K|^2(I)} \\
&\le \|f\|^2_{H_K(I)} + 2\|f\|^4_{H_K(I)}.
\end{aligned}$$

See [389, (17)]. It is worth noting that the right-hand side does not depend on $\{h_j\}_{j=1}^3$.

In some general linear transform of Hilbert spaces we could get essentially isometries between the input and the output function spaces. However, in nonlinear transforms, essentially we get norm inequalities. In general, even in the finite-dimensional case, it is involved, to determine the cases where the equalities hold in the inequalities. In each case we need arguments to determine the cases. See, for example, [373–377, 380, 381]. However, for many cases (not always), for the reproducing kernels $f = K(\cdot, q)$ $(q \in E)$ equalities hold in the inequalities. A. Yamada [485] discussed these general properties in depth for the equality problem and we included this entire argument in Sect. A.1.

8.1.4 Convolution Norm Inequalities

Based on the generalized convolution, we obtain some inequalities. As general reproducing kernels, represented by the Fourier integral containing the Paley Wiener spaces, finite and infinite orders Sobolev Hilbert spaces, we will consider, for any nonnegative integrable functions ρ_1, ρ_2 on $\mathbb{R}$ that are measurable functions and are not zero identically, for $j = 1, 2$, we define the positive definite quadratic form functions K_j by

$$K_j(x, y) = \frac{1}{2\pi} \int_{\mathbb{R}} \exp(i(x - y) \cdot t) \rho_j(t)\, dt. \tag{8.40}$$

Then we consider the induced integral transforms $L_j : L^2(\rho_j) \to H_{K_j}(\mathbb{R})$ by

$$f_j(x) = L_j F_j(x) = \frac{1}{2\pi} \int_{\mathbb{R}} F_j(t) \exp(it \cdot x)\, \rho_j(t) dt \tag{8.41}$$

for the functions F_j satisfying

$$\frac{1}{2\pi}\int_{\mathbb{R}} |F_j(t)|^2 \rho_j(t)dt < \infty, \tag{8.42}$$

respectively. Then, for the reproducing kernel Hilbert spaces $H_{K_j}(\mathbb{R})$ admitting the kernels K_j, we have the isometric identities:

$$\|f_j\|_{H_{K_j}(\mathbb{R})} = \sqrt{\frac{1}{2\pi}\int_{\mathbb{R}} |F_j(t)|^2 \rho_j(t)dt} \quad (j = 1, 2). \tag{8.43}$$

Now we will consider the nonlinear operator $\varphi(f_1, f_2)$, for$f_j \in H_{K_j}(\mathbb{R})(j = 1, 2)$ given by

$$\varphi(f_1, f_2) = (f_1 + \overline{f_1})(f_2 + \overline{f_2}) = f_1 f_2 + f_1\overline{f_2} + \overline{f_1}f_2 + \overline{f_1 f_2}.$$

We define, following [87, p. 708],

$$F *_1 G(t) \equiv \int_{\mathbb{R}} F(\xi)G(t-\xi)\,d\xi$$

$$F *_2 G(t) \equiv \int_{\mathbb{R}} F(\xi)\overline{G(\xi - t)}\,d\xi$$

$$F *_3 G(t) \equiv \int_{\mathbb{R}} \overline{F(\xi)}G(t+\xi)\,d\xi$$

$$F *_4 G(t) \equiv \int_{\mathbb{R}} \overline{F(\xi)G(-t-\xi)}\,d\xi$$

for $t \in \mathbb{R}$. Then we obtain the identity

$$\varphi(f_1, f_2)(x) = \frac{1}{(2\pi)^2}\sum_{j=1}^{4}\int_{\mathbb{R}} \exp(ix\cdot t)(F_1\rho_1) *_j (F_2\rho_2)(t)\,dt.$$

Following the operator $\varphi(f_1, f_2)$, we consider the identity

$$\begin{aligned}\mathbf{K}(x,y) &\equiv K_1(x,y)K_2(x,y) + K_1(x,y)\overline{K_2(x,y)} + \overline{K_1(x,y)}K_2(x,y) + \overline{K_1(x,y)K_2(x,y)} \\ &= \frac{1}{(2\pi)^2}\int_{\mathbb{R}} \exp(i(x-y)\cdot t)\cdot \Omega(t;\rho_1,\rho_2)\,dt \quad (x,y \in \mathbb{R})\end{aligned}$$

for

$$\Omega(t;\rho_1,\rho_2) \equiv \rho_1 *_1 \rho_2(t) + \rho_1 *_2 \rho_2(t) + \rho_1 *_3 \rho_2(t) + \rho_1 *_4 \rho_2(t) \quad (t \in \mathbb{R}). \tag{8.44}$$

Then, by the structure of the reproducing kernel Hilbert spaces of sum and product, we see that the image of the nonlinear operator $\varphi(f_1,f_2)$ belongs to the reproducing kernel Hilbert space $H_{\mathbf{K}}(\mathbb{R})$ with the kernel $\mathbf{K}$, and furthermore, we obtain the inequality

$$\|\varphi(f_1,f_2)\|_{H_{\mathbf{K}}(\mathbb{R})} \leq 4\|f_1\|_{H_{K_1}(\mathbb{R})}\|f_2\|_{H_{K_2}(\mathbb{R})}. \tag{8.45}$$

Meanwhile, note that the reproducing kernel Hilbert space $H_{\mathbf{K}}(\mathbb{R})$ itself is realized explicitly as we see from the representation of $\mathbf{K}$ in terms of the Fourier integral: We can represent any function $g \in H_{\mathbf{K}}(\mathbb{R})$ by the integral

$$g(x) \equiv \frac{1}{(2\pi)^2}\int_{\mathbb{R}} G(t)\exp(ix\cdot t)\Omega(t;\rho_1,\rho_2)\,dt \quad (x \in \mathbb{R}) \tag{8.46}$$

for a function G satisfying

$$\frac{1}{(2\pi)^2}\int_{\mathbb{R}} |G(t)|^2\Omega(t;\rho_1,\rho_2)\,dt < \infty, \tag{8.47}$$

and we obtain the isometric identity

$$\|g\|^2_{H_{\mathbf{K}}(\mathbb{R})} = \frac{1}{(2\pi)^2}\int_{\mathbb{R}} |G(t)|^2\Omega(t;\rho_1,\rho_2)\,dt \tag{8.48}$$

as we see from Theorem 2.28.

Therefor we obtain in the t-space the desired convolution inequality:

Theorem 8.4. *Let ρ_1 and ρ_2 be nonnegative and integrable functions on $\mathbb{R}$, which allows us to consider the spaces $\mathcal{F}(\rho_1) = L^2({\rho_1}^{-1})$ and $\mathcal{F}(\rho_2) = L^2({\rho_2}^{-1})$, respectively. Assume that*

$$\sum_{k=1}^{4} \rho_1 *_k \rho_2(t) > 0$$

for all $t \in \mathbb{R}$. Then the inequality for the generalized convolution

$$\int_{\mathbb{R}} \left|\sum_{k=1}^{4} F_1 *_k F_2(t)\right|^2 \left(\sum_{k=1}^{4} \rho_1 *_k \rho_2(t)\right)^{-1} dt \leq 4\prod_{i=1}^{2} \|F_i\|^2_{L^2(\rho_i^{-1})}$$

holds true, for functions $F_j \in \mathcal{F}(\rho_j)$, $j = 1, 2$.

We expanded this result for the usual convolution in various directions with applications to inverse problems and partial differential equations through $L^p (p > 1)$ versions and converse inequalities. See [133] for example and the references therein.

The theory of reproducing kernels yields a lot of entirely new inequalities. Some of the derived inequalities seem to be impossible to have been derived without the theory of reproducing kernels. We refer to [415] for typical examples. Furthermore, generally speaking, it is very difficult to specify when the equality occurs in the inequalities. See the theory of A. Yamada in [485] given in Sect. A.1. We can see from [133] that D. T. Duc and N. D. V. Nhan were able to fully generalize the results above and they applied their results to the boundedness of various integral transforms and to the estimates of solutions of integral equations that solved the equality problems completely.

Various convolutions $*_k, k = 1, 2, 3, 4$ are considered in [182]. See [94, 183, 184, 464, 466] for the further generalizations.

In particular, for each term we obtain the norm inequality going through a similar argument:

Corollary 8.2. *Let $i = 1, 2, 3, 4$ be fixed. Assume that*

$$\rho_1 *_i \rho_2(t) > 0$$

for $t \in \mathbb{R}$. One obtains the norm inequality

$$\int_{\mathbb{R}} \frac{|(F_1\rho_1) *_i (F_2\rho_2)(t)|^2}{\rho_1 *_i \rho_2(t)}\, dt \leq \int_{\mathbb{R}} |F_1(t)|^2 \rho_1(t)\, dt \cdot \int_{\mathbb{R}} |F_2(t)|^2 \rho_2(t)\, dt.$$

By a similar method, we can obtain modified versions. For example, by considering the reproducing kernel

$$(K_1 + K_2)^2 = K_1^2 + 2K_1K_2 + K_2^2$$

and the related operator

$$\psi(f_1, f_2) = (f_1 + f_2)^2 = f_1^2 + 2f_1f_2 + f_2^2,$$

we obtain the norm inequalities again.

Corollary 8.3. *Assume that*

$$\sum_{i,j=1}^{2} \rho_i *_2 \rho_j(t) > 0$$

for $t \in \mathbb{R}$. Then the norm inequalities

$$\|(f_1 + f_2)^2\|^2_{H_{|K_1+K_2|^2}(\mathbb{R})} \leq (\|f_1\|^2_{H_{K_1}(\mathbb{R})} + \|f_2\|^2_{H_{K_2}(\mathbb{R})})^2 \tag{8.49}$$

and

$$\int_{\mathbb{R}} \left| \sum_{i,j=1}^{2} (F_i\rho_i) *_2 (F_j\rho_j)(t) \right|^2 \left(\sum_{i,j=1}^{2} \rho_i *_2 \rho_j(t) \right)^{-1} dt \le \left(\sum_{i=1}^{2} \int_{\mathbb{R}} |F_i(t)|^2 \rho_i(t)\, dt \right)^2$$

hold.

We followed [87] in this section.

The typical inequality for convolution is the following:

Theorem 8.5 ([390, (7.2.8)]). *If two positive integrable functions ρ_1, ρ_2 on $\mathbb{R}$ satisfies $\rho_1 * \rho_2(t) > 0$ for all $t \in \mathbb{R}$ and if $F_1, F_2 : \mathbb{R} \to [0, \infty)$ are measurable functions, then we have*

$$\sqrt{\int_{\mathbb{R}} \frac{1}{\rho_1 * \rho_2(t)} \left| \int_{\mathbb{R}} F_1(\xi)\rho_1(\xi)F_2(t-\xi)\rho_2(t-\xi)\, d\xi \right|^2 dt} \le \|F_1\|_{L^2(\rho_1)} \|F_2\|_{L^2(\rho_2)}.$$

Remark 8.2. Remark that Theorem 8.5 can be proved in a couple of ways. Theorem 8.5 was expanded in various directions with applications to inverse problems and partial differential equations by D. T. Duc and N. D. V. Nhan. See the related references [129–132, 323–328, 330] and [329, 376, 380, 381, 395, 396, 402].

8.2 Inequalities for Gram Matrices

N. D. V. Nhan and D. T. Duc found interesting concrete inequalities for the Gram determinant inequality of the following form in [331]: For every $F_i \in H_{K_1}(E)$ and $G_j \in H_{K_2}(E)$ $(i, j = 1, 2, \ldots, n)$,

$$\det\{\langle F_iG_i, F_jG_j\rangle_{H_{K_1K_2}(E)}\}_{i,j=1}^n \le C \det\{\langle F_i, F_j\rangle_{H_{K_1}(E)} \langle G_i, G_j\rangle_{H_{K_2}(E)}\}_{i,j=1}^n,$$

where C is a positive constant which is independent of F_i and G_j, and $H_{K_j} = H_{K_j}(E)$ is a reproducing kernel Hilbert space defined on a set E with the reproducing kernel K_j for $j = 1, 2$.

However, Yamada gave more quantitative information of C [488]. Moreover, he established the equality condition for the general inequality. In this section, we will show his wonderful results. For concrete examples, we refer to [331]. By

$$G(x_1, x_2, \ldots, x_n) \equiv \{\langle x_i, x_j\rangle_{H_1}\}_{i,j=1,2,\ldots,n}$$

we denote the *Gram matrix* of the n vectors $\{x_1, x_2, \ldots, x_n\}$ in an inner product space H_1. Likewise, define the *Gram matrix in an inner product space* H_2.

Theorem 8.6. *Let $T: H_1 \to H_2$ be a non-zero bounded linear operator between inner product spaces H_1 and H_2.*

1. Let $x_1, x_2, \ldots, x_n \in H_1$. Then

$$\det G(Tx_1, Tx_2, \ldots, Tx_n) \leq \|T\|_{H_1 \to H_2}^{2n} \det G(x_1, x_2, \ldots, x_n). \tag{8.50}$$

Also, equality occurs if and only if one of the following conditions holds:

(a) The set $\{x_1, x_2, \ldots, x_n\}$ is linearly dependent in H_1.
(b) $T \neq 0$ and the operator $T/\|T\|$ is an isometry on $\mathrm{Span}\{x_1, x_2, \ldots, x_n\}$, that is,

$$\left\| T\left(\sum_{j=1}^{n} \xi_j x_j\right) \right\|_{H_2} = \|T\|_{H_1 \to H_2} \|\sum_{j=1}^{n} \xi_j x_j\|_{H_1} \tag{8.51}$$

for all scalars $\xi_1, \xi_2, \ldots, \xi_n$.

2. Let $x_1, x_2, \ldots, x_n \in H_1$. Then

$$G(Tx_1, Tx_2 \ldots, Tx_n) \leq \|T\|_{H_1 \to H_2}^{2} G(x_1, x_2, \ldots, x_n).$$

Also, equality occurs if and only if the above condition (8.51) *holds.*

Proof. Put the Gram matrices as

$$A \equiv \|T\|_{H_1 \to H_2}^{2} G(x_1, x_2, \ldots, x_n), \quad B \equiv G(Tx_1, Tx_2 \ldots, Tx_n).$$

Inequalities are easy to check.

1. We have, for $(\xi_1, \xi_2, \ldots, \xi_n) \in \mathbb{C}^n$,

$$\sum_{i,j=1}^{n} \xi_i \bar{\xi_j} \|T\|_{H_1 \to H_2}^{2} \langle x_i, x_j \rangle_{H_1} - \sum_{i,j=1}^{n} \xi_i \bar{\xi_j} \langle Tx_i, Tx_j \rangle_{H_2}$$
$$= \|T\|_{H_1 \to H_2}^{2} \left\| \sum_{i=1}^{n} \xi_i x_i \right\|_{H_1}^{2} - \left\| T\left(\sum_{i=1}^{n} \xi_i x_i\right) \right\|_{H_2}^{2} \geq 0.$$

Thus, we see that $B \ll A$, so that (8.50) holds.

2. Let us enumerate the eigenvalues of an $n \times n$ Hermitian matrix X as

$$\lambda_1(X) \geq \lambda_2(X) \geq \cdots \geq \lambda_n(X).$$

Then, by a general theory of Hilbert spaces or Weyl's monotonicity principle (cf. [52]) for eigenvalues of Hermitian matrices, we have $\lambda_j(B) \leq \lambda_j(A)$ $(j = 1, 2, \ldots, n)$. Since the determinant of a matrix is the product of its eigenvalues, and since the matrices A and B are positive semidefinite, we obtain the inequality immediately.

Next, we proceed to determine the equality condition.

1. It is well known that the set $\{x_1, x_2, \ldots, x_n\}$ is linearly independent if and only if the Gram matrix $G(x_1, x_2, \ldots, x_n)$ is nonsingular. Thus, equality holds whenever (8.51) holds or $\{x_1, x_2, \ldots, x_n\}$ is linearly dependent in H_1. Conversely, if equality holds, we need only to show that (8.51) holds by assuming that $T \neq 0$ and that $G(x_1, x_2, \ldots, x_n)$ is nonsingular. We may assume without loss of generality that $\|T\| = 1$. Then, since both Gram matrices are positive definite, we have $\lambda_j(A) = \lambda_j(B)$ $(j = 1, 2, \ldots, n)$. Since $A - B \gg 0$ and

$$\mathrm{tr}(A - B) = \sum_{j=1}^{n} \lambda_j(A) - \sum_{j=1}^{n} \lambda_j(B) = 0,$$

we conclude that $A = B$. Therefore, it follows that T is an isometry on the subspace $\mathrm{Span}\{x_1, x_2, \ldots, x_n\}$, so that (8.51) holds.
2. Similarly, it is easy to see that equality occurs if and only if (8.51) holds.

Let $H_{K_j}(E)$ be an RKHS on a set E with the reproducing kernel K_j $(j = 1, 2)$. Then their tensor product of the Hilbert space (direct product) $H_{K_1}(E) \otimes H_{K_2}(E)$ is an RKHS on $E \times E$ whose reproducing kernel $K_1 \otimes K_2$ is given by

$$(K_1 \otimes K_2)((x, y), (x', y')) \equiv K_1(x, x')K_2(y, y') \quad (x, y), (x', y') \in E \times E;$$

see Theorem 2.20.

Let $\iota\colon E \to E \times E$, $\iota(x) \equiv (x, x)$ be the diagonal embedding from the set E into $E \times E$. Denote by diag the image of ι. Then the operator range of the linear map

$$Tf = f \circ \iota = f|\mathrm{diag}$$

defined on $H_{K_1} \otimes H_{K_2}(E \times E)$ is an RKHS on $E \times E$ with the reproducing kernel K_1K_2. Moreover, recall that the induced operator $T\colon H_{K_1}(E) \otimes H_{K_2}(E) \to H_{K_1K_2}(E)$ is a contraction with $T(f \otimes g) \equiv fg$ thanks to (2.125) and (2.129):

$$\|fg\|_{H_{K_1K_2}(E)} \leq \|f \otimes g\|_{H_{K_1}(E) \otimes H_{K_2}(E)} = \|f\|_{H_{K_1}(E)} \|g\|_{H_{K_2}(E)}.$$

Applying Theorem 8.6 to the contraction T, we obtain

Theorem 8.7. *Let $H_{K_l}(E)$ be an RKHS on E with the reproducing kernel K_l for $l = 1, 2$. Then, for any $F_i \in H_{K_1}(E)$ and $G_j \in H_{K_2}(E)$ $(i, j = 1, 2, \ldots, n)$, we have*

$$\det\{\langle F_iG_i, F_jG_j\rangle_{H_{K_1K_2}(E)}\}_{i,j=1}^{n} \leq \det\{\langle F_i, F_j\rangle_{H_{K_1}(E)} \langle G_i, G_j\rangle_{H_{K_2}(E)}\}_{i,j=1}^{n}. \tag{8.52}$$

Equality holds in the above inequality if and only if one of the following conditions holds:

1. $\{F_i \otimes G_i : i = 1, 2, \ldots, n\}$ *is linearly dependent in* $H_{K_1}(E) \otimes H_{K_2}(E)$, *or*
2. $\{F_i \otimes G_i : i = 1, 2, \ldots, n\} \subset (H_{K_1}(E) \otimes H_{K_2}(E)) \ominus \ker T$.

Proof. We need only to prove the equality condition; the inequality (8.52) is clear from Theorem 8.6. The operator $T: H_{K_1}(E) \otimes H_{K_2}(E) \to H_{K_1 K_2}(E)$ is a coisometry (i.e., the adjoint T^* is an isometry) by definition of the operator range. Hence, the subspace on which T is an isometry is the orthogonal complement of $\ker T$. Thus, from Theorem 8.6 we easily conclude the equality conditions of Theorem 8.7.

From Theorem 8.7, we immediately obtain all the Gram determinant inequalities proved in [331].

8.3 Inversions

8.3.1 Inversion for Any Matrix by Tikhonov Regularization

As a simple application of Tikhonov regularization with the theory of reproducing kernels, we introduce a very simple algorithm constructing the Moore Penrose generalized inverse for an **arbitrary** matrix. The algorithm will be given in several lines and, the algorithm is effective and itself interesting for a small size matrix.

We consider a linear equation

$$A\mathbf{x} = \mathbf{y}. \tag{8.53}$$

The matrix A need not be invertible and A can be an $m \times n$ matrix. For simplicity, we consider everything on the real number field. In order to solve the equation (8.53), following the idea of Tikhonov regularization, we consider the extremal problem: any fixed $\lambda > 0$ and any $\mathbf{y} \in \mathbb{R}^m$

$$\inf_{\mathbf{x} \in \mathbb{R}^n} \left(\lambda \|\mathbf{x}\|^2_{\mathbb{R}^n} + \|A\mathbf{x} - \mathbf{y}\|^2_{\mathbb{R}^m} \right) \tag{8.54}$$

and we represent the extremal vector $\mathbf{x}^*_\lambda(\mathbf{y})$ in the form

$$\mathbf{x}^*_\lambda(\mathbf{y}) = B_\lambda \mathbf{y}. \tag{8.55}$$

Then, by letting $\lambda \downarrow 0$, we will be able to obtain an inverse of A and the natural solution of (8.53). Of course, B_λ is given by

$$B_\lambda = \{({}^t\!AA + \lambda I)^{-1}\}{}^t\!A \tag{8.56}$$

in (3.75). We are interested in some practical and effective construction of B_λ. On the construction $\mathbf{x}^*_\lambda$ of the extremal vector in the representation (8.53), following the idea of the theory of reproducing kernels, we give a constructive method by iteration.

Construction of Approximate Solutions by Iteration

Following the idea and method of the Tikhonov regularization, we consider the extremal problem

$$\inf_{\mathbf{x}\in\mathbb{R}^n}\left(\lambda\|\mathbf{x}\|_{\mathbb{R}^n}^2+\|A\mathbf{x}-\mathbf{y}\|_{\mathbb{R}^m}^2\right). \tag{8.57}$$

We view $\mathbb{R}^n=\mathscr{F}(\{1,2,\ldots,n\})$. We wish to construct the reproducing kernel for the inner product space $H_{A,\lambda}(\{1,2,\ldots,n\})$ with norm square

$$\lambda\|\mathbf{x}\|_{\mathbb{R}^n}^2+\|A\mathbf{x}\|_{\mathbb{R}^m}^2. \tag{8.58}$$

As seen from Example 1.1 note here that $\delta(i,j)=\delta_{ij}$, the Kronecker delta, is the reproducing kernel of the usual inner product space $\mathbb{R}^n$. According to Theorem 2.10, we can calculate this reproducing kernel $K_\lambda(i,j)$ simply as the solution of the equation

$$K_\lambda^{(0)}(i,j)\equiv\frac{1}{\lambda}\delta(i,j)=K_\lambda(i,j)+\frac{1}{\lambda}\langle A[K_\lambda(\cdot,j)],A[\delta(\cdot,i)]\rangle_{\mathbb{R}^m}. \tag{8.59}$$

Then the extremal function in (8.57) is given for each i-th component of the vector $\mathbf{x}_\lambda^*(\mathbf{y})$ by

$$\mathbf{x}_\lambda^*(\mathbf{y})(i)=\langle A[K_\lambda(\cdot,i)],\mathbf{y}\rangle_{\mathbb{R}^m}. \tag{8.60}$$

We can solve the equation (8.59) directly, of course, however, in order to avoid the inverse of a large size matrix of $n\times n$ and in order to obtain a constructive method, we introduce a new algorithm based on an iterative method. We rewrite A and $A\mathbf{x}$ as follows:

$$A=\begin{pmatrix}\mathbf{a}_1\\ \mathbf{a}_2\\ \vdots\\ \mathbf{a}_m\end{pmatrix},\quad A\mathbf{x}=\begin{pmatrix}\langle\mathbf{a}_1,\mathbf{x}\rangle_{\mathbb{R}^n}\\ \langle\mathbf{a}_2,\mathbf{x}\rangle_{\mathbb{R}^n}\\ \vdots\\ \langle\mathbf{a}_m,\mathbf{x}\rangle_{\mathbb{R}^n}\end{pmatrix}.$$

Then (8.58) reads

$$\|x\|_{(m)}\equiv\sqrt{\lambda\|\mathbf{x}\|_{\mathbb{R}^n}^2+\|A\mathbf{x}\|_{\mathbb{R}^m}^2}=\sqrt{\lambda\|\mathbf{x}\|_{\mathbb{R}^n}^2+\sum_{j=1}^m|\langle\mathbf{x},\mathbf{a}_j\rangle_{\mathbb{R}^n}|^2}. \tag{8.61}$$

What we propose here is to proceed by steps: We identify $\mathscr{F}(\{1,2,\ldots,n\})$ naturally with $\mathbb{R}^n$ and we use Example 2.12 with $E=\{1,2,\ldots,n\}$, $\ell(\mathbf{x})=\langle \mathbf{x},\mathbf{a}_j\rangle_{\mathbb{R}^n}$ and

$$K_\lambda^{(0)}(i,j) \equiv \frac{1}{\lambda}\delta(i,j)$$

for $i,j \in E$. Our task is to seek a reproducing kernel with respect to the norm

$$\|x\|_{(1)} \equiv \sqrt{\lambda\|\mathbf{x}\|_{\mathbb{R}^n}^2 + \sum_{j=1}^{1} |\langle \mathbf{x},\mathbf{a}_j\rangle_{\mathbb{R}^n}|^2}. \tag{8.62}$$

We apply Examples 2.6 and 2.12. According to Example 2.6, we have

$$\ell \bowtie K = \mathbf{a}_1.$$

Thus, by (2.63), we conclude that

$$K_\lambda^{(1)}(i,j) = \frac{1}{\lambda}\delta_{ij} - \frac{a_{1i}a_{1j}}{\lambda(1+\lambda\|\mathbf{a}_1\|_{\mathbb{R}^n}^2)}. \tag{8.63}$$

For the second step, for the space with the norm square

$$\lambda\|\mathbf{x}\|_{\mathbb{R}^n}^2 + \sum_{j=1}^{2} |\langle \mathbf{a}_j,\mathbf{x}\rangle_{\mathbb{R}^n}|^2, \tag{8.64}$$

its reproducing kernel $K_\lambda^{(2)}(i,j)$ is constructed from $K_\lambda^{(1)}(i,j)$ similarly as follows: We have to start from $H_{K_1}(\{1,2,\ldots,n\})$ with the kernel K_1 and the mapping $\mathbf{x} \in \mathbb{R}^n \mapsto \langle \mathbf{a}_2,\mathbf{x}\rangle_{\mathbb{R}^n} \in \mathbb{R}$. We write $\mathbf{a}_2 \equiv (a_{21},a_{22},\ldots,a_{2n})$.

First, we regard $K_\lambda^{(1)}(\cdot,i)$ as a vector in $\mathbb{R}^n$. Next we calculate $\ell \bowtie K_1$. With the understanding that the vector $\mathbf{e}_j$ denotes the function $k \mapsto \delta_{ij}$, we trivially have

$$K_1 = \sum_{j_1,j_2=1}^{n} K_\lambda^{(1)}(j_1,j_2)\mathbf{e}_{j_1} \otimes \mathbf{e}_{j_2}.$$

Thus, although $\{\mathbf{e}_j\}_{j=1}^n$ is not a CONS, we still have

$$\ell \bowtie K_1 = \sum_{j_1,j_2=1}^{n} a_{2j_1}K_\lambda^{(1)}(j_1,j_2)\mathbf{e}_{j_2}.$$

Hence

$$\ell(\ell \bowtie K_1) = \sum_{j_1,j_2=1}^{n} a_{2j_1}a_{2j_2}K_\lambda^{(1)}(j_1,j_2).$$

Thus, it follows from (2.12) that

$$K_\lambda^{(2)}(i,j) = K_\lambda^{(1)}(i,j) - \frac{\langle \mathbf{a}_2, K_\lambda^{(1)}(\cdot,i)\rangle_{\mathbb{R}^n} \langle \mathbf{a}_2, K_\lambda^{(1)}(\cdot,j)\rangle_{\mathbb{R}^n}}{1 + \sum_{j_1,j_2=1}^n a_{2j_1} a_{2j_2} K_\lambda^{(1)}(j_1,j_2)}. \tag{8.65}$$

We obtain, repeatedly, the desired matrix: In the end we obtain

$$K_\lambda^{(m)}(i,j) = K_\lambda^{(m-1)}(i,j) - \frac{\langle \mathbf{a}_m, K_\lambda^{(m-1)}(\cdot,i)\rangle_{\mathbb{R}^n} \langle \mathbf{a}_m, K_\lambda^{(m-1)}(\cdot,j)\rangle_{\mathbb{R}^n}}{1 + \sum_{j_1,j_2=1}^n a_{mj_1} a_{mj_2} K_\lambda^{(m-1)}(j_1,j_2)}, \tag{8.66}$$

by writing $\mathbf{a}_m = (a_{m1}, a_{m2}, \ldots, a_{mn})$.

Then we obtain the desired representation

$$\mathbf{x}_\lambda^*(\mathbf{y}) = \{K_\lambda^{(m)}(i,j)\}_{i,j=1}^n {}^t A\mathbf{y}.$$

By this simple algorithm, we were able to derive simply the inversions for 50×50 size dense matrices.

Size Estimates of the Algorithm

In order to obtain (8.63) for all i, j, the number of multiplications and additions is of order n^3. Further, by (8.66) with $m = 2, \ldots, N$, in order to obtain $K_\lambda^{(m)}(i,j)$, the number of the multiplications and the additions is of order Nn^4. Therefore, we need the multiplicities and additions of the order Nn^4 in our algorithm to calculate $K_\lambda^{(N)}(i,j)$.

For many matrices appearing from finite element methods and difference equations, the components of the vectors $\{{}^t\mathbf{a}_j\}$ are zero except around the diagonal part. So, in our algorithm, we can save calculations for such matrices. Indeed, we will assume that $m = n$ and

$${}^t\mathbf{a}_j = (0, 0, \ldots, a_{j,j-r}, \ldots, a_{j,j}, \ldots, a_{j,j+r}, 0, 0, \ldots, 0).$$

We fix such $r > 0$. Then we can estimate multiplicities less than n^3 for large n.

See [231] for the source of the above estimation of the size of the algorithm and the numerical experiments.

8.3.2 Representation of Inverse Functions

Using the theory of reproducing kernels, we can consider the problem **for an arbitrary mapping** and in addition we will be able to consider some representation of its inversion [387]. For example, we can consider the following concrete problems.

Let $\varphi : E' \to E$ be a bijection. Now from (2.52),

$$f \circ \varphi(p) = f(\varphi(p)) = \langle f, K_{\varphi(p)} \rangle_{H_K(E)}, \tag{8.67}$$

by Theorem 2.15, we obtain

$$\varphi^{-1}(p) = \langle \varphi^{-1}, K_p \rangle_{H_K(E)} = \langle \mathrm{id}, K(\varphi(\cdot), p) \rangle_{H_{\varphi^* K}(E')} \tag{8.68}$$

if $\varphi^{-1} \in H_K(E)$.

We derived the following simple result by representing the inverse of the Riemann mapping function on the unit disk in terms of the Bergman reproducing kernel and by the transform of the Riemann mapping function; however, the result may be derived also directly:

Example 8.4. Suppose that $\varphi : \Omega_1 \to \Delta(1)$ is a biholomorphic function. Then we have

$$\varphi^{-1}(z) = \frac{1}{\pi} \iint_{\Omega_1} \frac{z|\varphi'(x+iy)|^2}{(1 - z\overline{\varphi(x+iy)})^2} \, dx \, dy \tag{8.69}$$

for all $z \in \Delta(1)$.

Example 8.5 ([387, Section 4.3]). Let $K(x, y) \equiv \min(x, y)$ for $0 \le x, y < \infty$. Then we have

$$H_K[0, \infty) = \left\{ f \in W^{1,2}[0, \infty) : f(0) = 0 \right\}, \tag{8.70}$$

according to Theorem 1.6. Let $\varphi : [0, \infty) \to [0, \infty)$ be an increasing function such that $\varphi|(0, \infty) \in C^1(0, \infty)$ that $\varphi(0) = 0$, and that $\varphi'(x) > 0$ for almost every $x > 0$. First, observe that

$$\varphi^* K(x, y) = K(\varphi(x), \varphi(y)). \tag{8.71}$$

Thus, we have

$$H_{\varphi^* K} = \{ f \in AC[0, \infty) : \|f\|_{H_{\varphi^* K}} < \infty \}$$

and

$$\|f\|_{H_{\varphi^* K}} = \sqrt{\int_0^\infty f'(\xi)^2 \frac{d\xi}{\varphi'(\xi)}} < \infty \tag{8.72}$$

by Theorem 1.8. Observe also that $\displaystyle\int_0^\infty \frac{\sin at}{t} \, dt = \frac{\pi}{2} \mathrm{sgn}(a)$, which is known as the *Dirichlet integral*. Note that

$$\varphi^{-1}(x) = \langle \mathrm{id}, \varphi^* K(\cdot, \varphi^{-1}(x)) \rangle_{H_{\varphi^* K}[0,\infty)} = \langle \mathrm{id}, K(\varphi(\cdot), \varphi(\varphi^{-1}(x))) \rangle_{H_{\varphi^* K}[0,\infty)}$$

from (8.71). As a result, we obtain

$$\begin{aligned}\varphi^{-1}(x) &= \int_0^\infty \frac{d}{d\xi}\min(\varphi(\xi), x)\,\frac{d\xi}{\varphi'(\xi)} \\ &= \frac{2}{\pi}\int_0^\infty \left(\varphi'(\xi)\int_0^\infty \frac{\cos(\varphi(\xi)t)\sin x\,t}{t}\,dt\right)\frac{d\xi}{\varphi'(\xi)} \\ &= \frac{2}{\pi}\int_0^\infty \left(\int_0^\infty \frac{\cos(\varphi(\xi)t)\sin x\,t}{t}\,dt\right)d\xi.\end{aligned}$$

In particular, by letting $\varphi(x) = x^n$,

$$\sqrt[n]{x} = \frac{2}{\pi}\int_0^\infty \left(\int_0^\infty \frac{\cos(\xi^n\,t)\sin x\,t}{t}\,dt\right)d\xi \tag{8.73}$$

for all $n \in \mathbb{N}$ and $x > 0$.

Example 8.6 ([489, Section 3]). Suppose that $f : [a, b] \to \mathbb{R}$ is an increasing function such that $f' \in C[a, b]$. Then as is well known, f^{-1} belongs to the Sobolev class $W^{1,2}[a, b]$, whose inner product is given by

$$\langle f_1, f_2\rangle_{W^{1,2}[a,b]} = f_1(a)f_2(a)+f_1(b)f_2(b)+\int_a^b \left(f_1(x)f_2(x) + f_1'(x)f_2'(x)\right)dx. \tag{8.74}$$

Note that the kernel of this RHKS is given by $K(x, y) = \frac{1}{2}\exp(-|y_1 - y_2|)$. See Theorem 1.3. Using (8.74), we obtain

$$f^{-1}(y_0) = \langle f^{-1}, K_{y_0}\rangle_{H_K[a,b]}$$

$$= aK(f(a), y_0) + bK(f(b), y_0) + \int_{f(a)}^{f(b)} f^{-1}(y)K(y, y_0) + (f^{-1}(y))'\frac{\partial K_{y_0}}{\partial y}(y)\,dy$$

$$= \int_a^b \left(x\,K(f(x), y_0) + \frac{\partial K_{y_0}}{\partial y}(f(x))\right)f'(x)\,dx + a\,K(f(a), y_0) + b\,K(f(b), y_0).$$

Let us calculate the integral in the last term by setting it as I. Let us take $R \gg 1$ so that $-R < f(a) < f(b) < R$. Define the curve γ_R as the boundary of $\Delta(R) \cap \mathbb{H}$ with the usual orientation. Then we have

$$\begin{aligned}\mathrm{I} &= \int_a^b \left(xK(f(x), y_0) + \frac{\partial K_{y_0}}{\partial y}(f(x))\right)f'(x)\,dx \\ &= \lim_{R\to\infty}\int_a^b \left(\int_{-R}^R (xf'(x) + i\,\xi)\frac{\exp(i\xi(f(x) - y_0))}{\xi^2 + 1}\,d\xi\right)dx \\ &= \lim_{R\to\infty}\int_{-R}^R \left(\int_a^b (xf'(x) + i\,\xi)\frac{\exp(i\xi(f(x) - y_0))}{\xi^2 + 1}\,dx\right)d\xi.\end{aligned}$$

Using another limit, we have

$$\begin{aligned}
\mathrm{I} = & \lim_{R\to\infty}\lim_{\varepsilon\downarrow 0}\left[\int_{[-R,R]\setminus[-\varepsilon,\varepsilon]}\left(b\left(-\frac{1}{i\xi}\right)\frac{\exp(i\xi(f(b)-y_0))}{\xi^2+1}\right.\right.\\
& \left.\left. -a\left(-\frac{1}{i\xi}\right)\frac{\exp(i\xi(f(a)-y_0))}{\xi^2+1}\right)d\xi\right]\\
& +\left\{\int_a^b\left(i\,\xi-\frac{1}{i\xi}\right)\frac{\exp(i\xi(f(x)-y_0))}{\xi^2+1}\,dx\right\}\,d\xi\\
= & \frac{a+b}{2}+\lim_{R\to\infty}\lim_{\varepsilon\downarrow 0}\left[\int_{[-R,R]\setminus[-\varepsilon,\varepsilon]}\left(\int_a^b\left(i\,\xi-\frac{1}{i\xi}\right)\frac{\exp(i\xi(f(x)-y_0))}{\xi^2+1}\,dx\right)\,d\xi\right]\\
& +\lim_{R\to\infty}\lim_{\varepsilon\downarrow 0}\int_{[-R,R]\setminus[-\varepsilon,\varepsilon]}\left(a\frac{\exp(i\xi(f(a)-y_0))}{\xi^2+1}-b\frac{\exp(i\xi(f(b)-y_0))}{\xi^2+1}\right)\,d\xi.
\end{aligned}$$

As a result, we see that

$$f^{-1}(y_0)=\frac{a+b}{2}+\frac{1}{2}\int_a^b \operatorname{sgn}(y_0-f(x))\,dx. \tag{8.75}$$

This formula (8.75) can, however, be derived directly and simply, and we note that we do not need any smoothness assumptions on the function f. Indeed, we need only the strictly increasing assumption.

Denote by ω_n the volume of the unit ball in $\mathbb{R}^n$. We can naturally generalize the above observations to $\mathbb{R}^n$ and this may be considered as a counterpart to nonlinear simultaneous equations of Kramer's formula for regular matrices. Here and below $*$ denotes the Hodge star operator, G_n the fundamental solution of the Laplacian $\Delta=\sum_{i=1}^{n}\frac{\partial^2}{\partial x_i^2}$:

$$G_n(z)\equiv\frac{|z|^{2-n}}{n(n-2)\omega_n}.$$

Denote by x_i the coordingate function; $x_i : x=(x_1,x_2,\ldots,x_n)\in\mathbb{R}^n\mapsto x_i\in\mathbb{R}$.

Theorem 8.8. *Let $n\geq 3$. Let Ω be a bounded C^1-domain in $\mathbb{R}^n$ and let $f\in C^1(\overline{\Omega})$ that is, f is defined on an open set U containing $\overline{\Omega}$. Assume further that $f : U\mapsto f(U)$ is an orientation-preserving mapping. Then, for $y_0\in f(\Omega)$, we have*

$$f_i^{-1}(y_0)=\int_{\partial\Omega}x_i f^*[*dG_n(\cdot-y_0)](x)-\int_{\Omega}dx_i\wedge f^*[*dG_n(\cdot-y_0)](x). \tag{8.76}$$

Here, f_i^{-1} denotes the i component of f^{-1}.

Proof (of Theorem 8.8). Write $x_0 \equiv f^{-1}(y_0)$ and choose $r > 0$ so that $B(y_0, 2r) \subset f(\Omega)$. Denote by $S(y_0, 2r) = \partial B(x_0, 2r)$ and equip $f^{-1}(S(y_0, 2r))$ with the natural orientation. Then

$$\begin{aligned}&\int_{f^{-1}(S(y_0,r))} x_i f^*[*dG_n(\cdot - y_0)](x)\\&= \int_{\partial\Omega} x_i f^*[*dG_n(\cdot - y_0)] - \int_{\Omega\setminus f^{-1}(B(y_0,r))} \left[dx_i \wedge f^*[*dG_n(\cdot - y_0)]\right](x)\end{aligned}$$

since G_n is harmonic, $\Delta = \pm\delta d$ and $*\delta = (-1)^k d*$ for k forms.

Let $\tau_0 : \mathbb{R} \to \mathbb{R}$ be a function satisfying $\chi_{(1,\infty)} \leq \tau \leq \chi_{(2,\infty)}$. Define $\tau(x) \equiv \tau(2|x|)$ for $x \in \mathbb{R}^n$. Then we have

$$\begin{aligned}&\int_{f^{-1}(S(y_0,r))} x_i f^*[*dG_n(\cdot - y_0)](x)\\&= \int_{f^{-1}(S(y_0,r))} x_i \tau(r^{-1}(f(x) - y_0)) f^*[*dG_n(\cdot - y_0)](x)\\&= \int_{f^{-1}(B(y_0,r))} d[x_i \tau(r^{-1}(f(x) - y_0))] \wedge f^*[*dG_n(\cdot - y_0)](x)\\&= \int_{f^{-1}(B(y_0,r))} f^* d[x_i \circ f^{-1} \cdot \tau(r^{-1}(\cdot - y_0))] \wedge f^*[*dG_n(\cdot - y_0)](x),\end{aligned}$$

since G_n is harmonic. Since f is orientation preserving, we have

$$\begin{aligned}&\int_{f^{-1}(S(y_0,r))} x_i f^*[*dG_n(\cdot - y_0)](x)\\&= \int_{B(y_0,r)} d[x_i \circ f^{-1} \cdot \tau(r^{-1}(f^{-1}(\cdot) - x_0))] \wedge [*dG_n(\cdot - y_0)](x).\end{aligned}$$

If we keep track of the above calculation, then we have

$$\int_{f^{-1}(S(y_0,r))} x_i f^*[*dG_n(\cdot - y_0)](x) = \int_{S(y_0,r)} x_i \circ f^{-1}(y)[*dG_n(\cdot - y_0)](y).$$

In total, we have

$$\begin{aligned}&\int_{S(y_0,r)} x_i \circ f^{-1}(y)[*dG_n(\cdot - y_0)](y)\\&= \int_{\partial\Omega} x_i f^*[*dG_n(\cdot - y_0)] - \int_{\Omega\setminus f^{-1}(B(y_0,r))} \left[dx_i \wedge f^*[*dG_n(\cdot - y_0)]\right](x).\end{aligned}$$

It remains to let $r \to 0$.

In particular, for $n = 1$, we obtain (8.75), directly.

As a special case of $n = 2$, we obtain the following theorem:

Theorem 8.9 ([492]). *Let Ω be a bounded C^1-domain in $\mathbb{R}^2$ and let $f \in C^1(\overline{\Omega})$, that is, f is defined on an open set U containing $\overline{\Omega}$. Assume further that $f : U \mapsto f(U)$ is an orientation preserving mapping. Then we have*

$$f_1^{-1}(\hat{y}) = \frac{1}{2\pi}\oint_{\partial\Omega} x_1 d\left[\operatorname{Arc\,tan}\left(\frac{f_1(x)-\hat{y}_1}{f_2(x)-\hat{y}_2}\right)\right] - \frac{1}{2\pi}\iint_{\Omega} dx_1 \wedge d\left[\operatorname{Arc\,tan}\left(\frac{f_1(x)-\hat{y}_1}{f_2(x)-\hat{y}_2}\right)\right]$$

$$f_2^{-1}(\hat{y}) = \frac{-1}{2\pi}\oint_{\partial\Omega} x_2 d\left[\operatorname{Arc\,tan}\left(\frac{f_2(x)-\hat{y}_2}{f_1(x)-\hat{y}_1}\right)\right] + \frac{1}{2\pi}\iint_{\Omega} dx_2 \wedge d\left[\operatorname{Arc\,tan}\left(\frac{f_2(x)-\hat{y}_2}{f_1(x)-\hat{y}_1}\right)\right]$$

for any $\hat{y} = (\hat{y}_1, \hat{y}_2) \in f(\Omega)$.

Note that the differential forms

$$d\left[\operatorname{Arc\,tan}\left(\frac{f_1(x)-\hat{y}_1}{f_2(x)-\hat{y}_2}\right)\right], \quad d\left[\operatorname{Arc\,tan}\left(\frac{f_2(x)-\hat{y}_2}{f_1(x)-\hat{y}_1}\right)\right]$$

make sense despite the ambiguity of the choice of the branch of Arc tan.

Theorem 8.9 is represented more explicitly as follows: Let f be a one-to-one C^1 class mapping from $\overline{\Omega}$ into $\mathbb{R}^2$ and we assume that its Jacobian $J(x)$ is positive on Ω. We represent f as follows:

$$\begin{cases} y_1 = f_1(x) = f_1(x_1, x_2) \\ y_2 = f_2(x) = f_2(x_1, x_2) \end{cases} \tag{8.77}$$

and the inverse mapping f^{-1} of f as follows:

$$\begin{cases} x_1 = f_1^{-1}(y) = f_1^{-1}(y_1, y_2) \\ x_2 = f_2^{-1}(y) = f_2^{-1}(y_1, y_2). \end{cases} \tag{8.78}$$

Theorem 8.10 ([492]). *For the mappings* (8.77) *and* (8.78), *we have*

$$\begin{pmatrix} f_1^{-1}(y^*) \\ f_2^{-1}(y^*) \end{pmatrix} = \frac{1}{2\pi}\oint_{\partial\Omega} \begin{pmatrix} x_1 \\ x_2 \end{pmatrix} d\left[\operatorname{Arc\,tan}\left(\frac{f_2(x)-y_2^*}{f_1(x)-y_1^*}\right)\right] - \frac{1}{2\pi}\iint_{\Omega} \frac{\operatorname{adj}J(x)}{|f(x)-y^*|^2}\begin{pmatrix} f_1(x)-y_1^* \\ f_2(x)-y_2^* \end{pmatrix} dx_1 dx_2$$

for any $y^* = (y_1^*, y_2^*) \in f(\Omega)$, where $\operatorname{adj}J(x)$ *denotes the adjoint matrix of $J(x)$.*

The proof will be given in Sect. A.3, see [423] as well.

The fundamental application of Theorem 8.8 is identifying the solution space. Note that for the outer side of the solutions, the representations on the right-hand side in Theorem 8.8 are zero. By the implicit function theorem, the existence of explicit functions is guaranteed. Applications to the representation of implicit functions, see Sect. A.3.

8.3.3 Identifications of Nonlinear Systems

We assume a function f on a set E is an input function of a function space and a nonlinear mapping φ of f

$$\varphi : f \in \mathscr{F}(E) \mapsto \varphi(f) \in \mathscr{F}(E) \tag{8.79}$$

is given. For a finite number of points $\{p_j\}_{j=1}^N$ of the set E, we observe the data as follows:

$$\varphi(f)(p_j) = \alpha_j; \quad j = 1, 2, \ldots, N. \tag{8.80}$$

Then we wish to determine all the outputs of the system: For any $p \in E$, we want to know

$$\varphi(f)(p) \quad p \in E. \tag{8.81}$$

For example, for the typical nonlinear system

$$\varphi(f) = \sum_{j=0}^{\infty} C_j f^j, \tag{8.82}$$

from (8.80), we have to determine all the coefficients $\{C_j\}_{j=0}^{\infty}$ and so the identification problem will be quite involved. See, for example, [147] for the Volterra series idea for nonlinear systems. For this identification problem, we will be able to apply the results in this section in a very simple way. Recall that identification problems may be directly related to interpolation problems, approximations of functions, and the learning theory. See, for example, [117, 471, 502].

However, the true identification problem will mean that we should determine $\{C_j\}_{j=0}^{\infty}$ in (8.82) independently of the members of a function space of f, not a fixed function f. We referred for this more difficult problem to [147].

Since its general theory and a prototype method for (8.82) are almost identical, we will state the case of (8.82) clearly. We wish to determine $\{C_j\}_{j=0}^{\infty}$ from the data (8.80), however it seems that it is a difficult problem (see the idea of Volterra series in [147]). However, the image of (8.82) belongs to the reproducing kernel Hilbert space admitting the reproducing kernel

$$\mathbb{K}(p, q) \equiv \exp(K(p, q)) \quad (p, q \in E). \tag{8.83}$$

So, the function (8.82) is approximated in the form

$$\varphi(f)(p) = \sum_{j=1}^{n} A_j \mathbb{K}(p, p_j) \quad (p \in E), \tag{8.84}$$

where each A_j is a complex constant, pointwisely and in the norm topology in the reproducing kernel Hilbert space $H_{\mathbb{K}}(E)$. Then, since we can assume that the matrix $\{\mathbb{K}(p_i, p_j)\}_{i,j=1}^{n}$ is a positive definite Hermitian matrix (if it is not positive definite, then we can delete the points $\{p_j\}_{j=1}^{n}$ that are not independent of $\mathbb{K}(p, p_j)$ $j = 1, 2, \ldots, n$), mathematically we can determine the constants $\{A_j\}_{j=1}^{n}$ from the regular linear equations

$$\sum_{j=1}^{n} A_j \mathbb{K}(p_i, p_j) = \alpha_i \quad (i = 1, 2, \ldots, n). \tag{8.85}$$

So we can determine the system (8.84) and make a good approximation.

As another typical case, we consider the nonlinear system

$$\varphi(f) \equiv C_{1,1}|f|^2 + C_{2,1}f|f|^2 + C_{1,2}|f|^2\overline{f} + C_{2,2}|f|^4,$$

or more generally,

$$\varphi(f) \equiv \sum_{j=1,i=1}^{\infty} C_{j,i} f^j \overline{f^i}. \tag{8.86}$$

Let us assume that each $C_{j,i}$ is not so large. Then, note that the image (8.86) belongs to the Hilbert space admitting the reproducing kernel

$$\mathbb{K} \equiv \exp(K) \exp(\overline{K}) = \exp(K + \overline{K}). \tag{8.87}$$

Therefore we can determine similarly the approximate nonlinear system for (8.86). When the partial derivatives in a nonlinear system are contained, we can adjust the related reproducing kernel Hilbert space as in (8.87). However, when the nonlinear system contains the variable coefficients as in (8.39), we must assume that the coefficients are known as in (8.85).

8.4 Sampling Theory

We will introduce some general sampling theory based on the theory of reproducing kernels.

The sampling theory has a deep connection with the general theory of reproducing kernels. Following the survey article by Higgins [206] and the paper [204] which generalize both [388] and the Kramer sampling lemma [263] in a unified manner, we will introduce simply its essence. We refer to [205, Chapters 4–6 and 10] and [204, 206] for an explanation of how great the theory of sampling theory is. We recall the definition of frames; see Definition 3.1.

Meanwhile, for a general and wide survey article for sampling theory, see [234]. For the sampling theory with noisy sampled values, sampling theorems with a finite number of sampled values, best approximations, optimal noise suppressions, see [341] and its references. See [92, 93, 166, 319, 320] for some old and new type and general sampling theories with many numerical experiments.

For every t belonging to a domain D, let K_t belong to $\mathscr{H}$ (a separable Hilbert space with inner product denoted by $\langle , \rangle_{\mathscr{H}}$). Then

$$k(s,t) \equiv \langle K_s, K_t \rangle_{\mathscr{H}} \tag{8.88}$$

is defined on $D \times D$ and we denote by $\mathscr{K}$ the transformation $\mathscr{K} g(t) \equiv \langle g, K_t \rangle_{\mathscr{H}}$, $(g \in \mathscr{H})$.

Theorem 8.11 (A general sampling principle).

1. *Let $\mathbb{X}$ be an indexing set, and assume that there exists a sequence $\{t_n\}_{n\in\mathbb{X}} \subset D$, such that $\{K_{t_n}\}_{n\in\mathbb{X}}$ is a basis (resp. a frame) for $\mathscr{H}$. Let*

$$\nu_n \equiv \|K_{t_n}\|_{\mathscr{H}} = \nu_n \tag{8.89}$$

so that the system $\{\varphi_n, \tilde{\varphi}_n\}_{n\in\mathbb{X}}$, where $\varphi_n \equiv {\nu_n}^{-1} K_{t_n}$ with dual denoted by $\{\tilde{\varphi}_n\}$, is a normalized basis [291, p. 79] *(resp. frame) for $\mathscr{H}$. Let "reconstruction functions" $\{S_n\}_{n\in\mathbb{X}}$ be defined by*

$$S_n(t) \equiv \nu_n^{-1} \langle \tilde{\varphi}_n, K_t \rangle_{\mathscr{H}} = \nu_n^{-1} \mathscr{K} \varphi_n(t), \quad (t \in D). \tag{8.90}$$

Then

$$f = \sum_{n\in\mathbb{X}} f(t_n) S_n \tag{8.91}$$

converges in $H_k(D)$, pointwise over D, and uniformly over any compact subset of D on which $\|K_t\|_{\mathscr{H}}$ is bounded.

2. *If $\{\varphi_n\}_{n\in\mathbb{X}}$ is a frame or an unconditional basis for $\mathscr{H}$, then the convergence in* (8.91) *is absolute.*

Proof.

1. We give the proof for bases; the proof for frames is very similar and uses canonical frame expansions.

   Now $\mathscr{K} : \mathscr{H} \mapsto H_k(D)$ is one-to-one, since its null space consists of those $h \in \mathscr{H}$ that are orthogonal to every member of the complete system $\{K_{t_n}\}_{n\in\mathbb{X}}$; such an element h can only be null. It follows that $\mathscr{K}$ is an isometry according to Theorem 2.36. Furthermore, $\{\mathscr{K}\varphi_n, \mathscr{K}\tilde{\varphi}_n\}_{n\in\mathbb{X}}$ is a normalized basis (resp. frame) for $H_k(D)$. The basis (resp. canonical frame) expansion for $f = \mathscr{K}g \in H_k(D)$, where $g \in \mathscr{H}$, in the set $\{\mathscr{K}\tilde{\varphi}_n\}_{n\in\mathbb{X}}$ has coefficients

$$\langle f, \mathscr{K}\varphi_n\rangle_{H_k(D)} = \langle \mathscr{K}g, \mathscr{K}\varphi_n\rangle_{H_k(D)} = \langle g, \varphi_n\rangle_{\mathscr{H}}\nu_n^{-1} = \langle g, K_{t_n}\rangle_{\mathscr{H}} = \nu_n^{-1}f(t_n).$$

Therefore the expansion is

$$f(t) = \sum_{n\in\mathbb{X}} f(t_n)\nu_n^{-1}\mathscr{K}\tilde{\varphi}_n(t) = \sum_{n\in\mathbb{X}} f(t_n)S_n(t), \tag{8.92}$$

which proves (8.91). Pointwise and uniform convergence follows from the general properties for reproducing kernel Hilbert spaces; see Theorem 2.1.

2. As to the absolute convergence of (8.91), we have by the Schwarz inequality

$$\left\{\sum_{n\in\mathbb{X}} |f(t_n)S_n(t)|\right\}^2 \leq \sum_{n\in\mathbb{X}} |f(t_n)|^2 \sum_{n\in\mathbb{X}} |S_n(t)|^2. \tag{8.93}$$

Let $g \in \mathscr{H}$ be such that $f = \mathscr{K}g$. Now if $\{K_{t_n}\}_{n\in\mathbb{X}}$ is a frame, then its dual frame is $\{\tilde{K}_{t_n}\}_{n\in\mathbb{X}} = \{\nu_n^{-1}\tilde{\varphi}_n\}_{n\in\mathbb{X}}$. We have $f(t_n) = \langle g, K_{t_n}\rangle_{\mathscr{H}}$ and

$$\overline{S_n(t)} = \langle K_t, \tilde{K}_{t_n}\rangle_{\mathscr{H}}. \tag{8.94}$$

Therefore, by the frame condition (8.95) below both series in (8.93) converge.

We consider three special cases of Theorem 8.11 and some consequences.

1. **Special case 1** Let $\{t_n\}_{n\in\mathbb{X}}$ be such that $\{K_{t_n}\}_{n\in\mathbb{X}}$ is a frame (in particular, an *exact frame* or *Riesz basis*) for $\mathscr{H}$. That is, there exist positive constants a and b such that the frame condition

$$a\sum_{n\in X} |\langle g, K_{t_n}\rangle_{\mathscr{H}}|^2 \leq \|g\|^2_{\mathscr{H}} \leq b\sum_{n\in X} |\langle g, K_{t_n}\rangle_{\mathscr{H}}|^2 \tag{8.95}$$

holds. Since $\langle g, K_{t_n}\rangle = f(t_n)$, we can write (8.95) as

$$a\sum_{n\in X} |f(t_n)|^2 \leq \|g\|^2_{\mathscr{H}} \leq b\sum_{n\in X} |f(t_n)|^2.$$

This is the well-known condition for *stable sampling* (see, e.g., [204, Definition 10.2]).

2. **Special case 2** To specialize further, suppose that $\{K_{t_n}\}_{n\in\mathbb{X}}$ is an *orthonormal basis*. Then according to (8.89) and (8.94)

$$S_n(t) = \langle K_{t_n}, K_t\rangle_{\mathscr{H}} = k(t_n, t), \tag{8.96}$$

so that the reconstruction functions are just special values of the reproducing kernel. In view of the sampling series (8.91), we can think of $\{S_n(t)\}_{n\in\mathbb{X}}$ in this case as a *discrete reproducing kernel*.

3. **Special case 3** To specialize yet further, we can obtain the classical sampling theorem, usually associated with the names of Whittaker, Kotel'nikov and Shannon. Let f belong to the Paley Wiener space defined in Theorem 1.1 with $h = 1$;

$$\mathrm{PW} = H_{K_1}(\mathbb{R}) \equiv \{f \,:\, f \in L^2(\mathbb{R}) \cap C(\mathbb{R}),\ \mathrm{supp}(\mathscr{F}f) \subset [-\pi, \pi]\}. \tag{8.97}$$

Such functions are of the form

$$f(t) = \frac{1}{\sqrt{2\pi}} \int_{-\pi}^{\pi} g(x) \exp(ixt)\, dx, \quad g \in L^2(-\pi, \pi). \tag{8.98}$$

In the notation established above, we have effectively chosen $\mathscr{K}$ to be the inverse Fourier transform, $\mathscr{H}$ to be $L^2(-\pi, \pi)$ and $\varphi_n(x)$ to be $\exp(-inx)$.

Then from Theorem 8.11 and routine calculations, we obtain the following:

Theorem 8.12 (Classical Sampling Theorem). *Let $f \in$ PW. Then we have the orthogonal expansion*

$$f(t) = \sum_{n=-\infty}^{\infty} f(n) \frac{\sin \pi(t-n)}{\pi(t-n)} = \sum_{n=-\infty}^{\infty} f(n) \mathrm{sinc}(\pi(t-n)), \tag{8.99}$$

which converges in the norm of PW, *and absolutely and globally uniformly over* $\mathbb{R}$.

Two properties of PW should be noted. One of these is that PW is, via the Fourier transform, isometrically isomorphic to $L^2(-\pi, \pi)$. The other is that PW is a Hilbert space with reproducing kernel. These properties also follow from the general sampling Theorem 8.11, but are well known from earlier literatures; see, e.g., [204].

We followed [368, Section 3] for this section somehow. We can say that Rauhut and Ullrich generalized the classical theory obtained in [151–154]. For concrete sampling formulas, see also [55–57, 468].

8.4.1 *The Information: Loss Error in Sampling Theory*

In practical applications of Theorem 8.11, we use only a finite number of sample points of $\{t_j\}_{j\in I}$. In order to obtain an upper bound on the truncation error when we fix a finite number of sample points in Theorem 8.11, we need the following hypercircle inequality of Golomb and Weinberger [122, Theorem 9.4.7]:

Theorem 8.13 (hypercircle inequality). *Let $E > 0$. For a Hilbert space H, let L be a linear functional on H. For a complete orthonormal system $\{\mathbf{v}_j : j \in I\}$ of H and for some fixed complex constants $\{b_j : j \in I'\}$ for some subset I' of I, we take the element $\mathbf{x}_0$ of H with the minimum norm satisfying*

$$\langle \mathbf{x}, \mathbf{v}_j \rangle_H = b_j, \quad j \in I' \subset I. \tag{8.100}$$

Then, for $\mathbf{x} \in H'' \equiv \{\mathbf{y} \in H : \|\mathbf{y}\|_H^2 \le E\}$ we have the inequality

$$|L\mathbf{x} - L\mathbf{x}_0| \le \sqrt{(E - \|\mathbf{x}_0\|_H^2) \sum_{j \in I \setminus I'} |L\mathbf{v}_j|^2}. \tag{8.101}$$

Moreover, if $\|\mathbf{x}_0\|_H^2 \le E$, the equality in (8.101) can be attained by some element in H.

Proof. In the representation

$$\mathbf{x} = \sum_{j \in I} \langle \mathbf{x}, \mathbf{v}_j \rangle_H \mathbf{v}_j = \sum_{j \in I'} b_j \mathbf{v}_j + \sum_{j \in I \setminus I'} \langle \mathbf{x}, \mathbf{v}_j \rangle_H \mathbf{v}_j,$$

we have

$$\mathbf{x}_0 = \sum_{j \in I'} b_j \mathbf{v}_j \tag{8.102}$$

from (8.100). Hence,

$$\mathbf{x} - \mathbf{x}_0 = \sum_{j \in I \setminus I'} \langle \mathbf{x}, \mathbf{v}_j \rangle_H \mathbf{v}_j$$

and so,

$$L\mathbf{x} - L\mathbf{x}_0 = \sum_{j \in I \setminus I'} \langle \mathbf{x}, \mathbf{v}_j \rangle_H L\mathbf{v}_j.$$

By the Schwarz inequality we have

$$|L\mathbf{x} - L\mathbf{x}_0| \le \|\mathbf{x} - \mathbf{x}_0\|_H \sqrt{\sum_{j \in I \setminus I'} |L\mathbf{v}_j|^2}. \tag{8.103}$$

Since $\mathbf{x} - \mathbf{x}_0 \perp \mathbf{x}_0$,

$$\|\mathbf{x} - \mathbf{x}_0\|_H^2 = \|\mathbf{x}\|_H^2 - \|\mathbf{x}_0\|_H^2 \le E - \|\mathbf{x}_0\|_H^2. \tag{8.104}$$

Combining (8.103) and (8.104), we have (8.101).

Next we assume that $\|\mathbf{x}_0\|_H^2 \le E$. We set

$$\mathbf{z} \equiv \sum_{j \in I \setminus I'} \overline{L\mathbf{v}_j}\mathbf{v}_j. \tag{8.105}$$

If $\mathbf{z} = \mathbf{0}$, of course, $L\mathbf{v}_j = 0$ for $j \in I \setminus I'$. In this case, the element $\mathbf{x}' = \mathbf{x}_0$ satisfies the equality in (8.101) trivially.

If $\mathbf{z} \neq \mathbf{0}$, set

$$\lambda \equiv \frac{1}{\|\mathbf{z}\|_H}\sqrt{E - \|\mathbf{x}_0\|_H^2} \in [0, \infty), \quad \mathbf{x}' \equiv \mathbf{x}_0 + \lambda\mathbf{z} \in H.$$

Then, since $\{\mathbf{v}_j\}_{j \in I}$ is a complete orthonormal system,

$$\langle \mathbf{x}', \mathbf{v}_j \rangle_H = \langle \mathbf{x}_0, \mathbf{v}_j \rangle_H = b_j, \quad j \in I' \tag{8.106}$$

and since $\mathbf{x}_0$ is perpendicular to $\mathbf{z}$,

$$\|\mathbf{x}'\|_H^2 = \|\mathbf{x}_0\|_H^2 + \lambda^2\|\mathbf{z}\|_H^2 = E. \tag{8.107}$$

Furthermore, from (8.105) and the fact that $\{\mathbf{v}_j\}_{j \in I \setminus I'}$ is an orthonormal system,

$$|L(\mathbf{x}' - \mathbf{x}_0)| = \lambda|L(\mathbf{z})| = \lambda \sum_{j \in I \setminus I'} |L\mathbf{v}_j|^2 = \lambda\|\mathbf{z}\|_H^2,$$

which implies

$$|L(\mathbf{x}' - \mathbf{x}_0)| = \lambda\|\mathbf{z}\|_H \sqrt{\sum_{j \in I \setminus I'} |L\mathbf{v}_j|^2}. \tag{8.108}$$

Hence by inserting the definition of λ into (8.108), we have

$$|L(\mathbf{x}' - \mathbf{x}_0)| = \|\mathbf{z}\|_H\sqrt{E - \|\mathbf{x}_0\|_H^2} = \sqrt{(E - \|\mathbf{x}_0\|_H^2)\sum_{j \in I \setminus I'} |L\mathbf{v}_j|^2},$$

which implies (8.101).

Let the hypotheses of Theorem 8.11 stand except that $\{\mathscr{K}\tilde{\varphi}_n\}_{n \in \mathbb{Z}}$, defined in (8.90), is assumed to be a normalized basis for $\mathscr{H}$. Let $\mathbb{J} \subset \mathbb{Z}$ be a finite set. We adopt the following for $f_{\mathbb{J}}$:

Definition 8.2. Let $f \in H_k(D)$, where k is in (8.88). Put

$$f_{\mathbb{J}} \equiv \sum_{j \in \mathbb{J}} v_j^{-1} f(t_j)\mathscr{K}\tilde{\varphi}_j, \tag{8.109}$$

where ν_j is given by (8.89), and define the *information-loss error* E by

$$Ef(t) \equiv |f(t) - f_{\mathbb{J}}(t)| = \left| \sum_{j \in \mathbb{Z} \setminus \mathbb{J}} \nu_j^{-1} f(t_j) \mathscr{K} \tilde{\varphi}_j(t) \right| \quad (t \in D). \tag{8.110}$$

Example 8.7. Assume that $\{\nu_n \mathscr{K} \tilde{\varphi}_n\}_{n \in \mathbb{Z}}$ is a complete orthonormal system.

1. Let $\{b_j\}_{j \in \mathbb{J}}$ be a given coefficient, and put

$$w \equiv \sum_{j \in \mathbb{J}} b_j \mathscr{K} \tilde{\varphi}_j. \tag{8.111}$$

Consider the hyperplane P, the set of all elements which can be written

$$w + \sum_{j \in \mathbb{Z} \setminus \mathbb{J}} \langle p, \mathscr{K} \varphi_n \rangle_H \mathscr{K} \tilde{\varphi}_j \tag{8.112}$$

with some $p \in H$. Now consider the hypercircle $C \equiv P \cap B_r$, where $B_r \equiv \{f \in H : \|f\|_H \leq r\}$. For $f \in C$ define the evaluation functional $\mathscr{L}_t$ by

$$\mathscr{L}_t f \equiv f(t), \tag{8.113}$$

so that $Ef(t) = |\mathscr{L}_t f - \mathscr{L}_t f_{\mathbb{J}}|$. Then keeping in mind $\mathscr{L}_t \mathscr{K} \tilde{\varphi}_j = \mathscr{K} \tilde{\varphi}_j(t)$ and $f \in C$, we have

$$|\mathscr{L}_t f - \mathscr{L}_t w| = |f(t) - w(t)| \leq \left[r^2 - \|w\|_H^2\right] \sum_{n \in \mathbb{Z} \setminus \mathbb{J}} \left| \frac{\mathscr{K} \tilde{\varphi}_n(t)}{\nu_n} \right|^2 \tag{8.114}$$

from the Hypercircle Inequality (see Theorem 8.13 and [122, Theorem 9.4.7]).

When $b_j = \nu_j^{-1} f(t_j)$, we have

$$|Ef(t)| \leq \sqrt{(r^2 - \|f_{\mathbb{J}}\|_H^2) \sum_{n \in \mathbb{Z} \setminus \mathbb{J}} \left| \frac{\mathscr{K} \tilde{\varphi}_n(t)}{\nu_n} \right|^2} \tag{8.115}$$

from (8.110), (8.113) and (8.114).

2. We move on to a more specailized case. Recall that sinc is given by (1.5). In the classical case $\{\mathscr{K} \tilde{\varphi}_n\}_{n=-\infty}^{\infty} = \{\operatorname{sinc}(\pi \cdot - n\pi)\}_{n=-\infty}^{\infty}$ is a complete orthonormal basis for PW, and then

$$|Ef(t)|^2 \leq \left[r^2 - \sum_{n \in \mathbb{J}} |f(n)|^2 \right] \sum_{n \in \mathbb{Z} \setminus \mathbb{J}} \operatorname{sinc}^2(\pi t - n\pi) \tag{8.116}$$

from (8.115). In this case, we can make a further estimate which is simpler but not quite so sharp, by using the well-known series

$$\sum_{n\in\mathbb{Z}} \operatorname{sinc}^2(\pi t - n\pi) = 1. \tag{8.117}$$

Then combining (8.116) and (8.117) we conclude

$$|Ef(t)| \leq \sqrt{r^2 - \sum_{n\in\mathbb{J}} |f(n)|^2}.$$

See [34] for an application of such an expansion to signal analysis.

8.4.2 A General Kramer-Type Lemma in Sampling Theory

We now make the general sampling principle of Theorem 8.11 more concrete by considering the following specifications: Let (M, Σ, μ) be a measure space; that is, let M be a set, Σ a sigma algebra of subsets of M, and μ a countably additive set function defined on Σ. Furthermore, let K be a complex-valued kernel defined on $M \times D$ satisfying;

(i) take $\mathscr{H}$ to be $L^2(M, \mu)$;
(ii) $K_t \equiv K(\cdot, t) \in L^2(M, \mu)$ for all $t \in D$;
(iii) take $\mathscr{K}$ to be the integral transform

$$\mathscr{K} g(t) \equiv \int_M g(x)\overline{K(x,t)}d\mu(x) = \langle g, K_t\rangle_{L^2(M,\mu)}. \tag{8.118}$$

Then we will have a kernel k given by

$$k(t,s) = \int_M K(x,t)\overline{K(x,s)}d\mu(x) = \mathscr{K} K_t(s) = \langle K_t, K_s\rangle_{L^2(M,\mu)} \quad (t, s \in M).$$

The set D is often taken to be $\mathbb{R}$.

Theorem 8.14 (A General Kramer–type Lemma). *With the notation as before, let D be a set, and let K be a complex-valued kernel defined on $M \times D$, where $K_t = K(\cdot, t) \in L^2(M, \mu)$ for all $t \in D$. Assume that there exists $\{t_n\}_{n\in\mathbb{X}} \subset D$ such that $\{K_{t_n}\}_{n\in\mathbb{X}}$ forms a basis for $L^2(M, \mu)$, with dual basis denoted by $\{\tilde{\varphi}_n\}_{n\in\mathbb{X}}$. That is, for all $g \in L^2(M, \mu)$,*

$$g = \sum_{n\in\mathbb{X}} \langle g, K_{t_n}\rangle_{L^2(M,\mu)}\tilde{\varphi}_n$$

holds in the topology of $L^2(M, \mu)$ *and there exists a constant* $C > 0$ *independent of* g *such that*

$$\sum_{n\in\mathbb{X}} |\langle g, K_{t_n}\rangle_{L^2(M,\mu)}|^2 + \sum_{n\in\mathbb{X}} |\langle g, \tilde{\varphi}_n\rangle_{L^2(M,\mu)}|^2 \le C\|g\|_{L^2(M,\mu)}. \tag{8.119}$$

We set

$$S_n(t) \equiv \int_M \tilde{\varphi}_n(x)\overline{K(x,t)}\, d\mu(x) = \langle \tilde{\varphi}_n, K_t\rangle = \mathscr{K}\tilde{\varphi}(t) \quad (t \in D), \tag{8.120}$$

where $\mathscr{K}$ *is given by* (8.118)*.*

1. *The range* $\mathrm{Ran}(\mathscr{K})$ *is characterized as an RKHS* $H_k(D)$*, and it is uniquely determined by the kernel* k*. For every* $g \in L^2(M, \mu)$ *we have*

$$\|\mathscr{K}g\|_{H_k(D)} = \|g\|_{L^2(M,\mu)}. \tag{8.121}$$

2. *For all* $f \in H_k(D)$*, we have the sampling series representation:*

$$f = \sum_{n\in\mathbb{X}} f(t_n)S_n \tag{8.122}$$

   in $H_k(D)$*.*
3. *About the convergence of* (8.122)*, we have the following:*
   (a) *The convergence is absolute, that is,*

$$\sum_{n\in\mathbb{X}} |f(t_n)S_n(t)| < \infty \quad (t \in D).$$

   (b) *Let* D *be a compact topological space. Assume that*

$$s, t \in D \mapsto k(s,t) \equiv \sum_{n=1}^{\infty} S_n(t)\overline{S_n(s)} \in \mathbb{C}$$

   is continuous. Then the convergence of (8.122) *is uniform.*

Proof.

1. We apply Theorem 2.36. In (2.228), we let $L = S$ with $\mathscr{H} = L^2(M, \mu)$. Since we are assuming that $\{K_{t_n}\}_{n\in\mathbb{X}}$ spans a dense space, we can use Theorem 2.36.
2. Let $t \in D$. Since $f \in H_k(D)$, we can write $f = \mathscr{K}g$ for some $g \in L^2(M, \mu)$. Observe

$$g = \sum_{n\in\mathbb{X}} \langle g, K_{t_n}\rangle_{L^2(M,\mu)}\tilde{\varphi}_n = \sum_{n\in\mathbb{X}} \mathscr{K}g(t_n)\tilde{\varphi}_n = \sum_{n\in\mathbb{X}} f(t_n)\tilde{\varphi}_n$$

in the topology of $L^2(M,\mu)$. Since $\mathscr{K}$ is continuous from $L^2(M,\mu)$ to $H_k(D)$, we have

$$f = \mathscr{K} g = \sum_{n \in \mathbb{X}} f(t_n) \mathscr{K} \tilde{\varphi}_n = \sum_{n \in \mathbb{X}} f(t_n) S_n.$$

Thus, (8.122) is proved.

3. (a) Note that

$$\sum_{n \in \mathbb{X}} |f(t_n)|^2 = \sum_{n \in \mathbb{X}} |\langle g, K_{t_n} \rangle_{L^2(M,\mu)}|^2 \leq C \|g\|^2_{L^2(M,\mu)} \tag{8.123}$$

and that

$$\sum_{n \in \mathbb{X}} |S_n(t)|^2 = \sum_{n \in \mathbb{X}} |\langle K_t, \tilde{\varphi}_n \rangle_{L^2(M,\mu)}|^2 \leq C \|K_t\|^2_{L^2(M,\mu)} = Ck(t,t), \tag{8.124}$$

both of which follows from (8.119). Thus, the convergence of (8.122) is absolute by the Cauchy-Schwarz inequality.

(b) This follows again from (8.123) and (8.124). In fact, the function k is continuous. Thus, from Theorem 2.3 and Theorem 2.17 we see that each S_n belongs to $H_k(D,D)$ and hence is continuous.

The sampling theorem is famous and we have great references; we can find a motivation for this theorem in [205, Section 1.1] and we refer to [446, Section 1.10] and [204] as well. However, as we see in the typical Shannon theorem, sampling points may be determined with strictly strong conditions; this viewpoint will be easily understood. One more very important point is: we see from the Shannon theorem, the function space, in the Shannon sampling theorem, the Paley Wiener space will contain very bad functions such that for any given finite point set, functions with any given values exist. Therefore, practically, the sampling theorem is valid among some good functions only not mathematically, but numerically. Therefore, in the sampling theorem, we wish to select sampling points following the function property that we are concerned about, because we wish to collect useful information regarding functions from the sampling points. In this sense the sampling theory will have weak points, because the sampling points are determined by the function spaces, as in the Paley Wiener space. From these viewpoints, we will be able to apply *an ultimate sampling theorem* which is introduced in [92, 93] and clearly stated in Theorem 2.33. For these comments and many numerical experiments, see [166].

For the recent articles in sampling theory and reproducing kernels, see also [1–5].

8.5 Error and Convergence Rate Estimates in Statistical Learning Theory

In this section, that each function assumes its value in $\mathbb{R}$. We work on a compact metric space X. For a function $\varphi : X \times X \to \mathbb{C}$ we let $\varphi_x(\cdot) \equiv \varphi(\cdot, x)$. We begin with the general theorem.

Theorem 8.15. *Let ν be a Borel measure on a compact metric space X. Let $h : X \times X \to \mathbb{R}$ a continuous function such that $\{h_x = h(\cdot, x) \,:\, x \in X\}$ and $\{h(t, \cdot) \,:\, t \in X\}$ are complete in $L^2(X, \nu)$. Assume in addition that*

$$\iint_{X \times X} |h(x, y)|^2 \, d\nu(x) d\nu(y) < \infty. \tag{8.125}$$

1. *Define the linear mapping defined by*

$$L_h f(x) = F(x) = \int_X f(t) h(t, x) d\nu(t) \quad (x \in X) \tag{8.126}$$

for $f \in L^2(X, \nu)$. Then L_h is a Hilbert-Schmidt operator on $L^2(X, \nu)$.

2. *We write*

$$K^*(x, y) \equiv \int_X h(x, \xi) h(y, \xi) d\nu(\xi) \quad (x, y \in X). \tag{8.127}$$

Then the family $\{K^*(\cdot, x) \,:\, x \in X\}$ is complete in $L^2(X, \nu)$.

3. *We define the associated reproducing kernel by*

$$\mathbb{K}(x, y) \equiv \int_X K^*(t, y) K^*(t, x) d\nu(t) = \langle K_y^*, K_y^* \rangle_{L^2(X,\nu)} \quad (x, y \in X). \tag{8.128}$$

Then, for any member G of $H_{\mathbb{K}}(X)$, we have the estimate

$$\min\{\|G - L_h^* L_h L_h^* L_h f\|_{L^2(X,\nu)} \,:\, \|f\|_{L^2(X,\nu)} \leq R\} \leq \frac{1}{R} \|G\|_{H_{\mathbb{K}}(X)}^2. \tag{8.129}$$

The proof of Theorem 8.15 requires us (8.133) below. To obtain (8.133) we need the following fundamental lemma:

Lemma 8.1. *Let A be a compact and self-adjoint operator on a Hilbert space H such that $\dim \mathrm{Ran}(A) = \infty$. Then there exists a sequence $\{r_j\}_{j=1}^{\infty} \subset \mathbb{R} \setminus \{0\}$ decreasing to 0 and a sequence of projections to finite-dimensional subspaces such that*

$$A = \sum_{j=1}^{\infty} r_j E_j. \tag{8.130}$$

Proof. We may assume A to be positive by decomposing A into A_+ and A_-. Let $\{E_\lambda\}_{\lambda \in \mathbb{R}}$ be a spectral decomposition of A. First, we observe that

$$A = \lim_{\varepsilon \to 0} A(1 - E_\varepsilon). \tag{8.131}$$

Note that $\ker (A - \lambda)_+{}^{\perp} = \bigcap_{\rho > \lambda} \ker (A - \rho)^{\perp}$ is finite dimensional, if $\lambda \in \mathbb{R}$. Using this, we see that there exists a sequence $r_1 > r_2 > \ldots > r_j > \ldots \to 0$ such that $E_{r_j} - \lim_{\rho \uparrow a_j} E_{r_j}$ is a nonzero projection and that $E_r - \lim_{\rho \uparrow r} E_r$ if $r > 0$ and $r \notin \{r_1, r_2, \ldots, r_j, \ldots\}$. Using this projection, we obtain

$$A = \sum_{j=1}^{\infty} r_j E_j \tag{8.132}$$

as is described in (8.130). This is the desired result.

We need the next proposition:

Proposition 8.2. *Let H be a Hilbert space and A a self-adjoint, positive compact operator on H and let $s, r \in \mathbb{R}$ such that $s > r > 0$. Then, for $R > 0$ and for $a \in H$,*

$$\min\{\|A^r a - A^s b\|_H \, : \, \|b\|_H \le R\} \le \left\{ \left(\frac{r}{s}\right)^{\frac{r}{s-r}} - \left(\frac{r}{s}\right)^{\frac{s}{s-r}} \right\} R^{-\frac{r}{s-r}} \|a\|_H{}^{\frac{s}{s-r}}. \tag{8.133}$$

Proof. The ball $\{b \in H \, : \, \|b\|_H \le R\}$ being weakly compact, and the mapping $T : H \to H$ being weak-strong continuous, the minimum in (8.133) does exist. Here and below we suppose that $\mathrm{Ran}(A)$ is finite dimensional. However, this is not immaterial because we can modify our arguments below to the case where $\mathrm{Ran}(A)$ is infinite dimensional.

We prove (8.133) by induction on $d = \mathrm{Ran}(A)$. Let $d = 1$. Then $Ax = \lambda \langle x, e \rangle_H e$ for some unit vector e and $\lambda > 0$ by Lemma 8.1. Set $\alpha = \langle a, e \rangle_H$. With this setup, we have

$$\min\{\|A^r a - A^s b\|_H \, : \, \|b\|_H \le R\} = \begin{cases} |\lambda^r \alpha - \lambda^s R|, & R \le \lambda^{-s+r} \alpha, \\ 0, & \text{otherwise.} \end{cases} \tag{8.134}$$

Let us consider the function given by

$$\Theta : \lambda \in \left(0, (\alpha/R)^{\frac{1}{s-r}}\right) \mapsto \lambda^r \alpha - \lambda^s R \in (0, \infty).$$

Then Θ attains its maximum at

$$\lambda_0 = \left(\frac{r\alpha}{sR}\right)^{\frac{1}{s-r}} \in \left(0, (\alpha/R)^{\frac{1}{s-r}}\right),$$

and hence, for all $\lambda \in \left(0, (\alpha/R)^{\frac{1}{s-r}}\right)$,

$$\Theta(\lambda) \le \Theta(\lambda_0) \le \left\{ \left(\frac{r}{s}\right)^{\frac{r}{s-r}} - \left(\frac{r}{s}\right)^{\frac{s}{s-r}} \right\} \left(\frac{\alpha}{R}\right)^{\frac{r}{s-r}} \alpha. \tag{8.135}$$

Thus (8.134) and (8.135) prove (8.133) when $d = 1$.

Assume that (8.133) is true when $d \le d_0$ with $d_0 \ge 1$. Suppose that we are given a self-adjoint compact operator A with $\dim(\mathrm{Ran}(A)) = d_0 + 1$. By Lemma 8.1, we have the decomposition $\mathrm{id}_H = P_1 + P_2$, where P_1 and P_2 are projections, AP_1 and AP_2 are self-adjoint compact operators with commutative ranges and

$$\max(\dim(\mathrm{Ran}(AP_1)), \dim(\mathrm{Ran}(AP_2))) \le d_0.$$

Let $\theta \in [0, \pi/2]$ and set

$$\beta_\theta \equiv \min\{\|A^r P_1 a - A^s P_1 b\|_H \,:\, \|b\|_H \le R\cos\theta\}$$
$$\gamma_\theta \equiv \min\{\|A^r P_2 a - A^s P_2 b\|_H \,:\, \|b\|_H \le R\sin\theta\}.$$

Then

$$\min\{\|A^r a - A^s b\|_H \,:\, \|b\|_H \le R\} = \min_{\theta\in[0,\pi/2]} \sqrt{\beta_\theta^2 + \gamma_\theta^2}.$$

By the induction assumption, we have

$$\beta_\theta \le \left\{ \left(\frac{r}{s}\right)^{\frac{r}{s-r}} - \left(\frac{r}{s}\right)^{\frac{s}{s-r}} \right\} (\|P_1 a\|_H)^{\frac{s}{s-r}} (R\cos\theta)^{1-\frac{s}{s-r}}$$

and

$$\gamma_\theta \le \left\{ \left(\frac{r}{s}\right)^{\frac{r}{s-r}} - \left(\frac{r}{s}\right)^{\frac{s}{s-r}} \right\} (\|P_2 a\|_H)^{\frac{s}{s-r}} (R\sin\theta)^{1-\frac{s}{s-r}}.$$

Thus,

$$\min\{\|A^r a - A^s b\|_H \,:\, \|b\|_H \le R\} \le \left\{ \left(\frac{r}{s}\right)^{\frac{r}{s-r}} - \left(\frac{r}{s}\right)^{\frac{s}{s-r}} \right\}$$
$$\min_{\theta\in[0,\pi/2]} \sqrt{(\|P_1 a\|_H)^{\frac{2s}{s-r}} (R\cos\theta)^{2-\frac{2s}{s-r}} + (\|P_2 a\|_H)^{\frac{2s}{s-r}} (R\sin\theta)^{2-\frac{2s}{s-r}}}.$$

Let $t > 2$ and $A, B > 0$ be a fixed constant. Observe that the function

$$\theta \in \left[0, \frac{\pi}{2}\right] \mapsto \frac{A^t}{\cos^{t-2}\theta} + \frac{B^t}{\sin^{t-2}\theta}$$

attains mininum $(A^2+B^2)^{\frac{t}{2}}$. Thus,

$$\begin{aligned}
&\min\{\|A^r a - A^s b\|_H \ : \ \|b\|_H \le R\} \\
&\le \left\{\left(\frac{r}{s}\right)^{\frac{r}{s-r}} - \left(\frac{r}{s}\right)^{\frac{s}{s-r}}\right\} \sqrt{(\|P_1 a\|_H^2 + \|P_2 a\|_H^2)^{\frac{2s}{s-r}} R^{2-\frac{2s}{s-r}}} \\
&\le \left\{\left(\frac{r}{s}\right)^{\frac{r}{s-r}} - \left(\frac{r}{s}\right)^{\frac{s}{s-r}}\right\} (\|a\|_H)^{\frac{s}{s-r}} R^{1-\frac{s}{s-r}}.
\end{aligned}$$

Consequently, we obtain (8.133).

Proof (of Theorem 8.15).

1. From Example 3.3, L_h is a Hilbert-Schmidt operator from $L^2(X,\nu)$ into $L^2(X,\nu)$.
2. Suppose that g is perpendicular to any element in $\{K^*(\cdot,x) \ : \ x \in X\}$. Then we have

$$\int_X g(y)K^*(y,x)\,d\nu(y) = 0 \quad (x \in X).$$

   Since h satisfies (8.125), we can use the Fubini theorem. Hence

$$\int_X \left(\int_X g(x)h(x,\xi)\,d\nu(x)\right)^2 d\nu(\xi) = \int_{X\times X} g(x)g(y)K^*(y,x)\,d\nu(y) = 0.$$

   Consequently,

$$\int_X g(x)h(x,\xi)\,d\nu(x) = 0 \quad (\xi \in X).$$

   Now that $\{h(\cdot,\xi)\}_{\xi\in X}$ is complete in X, we have $g = 0 \in L^2(X,\nu)$.
3. Now, in order to examine the integral transform (8.126) for $L^2(X,\nu)$, we consider the function K on $X\times X$ defined by

$$K(x,y) = \int_X h(t,y)h(t,x)\,d\nu(t) \quad (x,y \in X). \tag{8.136}$$

   Then we have the isometric identity

$$\|f\|_{L^2(X,\nu)} = \|F\|_{H_K(X)}$$

   by Theorem 2.36.

   The adjoint operator $L_h^* : L^2(X,\nu) \to L^2(X,\nu)$ of $L_h : L^2(X,\nu) \to L^2(X,\nu)$ is realized by the integral transform

$$L^*h(x) = \int_X f(t)h(x,t)d\nu(t) \quad (x \in X) \tag{8.137}$$

on $L^2(X,\nu)$. Then the composition operator $L_h^*L_h$ is realized in the form

$$F(x) = \int_X f(t)K^*(t,x)d\nu(t) \tag{8.138}$$

by Fubini's theorem. We consider the associated reproducing kernel Hilbert space $H_{\mathbb{K}}(X)$. We have the isometric identity

$$\|f\|_{L^2(X,\nu)} = \|F\|_{H_{\mathbb{K}}(X)}, \tag{8.139}$$

in (8.138).

Since the operator $L_h^*L_h$ is a self-adjoint, positive compact operator on $L^2(X,\nu)$, we obtain for $G \in L^2(X,\nu)$ such that G is an image of $L_h^*L_h$,

$$\min\{\|G - L_h^*L_hL_h^*L_hf\|_{L^2(X,\nu)} : \|f\|_{L^2(X,\nu)} \le R\} \le \frac{1}{R}\|(L_h^*L_h)^{-1}G\|_{L^2(X,\nu)}^2 \tag{8.140}$$

by Proposition 8.2, with $A = L_h^*L_h, s = 2, r = 1$. Here, we see that G belongs to the reproducing kernel Hilbert space $H_{\mathbb{K}}(X)$ with the meaning $(L_h^*L_h)^{-1}G$, and furthermore,

$$\|(L_h^*L_h)^{-1}G\|_{L^2(X,\nu)} = \|G\|_{H_{\mathbb{K}}(X)} \tag{8.141}$$

by (2.233). Combining (8.140) and (8.141), we obtain (8.129).

In general, if $k : X \times X \to \mathbb{R}$ is a positive definite quadratic function, continuous and symmetric, then k is called *Mercer kernel* which has appeared in many cases (see Theorem 2.30). For the interesting theory and results for Mercer kernels, see [62, 63]. In our context, K^* is a Mercer kernel and when we are given the linear mapping $L = L_h^*L_h$ defined by (8.138), then the isometry (8.139) yields:

Corollary 8.4. *For any member $G \in H_{\mathbb{K}}(X)$, one obtains the estimate*

$$\min\{\|G - LF\|_{L^2(X,\nu)} : \|F\|_{H_{\mathbb{K}}(X)} \le R\} \le \frac{1}{R}\|G\|_{H_{\mathbb{K}}(X)}^2. \tag{8.142}$$

Proof. Let $F \in H_{\mathbb{K}}(X)$. We use Theorem 2.37. Since $G \in H_{\mathbb{K}}(X)$, we have $G = Lg$ for some $g \in L^2(X,\nu)$ such that $\|g\|_{L^2(X,\nu)} = \|G\|_{H_{\mathbb{K}}(X)}$. Likewise, we have $F = Lf$ for some $f \in L^2(X,\nu)$ such that $\|f\|_{L^2(X,\nu)} = \|F\|_{H_{\mathbb{K}}(X)}$. Since L is compact, we are again in the position of using Proposition 8.2 to obtain

$$\min\{\|Lg - L^2f\|_{L^2(X,\nu)} : \|f\|_{L^2(X,\nu)} \le R\} \le \frac{1}{R}\|g\|_{L^2(X,\nu)}^2.$$

If we combine the above observation with $\|Lg - L^2f\|_{L^2(X,\nu)} = \|G - LF\|_{H_{\mathbb{K}}(X)}$, then we have (8.142).

The estimate (8.142) just meets the approximation in statistical learning theory in [117] and approximations in reproducing kernel Hilbert spaces in [502].

Therefore, without loss of generality, we can consider $s = 1$, because we can consider A^s as A.

Meanwhile, for $r > 0$, consider the relation between the integral transforms (8.137) and (8.138) for the symmetric kernel $h(x,t)$. Then, we see how to realize the $1/2$ operator from (8.138) to (8.137). In this way, we are convinced of how to deal with the general power 2^{-n}, $n \in \mathbb{N}$ of a positive operator, by repeating the procedure.

As a typical example, we consider the integral transform

$$F(x) = \int_0^T f(\xi)U(x, T - \xi)d\xi \tag{8.143}$$

for

$$U(x,t) = \chi_{\{(x,t)\,:\,|x| \le t\}}(x,t) = \begin{cases} 1 & (|x| \le t), \\ 0 & (|x| > t), \end{cases} \tag{8.144}$$

which appeared in the wave equation in one-dimensional space [388, pp. 143–146]. We apply Theorem 1.9 and Corollary 8.4 to

$$K(x_1, x_2) = \min(T - x_1, T - x_2) \quad (x_1, x_2 \in [0, T]) \tag{8.145}$$

is the associated reproducing kernel.

In (1.75), we have the isometric identity

$$\int_0^T f(\xi)^2 d\xi = \int_0^T F''(x)^2 dx.$$

Then, from Corollary 8.4, we obtain the following theorem:

Theorem 8.16. *Let $T > 0$ and let K be given by* (8.145). *Define*

$$\mathbb{K}(x,y) \equiv \int_0^T K(x,\xi)K(y,\xi)\,d\xi = \int_0^T \min(T - x, T - \xi)\min(T - y, T - \xi)\,d\xi$$

for $x, y \in [0, T]$. For any $G \in H_{\mathbb{K}}[0, T]$, we have the estimate

$$\min\left\{ \left\| G - \int_0^T F(\xi)K_\xi d\xi \right\|_{L^2[0,T]} : \|F\|_{H_{\mathbb{K}}[0,T]} \le R \right\} \le \frac{1}{R}\|G\|^2_{H_{\mathbb{K}}[0,T]}. \tag{8.146}$$

Proof. In the statement of Theorem 8.15, the function $h : [0,T] \times [0,T]$ is given by

$$h(x,t) \equiv \begin{cases} 1 & x < T - t, \\ 0 & x > T - t, \end{cases} \qquad (x, t \in [0, T]).$$

Therefore, if we consider the mapping

$$K^*(x, y) \equiv \int_0^T h(x, t)h(y, t)\, dt \quad (x, y \in [0, T]),$$

then a simple calculation shows that this mapping K^* is exactly the mapping K given by (8.145). In the statement of Corollary 8.4, we have

$$LF(x) = \int_0^T F(y)K^*(x, y)\, dy \quad (x \in [0, T]),$$

or in terms of the Bochner integral,

$$LF = \int_0^T F(y)K^*(\cdot, y)\, dy.$$

Thus, if we apply Corollary 8.4, then (8.146) follows.

We followed [117] in this section.

8.6 Membership Problems for RKHSs

Could we obtain some smoothing properties and analyticity of functions by using computers? In this section, we will consider such general problems based on [93].

8.6.1 *How to Obtain Smoothing Properties and Analyticity of Functions Using Computers?*

Can we realize some smoothing properties and analyticity of functions by using computers? Clearly, in a strict mathematical sense, such methods will be impossible. Anyway, from a certain practical point of view, we can propose a method that may be applied to many cases where some good functions are involved, despite the fact that it will not solve other situations where "bad functions" occur. Main arguments will be therefore concerned with numerical experiments to meet the challenge using a new concept.

In order to consider analyticity or smoothing properties, we will represent them by the functions of reproducing kernel Hilbert spaces. Indeed, using Sobolev spaces, we can classify smoothness properties by their corresponding orders in a quite useful way. On the unit disk in the complex plane, by the Bergman Selberg spaces, we can classify the family of analytic functions, for example. In this way, many properties of analytic functions may be classified similarly by reproducing kernel Hilbert spaces. So, our interest turns to the problem that we want to determine whether or not any given function belongs to a certain reproducing kernel Hilbert space.

In connection with the membership problem, we note that for any large number of points $\{p_j\}_{j=1}^n$ on the set E, we cannot apply the result by positive definiteness.

Indeed, we will assume that without loss of generality, $\{K(\cdot, p_j)\}_{j=1}^n$ are linearly independent in $H_K(E)$. Then for any given values $\{\alpha_j\}_{j=1}^n$, there exists a uniquely determined $f \in H_K(E)$ satisfying

$$f(p_j) = \alpha_j, \quad j = 1, 2, 3, \ldots, n, \tag{8.147}$$

as follows:

$$f = \sum_{j=1}^{n} C_j K_{p_j} \tag{8.148}$$

where the constants $\{C_j\}_{j=1}^n$ are determined by the equations

$$\sum_{j=1}^{n} C_j K(p_{j'}, p_j) = \alpha_{j'}, \quad j' = 1, 2, \ldots, n, \tag{8.149}$$

and we obtain the inequality in Corollary 2.1 for

$$\|f\|_{H_K(E)} = \sqrt{\sum_{j=1}^{n} \sum_{j'=1}^{n} C_j \overline{C_{j'}} K(p_{j'}, p_j)}, \tag{8.150}$$

that is the square of the norm of $f \in H_K(E)$.

Note that the function f satisfying (8.147) is not uniquely determined, but we can check the function f given by (8.148) has the minimum norm among the functions f satisfying (8.147).

For any finite number of points $\{p_j\}_{j=1}^n$ and for any given values $\{\alpha_j\}_{j=1}^n$, there certainly exists a function $f \in H_K(E)$ satisfying (8.147) as long as (8.149) admits the unique solution. However, for many points $\{p_j\}_{j=1}^n$ and bad values $\{\alpha_j\}_{j=1}^n$, the calculations in (8.149) in search of $\{C_j\}_{j=1}^n$ will be numerically and practically difficult. The difficulty in calculating the solution in (8.149) will depend on the given data and the function space $H_K(E)$. We saw such phenomena for the Paley Wiener spaces in some cases (cf. [91]). However, to represent such deep and delicate phenomena exactly will be difficult. Anyway, we may expect that the smoothness property of functions may be reflected in some properties on a large point set.

The goodness of a function in the reproducing kernel $H_K(E)$ may be given by

($g1$) the number of the points $\{p_j\}_{j=1}^n$ in (8.148),
($g2$) the distribution of the coefficients $\{C_j\}_{j=1}^n$ in (8.148), and
($g3$) the distribution of the points $\{p_j\}_{j=1}^n$ on the set E.

The factors ($g1$) and ($g2$) may be considered in a general setting, however, ($g3$) will depend on the reproducing kernel Hilbert space $H_K(E)$.

Now, from Theorem 2.33, we see that the membership property for a general reproducing kernel Hilbert space is faithfully reflected in the uniqueness point set $\{p_j\}_{j=1}^n$ by the norm in Theorem 2.33. Then, Fujiwara's infinite precision algorithm and computer power will be able to calculate the norm in Theorem 2.33 for many practical cases; see the case of numerical and real inversion formula of the Laplace transform that is a repetitive difficult problem (cf. [91, 92, 162–164]). Fujiwara gave the solution for some Fredholm integral equation containing a parameter $\alpha = 10^{-400}$ by 6000 discretization (6000 linear equations) with **600 digits precision**.

Furthermore, Fujiwara already gave numerical experiments for the norms for many points (cf. [91–93]), and we see that when increasing the number of points we are requested to calculate with more precision and more calculation costs. Therefore, we will need more precision and computer costs for realizing our method. However, this has succeeded already for the present method, see [92, 93] for numerical experiments for several typical examples.

8.6.2 Band Preserving, Phase Retrieval and Related Problems

The membership problem is applied in [363]. The corresponding fundamental problems, with many applications to analytical signals, are stated in the general theory as follows: For any fixed member $f \in H_K(E)$, look for a function g satisfying

$$f \cdot g \in H_K(E). \tag{8.151}$$

Indeed, the paper [363] examined the special Paley Wiener space for the images by Fourier inversion for the L^2 functions on an interval $[0, A]$.

If we are concerned with the functions f and g on a finite number of points $\{p_j\}_{j=1}^n$, then for the nonvanishing points $\{p_j\}_{j=1}^n$ of the function f, the values $g(p_j)$ are arbitrary, and we obtain the representation

$$f \cdot g = \sum_{j=1}^{n} C_j K_{p_j}, \tag{8.152}$$

as in (8.148) and we can obtain the meromorphic function g completely, because the function f and the right-hand side are analytic functions. We can assume, without loss of generality, that the functions $\{K_{p_j}\}_{j=1}^n$ are linearly independent and in this case, the function (8.152) is uniquely determined.

The above logic is very interesting, because in looking for the functions g, we have a great freedom. Furthermore, we note that, indeed, for the function g, we can take the functions:

$$g(p) = \frac{h(p)}{f(p)} \tag{8.153}$$

for any given function $h \in H_K(E)$ and $p \in E$, that are meromorphic functions. For the membership problem, we can conclude a simple result. We note here that all the logics in the paper [363] depend on the restriction property for the functions g that belong to L_p-spaces on the real line. We will not be able to consider in a simpler manner such conditions on g in the above idea. Therefore, we can examine if in the future it will be possible to expose some connection between the above theory and the hard analysis obtained in [363].

Moreover, it is also significant to notice that of course all the analytical and delicate theory of [363] depends on the zero point properties of the function f. Therefore, consequently, the construction of the desired functions g is not simple. When we consider the properties supporting the related Fourier inversions in connection with the convolution property, the results will be *mysteriously deep* and many concrete problems may happen. For the sake of *hard analysis*, the authors of [363] were able to solve surprisingly and perfectly the phase retrieval problem that asks, however, the strong condition $|g| = 1$ to be on the real line. As it was exposed above, for the sake of freedom for the functions g, we can construct the desired function g even in the case of phase retrieval problems.

8.7 Eigenfunctions, Initial Value Problems, and Reproducing Kernels

In this section, a new global theory combining the fundamental relations among eigenfunctions, initial value problems in general linear partial differential operators, and reproducing kernels will be given. We have various applications for the theory, however, here, we will see the examples of the prototype.

8.7.1 Formulation of the Problem

For some general linear operator L_x(and differential operator ∂_t), for some function space on a certain domain (to be specified later on), we will consider the problem

$$(\partial_t + L_x)u_f(t,x) = 0, \quad t > 0, \tag{8.154}$$

for an unknown u_f satisfying the initial value condition

$$u_f(0,x) = f(x). \tag{8.155}$$

For this global problem, we have to consider a general method which includes the analysis of the existence and construction of the solution of that type of initial value problem using the theory of reproducing kernels. Furthermore, the method will have the power to completely characterize the solutions under each specific conditions.

One of the basic procedures in the method is to use some eigenvalues λ on a set I, and eigenfunctions W_λ satisfying

$$L_x W_\lambda = \lambda W_\lambda.$$

The case of discrete eigenvalues may be dealt with similarly and so we will assume that the eigenvalues are continuous on an interval I for $\lambda > 0$. In this way, we note that the functions

$$\exp(-\lambda t) W_\lambda(x) \tag{8.156}$$

are the solutions for the operator equation

$$(\partial_t + L_x) u(t,x) = 0. \tag{8.157}$$

We will consider some general solution of (8.157) by a suitable sum of the elements in (8.156). In order to consider a convenient sum, we will use the following kernel form, with a continuous nonnegative weight function ρ over the interval I,

$$\int_I \exp(-\lambda t) W_\lambda(x) W_\lambda(y) \rho(\lambda) d\lambda \tag{8.158}$$

(where we are naturally considering the integral with absolute convergence for the kernel form). Moreover, here we assume λ to be real-valued and also the eigenfunctions $W_\lambda(x)$ are real-valued. Then, fully general solutions of the equation (8.157) may be represented in the integral form

$$u(t,x) = \int_I \exp(-\lambda t) W_\lambda(x) F(\lambda) \rho(\lambda) d\lambda \tag{8.159}$$

for the functions F satisfying

$$\int_I \exp(-\lambda t) |F(\lambda)|^2 \rho(\lambda) d\lambda < \infty. \tag{8.160}$$

Therefore, we can obtain the solution $u(t,x)$ of (8.157) satisfying the initial condition

$$u(0,x) = F(x) \tag{8.161}$$

by taking $t \to 0$ in (8.159). However, this point will be delicate and we will need to consider some more intricate structure. Here, (8.158) is a reproducing kernel and in order to analyze in detail the strategy above, we will need the theory of reproducing kernels. In particular, in order to construct certain natural solutions of (8.158) we will need a new framework and function space.

In order to analyze the integral transform (8.159) and to set the basic background for our purpose, we will need the essence of the theory of reproducing kernels.

Following the general framework of Sect. 2.5.9, we will now built our general theorem. For this purpose, we will assume that I is a positive interval, $\lambda > 0$, and that this parameter λ represents the eigenvalues satisfying

$$L_x \overline{h_\lambda} = \lambda \overline{h_\lambda}, \quad x \in E, \quad \lambda \in I. \tag{8.162}$$

Here, $\overline{h_\lambda}$ is the eigenfunction and in order to set down our notation in a consistent way, we put the complex conjugate on the function h_λ.

We form the reproducing kernel

$$K_t(x, y) = \int_I \exp(-\lambda t) h_\lambda(y) \overline{h_\lambda}(x)\, dm(\lambda), \quad t > 0, \tag{8.163}$$

and consider the reproducing kernel Hilbert space $H_{K_t}(E)$ admitting the kernel K_t. In particular, note that for $K_0(x, y) = K(x, y)$;

$$(K_t)_y \in H_K(E), \quad y \in E.$$

Then we have the following main theorem:

Theorem 8.17. *For any element $f \in H_K(E)$, the solution $u_f(t, x)$ of the initial value problem* (8.154)–(8.155) *exists and it is given by*

$$u_f(t, x) = \langle f, (K_t)_x \rangle_{H_K(E)}. \tag{8.164}$$

In fact, the solution (8.164) *satisfies* (8.155) *in the sense that*

$$\lim_{t \to +0} u_f(t, x) = \lim_{t \to +0} \langle f, K_t(\cdot, x) \rangle_{H_K(E)} = \langle f, K_x \rangle_{H_K(E)} = f(x), \tag{8.165}$$

whose existence is ensured and the limit is given in the sense of uniform convergence on any subset of E such that $K|\mathrm{diag} \cap (E \times E)$ is bounded.

The uniqueness property of the initial value problem depends on the completeness of the family of functions

$$\{(K_t)_x ; x \in E\} \tag{8.166}$$

in $H_K(E)$.

Proof. At first, note that the kernel $K_t(x, y)$ satisfies the operator equation (8.154) for any fixed y, because the functions

$$\exp(-\lambda t) \overline{h_\lambda}$$

satisfy the operator equation and it is the summation. Similarly, the function $u_f(t,x)$ defined by (8.164) is the solution of the operator equation (8.154).

In order to see the initial value problem, we note the important general property

$$K_t(x,y) \ll K(x,y); \tag{8.167}$$

and we have

$$H_{K_t} \subset H_K(E)$$

and for any function $f \in H_{K_t}$, it holds that

$$\|f\|_{H_K(E)} = \lim_{t\to+0} \|f\|_{H_{K_t}}$$

in the sense of nondecreasing norm convergence. In order to verify the crucial point in (8.165), note that

$$\begin{aligned}\|K(x,y) - K_t(x,y)\|^2_{H_K(E)} &= K(y,y) - 2K_t(y,y) + \|K_t(x,y)\|^2_{H_K(E)}\\ &\leq K(y,y) - 2K_t(y,y) + \|K_t(x,y)\|^2_{H_{K_t}}\\ &= K(y,y) - K_t(y,y),\end{aligned}$$

that converges to zero as $t \to +0$. We thus obtain the desired limit property in the theorem.

The uniqueness property of the initial value problem follows directly from (8.164).

In Theorem 8.17, we can derive the characteristic property of the solutions u_f of (8.154)–(8.155) satisfying the initial value f by the reproducing kernel Hilbert space admitting the kernel

$$k(x,t;y,\tau) \equiv \langle (K_\tau)_y, (K_t)_x \rangle_{H_K(E)}. \tag{8.168}$$

In this method, we see that the existence problem of the initial value problem is based on the eigenfunctions and we are constructing the desired solution satisfying the considered initial condition. In view of this, for a larger knowledge of the eigenfunctions we can consider a more general initial value problem. Furthermore, by considering the linear mapping of (8.164) in difference situations, we will be able to obtain many inverse problems which may be described by looking for the initial values f in different output data of $u_f(t,x)$.

We also would like to remark that in the stationary case of

$$L_x u(x) = 0, \tag{8.169}$$

the method is simpler and more direct. We may consider the family of solutions $\overline{h_\lambda}$ such that

$$L_x \overline{h_\lambda} = 0. \tag{8.170}$$

Then the general solutions are constructed by the integral form

$$u(x) = LF(x) \equiv \langle F, h_x \rangle_{L^2(I,dm)} = \int_I F(\lambda)\overline{h_\lambda}\, dm(\lambda), \quad x \in E, \tag{8.171}$$

for $F \in \mathcal{H} = L^2(I, dm)$.

8.7.2 The Derivative Operator and the Corresponding Exponential Function Case

For the simplest derivative operator D, we obtain, of course,

$$De^{\lambda x} = \lambda e^{\lambda x}. \tag{8.172}$$

We will be able to see that we can consider initial value problems in various situations by considering λ and the variable x. As typical cases we can point out the weighted Laplace transforms, the Paley Wiener spaces and the Sobolev spaces depending on $\lambda > 0$, λ being on a symmetric interval or λ on the whole real space.

The Laplace transform may be taken into account in many situations by considering various weights (see [388]), and so we will consider the simplest case:

$$K(z, \overline{u}) = \int_0^\infty e^{-\lambda z} e^{-\lambda \overline{u}} d\lambda = \frac{1}{z + \overline{u}}, \quad z = x + iy, \tag{8.173}$$

on the right-half complex plane. The reproducing kernel is the Szegö kernel and we have the image of the integral transform

$$f(z) = \int_0^\infty e^{-\lambda z} F(\lambda) d\lambda \tag{8.174}$$

for $L^2(0, \infty)$-functions $F(\lambda)$. Thus, we obtain the isometric identity

$$\frac{1}{2\pi} \int_{-\infty}^{+\infty} |f(iy)|^2 dy = \int_0^\infty |F(\lambda)|^2 d\lambda. \tag{8.175}$$

Here, $f(iy)$ mean the Fatou's non-tangential boundary values of the Szegö space of analytic functions on the right-hand half complex plane.

Now, we will consider the reproducing kernel $K_t(z, \overline{u})$ and the corresponding reproducing kernel Hilbert space H_{K_t} by taking

$$K_t(z, \overline{u}) = \int_0^{\infty} e^{-\lambda t} e^{-\lambda z} e^{-\lambda \overline{u}} d\lambda. \tag{8.176}$$

Note that the reproducing kernel Hilbert space H_{K_t} is the Szegö space on the right-hand complex plane $x > \frac{-t}{2}$.

For any Szegö kernel function space, $f(z)$ on the right-half complex plane, the function

$$U_f(t, z) = \langle f, (K_t)_{\overline{z}} \rangle_{H_K} = \frac{1}{2\pi} \int_{-\infty}^{+\infty} f(iy) \overline{K_t(iy, \overline{z})} dy$$

satisfies the partial differential equation

$$(\partial_t - D_z) U(t, z) = 0. \tag{8.177}$$

For the sake of the monotonicity of the reproducing kernels, it holds that

$$K_t(z, \overline{u}) \ll K(z, \overline{u}); \tag{8.178}$$

we obtain the desired initial condition

$$\lim_{t \to 0} U_f(t, z) = \lim_{t \to 0} \langle f, K_t(\cdot, \overline{z}) \rangle_{H_K} = \langle f, K(\cdot, \overline{z}) \rangle_{H_K} = f(z).$$

From the general property of the reproducing kernels, we see that the above convergence is uniform on any compact subset of the right-half complex plane.

In order to see the characteristic property of the solutions $U(t, z)$, we consider the kernel form

$$k(t, z; \tau, \overline{u}) = \langle K_\tau(\cdot, \overline{u}), K_t(\cdot, \overline{z}) \rangle_{H_K} = \frac{1}{t + \tau + z + \overline{u}}.$$

From this representation we see that for any fixed $t > 0$, the solutions $U(t, z)$ belong to the Szegö space on the right-hand complex plane

$$\text{Re}\, z > -t,$$

and for any fixed z, $\text{Re}\, z > 0$, the solutions $U_f(t, z)$ may be continued analytically onto the half-complex plane with respect to t,

$$\text{Re}\, t > -\text{Re}\, z.$$

In particular, note that the solutions $U_f(t, z)$ can also be interpreted for *negative time*.

For any fixed time t, we can obtain the inversion formula in the complex version in (8.178) by our general formula in Sect. 2.5.9, because the required situations are concretely given. Meanwhile, for any fixed space point z, we will be able to see that the situation is similar, and we naturally consider the inversion with the "complex time" t. When we wish to establish the real inversion, we can consider the inversion formulas by the *Aveiro Discretization Method in Mathematics* [92, 93] or by applying the *Tikhonov Regularization Method* [415], as in the numerical real inversion formula of the Laplace transform. Then, the analytical inversion formula is deep and complicated (cf. e.g., [388] for the real inversion formula of the Laplace transform).

8.7.3 The Case of the Operator $L(D) = D^2$

In the case of $L(D) = D^2$, we have

$$\sin \lambda x, \quad \cos \lambda x \tag{8.179}$$

as eigenfunctions with λ eigenvalues. Therefore, we can consider the sine and cosine transforms with suitable weights on the positive real line.

On the half line, in the cosine case, we have the following important reproducing kernels given by (1.40):

$$K(s,t) \equiv \int_0^\infty \frac{\cos(s\,\lambda)\cos(t\,\lambda)}{\lambda^2+1}\,d\lambda = \frac{\pi}{4}\left(\exp(-|s-t|) - \exp(-s-t)\right)$$

for $s, t > 0$. Then we have $H_K(0,\infty) = W^{1,2}(0,\infty)$ as a set and the norm is given by (1.41) according to Theorem 1.4.

Meanwhile, for K given by (1.43), $H_K(0,\infty) = \{f \in W^{1,2}(0,\infty) : f(0) = 0\}$ as a set and the norm $\|\cdot\|_{H_K(0,\infty)}$ is still given by (1.41) according to Theorem 1.5.

In addition, for the hyperbolic sine and cosine functions, we note the beautiful formulas

$$\int_0^\infty \frac{\sinh ax \sinh bx}{\cosh cx}\,dx = \frac{\pi}{c}\left(\sin\frac{a\pi}{2c}\sin\frac{b\pi}{2c}\right)\left(\cos\frac{a\pi}{c} + \cos\frac{b\pi}{c}\right)^{-1},$$

and

$$\int_0^\infty \frac{\cosh ax \cosh bx}{\cosh cx}\,dx = \frac{\pi}{c}\left(\cos\frac{a\pi}{2c}\cos\frac{b\pi}{2c}\right)\left(\cos\frac{a\pi}{c} + \cos\frac{b\pi}{c}\right)^{-1}$$

for $c > |a| + |b|$; see [189, pp. 372].

From these integral forms, we can examine the integral transforms, analytically. However, their details are needed in a separate paper especially devoted to these cases.

8.7.4 How to Obtain Eigenfunctions

We will refer to how to look for various eigenfunctions.

First, from general solutions of homogeneous equations, we can look for eigenfunctions. For example, we consider modified Bessel differential equations

$$x^2 \frac{d^2 y}{dx^2} + x \frac{dy}{dx} - (x^2 + \nu^2) y = 0 \tag{8.180}$$

which have as solutions modified Bessel functions of the second kind, i.e., $y = K_\nu$. And consequently, K_ν are eigenfunctions of the second-order differential, i.e.,

$$L_x K_\nu(x) = \nu^2 K_\nu(x) \tag{8.181}$$

with an eigenvalue ν^2, and where

$$L_x = x^2 \frac{d^2}{dx^2} + x \frac{d}{dx} - x^2. \tag{8.182}$$

Then, note that in the case that ν are positive reals, the functions

$$\exp(-\nu^2 t) K_\nu(x)$$

are the solutions of the operator equation

$$(\partial_t + L_x)\, u(t, x) = 0.$$

In addition, for pure imaginary values of ν, the functions

$$\exp(\nu^2 t) K_\nu(x) \tag{8.183}$$

are the solutions of the operator equation

$$(\partial_t - L_x)\, u(t, x) = 0. \tag{8.184}$$

Furthermore, for a pure imaginary $\nu = i\tau$, we have as eigenvalues associated to the second-order differential (8.182) $\nu^2 = -\tau^2$. For these eigenfunctions, we can discuss the detailed results following the general theory.

Next, we consider the Euler differential equations, for constants a, b,

$$x^2\frac{d^2y}{dx^2} + ax\frac{dy}{dx} = -by, \tag{8.185}$$

whose solutions are completely known as follows: We set

$$\mu \equiv \frac{1}{2}\sqrt{|(1-a)^2 - 4b|}$$

and constants as C_1 and C_2. A direct calculation shows that the solution y is given by

$$y = \begin{cases} |x|^{\frac{1-a}{2}}(C_1|x|^{\mu} + C_2|x|^{-\mu}), & (1-a)^2 > 4b \\ |x|^{\frac{1-a}{2}}[C_1 \sin(\mu \log|x|) + C_2 \cos(\mu \log|x|)], & (1-a)^2 < 4b \\ |x|^{\frac{1-a}{2}}(C_1 + C_2 \log|x|), & (1-a)^2 = 4b; \end{cases}$$

see [360, p. 248] for example. From the above general solutions, we find four eigenfunction groups: Without loss of generality, we will consider the problem on the positive real line.

Lemma 8.2. *Let* $P_a(D)y \equiv x^2y'' + axy'$.

1. *For* $(1-a)^2 > 4\lambda$,

$$y_\lambda(x) = x^{\frac{1-a}{2} + \frac{1}{2}\sqrt{(1-a)^2 - 4\lambda}} \tag{8.186}$$

and

$$y_\lambda(x) = x^{\frac{1-a}{2} - \frac{1}{2}\sqrt{(1-a)^2 - 4\lambda}} \tag{8.187}$$

are eigenfunctions.

2. *For* $(1-a)^2 < 4\lambda$,

$$y_\lambda(x) = x^{\frac{1-a}{2}} \sin\left(\frac{\sqrt{4\lambda - (1-a)^2}}{2}\log x\right), \tag{8.188}$$

and

$$y_\lambda(x) = x^{\frac{1-a}{2}} \cos\left(\frac{\sqrt{4\lambda - (1-a)^2}}{2}\log x\right). \tag{8.189}$$

Proof. We suppose $(1-a)^2 > 4\lambda$. Another case is similar. Indeed, for the operator $P_a(D)$ we have

$$P_a(D)y_\lambda = -\lambda y_\lambda. \tag{8.190}$$

Then, note that the functions

$$\exp(-\lambda t)y_\lambda$$

are the solutions of the operator equation

$$(\partial_t - P_a(D))\, u(t,x) = 0. \tag{8.191}$$

Next, we will consider the typical and famous Lalesco Picard integral equation:

$$y(x) - \lambda \int_{-\infty}^{\infty} e^{-|x-t|} y(t)dt = y(x) - \lambda \int_{-\infty}^{x} e^{x-t} y(t)dt - \lambda \int_{x}^{\infty} e^{-x+t} y(t)dt = 0 \tag{8.192}$$

for $\lambda \in (0, 1/2) \cup (1/2, \infty)$. Then, we have the general solution

$$y(x) = C_1 \exp(x\sqrt{1-2\lambda}) + C_2 \exp(-x\sqrt{1-2\lambda}), \quad 0 < \lambda < \frac{1}{2}, \tag{8.193}$$

and

$$y(x) = C_1 \cos(x\sqrt{2\lambda - 1}) + C_2 \sin(x\sqrt{2\lambda - 1}), \quad \frac{1}{2} < \lambda; \tag{8.194}$$

see [217, 360].

From the results, we can consider the four eigenfunctions and eigenvalues; therefore, without loss of generality, we will consider the case, for

$$y_\lambda(x) = \exp(-x\sqrt{1-2\lambda}), \tag{8.195}$$

and we have

$$\int_{-\infty}^{\infty} e^{-|x-t|} y_\lambda(t)dt = \frac{1}{\lambda} y_\lambda(x), \quad 0 < \lambda < \frac{1}{2}. \tag{8.196}$$

Therefore, by a suitable weight ρ, we will consider the reproducing kernel

$$\int_0^{1/2} \exp(-(x+y)\sqrt{1-2\lambda})\rho(\lambda)d\lambda. \tag{8.197}$$

We can consider many weights ρ, however, as the simplest case, we obtain the reproducing kernel

$$K(x,y) = \int_0^{1/2} \frac{\exp(-(x+y)\sqrt{1-2\lambda})}{\sqrt{1-2\lambda}} d\lambda = \frac{1 - e^{-x}e^{-y}}{x+y}$$

by letting $\rho(\lambda) \equiv \dfrac{1}{\sqrt{1-2\lambda}}$. Now, we are interested in the integral transform

$$f(x) = \int_0^{1/2} F(\lambda) \exp(-x\sqrt{1-2\lambda}) \frac{1}{\sqrt{1-2\lambda}} d\lambda \tag{8.198}$$

for the functions F satisfying the conditions

$$\int_0^{1/2} |F(\lambda)|^2 \frac{1}{\sqrt{1-2\lambda}} d\lambda < \infty. \tag{8.199}$$

Then, we can realize the image domain and the inversion formula using the Szegö kernel. For other eigenfunctions, we can obtain similar results. For other weighted functions, we can obtain more complicated results, see [388] for the details.

Next, as a typical example of Volterra integral equations, we will consider, Dixon's equation

$$y(x) - \lambda \int_0^x \frac{y(t)dt}{x+t} = f(x), \quad \lambda > 0. \tag{8.200}$$

For the homogeneous case of $f \equiv 0$, we know the solutions

$$y(x) = Cx^{\beta}; \quad \beta > -1, \tag{8.201}$$

where

$$\lambda = \frac{1}{I(\beta)}; \quad I(\beta) = \int_0^1 \frac{\xi^{\beta} d\xi}{1+\xi}, \tag{8.202}$$

[360, p. 136]. Therefore, for the integral operator

$$\int_0^x \frac{y(t)dt}{x+t} \tag{8.203}$$

we have the eigenfunctions and eigenvalues

$$y(t) = I(\lambda)t^{\lambda}, \lambda > -1. \tag{8.204}$$

Therefore, we obtain the related reproducing kernel, for $0 < x, y < 1$, for example,

$$K(x,y) = \int_{-1}^{\infty} x^{\lambda} y^{\lambda} d\lambda = -\frac{1}{xy \log xy}. \tag{8.205}$$

The property of the integral transform

$$f(x) = \int_{-1}^{\infty} x^{\lambda} F(\lambda) d\lambda \tag{8.206}$$

for the functions F satisfying

$$\int_{-1}^{\infty} |F(\lambda)|^2 d\lambda < \infty \tag{8.207}$$

will be involved, however, we see that the image f is extended analytically onto the complex plane cut by the real half line $(-\infty, 0]$ and the property at infinity.

By the complex conformal mapping $W = \log z$, the image $f(z) = f(e^w)$ of (8.206) may be discussed by the Szegö space on the strip domain

$$D \equiv \left\{ w \in \mathbb{C} \, : \, \mathrm{Im}(w) < \frac{\pi}{2} \right\}.$$

At last, as a typical example of singular integral equations, we will consider the Tricomi equations

$$y(x) - \lambda \int_0^1 \left(\frac{1}{t-x} + \frac{1}{x+t-2xt} \right) y(t) dt = f(x), \quad \lambda > 0. \tag{8.208}$$

For the homogeneous case of $f \equiv 0$, we know the solutions from [360, page 769]

$$y(x) = C \frac{(1-x)^{\beta}}{x^{1+\beta}}; \quad \tan \frac{\beta \pi}{2} = \lambda \pi, -2 < \beta < 0. \tag{8.209}$$

Therefore, for the integral operator

$$\int_0^1 \left(\frac{1}{t-x} + \frac{1}{x+t-2xt} \right) y(t) dt, \tag{8.210}$$

we have the eigenfunctions and eigenvalues

$$y(t) = \frac{(1-x)^{\frac{2}{\pi} \arctan \pi \lambda}}{\lambda x^{1+\frac{2}{\pi} \arctan \pi \lambda}}. \tag{8.211}$$

Therefore, we obtain the related reproducing kernel

$$\begin{aligned} K(x,y) &= \int_0^{\infty} \frac{(1-x)^{\frac{2}{\pi} \arctan \pi \lambda} (1-y)^{\frac{2}{\pi} \arctan \pi \lambda}}{x^{1+\frac{2}{\pi} \arctan \pi \lambda} y^{1+\frac{2}{\pi} \arctan \pi \lambda}} d\lambda \\ &= \int_0^1 \frac{(1-x)^{\xi} (1-y)^{\xi}}{2x^{1+\xi} y^{1+\xi}} \sec^2 \frac{\pi \xi}{2} d\xi \end{aligned}$$

for $0 < x, y < 1$ and for example, for $0 < \lambda < \infty$.

As typical integral equations, we stated three integral equations; however, we can consider many more integral equations, and the structures of the eigenfunctions will be mysteriously deep so that we have to analyze their structures and the integral transforms. Furthermore, we are interested in the formed reproducing kernels. See the book [360].

As a concluding remark, we stress that from the books [189, 358–360, 362] one can find many concrete problems among partial differential equations, eigenfunctions, integral transforms and reproducing kernels to which the present method can be applied. In view of this, in order to obtain some analytical results for concrete cases, we have to analyze and realize the corresponding reproducing kernel Hilbert spaces. However, for this part of the process, in most of the cases, we can also benefit from the *Aveiro Discretization Method in Mathematics* [92, 93]. For this session, the source was given by [96, 97].

Here, we discussed the initial value problems for the initial value functions belonging to the naturally determined reproducing kernel Hilbert spaces. However, by the concept of general reproducing kernels in the next section. we can discuss the initial value problems in the framework of some naturally determined Hilbert spaces. See [417, 418].

8.8 Generalized Reproducing Kernels and Generalized Delta Functions

We introduce the concept of generalized reproducing kernels and at the same time, we will be able to answer clearly the general and essential question that: what are reproducing kernels? By considering the basic problem, we will be able to obtain a general concept of the generalized delta functions as generalized reproducing kernels and, as general reproducing kernel Hilbert spaces, in particular, we can consider all separable Hilbert spaces consisting of functions.

8.8.1 *What Is a Reproducing Kernel?*

We shall consider a family of **any complex-valued functions** $\{U_n\}_{n=0}^{\infty}$ defined on an abstract set E that are linear independent. Let $N \in \mathbb{N}$. Then, we consider the form

$$K_N = \sum_{n=0}^{N} U_n \otimes \overline{U_n}. \tag{8.212}$$

Then, K_N is a *reproducing kernel* of a Hilbert space.

We consider the family of all the functions, for arbitrary complex numbers $\{C_n\}_{n=0}^{N}$

$$F = \sum_{n=0}^{N} C_n U_n \tag{8.213}$$

and we introduce the norm

$$\|F\| \equiv \sqrt{\sum_{n=0}^{N} |C_n|^2}. \tag{8.214}$$

The function space forms a Hilbert space $H_{K_N}(E)$ determined by the kernel K_N with the inner product induced from the norm (8.214), as usual. Then we note that for any $y \in E$

$$(K_N)_y = K_N(\cdot, y) \in H_{K_N}(E) \tag{8.215}$$

and for any $F \in H_{K_N}(E)$ and for any $y \in E$

$$F(y) = \langle F, (K_N)_y \rangle_{H_{K_N}(E)} = \sum_{n=0}^{N} C_n U_n(y). \tag{8.216}$$

The properties (8.215) and (8.216) is a reproducing property of the kernel $K_N(x, y)$ for the Hilbert space $H_{K_N}(E)$.

8.8.2 A Generalized Reproducing Kernel

We wish to introduce a pre-Hilbert space by

$$H_{K_\infty} \equiv \bigcup_{N \geqq 0} H_{K_N}(E),$$

where K_N is given by (8.212). For any $F \in H_{K_\infty}$, there exists a space $H_{K_M}(E)$ containing the function F for some $M \geqq 0$. Then, for any N such that $M < N$

$$H_{K_M}(E) \subset H_{K_N}(E)$$

and for the function $F \in H_{K_M}(E)$

$$\|F\|_{H_{K_M}(E)} = \|F\|_{H_{K_N}(E)}.$$

Therefore, there exists the limit

$$\|F\|_{H_{K_\infty}} \equiv \lim_{N \to \infty} \|F\|_{H_{K_N}(E)}.$$

Denote by H_∞ the completion of H_{K_∞} with respect to this norm. Note that for any $M < N$ and for any $F_M \in H_{K_M}(E)$, $F_M \in H_{K_N}(E)$ and furthermore, in particular, that

$$\langle f, g \rangle_{H_{K_M(E)}} = \langle f, g \rangle_{H_{K_N(E)}}$$

for all $N > M$ and $f, g \in H_{K_M}(E)$.

Theorem 8.18. *Under the above conditions, for any function $F \in H_\infty$ and for F_N^* defined by*

$$F_N^*(x) \equiv \langle F, K_N(\cdot, x) \rangle_{H_\infty} \quad x \in E,$$

$F_N^* \in H_{K_N}(E)$ for all $N > 0$, and as $N \to \infty$, $F_N^* \to F$ in the topology of H_∞.

Proof. Just observe that

$$|F_N^*(x)|^2 \leq \|F\|_{H_\infty}^2 \|K_N(\cdot, x)\|_{H_\infty}^2 \leq \|F\|_{H_\infty}^2 \|K_N(\cdot, x)\|_{H_{K_N}(E)}^2 = \|F\|_{H_\infty}^2 K_N(x, x).$$

Therefore, we see that $F_N^* \in H_{K_N}(E)$ and that $\|F_N^*\|_{H_{K_N}(E)} \leq \|F\|_{H_\infty}$.

Indeed, note the identity

$$K_N(x, y) = \langle K_N(\cdot, y), K_N(\cdot, x) \rangle_{H_\infty}.$$

The mapping $F \mapsto F_N^*$ being uniformly bounded, and so, we can assume that $F \in H_{K_L}(E)$ for any fixed L. However, in this case the result is clear, since $F \in H_{K_N}(E)$ and for $L < N$

$$\lim_{N \to \infty} F_N^*(x) = \lim_{N \to \infty} \langle F, K_N(\cdot, x) \rangle_{H_\infty} = \lim_{N \to \infty} \langle F, K_N(\cdot, x) \rangle_{H_{K_N}(E)} = F(x).$$

Theorem 8.18 may be looked as a reproducing kernel in the natural topology and in the sense of this theorem and the reproducing property may be written as follows

$$F(x) = \langle F, K_\infty(\cdot, x) \rangle_{H_\infty},$$

with

$$K_\infty(\cdot, x) \equiv \lim_{N \to \infty} K_N(\cdot, x) = \sum_{n=0}^{\infty} U_n(\cdot) \overline{U_n(x)}. \tag{8.217}$$

Here **the limit does, in general, not need to exist**, however, the series are non-decreasing in the sense that for any $N > M$, $K_N(y, x) \gg K_M(y, x)$.

8.8.3 *Generalized Reproducing Kernels*

Any reproducing kernel (separable case) may be considered as the form (8.217) by arbitrary linear independent functions $\{U_n\}_{n=0}^{\infty}$ on an abstract set E, here, the sum does not need converge. Furthermore, the property of linear independent is not essential. We call the form (8.217) a *general or generalized reproducing kernel* in the sense of Theorem 8.18. Recall the *double helix structure of gene* for the form (8.217). Furthermore, the form may be looked as **generalized delta functions**. The completion H_∞ may be found, in concrete cases, from the realization of the spaces $H_{K_N}(E)$.

The typical case is that the family $\{U_n\}_{n=0}^{\infty}$ is a complete orthonormal system in a Hilbert space with the norm

$$\|F\| \equiv \sqrt{\int_E |F(x)|^2 dm(x)} \tag{8.218}$$

with a dm measurable set E in the usual form $L^2(E, dm)$. Then, the functions (8.213) and the norm (8.214) are realized by this norm and the completion of the space $H_{K_\infty}(E)$ is given by this Hilbert space with the norm (8.218).

For any separable Hilbert space consisting of functions, there exists a complete orthonormal system, and therefore, in the generalized sense, for the Hilbert space there exists an approximating reproducing kernel Hilbert spaces and, the Hilbert space is the generalized reproducing kernel Hilbert space in the sense of this section.

The results will mean that the concept of the classical reproducing kernels [28, 46] was extended beautifully and completely. See [417, 418] for the fundamental applications to initial value problems using eigenfunctions and reproducing kernels.

8.9 General Integral Transforms by the Concept of Generalized Reproducing Kernels

The general integral transforms in the framework of Hilbert spaces were combined with the general theory of reproducing kernels and the basic assumption was that the integral kernels belong to some Hilbert spaces. However, as a very typical integral transform, in the case of Fourier integral transform, the integral kernel does not belong to $L^2(\mathbf{R})$; however, we can establish the isometric identity and inversion formula.

Therefore we will develop some general integral transform theory containing the Fourier integral transform that the integral kernel does not belong to any Hilbert space, based on the general concept of generalized reproducing kernels in Sect. 8.8.

8.9.1 Formulation of a Fundamental Problem

In order to formulate concretely general integral transforms, we will consider the situation of Theorem 2.47 for $T = I$.

Our basic assumption was that $h : I \times E \to \mathbb{C}$ satisfies $h_y = h(\cdot, y) \in L^2(I, dm)$ for all $y \in E$; that is, the integral kernel or linear mapping is in the framework of Hilbert spaces. Here, we assume that the integral kernel $h_y = h(\cdot, y)$ does not belong to $L^2(I, dm)$, however, for any exhaustion $\{I_t\}_{t>0}$ such that $I_t \subset I_{t'}$ for $t \leq t'$, $\bigcup_{t>0} I_t = I$, $h_y = h(\cdot, y) \in L^2(I_t, dm)$ for all $y \in E$ and $\{h_y : y \in E\}$ is complete in $L^2(I_t, dm)$ for any $t > 0$.

In the setting of Theorem 2.47, we will consider the integral transform

$$f_t(x) = \langle F, h_x \rangle_{L^2(I_t, dm)} \text{ for } \quad F \in L^2(I, dm) \tag{8.219}$$

and the corresponding reproducing kernel

$$K_t(x, y) = \langle h_y, h_x \rangle_{L^2(I_t, dm)} \quad (x, y \in E). \tag{8.220}$$

Here, we assume that $\mathcal{H}_t$ is the Hilbert space $L^2(I_t, dm)$ and $h_x \in \mathcal{H}_t$ for any x. We assume that the nondecreasing reproducing kernels $K_t(x, y)$, in the sense: for any $t' > t$, $K_{t'}(y, x) \gg K_t(y, x)$, do, in general, not converge for $\lim_{t \uparrow \infty} K_t(x, y)$. We will write, however, the limit by $K_\infty(x, y)$ formally, that is,

$$K_\infty(x, y) \equiv \lim_{t \uparrow \infty} K_t(x, y) = \langle h_y, h_x \rangle_{L^2(I, dm)} \quad (x, y \in E). \tag{8.221}$$

This integral does, in general, not exist and the limit is a special meaning. We are interesting, however, in the relationship beteween the spaces $L^2(I_t, dm)$ and $L^2(I, dm)$ associating the kernels $K_t(x, y)$ and $K_\infty(x, y)$, respectively.

At first, for the space $\mathcal{H}_t$ and the reproducing kernel Hilbert space $H_{K_t}(E)$, we recall the isometric identity in (8.219), assuming that $\{h_x : x \in E\}$ is complete in the space $\mathcal{H}_t$

$$\|f_t\|_{H_{K_t}(E)} = \|F\|_{L^2(I_t, dm)}. \tag{8.222}$$

Next note that for any $F \in L^2(I, dm)$,

$$\lim_{t \uparrow \infty} \|F\|_{L^2(I_t, dm)} = \|F\|_{L^2(I, dm)}. \tag{8.223}$$

Here, of course, the norms are nondecreasing.

As the corresponding function to $f_t \in H_{K_t}(E)$, we will consider the function, in the view point of (8.220)

$$f(x) = \langle F, h_x \rangle_{L^2(I, dm)} \text{ for } \quad F \in L^2(I, dm). \tag{8.224}$$

However, this function is not defined, because the above integral does, in general, not exist. So, we consider the function formally, tentatively. However, we are considering the correspondence

$$f_t \longleftrightarrow f \tag{8.225}$$

and

$$H_{K_t}(E) \longleftrightarrow H_{K_\infty}(E). \tag{8.226}$$

However, for the space $H_{K_\infty}(E)$, we have to give its meaning; here, when the kernel K_∞ exists by the condition $h_x \in L^2(I, dm), x \in E, H_{K_\infty}(E)$ is the reproducing kernel Hilbert space admitting the kernel K_∞.

8.9.2 Completion Property

We note the general and fundamental property.

We introduce a pre-Hilbert space by

$$H_{K_\infty} \equiv \bigcup_{t>0} H_{K_t}(E).$$

For any $f \in H_{K_\infty}$, there exists a space $H_{K_t}(E)$ containing the function f for some $t > 0$. Then, for any t' such that $t < t'$,

$$H_{K_t}(E) \subset H_{K_{t'}}(E)$$

and, for the function $f \in H_{K_\infty}$,

$$\|f\|_{H_{K_t}(E)} \geq \|f\|_{H_{K_{t'}}(E)}.$$

(Here, inequality in general holds; however, in this case equality holds, for the completeness of the integral kernel.) Therefore, there exists the limit

$$\|f\|_{H_{K_\infty}} \equiv \lim_{t'\uparrow\infty} \|f\|_{H_{K_{t'}}(E)}.$$

Denote by H_∞ the completion of H_{K_∞}.

Then, as in Theorem 8.18 we can obtain

Theorem 8.19. *For the general situation such that $K_t(x, y)$ exists for all $t > 0$ and $K_\infty(x, y)$ does, in general, not exist, for any function $f \in H_\infty$*

$$\lim_{t\uparrow\infty} \langle f, (K_t)_x \rangle_{H_\infty} = f(x) \quad (x \in E), \tag{8.227}$$

in the space H_∞.

Theorem 8.19 may be looked as a reproducing kernel in the natural topology and by the sense of Theorem 8.19, and the reproducing property may be written as follows:

$$f(x) = \langle f, K_\infty(\cdot, x) \rangle_{H_\infty} \quad (x \in E),$$

with (8.227). Here **the limit** $K_\infty(\cdot, x)$ **does, in general, not need exist**; however, the series are nondecreasing.

The completion space H_∞ will be determined in many concrete cases, from the realizations of the spaces $H_{K_t}(E)$ by case by case.

8.9.3 Convergence of $f_t(x) = \langle F, h_x \rangle_{L^2(I_t, dm)}$ for $F \in L^2(I, dm)$

As in the case of Fourier integral, we will prove the convergence of (8.219) in the completion space H_∞. Indeed, for any $t, t' > 0 \quad (t < t')$ we have

$$\begin{aligned}\lim_{t,t'\to\infty} \|f_t - f_{t'}\|^2_{H_\infty} &= \lim_{t,t'\to\infty} \|f_t - f_{t'}\|^2_{H_{K_t}(E)} \\ &\le \lim_{t,t'\to\infty} \left(\|f_t\|^2_{H_{K_t}(E)} + \|f_{t'}\|^2_{H_{K_{t'}}(E)} - 2\|f_t\|^2_{H_{K_t}(E)} \right) \\ &= 0.\end{aligned}$$

In this sense, as in the Fourier integral of the cace $L^2(\mathbb{R}, dx)$ we will write for

$$\lim_{t\uparrow\infty} f_t = f \quad \text{in} \quad H_\infty$$

as follows:

$$f(x) = \lim_{t\uparrow\infty} \langle F, h(\cdot, x) \rangle_{L^2(I_t, dm)} = \langle F, h(\cdot, x) \rangle_{L^2(I, dm)}. \tag{8.228}$$

8.9.4 Inversion of Integral Transforms

We will consider the inversion of the integral transform (8.228) from the space H_∞ onto $L^2(I, dm)$. For any $f \in H_\infty$, we take functions $f_t \in H_{K_t}(E)$ such that

$$\lim_{t\uparrow\infty} f_t = f$$

in the space H_∞. This is possible, because the space H_∞ is the completion of the spaces $H_{K_t}(E)$. However, f_t may be constructed by Theorem 8.19 in the form

$$f_t(x) \equiv \langle f, K_t(\cdot, x) \rangle_{H_\infty}.$$

For the functions $f_t \in H_{K_t}(E)$, by Theorem 2.47–we are assuming the conditions in Theorem 2.47 – we can construct the inversion in the following way

$$F_t(\lambda) = \lim_{N\to\infty} \int_{E_N} f_t(x) h(\lambda, x)\, d\mu_t(x) \tag{8.229}$$

in the topology of $L^2(I, dm)$ satisfying

$$f_t(x) = \langle F_t, h_x \rangle_{L^2(I_t, dm)} \quad (x \in E). \tag{8.230}$$

Here, of course, the function F_t of $L^2(I, dm)$ is the zero extension of a function F_t of $L^2(I_t, dm)$. Note that the isometric relation that for any $t < t'$

$$\|f_t - f_{t'}\|_{H_0} = \|F_t - F_{t'}\|_{L^2(I, dm)}. \tag{8.231}$$

Then, we see the desired result: The functions F_t converse to a function F in $L^2(I, dm)$ and

$$f(x) = \langle F, h_x \rangle_{L^2(I, dm)} \quad (x \in E) \tag{8.232}$$

in our sense. We can write down the inversion formula as follows

$$F(\lambda) = \lim_{t \uparrow \infty} \lim_{N \to \infty} \int_{E_N} \langle f, (K_t)_x \rangle_{H_\infty} h(\lambda, x)\, d\mu_t(x), \tag{8.233}$$

where the both limits $\lim_{N \to \infty}$ and $\lim_{t \uparrow \infty}$ are taken in the sense of the space $L^2(I, dm)$.

Of course, the correspondence $f \in H_\infty$ and $F \in L^2(I, dm)$ is one to one. Indeed, we assume that $f \in H_\infty, f \equiv 0$, then

$$0 \equiv f(x) = \lim_{t \uparrow \infty} \langle f, K_t(\cdot, x) \rangle_{H_{K_t}(E)} = \lim_{t \uparrow \infty} f_t(x) \tag{8.234}$$

in the space H_∞ ; that is,

$$\lim_{t \uparrow \infty} \|f_t\|_{H_{K_t}(E)} = 0 = \lim_{t \uparrow \infty} \|F\|_{L^2(I, dm)}; \tag{8.235}$$

this implies the desired result that $F = 0$ on the space $L^2(I, dm)$.

8.9.5 Discrete Versions

We will refer to a typical discrete version under a very general situation. Let the family $\{U_n(x)\}_{n=0}^{\infty}$ be a complete orthonormal system in a Hilbert space with the norm

$$\|F\| \equiv \sqrt{\int_E |F(x)|^2 dm(x)} \tag{8.236}$$

with a dm measurable set E in the usual form $L^2(E, dm)$. We will consider the family of all functions, for arbitrary complex numbers $\{C_n\}_{n=0}^N$

$$F = \sum_{n=0}^{N} C_n U_n \tag{8.237}$$

and we introduce the norm

$$\|F\| \equiv \sqrt{\sum_{n=0}^{N} |C_n|^2}. \tag{8.238}$$

Then, the function space forms a Hilbert space $H_{K_N}(E)$ determined by the reproducing kernel $K_N(x, y)$

$$K_N = \sum_{n=0}^{N} U_n \otimes \overline{U_n} \tag{8.239}$$

with the inner product induced from the norm (8.238), as usual. Then, the functions in the Hilbert space $L^2(E, dm)$ and the norm (8.238) are realized as the completion $H_{K_\infty}(E)$ of the spaces $H_{K_N}(E)$. In this case, for the correspondence:

$$\ell^2 : \{C_n\}_{n=0}^N \leftrightarrow F = \sum_{n=0}^{N} C_n U_n, \tag{8.240}$$

we obtain the same results in the classical analysis and in this section.

We can consider such linear mappings for arbitrary functions $\{U_n\}_{n=0}^N$ which are linear independent and by considering the kernel forms (8.239); however, the realization of the completion space H_∞ becomes the crucial problem, in the new approach.

Appendices

A.1 Equality Problems for Norm Inequalities

In this section, we will introduce the general theory of reproducing kernels that may be considered as the deepest theory in the general theory of reproducing kernels.

A.1.1 *Introduction of the Results by Akira Yamada*

We will introduce the quite general theory by A. Yamada [485] for some general conditions for norm inequalities derived from the theory of reproducing kernels.

In 1965, Lebedev and Milin [281] found the following inequality: If $f \in \mathscr{O}(\Delta(1))$ and e^f have Taylor expansion $f(z) = \sum_{n=1}^{\infty} a_n z^n$ and $e^{f(z)} = \sum_{n=0}^{\infty} b_n z^n$, respectively, then

$$\sum_{n=0}^{\infty} |b_n|^2 \leq \exp\left(\sum_{n=1}^{\infty} n|a_n|^2\right),$$

or equivalently

$$\sum_{n=1}^{\infty} |b_n|^2 \leq \exp\left(\sum_{n=1}^{\infty} n|a_n|^2\right) - 1, \tag{A.1}$$

where equality occurs if and only if there exists $\rho \in \Delta(1)$ with $a_n = \rho^n/n$ for all n. This is a prototype of the inequalities treated in Sect. A.1; that is, we can rewrite the above inequality as

$$\|e^f\|_{H^2}^2 \leq \exp(\|f\|_D^2), \tag{A.2}$$

S. Saitoh, Y. Sawano, *Theory of Reproducing Kernels and Applications*, Developments in Mathematics 44, DOI 10.1007/978-981-10-0530-5

where

$$\|f\|_D = \sqrt{\frac{1}{\pi}\iint_{\Delta(1)} |f'(z)|^2 dxdy}$$

is the Dirichlet norm, and

$$\|e^f\|_{H^2} = \sqrt{\sum_{n=0}^{\infty} |b_n|^2}$$

is the Hardy H^2-norm of the function e^f. See Proposition 1.2. Also we remark that in this case we have the identity

$$k_{H^2}(z,w) = e^{k_D(z,w)} = \frac{1}{1 - z\bar{w}}, \quad z, w \in \Delta(1), \tag{A.3}$$

where k_{H^2} and k_D are the reproducing kernels for the Hardy H^2-space and the Dirichlet space D on $\Delta(1)$ normalized by $f(0) = 0$, respectively. See Theorem 1.15 for the definition of k_D. Moreover, the above equality condition is equivalent to $f(z) = k_D(z,q)$ for some $q \in \Delta(1)$. Therefore, there exists a deep connection between the Lebedev Milin inequality and reproducing kernels.

Before we start to develop the problem, we recall a generality: Let $\psi(z) = \sum_{n=1}^{\infty} p_n z^n$ be an entire function such that $p_n \geq 0$ for all $n \in \mathbb{N}$. If $H_K(E)$ is a complex RKHS on a set E with the reproducing kernel K and the norm $\|\cdot\|_K$, then there exists a unique RKHS $H_{\psi^* K}(E)$ on E with the reproducing kernel $\psi^* K$, which is given by

$$\psi^* K = \psi(K(\cdot,\cdot)).$$

See Proposition 2.6 and Corollary 2.4 for the details.

Keeping this in mind, we consider another entire function φ such that $\varphi(0) = 0$. We expand

$$\varphi(z) = \sum_{n=1}^{\infty} c_n z^n$$

and we assume:

1. $p_n \geq 0$ for all n
2. $c_n = 0$ whenever $p_n = 0$.

Putting

$$\varphi_\psi(z) \equiv \sum_{p_n > 0} \frac{|c_n|^2}{p_n} z^n, \tag{A.4}$$

we have the following norm inequality which generalizes the Lebedev Milin inequality (A.1) [65, 388]: For all $f \in H_K(E)$,

$$\|\varphi(f)\|_{H_{\psi * K}(E)} \leq \sqrt{\varphi_\psi(\|f\|^2_{H_K(E)})}. \tag{A.5}$$

In the special case where $\varphi = \psi$, it is easy to see that if $f = K_q$ for some $q \in E$, then equality holds in (A.5) [65]. The converse of this fails in general. See [381] for counterexamples. However, by studying various special but important RKHSs, there are many papers [65–68, 373–376, 380, 381, 500] asserting that equality occurs in (A.5) if and only if $f = K_q$ for some $q \in E$. The most concrete case was dealt with [373, Section 31] in detail. In order to investigate the condition for equality, all these papers relied on case-by-case arguments. Yamada gave a general and satisfactory theory of equality conditions for such norm inequalities. In order to see his theory, we will introduce a class of RKHSs called *algebra-dense* and study relations between equality conditions and $\mathbb{C}$-algebra homomorphisms.

Let $m \geq 2$. Let $H_j(E) = H_{K^{(j)}}(E)$ be a complex RKHS on the set E with the reproducing kernel $K^{(j)}$ for $j = 1, 2, \ldots, m$. Then, according to Theorem 2.20, the Hilbert tensor product

$$H \equiv \otimes_{j=1}^m H_j(E)$$

is an RKHS on

$$E^m \equiv \prod_{j=1}^m E.$$

Definition A.1.

1. The set $E^m_{\rm d}$ denotes the *diagonal* $\{(x, x, \ldots, x) : x \in E\}$ of the set E^m.
2. Denote by H_0 the subspace of H defined by $\{f \in H : f|_{E^m_{\rm d}} = 0\}$.
3. For $f, g \in H$ define an equivalence relation "$\sim$" by $f \sim g$ if and only if f and g agree on the diagonal $E^m_{\rm d}$.

Definition A.2. An element $\phi \in H$ is said to be *extremal* if $\phi \in (H_0)^\perp$, or equivalently, ϕ is extremal if and only if $f \sim g$ implies $\langle f, \phi \rangle_H = \langle g, \phi \rangle_H$.

Remark A.1. The definition of the extremality above is closely related to equality conditions of norm inequalities for the tensor product. As we see easily, if H' denotes the unique RKHS with the kernel $\prod_{j=1}^m K^{(j)}_x$, $x \in E$, then H' consists of functions on E induced from the restrictions of functions in H to the diagonal $E^m_{\rm d}$. For $\phi_j \in H_j(E)$ with $j = 1, 2 \ldots, m$, we have

$$\|\phi_1\phi_2 \cdots \phi_m\|_{H'} \leq \|\phi_1 \otimes \phi_2 \otimes \cdots \otimes \phi_m\|_H = \|\phi_1\|_{H_1(E)}\|\phi_1\|_{H_2(E)} \cdots \|\phi_m\|_{H_m(E)}, \tag{A.6}$$

where equality occurs if and only if $\otimes_{j=1}^m \phi_j$ is extremal (cf. Sect. A.1.2 as well as Remark 2.7).

Here and below we denote $\mathbb{C}^E \equiv \mathscr{F}(E)$.

Lemma A.1. *Let R be a subalgebra of $\mathbb{C}^E$. For $j = 1, 2, \ldots, m$, suppose that we are given $f_j, g_j, \phi_j \in H_j(E)$. Assume that $\phi \equiv \otimes_{j=1}^m \phi_j \in (H_0)^\perp \setminus \{0\} \subset \otimes_{j=1}^m H_j(E)$ is nonzero extremal.*

Then

$$\prod_{j=1}^m \langle f_j, \phi_j \rangle_{H_j(E)} = \prod_{j=1}^m \langle g_j, \phi_j \rangle_{H_j(E)}, \tag{A.7}$$

if $\otimes_{j=1}^m f_j \sim \otimes_{j=1}^m g_j$, namely

$$\prod_{j=1}^m f_j = \prod_{j=1}^m g_j \tag{A.8}$$

on E.

Proof. Recall that the inner product of the tensor product satisfies

$$\langle \otimes_{j=1}^m \tilde{\psi}_j, \otimes_{j=1}^m \psi_j \rangle_H = \prod_{j=1}^m \langle \tilde{\psi}_j, \psi_j \rangle_{H_j(E)} \tag{A.9}$$

for all elements $\{\tilde{\psi}_j\}_{j=1}^m$, and $\{\psi_j\}_{j=1}^m \in H_1(E) \otimes H_2(E) \otimes \ldots \otimes H_m(E)$. Observe also that (A.8) implies $\otimes_{j=1}^m f_j \sim \otimes_{j=1}^m g_j$. Since ϕ is extremal, (A.7) follows.

In words of the paper [377], we recall several definitions:

Definition A.3. Let $H_j(E)$ be RKHSs on E for $j = 1, 2, \ldots, m$. Then the tensor product $H = \otimes_{j=1}^m H_j(E)$ is called *regular*, if for every nonzero extremal $\otimes_{j=1}^m \phi_j \in H$, there exists a point $q \in E$ such that $\phi_j \in \mathrm{Span}\{K_q^{(j)}\}$ for all $j = 1, 2, \ldots, m$. Also, H is called *weakly regular*, if for every nonzero extremal $\otimes_{j=1}^m \phi_j \in H$, there exists a point $q \in E$ such that for each $j = 1, 2, \ldots, m$, either of the following holds:

1. q is a common zero of the functions in $H_j(E)$,

$$f(q) = 0 \tag{A.10}$$

for all $f \in H_j(E)$ with $j = 1, 2, \ldots, m$, or
2. $\phi_j \in \mathrm{Span}\{K_q^{(j)}\}$, that is, ϕ_j is a constant multiple of $K_q^{(j)}$.

For later references, we will call (A.10) above the *exceptional case*; this is independent of the functions ϕ_j.

In what follows, we always assume that R denotes a $\mathbb{C}$-subalgebra of $\mathbb{C}^E$, an algebra made up of complex-valued functions on E. The existence of the identity of R is not assumed. For a complex subspace H of $\mathbb{C}^E$, let $R^{-1}H$ denote the subspace of H defined by

$$R^{-1}H \equiv \{f \in H : rf \in H \text{ for all } r \in R\}.$$

Remark that under this definition $R^{-1}H$ is always a subset of H.

Lemma A.2. *Let R be a subalgebra of $\mathbb{C}^E$. Assume that $\phi = \otimes_{j=1}^m \phi_j \in (H_0)^\perp \setminus \{0\} \subset \otimes_{j=1}^m H_j(E)$ is nonzero extremal. If each $R^{-1}H_j(E)$ is dense in $H_j(E)$ ($j = 1, 2, \ldots, m$), then there exists a unique $\mathbb{C}$-algebra homomorphism $\Lambda_\phi : R \to \mathbb{C}$ satisfying, for every $j = 1, 2, \ldots, m$,*

$$\langle fu, \phi_j \rangle_{H_j(E)} = \Lambda_\phi(f) \langle u, \phi_j \rangle_{H_j(E)} \tag{A.11}$$

for every $f \in R$ and for every $u \in R^{-1}H_j(E)$ ($j = 1, 2, \ldots, m$).

Proof. Since $R^{-1}H_j(E)$ is assumed to be dense in $H_j(E)$ and $\phi_j \neq 0$, we can find an element $u_j \in R^{-1}H_j(E)$ with

$$\langle u_j, \phi_j \rangle_{H_j(E)} \neq 0. \tag{A.12}$$

Fixing such an element u_j for each j below, we defined $\Lambda_\phi(f)$ by

$$\Lambda_\phi(f) \equiv \frac{\langle fu_j, \phi_j \rangle_{H_j(E)}}{\langle u_j, \phi_j \rangle_{H_j(E)}} \tag{A.13}$$

for $f \in R$. Now we show that $\Lambda_\phi(f)$ is well defined, that is, Λ_ϕ is determined despite the ambiguity of the choice of j and u_j. For $f \in R$ define $f_k,\ g_k \in H_k(E)$ ($k = 1, 2, \ldots, m$) by

$$f_k \equiv \begin{cases} fu_i & (k = i) \\ u_k & (k \neq i) \end{cases}, \quad g_k \equiv \begin{cases} fu_j & (k = j) \\ u_k & (k \neq j) \end{cases}. \tag{A.14}$$

From (A.7) and (A.12), we have

$$\langle fu_i, \phi_i \rangle_{H_i(E)} \langle u_j, \phi_j \rangle_{H_j(E)} = \langle u_i, \phi_i \rangle_{H_i(E)} \langle fu_j, \phi_j \rangle_{H_j(E)} \tag{A.15}$$

from (A.12), since

$$u_1 \otimes \cdots \otimes u_{i-1} \otimes fu_i \otimes u_{i+1} \otimes \cdots \otimes u_m \sim u_1 \otimes \cdots \otimes u_{j-1} \otimes fu_j \otimes u_{j+1} \otimes \cdots \otimes u_m$$

and $\otimes_{j=1}^m \phi_j$ is extremal. Thus, for all $f \in R$ and i, j, from (A.12) and (A.15), we deduce that

$$\frac{\langle fu_i, \phi_i\rangle_{H_i(E)}}{\langle u_i, \phi_i\rangle_{H_i(E)}} = \frac{\langle fu_j, \phi_j\rangle_{H_j(E)}}{\langle u_j, \phi_j\rangle_{H_j(E)}}. \tag{A.16}$$

Similarly, for $f, g \in R$, setting

$$f_k \equiv \begin{cases} fu_i & (k = i) \\ gu_j & (k = j) \\ u_k & (k \neq i, j) \end{cases}, \quad g_k \equiv \begin{cases} fgu_i & (k = i) \\ u_k & (k \neq i) \end{cases},$$

we have

$$f_1 \otimes f_2 \otimes \cdots \otimes f_k \sim g_1 \otimes g_2 \otimes \cdots \otimes g_k$$

and hence

$$\langle fu_i, \phi_i\rangle_{H_i(E)} \langle gu_j, \phi_j\rangle_{H_j(E)} = \langle fgu_i, \phi_i\rangle_{H_i(E)} \langle u_j, \phi_j\rangle_{H_j(E)}.$$

Hence, from (A.16), we deduce that

$$\frac{\langle fgu_i, \phi_i\rangle_{H_i(E)}}{\langle u_i, \phi_i\rangle_{H_i(E)}} = \frac{\langle fu_i, \phi_i\rangle_{H_i(E)} \langle gu_j, \phi_j\rangle_{H_j(E)}}{\langle u_i, \phi_i\rangle_{H_i(E)} \langle u_j, \phi_j\rangle_{H_j(E)}} = \frac{\langle fu_i, \phi_i\rangle_{H_i(E)} \langle gu_i, \phi_i\rangle_{H_i(E)}}{\langle u_i, \phi_i\rangle_{H_i(E)} \langle u_i, \phi_i\rangle_{H_i(E)}}. \tag{A.17}$$

Therefore, we have proved by (A.17) that the linear functional Λ_ϕ on R is multiplicative and that its definition makes sense by (A.16). Since the right-hand side of (A.13) remains unchanged if we multiply ϕ_j by any nonzero constant, Λ_ϕ is dependent only on the tensor product ϕ. Hence $\Lambda_\phi \colon R \to \mathbb{C}$ is a well-defined $\mathbb{C}$-algebra homomorphism. The uniqueness of Λ_ϕ is clear from the definition and the assumption that $R^{-1}H_j$ is dense in H_j.

In view of the definition (A.13), (A.11) holds when $\langle u, \phi_j\rangle_{H_j(E)} \neq 0$. Thus, to show that the identity (A.11) holds for every $u \in R^{-1}H_j(E)$, it suffices only to show that if $\langle u, \phi_j\rangle_{H_j(E)} = 0$, then $\langle fu, \phi_j\rangle_{H_j(E)} = 0$. Indeed, from (A.15) we have, for $k \neq j$,

$$\langle fu, \phi_j\rangle_{H_j(E)} \langle u_k, \phi_k\rangle_{H_k} = \langle u, \phi_j\rangle_{H_j(E)} \langle fu_k, \phi_k\rangle_{H_k} = 0.$$

Thus $\langle fu, \phi_j\rangle_{H_j(E)} = 0$, since $\langle u_k, \phi_k\rangle_{H_k} \neq 0$.

Remark A.2. If R has the identity, then from (A.11) we have $\Lambda_\phi(1) = 1$.

Given complex subspaces R_1 and R_2 of R, let $R_1 \cdot R_2$ denote the complex subspace of R given by

$$R_1 \cdot R_2 \equiv \bigcup_{N=1}^{\infty} \left\{ \sum_{i=1}^{N} a_i b_i \,:\, a_i \in R_1 \text{ and } b_i \in R_2 \right\}. \tag{A.18}$$

Definition A.4. Let R be a subalgebra of $\mathbb{C}^E$. An RKHS H on E is called *R-dense* if $R \cdot (R \cap H)$ is densely contained in H. If H is R-dense for some $\mathbb{C}$-algebra R on E, H is called *algebra-dense*.

Remark A.3. If H is R-dense, then $R \cap H$ is both (1) a dense subspace of H and (2) an ideal of R. Moreover, if $1 \in R$, then H is R-dense if and only if (1) and (2) hold.

Indeed, if H is R-dense, then $R \cdot (R \cap H) \subset R \cap H \subset H$. Hence $R \cap H$ is dense in H. Also, since H is R-dense, we have $R \cdot (R \cap H) \subset H$ and since R is an algebra, we have $R \cdot (R \cap H) \subset R$. Thus, $R \cap H$ is an ideal of R. Conversely, suppose $1 \in R$ and that (1) and (2) hold. Then $R \cdot (R \cap H) = R \cap H$ is a dense subspace of H, which implies H is R-dense.

Definition A.5. Let R be a subalgebra of $\mathbb{C}^E$ and let H be an R-dense RKHS on E. Let $\chi: R \to \mathbb{C}$ be a $\mathbb{C}$-algebra homomorphism. If there exists a constant $C > 0$ such that

$$|\chi(f)| \leq C\|f\|_H \text{ for all } f \in R \cap H, \tag{A.19}$$

χ is called an *H-bounded homomorphism* of R. The set of nonzero H-bounded homomorphisms of R is called an *H-hull* of E and is denoted by $\hat{E}_H$.

For example, $\chi = \mathrm{ev}_q$ is an *H-bounded homomorphism* of R; see Lemma 2.1.

If $H_j(E)$ $(j = 1, 2, \ldots, m)$ is R-dense, then $R \cap H_j(E)$ is a dense R-invariant subspace of $H_j(E)$ and we can apply Lemma A.2 to the tensor product $\otimes_{j=1}^m H_j(E)$. In fact, $R^{-1}H_j(E)$ contains $R \cap H_j(E)$ and $R \cap H_j(E)$ is assumed dense in $H_j(E)$. Then, we have the main result:

Theorem A.1. *Let R be a subalgebra of $\mathbb{C}^E$. For each $j = 1, 2, \ldots, m$, assume that $H_j(E)$ is an R-dense RKHS on E. If $\phi = \otimes_{j=1}^m \phi_j \in \otimes_{j=1}^m H_j(E)$ is nonzero extremal, then there exists a unique nonzero $\mathbb{C}$-algebra homomorphism $\Lambda_\phi \in \bigcap_{j=1}^m \hat{E}_{H_j(E)}$ satisfying* (A.11). *Furthermore, for each $j = 1, 2, \ldots, m$, either of the following holds:*

1. *$\Lambda_\phi | R \cap H_j(E) = 0$.*
2. *For every $j = 1, 2, \ldots, m$, there exists a constant $C_j \neq 0$ such that*

$$\langle f, \phi_j \rangle_{H_j(E)} = C_j \Lambda_\phi(f) \tag{A.20}$$

for each $f \in R \cap H_j(E)$.

Proof. Let Λ_ϕ be a mapping from Lemma A.2. We claim this Λ_ϕ is the one we are looking for.

1. We must show that $\Lambda_\phi \neq 0$. Since $R \cdot (R \cap H_1(E))$ is dense in $H_1(E)$, there exists finite subsets $\{f_k\}_{k=1}^N \in R$ and $\{g_k\}_{k=1}^N \in R \cap H_1(E)$ such that

$$\left\langle \sum_{k=1}^N f_k g_k, \phi_1 \right\rangle_{H_1(E)} = \sum_{k=1}^N \langle f_k g_k, \phi_1 \rangle_{H_1(E)} \neq 0.$$

Thus there exists an index k_0 such that

$$\langle f_{k_0} g_{k_0}, \phi_1 \rangle_{H_1(E)} = \Lambda_\phi(f_{k_0}) \langle g_{k_0}, \phi_1 \rangle_{H_1(E)} \neq 0.$$

Hence $\Lambda_\phi(f_{k_0}) \neq 0$ and so $\Lambda_\phi \neq 0$.

2. We look for C_j satisfying (A.20) assuming that $\Lambda_\phi | R \cap H_j(E) \neq 0$. Then there exists an element $f_0^* \in R \cap H_j(E)$ such that $\Lambda_\phi(f_0^*) \neq 0$. Since $R \cdot R \cap H_j(E)$ is dense in $H_j(E)$, there exist elements $f_0 \in R$ and $g \in R \cap H_j(E) \subset R^{-1}H_j(E)$ with $\langle f_0 \cdot g, \phi_j \rangle_{H_j(E)} \neq 0$. For every $f \in R \cap H_j(E)$, Lemma A.2 implies the identity

$$\langle fg, \phi_j \rangle_{H_j(E)} = \Lambda_\phi(f_0) \langle g, \phi_j \rangle_{H_j(E)}. \tag{A.21}$$

Note that

$$\Lambda_\phi(g) \langle f, \phi_j \rangle_{H_j(E)} = \Lambda_\phi(f) \langle g, \phi_j \rangle_{H_j(E)} = \langle fg, \phi_j \rangle_{H_j(E)}$$

from (A.21) when f belongs to $R \cap H_j(E)$ as well. Putting $f = f_0 \in R$ above, we obtain $\Lambda_\phi(g) \neq 0$. Hence, (A.20) holds if we set

$$C_j \equiv \frac{\langle g, \phi_j \rangle_{H_j(E)}}{\Lambda_\phi(g)} \neq 0.$$

3. Now it is clear that Λ_ϕ is an $H_j(E)$-bounded homomorphism of R. For, if $\Lambda_\phi|_{R \cap H_j(E)} = 0$, then this is trivial, otherwise this follows from Schwarz's inequality. Thus, $\Lambda_\phi \in \hat{E}_{H_j(E)}$ for every $j = 1, 2, \ldots, m$.

Definition A.6. Let R be a subalgebra of $\mathbb{C}^E$ and let H be an R-dense RKHS on E. Then H is called *maximal* if every nonzero H-bounded homomorphism of R is a point evaluation of R at some point in E. If one needs to specify the algebra R when H is maximal, then H is called *R-maximal*.

We defined the generalized ℓ^2-space over any set E in Example 2.4. Here we can prove that this is maximal.

Example A.1. Let E be a set. Let $\ell^2(E)$ be the complex Hilbert space given by

$$\ell^2(E) \equiv \left\{ f \in \mathbb{C}^E : \sum_{x \in E} |f(x)|^2 < \infty \right\},$$

whose inner product is given by

$$\langle f, g \rangle_{\ell^2(E)} = \sum_{x \in E} f(x) \overline{g(x)},$$

for $f, g \in \ell^2(E)$. It is easy to see that $\ell^2(E)$ is a $\mathbb{C}$-subalgebra of $\mathbb{C}^E$(without the identity). Putting $R \equiv \ell^2(E)$ we will show that $\ell^2(E)$ is R-dense and maximal. From $\delta_x^2 = \delta_x$ we see that $R \cdot R$ is dense in R, which implies that $\ell^2(E)$ is R-dense. Let $\chi: R \to \mathbb{C}$ be a nonzero $\mathbb{C}$-algebra homomorphism. From $\delta_x^2 = \delta_x$, $\chi(\delta_x)$ is equal to 0 or 1. Also, $\delta_x\delta_y = 0$ $(x \neq y)$ implies that $\chi(\delta_x)\chi(\delta_y) = 0$ $(x \neq y)$. From $\chi \not\equiv 0$, we conclude that there uniquely exists a point $q \in E$ such that $\chi(\delta_x) = \delta_x(q)$ for $x \in E$. Since the span of δ_x's is dense in R, $\chi(f) = f(q)$ for all $f \in R$. Thus $\ell^2(E)$ is maximal.

As a corollary to Theorem A.1, we have:

Corollary A.1. *Let $H_j(E)$ $(j = 1, 2, \ldots, m)$ be R-dense RKHSs on E. If $H_{j_0}(E)$ is maximal for some $j_0 = 1, 2, \ldots, m$, then their tensor product $\otimes_{j=1}^m H_j(E)$ is weakly regular.*

Proof. Let $\phi_1 \in H_1(E)$, $\phi_2 \in H_2(E)$, ..., $\phi_m \in H_m(E)$ and suppose that $\phi \equiv \otimes_{j=1}^m \phi_j$ is nonzero extremal in $\otimes_{j=1}^m H_j(E)$. Then by Theorem A.1 the algebra homomorphism Λ_ϕ is an $H_j(E)$-bounded homomorphism of R $(j = 1, 2, \ldots, m)$. Since $H_{j_0}(E)$ is maximal, Λ_ϕ is a point evaluation of R at some point $q \in E$; $\Lambda_\phi(f) = f(q)$.

We fix $j = 1, 2, \ldots, m$ until the end of the proof. Assume that q is not a common zero of $H_j(E)$, that is, $\tilde{f}(q) \neq 0$ for $\tilde{f} \in H_j(E)$. We prove $\phi_j \in \mathrm{Span}\{K_q^{(j)}\}_{q \in E}$. Since $R \cap H_j(E)$ is dense in $H_j(E)$, $\Lambda_\phi | R \cap H_j(E) \neq 0$. Thus, there exists a constant $C_j \neq 0$ such that $\langle f, \phi_j \rangle_{H_j(E)} = C_j f(q)$ for all $f \in R \cap H_j(E)$. Since $H_j(E)$ is R-dense, for each $f \in H_j(E)$ there exists a sequence $f_n \in R \cap H_j(E)$ $(n = 1, 2, \ldots)$ such that f_n converges strongly to f as $n \to \infty$. From $\langle f_n, \phi_j \rangle_{H_j(E)} = C_j f_n(q)$, letting $n \to \infty$ we have $\langle f, \phi_j \rangle_{H_j(E)} = C_j f(q)$ for $f \in H_j(E)$. Since $C_j \neq 0$, it follows that ϕ_j induces a constant multiple of the point evaluation of $H_j(E)$ at $q \in E$. Thus, ϕ_j is the reproducing kernel of $H_j(E)$ at q up to a nonzero multiplicative constant. Hence, $\otimes_{j=1}^m H_j(E)$ is weakly regular.

A.1.2 Equality Conditions for the Norm Inequalities

Before studying equality conditions we first recall some facts from the theory of reproducing kernels. Let $A : H_1 \to H_2$ be a linear map from a Hilbert space H_1 into a linear space H_2 with closed kernel $\ker(A) = A^{-1}(0)$. The *range norm* of Ran(A) is the norm which makes A a partial isometry from H_1 onto Ran(A):

$$\|Ax\|_{\mathrm{Ran}(A)} = \|x\|_{H_1}$$

for $x \in H_1 \cap \ker(A)^\perp$. In fact, the range Ran(A) equipped with this range norm is a Hilbert space isomorphic to $H_1 \ominus \ker(A)$ and is called the *operator range* of the map A. With these terminologies, the RKHS $H_{K_1+K_2}(E)$ is the operator range of the map

$$(f,g) \in H_{K_1}(E) \oplus H_{K_2}(E) \mapsto f+g \in H_{K_1+K_2}(E).$$

See (2.88). Hence, we have the Pythagorean inequality (see (2.95))

$$\|f+g\|^2_{H_{K_1+K_2}(E)} \leq \|f\|^2_{H_{K_1}(E)} + \|g\|^2_{H_{K_2}(E)}, \tag{A.22}$$

for all $f_1 \in H_{K_1}(E)$ and $f_2 \in H_{K_2}(E)$, where equality holds if and only if

$$\langle f_1, h\rangle_{H_{K_1}(E)} = \langle f_2, h\rangle_{H_{K_2}(E)} \tag{A.23}$$

for all $h \in H_{K_1}(E) \cap H_{K_2}(E)$ according to (2.89). Also, the RKHS $H_{K_1K_2}(E)$ is the operator range of the mapping

$$H_{K_1}(E) \otimes H_{K_2}(E) \ni \sum_i f_i \otimes g_i \mapsto \sum_i f_i g_i \in H_{K_1K_2}(E)$$

which is induced by the restriction map from $E \times E$ to its diagonal. Hence, for all $f_1 \in H_{K_1}(E)$ and $f_2 \in H_{K_2}(E)$

$$\|fg\|_{H_{K_1K_2}(E)} \leq \|f \otimes g\|_{H_{K_1 \otimes K_2}(E)} = \|f\|_{H_{K_1}(E)}\|g\|_{H_{K_2}(E)}, \tag{A.24}$$

where equality holds if and only if $f \otimes g \in (H_{K_1}(E) \otimes H_{K_2}(E))_0^{\perp}$. See Example 2.3 and (2.129). We note that if $\otimes_{j=1}^J \phi_j$ is extremal in $\otimes_{j=1}^J H_j$, then by the basic properties of the range norm and the tensor product, for any $\otimes_{j=1}^J g_j \in \otimes_{j=1}^J H_j$ we have

$$\langle \Pi_{j=1}^J g_j, \Pi_{j=1}^J \phi_j\rangle_{H_{\Pi_{j=1}^J K_j}(E^m)} = \left\langle \otimes_{j=1}^J g_j, \otimes_{j=1}^J \phi_j\right\rangle_{H_{\otimes_{j=1}^J K_j}(E)} = \Pi_{j=1}^J \langle g_j, \phi_j\rangle_{H_{K_j}(E)}$$

from the definition of the operator range. As a special case of what we have obtained we note that

$$\|f\|_{H_{cK}(E)} = \sqrt{c^{-1}}\|f\|_{H_K(E)} \tag{A.25}$$

for any positive constant c and hence

$$\sum_{p_n>0} \|c_n f^n\|^2_{H_{p_nK^n}(E)} = \sum_{p_n>0} \frac{|c_n|^2}{p_n}\|f^n\|^2_{H_{K^n}(E)}.$$

See Corollary 2.5.

Applying (A.22), (A.24) and (A.25), for $f \in H_K(E)$ we have

$$\|\varphi(f)\|^2_{H_\psi * K(E)} \leq \sum_{p_n>0} \|c_n f^n\|^2_{H_{p_n K^n}(E)} \leq \sum_{p_n>0} \frac{|c_n|^2}{p_n} \|f\|^{2n}_{H_K(E)}. \tag{A.26}$$

Thus we obtain the inequality (A.5) stated in the beginning of Sect. A.1.1.

Theorem A.2 below asserts that if the RKHS $H_K(E)$ is algebra-dense and maximal, then "usually" its reproducing kernels are, up to constants, the only functions which attain equality in (A.5), and hence the exceptional case does not occur.

Theorem A.2. *Let $H_K(E)$ be an RKHS on E which is R-dense and maximal. Assume that $\varphi(z) = \sum_{n=1}^{\infty} c_n z^n$ and $\psi(z) = \sum_{n=1}^{\infty} p_n z^n$ are entire functions with the properties:*

1. $p_n \geq 0$ *for all n.*
2. $c_n = 0$ *whenever* $p_n = 0$.
3. *At least two different elements c_i and c_j are nonzero.*

Then, equality in inequality (A.5) *holds if and only if there exist a point $q \in E$ and constants C, C' such that $\varphi(Cz) = C'\psi(z)$ for all $z \in \mathbb{C}$ and that $f = CK_q$.*

Proof. If $\varphi(Cz) = C'\psi(z)$ for all $z \in \mathbb{C}$, then $c_n C^n = C' p_n$ for all $n \geq 1$. Thus $\varphi_\psi(|C|^2 z) = |C'|^2 \psi(z)$, where φ_ψ is given by (A.4). Observe also that

$$\|C'\psi(K_q)\|^2_{H_\psi * K(E)} = |C'|^2 \|\psi(K_q)\|^2_{H_\psi * K(E)} = |C'|^2 \sum_n p_n K^n(q, q).$$

Hence we have

$$\|\varphi(CK_q)\|^2_{H_\psi * K(E)} = \|C'\psi(K_q)\|^2_{H_\psi * K(E)} = |C'|^2 \psi(\|K_q\|^2_{H_K(E)}) = \varphi_\psi(\|CK_q\|^2_{H_K(E)}),$$

proving the sufficiency part of Theorem A.2.

To prove the necessity, assume that equality holds in (A.5) for $f \in H_K(E)$. Noting that the case $f = 0$ corresponds to the choice with $C = C' = 0$, we may assume that $f \neq 0$. By the hypotheses above there exist indices i and j with $p_i p_j \neq 0$, $1 \leq i < j$. From the chain of inequalities (A.26) and since $j \geq 2$, $f^{\otimes j}$ must be nonzero extremal in $\otimes^j H_K(E)$. Meanwhile, by Corollary A.1 $\otimes^j H_K(E)$ is weakly regular. Thus, there exists a point $q \in E$ such that

$$\Lambda_{f^{\otimes j}}(g) = g(q) \tag{A.27}$$

for all $g \in R$, and that q is either a common zero of $H_K(E)$ or $f = CK_q$ for some nonzero constant C.

Next we disprove that q is a common zero of $H_K(E)$. Since $H_K(E)$ is R-dense and $f \neq 0$, there exists $u \in R \cap H_K(E)$ with $\langle f, u \rangle_{H_K(E)} \neq 0$ according to Remark A.3.

Since $f^{\otimes j}$ is extremal in $H_{K^{\otimes j}}(E)$, we have

$$\langle c_j f^j, u^j\rangle_{H_{p_jK^j}(E)} = \frac{c_j}{p_j}\langle f^j, u^j\rangle_{H_{K^j}(E)} = \frac{c_j}{p_j}\langle f, u\rangle^j_{H_K(E)} \tag{A.28}$$

from (A.25). On the other hand, if $i \geq 2$, $f^{\otimes i}$ is extremal in $\otimes^i H_K(E)$, and so by decomposing u^j as a product $u^{j-i+1} \cdot u^{i-1}$, we have

$$\langle c_i f^i, u^j\rangle_{H_{p_iK^i}(E)} = \frac{c_i}{p_i}\langle f, u^{j-i+1}\rangle_{H_K(E)}\langle f^{i-1}, u^{i-1}\rangle_{H_{K^{i-1}}(E)} \tag{A.29}$$

from (A.27) Obviously, (A.29) also holds for $i = 1$. Since $f^{\otimes j}$ is nonzero extremal, it follows from Lemma A.2

$$\langle u^{j-i+1}, f\rangle_{H_K(E)} = u^{j-i}(q)\langle u, f\rangle_{H_K(E)}. \tag{A.30}$$

Decomposing u^j as above we see that

$$u^j \in H_{p_iK^i}(E) \cap H_{p_jK^j}(E)$$

since $u \in R \cap H_K(E)$ and $H_K(E)$ is R-dense. The equality condition (A.23) implies, for any k, l with $p_k p_l \neq 0$,

$$\langle c_k f^k, u^l\rangle_{H_{p_kK^k}(E)} = \langle c_l f^l, u^l\rangle_{H_{p_lK^l}(E)}. \tag{A.31}$$

Combining (A.28), (A.29), (A.30), and (A.31), we have

$$\overline{u^{j-i}(q)} = \frac{c_j p_i}{c_i p_j}\langle f, u^{j-i}\rangle_{H_K(E)} \neq 0. \tag{A.32}$$

Since $j > i$, the point q is not a common zero of $H_K(E)$, as desired.

Thus, $f = CK_q$ for some constant $C \neq 0$, see Definition A.3. By (A.31) the reproducing property of f yields, for any k, l with $p_k p_l \neq 0$,

$$\frac{c_k C^k u(q)^k}{p_k} = \frac{c_l C^l u(q)^l}{p_l}. \tag{A.33}$$

Putting $C' = c_k C^k u(q)^k / p_k$, we immediately obtain the identity $\varphi(Cz) = C'\psi(z)$.

A.1.3 The Case of Polynomial Ring

As an important example, we first consider the case where E is a subset of the complex n-dimensional space $\mathbb{C}^n$, and R is a restriction to E of the polynomial ring $\mathbb{C}[z_1, z_2, \ldots, z_n]$. A power series with center at the origin is denoted by $\sum_{\alpha} a_\alpha z^\alpha$.

Definition A.7. Let H be an RKHS on a subset E of $\mathbb{C}^n$. If H is $\mathbb{C}[z_1, z_2, \ldots, z_n]|_E$-*dense*, then H is called *polynomially dense*.

Let $\chi\colon \mathbb{C}[z_1, z_2, \ldots, z_n] \to \mathbb{C}$ be a $\mathbb{C}$-algebra homomorphism with $\chi(1) = 1$. For any polynomial $f(z) = \sum_\alpha a_\alpha z^\alpha$, we have

$$\chi(f) = \sum_{\alpha} a_\alpha \chi(z)^\alpha = f(w), \tag{A.34}$$

where $w \equiv \chi(z) = (\chi(z_1), \chi(z_2), \ldots, \chi(z_n)) \in \mathbb{C}^n$. Hence we conclude that any nonzero $\mathbb{C}$-algebra homomorphism of $\mathbb{C}[z_1, z_2, \ldots, z_n]|_E$ is a point evaluation at some point of $\mathbb{C}^n$. Thus we have immediately:

Proposition A.1. *Let H be a polynomially dense RKHS on E. Then H is maximal if and only if the following holds: Given a point $q \in \mathbb{C}^n$, if there exists a constant $C > 0$ with $|f(q)| \leq C\|f\|$ for all $f \in \mathbb{C}[z_1, z_2, \ldots, z_n] \cap H$, then $q \in E$.*

We next give an example of polynomially dense RKHSs and provide a sufficient condition for these RKHSs to be maximal.

Example A.2 ([68]). For $z, \zeta \in \mathbb{C}^n$ we put

$$z\zeta \equiv (z_1\zeta_1, z_2\zeta_2, \ldots, z_n\zeta_n) \in \mathbb{C}^n.$$

Fix a power series with positive coefficients $\eta(z) = \sum_\alpha c_\alpha z^\alpha$, $(c_\alpha > 0,\ \alpha \in \mathbb{Z}_+^n)$, and assume that the domain D of convergence of the function η is nonempty. A function f holomorphic in the domain D has a power series expansion $f(z) = \sum_{\alpha} a_\alpha z^\alpha$ on D. Define the norm $\|f\|$ of f by

$$\|f\| \equiv \sqrt{\sum_{\alpha \in \mathbb{Z}_+^n} \frac{|a_\alpha|^2}{c_\alpha}}, \tag{A.35}$$

and let $\mathscr{H}_\eta$ denote the space of holomorphic functions in D with $\|f\| < \infty$. Define an inner product of f and $g \in \mathscr{H}_\eta$ by

$$\langle f, g\rangle_{H_\eta} = \sum_{\alpha} \frac{a_\alpha \bar{b}_\alpha}{c_\alpha}, \quad g(z) = \sum_{\alpha} b_\alpha z^\alpha, \tag{A.36}$$

then $\mathscr{H}_\eta$ is a Hilbert space. For $\zeta \in D$ let $k_\zeta(z)$ denote the function $\eta(z\bar{\zeta})$. Then we easily see that $k_\zeta \in \mathscr{H}_\eta$ and that $f(\zeta) = \langle f, k_\zeta\rangle_{H_\eta}$ for all $f \in \mathscr{H}_\eta$. Thus, k_ζ is the reproducing kernel at ζ for the space $\mathscr{H}_\eta$, and hence $\mathscr{H}_\eta$ is an RKHS on D. By definition of the norm, $\mathscr{H}_\eta$ is clearly polynomially dense.

Proposition A.2. *If $\eta(z\bar{z}) = \infty$ for every $z \in \partial D$, then $\mathscr{H}_\eta$ is polynomially dense and maximal.*

Proof. Since $\eta(z\bar{z})$ is a series with nonnegative terms, if $\eta(z\bar{z}) < \infty$, then $\eta(tz\overline{tz}) < \infty$ for all t with $0 < t < 1$. Thus, from the hypothesis it is easy to see that $\eta(z\bar{z}) = \infty$ for every $z \notin D$. For $\zeta \notin D$ and $n \in \mathbb{N}$, let $k_\zeta^{(n)}(z) = \sum_{|\alpha|\le n} c_\alpha \bar{\zeta}^\alpha z^\alpha \in \mathscr{H}_\eta$ be the n-th partial sum of $k_\zeta(z)$. Then

$$\frac{|k_\zeta^{(n)}(\zeta)|}{\|k_\zeta^{(n)}\|} = \sqrt{k_\zeta^{(n)}(\zeta)} \to \sqrt{\eta(\zeta\bar{\zeta})} = \infty \ (n \to \infty). \tag{A.37}$$

Thus, the point evaluation at $\zeta \notin D$ is not $\mathscr{H}_\eta$-bounded. In view of Proposition A.1 this implies that $\mathscr{H}_\eta$ is maximal.

Remark A.4. All the theorems in [68, Sections 5,6] are immediate consequences of our Theorem A.2 and Proposition A.2.

A.1.4 Algebra of Meromorphic Functions

Throughout Sect. A.1.4, let E be a regular subregion of a compact Riemann surface S. Here a proper subregion E of S is called *regular* if E and its exterior have the same boundary consisting of a finite number of analytic Jordan curves. Let $\mathscr{R}_E$ denote the complex algebra of meromorphic functions on S which are holomorphic on $\overline{E}$.

Definition A.8. An RKHS H on E is called *meromorphically dense* if H is $\mathscr{R}_E$-dense.

Lemma A.3. *For any $f \in \mathscr{R}_E$, $\chi(f) \in f(\overline{E})$.*

Proof. If $f = 0$, then this is obvious. If f never vanishes on $\overline{E}$, then $1/f \in \mathscr{R}_E$, and from the identity $\chi(f)\chi(1/f) = \chi(1) = 1$ we have $\chi(f) \neq 0$. Therefore, $\chi(f - \chi(f)) = 0$ implies that $f - \chi(f)$ vanishes for some point in $\overline{E}$. Thus we have $\chi(f) \in f(\overline{E})$.

Let $P_1, P_2, \ldots, P_m$ and $Q_1, Q_2, \ldots, Q_k$ be finite collections of points in S. Suppose that we are given positive integers $p_1, p_2, \ldots, p_m$ and $q_1, q_2, \ldots, q_k$. Denote by g the genus of S. Assume that

$$1 - g - \sum_{j=1}^{m} p_j + \sum_{j=1}^{k} q_j \ge 1.$$

According to the Riemann Roch theorem, there exists a (nonconstant) meromorphic function f such that f vanishes P_j with order p_j for each $j = 1, 2, \ldots, m$ and that f has pole Q_j with order q_j for each $j = 1, 2, \ldots, k$. See [145] for details, which we admit in this book.

Lemma A.4. *Let $P \in S$ and $Q_1, Q_2, \ldots, Q_k \in S \setminus \{P\}$. Then there exists a function F such that $F(P) = 1$ and that $F(Q_1) = F(Q_2) = \cdots = F(Q_k) = 0$.*

Proof. Let $j = 1, 2, \ldots, k$. By the Riemann Roch theorem, we can find a function G_j such that G_j has a sole pole at Q_j and that G_j never vanishes at $\{Q_1, Q_2, \ldots, Q_k\} \setminus \{Q_j\}$. Define $H \equiv G_1 G_2 \cdots G_k - G_1(P)G_2(P) \cdots G_k(P) + 1$. Then $1/H = F$ has the desired result.

Lemma A.5. *Let $E \subset S$ be a compact set and $Q_1, Q_2, \ldots, Q_k \in S \setminus E$. Assume that we are given positive integers $q_1, q_2, \ldots, q_k$ such that*

$$\sum_{j=1}^{k} q_j > g,$$

where g denotes the genus of S. Then we can find a meromorphic function F such that $F^{-1}(0) = \{Q_1, Q_2, \ldots, Q_k\}$.

Proof. In fact, by the Riemann Roch theorem, we can find a function F_j such that F_j has a pole only at Q_j. If necessary, by adding a constant we can suppose that F_j never vanishes on $\{Q_1, Q_2, \ldots, Q_k\} \setminus \{Q_j\}$. Let $G \equiv F_1 \cdot F_2 \cdots F_k$. Then, the set of the poles of G is $\{Q_1, Q_2, \ldots, Q_k\}$. Since E is compact, we can find M such that $|G(z)| \leq M$ for all $z \in E$. Thus, if we set $F \equiv \dfrac{1}{G + 3M}$, then we have the desired function.

We prepare the following proposition which is useful for testing the maximality of meromorphically dense RKHSs.

Proposition A.3. *Let $\chi : \mathscr{R}_E \to \mathbb{C}$ be a $\mathbb{C}$-algebra homomorphism with $\chi(1) = 1$. Then there exists a unique point $q \in \overline{E}$ such that $\chi(f) = f(q)$ for all $f \in \mathscr{R}_E$.*

Proof. Choose any $f \in \mathscr{R}_E \setminus \mathbb{C}$ and set $\zeta = \chi(f)$. Let $f^{-1}(\zeta)$ consist of distinct points $q_1, q_2, \ldots, q_r$ $(1 \leq r \leq n)$, where n is the degree of the meromorphic function f on the compact Riemann surface S. From the above remark, $f^{-1}(\zeta) \cap \overline{E} \neq \phi$. Let us choose a point $\zeta_0 \in \mathbb{C}$ such that $\zeta_0 \neq \zeta$ and $f^{-1}(\zeta_0)$ consists of n distinct points. Applying Lemmas A.4 and A.5, one verifies easily that there exists $g_0 \in \mathscr{R}_E$ with the following properties:

1. g_0 takes different values at different points of $f^{-1}(\zeta_0)$.
2. $g_0(p) \neq 0$ for any $p \in \overline{E}$.
3. $g_0(q_j) = 0$ for all $q_j \notin \overline{E}$.

Indeed, by invoking Lemma A.5, we can construct a function g_1 such that $g_0(p) \neq 0$ for any $p \in \overline{E}$ and that $g_0(q_j) = 0$ for all $q_j \notin \overline{E}$. By using g_1 and Lemma A.4, we can find a function g_2 such that g_0 takes different values at different points of $f^{-1}(\zeta_0)$

and that $g_0(q_j) = 0$ for all $q_j \notin \overline{E}$. Since E is compact, if we choose M large enough, we see that $g_0 = g_2 + M$ is the desired function.

For sufficiently small $\varepsilon \neq 0$, the function $g \equiv 1/(g_0 + \varepsilon)$ is an element of $\mathscr{R}_E$ holomorphic at all points q_j $(j = 1, 2, \ldots, r)$ and satisfies the inequality

$$\sup_{z \in \overline{E}} |g(z)| < |g(q_j)| \quad \text{for any } q_j \notin \overline{E}. \tag{A.38}$$

By the general theory of compact Riemann surfaces [145], there exist rational functions a_k $(k = 0, \ldots, n)$ such that

$$\sum_{k=0}^{n} a_k(f) g^k = 0, \quad (a_n(z) = 1). \tag{A.39}$$

Since g is holomorphic at all points $q_1, q_2, \ldots, q_r$, it is clear from well-known constructions of (A.39) that the rational functions $a_k(z)$ $(k = 0, 1, 2, \ldots, n)$ are holomorphic at $z = \zeta$. Thus, we may assume that a_k is of the form b_k/c_k such that $b_k, c_k \in \mathbb{C}[z]$ and $c_k(\zeta) \neq 0$. Put $s = \prod_{k=0}^{n} c_k$ and multiply the identity (A.39) by $s(f)$. Since $g \in \mathscr{R}_E$ and each $sa_k (k = 0, \ldots, n)$ is a polynomial, we can apply the homomorphism χ to the resulting identity. Thus

$$\sum_{k=0}^{n} sa_k(\zeta) \chi(g)^k = 0. \tag{A.40}$$

Since $sa_k(\zeta) = s(\zeta) a_k(\zeta)$ with $s(\zeta) \neq 0$, we have $\sum_{k=0}^{n} a_k(\zeta)\chi(g)^k = 0$. Hence by construction of (A.39), $\chi(g) = g(q_j)$ for some j $(1 \leq j \leq r)$. Therefore, we have proved that there exists a point $q = q_j$ such that $\chi(f) = f(q)$ and $\chi(g) = g(q)$. From (A.38) we conclude that $q \in \overline{E}$, since $\chi(g) \in g(\overline{E})$ and the algebra of functions $\mathscr{R}_E$ separates points of S.

We now show that the homomorphism χ is the point evaluation at q. By the property (i) above, the meromorphic functions f and g form a primitive pair [145, p. 233]. Thus, it is well known that for any meromorphic function h on S, there exist rational functions A_k $(k = 0, \ldots, n-1)$ such that

$$h = \sum_{k=0}^{n-1} A_k(f) g^k. \tag{A.41}$$

First, consider the case where $h \in \mathscr{R}_E$ is holomorphic at each point of the set $f^{-1}(\zeta)$. Then, from well-known constructions of (A.41), every coefficient A_k is holomorphic at ζ. Again, by multiplying suitable polynomial as above, we obtain

$$\chi(h) = \sum_{k=0}^{n-1} A_k(\zeta) g(q)^k = h(q). \tag{A.42}$$

Second, for general $h \in \mathscr{R}_E$, choose a function $h_2 \in \mathscr{R}_E$ such that $h_2(q) \neq 0$ and that the product $h_2 h$ is holomorphic at each point of $f^{-1}(\zeta)$. This is possible if h_2 has sufficiently large order of zeros at every $q_j \notin \overline{E}$. From

$$h_2(q)h(q) = \chi(h_2 h) = \chi(h_2)\chi(h) = h_2(q)\chi(h), \tag{A.43}$$

we conclude that $\chi(h) = h(q)$ for all $h \in \mathscr{R}_E$, as desired. The uniqueness of the point q is obvious, since the algebra of functions $\mathscr{R}_E$ separates points of S.

From Definition A.6 and Proposition A.3, we immediately have

Proposition A.4. *Let H be a meromorphically dense RKHS on E. Then H is maximal if and only if the following holds: Given a point $q \in \overline{E}$, if there exists a constant $C > 0$ with $|f(q)| \leq C\|f\|_H$ for all $f \in \mathscr{R}_E \cap H$, then $q \in E$.*

A.1.5 Applications

As an application of the results obtained in the previous sections we study the regularity of tensor products of RKHSs consisting of analytic functions or analytic differentials on a compact bordered Riemann surface. Let E be the interior of a compact bordered Riemann surface $\overline{E} = E \cup \mathbf{E}$ with nonempty boundary $\mathbf{E}$. Let $\hat{E}$ be the Schottky double of $\overline{E}$ [425]. Then E can be viewed as a regular subregion of the compact Riemann surface $\hat{E}$. Define the $\mathbb{C}$-algebra $\mathscr{R}_E$ in this context. Consider the following RKHSs on E:

1. $\mathscr{H}_1(E, \rho)$: (Weighted Szegö space) Let ρ be a positive measurable function on E. The Hardy H^2 space of analytic functions f on E with norm

$$\|f\|_{\mathscr{H}_1(E,\rho)} \equiv \sqrt{\int_{\mathbf{E}} |f(z)|^2 \rho(z)|dz|} \tag{A.44}$$

   where $\rho|dz|$ is a positive continuous metric on $\mathbf{E}$. In the integrand f denotes the nontangential boundary value of f on $\mathbf{E}$.
2. $\mathscr{H}_2(E, \rho)$: (Weighted Dirichlet space) The space of analytic functions f on E with finite Dirichlet norm

$$\|f\|_{\mathscr{H}_2(E,\rho)} \equiv \sqrt{\frac{i}{2\pi} \iint_E \rho(z)\, df(z) \wedge \overline{df}(z)} \tag{A.45}$$

   satisfying $f(a) = 0$ for a fixed point $a \in E$, where ρ is a positive continuous function on $\overline{E}$.
3. $\mathscr{H}_3(E, \rho)$: (Weighted Bergman space) The *weighted Bergman space of analytic differentials* f on E with norm

$$\|f\|_{\mathscr{H}_3(E,\rho)} \equiv \sqrt{\frac{i}{2\pi}\iint_E \rho(z) f(z) \wedge \overline{f(z)}}, \tag{A.46}$$

where ρ is a positive continuous function on $\overline{E}$.

Theorem A.3. *The following hold:*

1. *The RKHSs* $\mathscr{H}_j(E,\rho)$ $(j = 1, 2, 3)$ *are meromorphically dense and maximal.*
2. *For any integer* $n \geq 2$, $\mathscr{H}_j(E,\rho)^{\otimes n}$ $(j \neq 2)$ *is regular and* $\mathscr{H}_2(E,\rho)^{\otimes n}$ *is weakly regular.*
3. *Let* $a = 0$. *Then* $\phi^{\otimes 2}$ *is extremal in* $\mathscr{H}_2(\Delta(1), 1)^{\otimes 2}$ *if and only if* $\phi(z) = cz$ *or* $\phi = ck_q$ $(q \in \Delta(1) \setminus \{0\})$ *for some* $c \in \mathbb{C}$, *where*

$$k_q(z) \equiv -\log(1 - \bar{q}z)$$

is the reproducing kernel of $\mathscr{H}_2(\Delta(1), 1)$ *at* q. *Thus,* $\mathscr{H}_2(\Delta(1), 1)^{\otimes 2}$ *is not regular.*

For the proof of Theorem A.3 above we need a weaker form of the result in [424, Theorem 8] on uniform approximation.

Proposition A.5 (S. Scheinberg). *Let* E *be a regular subregion of a compact Riemann surface* S. *For every holomorphic function* f *on* $\overline{E}$ *and for every positive constant* ε, *there exists a function* $g \in \mathscr{R}_E$ *such that* $\|f - g\|_\infty < \varepsilon$ *on* $\overline{E}$, *where* $\|\cdot\|_\infty$ *denotes the sup-norm on* $\overline{E}$.

Proof (of Theorem A.3).

1. First, we remark that the norm of the RKHS $\mathscr{H}_j(E,\rho)$ is equivalent for each weight ρ and, as a set, the space $\mathscr{H}_j(E,\rho)$ is independent of ρ, since ρ is positive and continuous on compact set $\overline{E}$. Thus we may assume $\rho = 1$ without loss of generality. Let $\mathscr{H}_j(E) = \mathscr{H}_j(E, 1)$ for simplicity.

   Since $1 \in \mathscr{R}_E$, to prove that $\mathscr{H}_j(E)$ is meromorphically dense, it suffices to show that $\mathscr{R}_E \cap \mathscr{H}_j(E)$ is dense in $\mathscr{H}_j(E)$ and is an ideal of $\mathscr{R}_E$. To establish that the space $\mathscr{R}_E \cap \mathscr{H}_j(E)$ is dense in $\mathscr{H}_j(E)$ $(j = 1, 2, 3)$, we recall that the Szegö kernels, the exact Bergman kernels and the Bergman kernels are analytically continued to a neighborhood of $\overline{E}$ [388, 425]. Since the linear span of the reproducing kernels is dense, the set of functions holomorphic on $\overline{E}$ are dense in $\mathscr{H}_j(E)$, and by Proposition A.5 we see easily that $\mathscr{R}_E \cap \mathscr{H}_j(E)$ is dense in $\mathscr{H}_j(E)$ $(j = 1, 2, 3)$. On the other hand, it is clear that $\mathscr{R}_E \cap \mathscr{H}_j(E)$ is an ideal of $\mathscr{R}_E$. Thus, $\mathscr{H}_j(E)$ $(j = 1, 2, 3)$ is meromorphically dense.

   Next, we show that all the $\mathscr{H}_j(E)$ $(j = 1, 2, 3)$ are maximal. In view of Proposition A.4 it suffices to show that, for fixed $b \in \mathbf{E}$, there exists a family of functions $\{f_p\}$, $f_p \in \mathscr{R}_E \cap \mathscr{H}_j(E)$ such that $|f_p(b)|/\|f_p\|$ tends to ∞ as $p \to b$. Consider the case of the Dirichlet space $\mathscr{H}_2(E)$. The proof of the other cases is similar but easier. Now we claim that we need only to show that there exists a family of functions $\{f_p\}$ holomorphic on $\overline{E}$ with the above property. This is seen as follows. Given a holomorphic function f on $\overline{E}$, there exists a regular subregion E_1 with $\overline{E} \subset E_1$ such that f is holomorphic on $\overline{E_1}$. Applying Cauchy's integral formula we see that there exists a constant $C > 0$ with $\|f\| \leq C\|f\|_{\infty,E_1}$ where $\|f\|_{\infty,E_1}$ denotes $\sup_{x\in E_1} |f(x)|$. From Proposition A.5 there exists $g \in \mathscr{R}_{E_1}$ such that $\|f - g\|_{\infty,E_1} \leq \varepsilon\|f\|$. Then

$$|g(b)| \geq |f(b)| - \|f - g\|_{\infty,E_1} \geq |f(b)| - \varepsilon\|f\|,$$
$$\|g\| \leq \|f\| + \|f - g\| \leq \|f\| + C\|f - g\|_{\infty,E_1} \leq (1 + \varepsilon C)\|f\|.$$

By choosing $\varepsilon(0, \min\{1, C^{-1}\})$, we have

$$\frac{|g(b)|}{\|g\|} \geq \frac{|f(b)|}{(1 + \varepsilon C)\|f\|} - \varepsilon > \frac{|f(b)|}{2\|f\|} - 1, \tag{A.47}$$

which implies our claim, as desired.
By definition we have the identity

$$k_B(x, y) = \frac{\partial^2 k_D}{\partial x \partial \bar{y}}(x, y), \tag{A.48}$$

where $k_D(x, y)$ is the reproducing kernel for the Dirichlet space $\mathscr{H}_2(E)$ and $k_B(x, y)$ is the exact Bergman kernel for E. Let ϕ be the canonical anticonformal involution for the double $\hat{E}$ fixing $\mathbf{E}$. It is well known [425, p. 118] that the exact Bergman kernel $k_B(x, y)$ is extended to a meromorphic bilinear differential on $\hat{E}$ with double pole only at $x = \phi(y)$, and that $k_B(x, y)$ has the expansion

$$k_B(x, y) = -\frac{1}{\pi(x - \phi(y))^2} + \text{regular terms} \tag{A.49}$$

for x, y in a coordinate neighborhood U centered at $b \in \mathbf{E}$. Integrating (A.49), we see that for $p \in E$ the Dirichlet kernel $k_D(x, p)$ is extended holomorphically to a neighborhood of $\overline{E}$ with the expansion

$$k_D(x, y) = \frac{1}{\pi} \log \frac{1}{x - \phi(y)} + \text{regular terms} \tag{A.50}$$

for $x, y \in U \cap E$. Setting $f_p(x) = k_D(x, p)$ for $p \in E$ near b, we have

$$\frac{|f_p(b)|}{\|f_p\|} = \frac{|k_D(b, p)|}{\sqrt{k_D(p, p)}}.$$

By (A.50) this tends to ∞ as $p \to b \in \mathbf{E}$ nontangentially. Thus $\mathscr{H}_2(E)$ is maximal.

2. By Corollary A.1 and (A.44) all the tensor products of these spaces are weakly regular. Moreover, since both $\mathscr{H}_1(E,\rho)$ and $\mathscr{H}_3(E,\rho)$ have no common zeros, the products $\mathscr{H}_j(E,\rho)^{\otimes n}$ ($j = 1, 3$; $n \geq 2$) are regular.
3. By definition, if $\phi = ck_q$, it is clear that $\phi^{\otimes 2}$ is extremal. We will show that $z^{\otimes 2} \in (\mathscr{H}_2(\Delta(1), 1)^{\otimes 2})_0^{\perp}$. Since the set of functions $\{z^i/\sqrt{i}\}_{i=1}^{\infty}$ is a complete orthonormal system (CONS) for $\mathscr{H}_2(\Delta(1), 1)$, the set $\{z^i \otimes z^j/\sqrt{ij}\}_{i,j=1}^{\infty}$ is a CONS for the tensor product $\mathscr{H}_2(\Delta(1), 1)^{\otimes 2}$. Hence, any $f \in \mathscr{H}_2(\Delta(1), 1)^{\otimes 2}$ is given by

$$f = \sum_{i,j=1}^{\infty} \frac{c_{ij}}{\sqrt{ij}} z^i \otimes z^j, \quad \text{with} \quad \sum_{i,j=1}^{\infty} |c_{ij}|^2 < \infty. \tag{A.51}$$

Then we see that $f \in (\mathscr{H}_2(\Delta(1), 1)^{\otimes 2})_0$ if and only if $\sum_{i+j=n} c_{ij}/\sqrt{ij} = 0$ for all $n \geq 2$. In particular, $f \in (\mathscr{H}_2(\Delta(1), 1)^{\otimes 2})_0$ implies $c_{11} = 0$. Since $\langle f, z^{\otimes 2}\rangle_{H_\eta} = c_{11}$, $z^{\otimes 2} \in (\mathscr{H}_2(\Delta(1), 1)^{\otimes 2})_0^{\perp}$, that is, $z^{\otimes 2}$ is extremal.

Contrarily, suppose that $\phi^{\otimes 2} \in (\mathscr{H}_2(\Delta(1), 1)^{\otimes 2})_0^{\perp}$. We may assume without loss of generality $\phi \neq 0$. From Lemma A.2 there exists a point $q \in \Delta(1)$ such that

$$\langle zf, \phi\rangle_{H_\eta} = f(q)\langle z, \phi\rangle_{H_\eta}. \tag{A.52}$$

If $q \neq 0$, then q is not a common zero of $\mathscr{H}_2(\Delta(1), 1)$. Since $\mathscr{H}_2(\Delta(1), 1)^{\otimes 2}$ is weakly regular, $\phi = ck_q$ for some constant $c \in \mathbb{C}$. On the other hand, if $q = 0$, then by (A.52) $\phi \perp z^n$ for all $n \geq 2$. Therefore, ϕ is a constant multiple of the function z. Thus the first assertion of (A.46) is proved. Finally observe that $\mathscr{H}_2(\Delta(1), 1)^{\otimes 2}$ is not regular, since the function z cannot be a constant multiple of the reproducing kernel k_q for any $q \in \Delta(1)$.

A.2 Generalizations of Opial's Inequality

In 1995, natural norm inequalities hold for a wide class of nonlinear maps between reproducing kernel Hilbert spaces in [386, 388]. This can be shown also by the theory of reproducing kernels. The method has proved to be very important for applications such as identifications of nonlinear systems [490]. Among several concrete inequalities, the most beautiful one is the following: For a real-valued function $f \in \mathrm{AC}[0, 1]$ with $f(0) = 0$ and $\int_0^1 f'(x)^2 dx < 1$, where $\mathrm{AC}[0, 1]$ is the set of all absolutely continuous functions on $[0, 1]$, we have

$$\int_0^1 \left(\frac{f(x)}{1 - f(x)}\right)'^2 (1 - x)^2 dx \leq \frac{\int_0^1 f'^2(x)\,dx}{1 - \int_0^1 f'^2(x)\,dx}. \tag{A.53}$$

Equality holds in (A.53), if f is of form $f(x) = \min(x, y)$, $x \in [0, 1]$ for some $y \in [0, 1)$. This is stated in [389, (22)]. But the equality condition was not completely determined.

Meanwhile, the following Opial's inequality [349] is famous and there are many papers extending it: For $f \in \mathrm{AC}[0, a]$ with $f(0) = 0$, we have

$$\int_0^a |f(x)f'(x)|\, dx \leq \frac{a}{2} \int_0^a |f'(x)|^2\, dx. \tag{A.54}$$

For Opial-type inequalities see, e.g., [38, 39, 307].

A function $f(x)$ positive and continuous on an interval $(0, R)$ is called *geometrically convex* if f satisfies the inequality, for all $x, y \in (0, R)$

$$f(\sqrt{xy}) \leq \sqrt{f(x)f(y)}, \tag{A.55}$$

that is, $\log \circ f \circ \exp$ is convex on $(-\infty, \log R)$ [290].

Following Akira Yamada [486], we will introduce a generalization of the inequality (A.53) with an elementary and direct proof, and for the generalization, we will be able to generalize Opial's inequality (Theorem A.6), by using geometrically convex functions (Theorem A.4): The tool is Hölder's inequality, and so the proof is elementary.

Very interestingly, similar, but many different generalizations of the results obtained by the theory of reproducing kernels were independently published by Nguyen Du Vi Nhan, Dinh Thanh Duc, and Vu Kim Tuan, in the same year, in [329].

A.2.1 Main Inequality

We now state the main theorem. Recall that $\mathrm{AC}[a, b]$ denotes the set of all absolutely continuous function on $[a, b]$.

Theorem A.4. *Let $p > 1$ and G be a function of class $C^1(-R, R)$ satisfying the conditions*

$$G(0) = 0, \tag{A.56}$$

$$|G'(x)| \leq G'(|x|) \tag{A.57}$$

for all $x \in (-R, R)$ and

$$0 < G'(x)^2 \leq G'(y)G'(z) \tag{A.58}$$

for all $x, y, z \in (0, R)$ satisfying $x^2 \leq yz$. Assume that functions F, $f \in \mathrm{AC}[a, b]$ with $F(a) = f(a) = 0$ satisfy

$$F'(x) > 0, F(b) \leq R, \tag{A.59}$$

for almost all $x \in [a, b]$ and

$$\int_a^b |f'(t)|^p F'(t)^{1-p}\, dt < R. \tag{A.60}$$

Then,

$$\int_a^b \frac{|(G \circ f)'(x)|^p}{(G \circ F)'(x)^{p-1}}\, dx \leq G\Big(\int_a^b \frac{|f'(x)|^p}{F'(x)^{p-1}}\, dx\Big). \tag{A.61}$$

If $f(x) = F(\min(x, y))$ for some $y \in (a, b]$, then equality holds in (A.61).

Before the proof, a helpful remark may be in order.

Remark A.5. Under the assumptions above, one can obtain $|f(x)| < R$. In fact, $f(a) = 0$. Thus, applying Hölder's inequality with conjugate exponent $1/p + 1/q = 1$ to the identity $f(x) = \int_a^x f'(t)\, dt$, we have

$$\begin{aligned} |f(x)| &\leq \sqrt[q]{F(x) - F(a)}\sqrt[p]{\int_a^x \frac{|f'(t)|^p}{F'(t)^{p-1}}\, dt} \\ &= \sqrt[q]{F(x)}\sqrt[p]{\int_a^x \frac{|f'(t)|^p}{F'(t)^{p-1}}\, dt}. \end{aligned} \tag{A.62}$$

Thus $|f(x)| < R$ from (A.60).

Proof. Since G' is continuous, we remark that from (A.57) and (A.58), the function $G'(x)$ is positive, monotone increasing and geometrically convex on the interval $(0, R)$. Hence we see that for $0 \leq x, y, z < R$,

$$x \leq \sqrt[p]{y}\sqrt[q]{z} \implies G'(x) \leq \sqrt[p]{G'(y)}\sqrt[q]{G'(z)}. \tag{A.63}$$

Hence, by (A.63) and (A.62), we obtain for $a \leq x < b$

$$G'(|f(x)|) \leq \sqrt[q]{G'(F(x))}\sqrt[p]{G'\left(\int_a^x \frac{|f'(t)|^p}{F'(t)^{p-1}}\, dt\right)}.$$

Multiplying $|f'(x)|^p/(G \circ F)'(x)^{p-1}$ (≥ 0) to the p-th power of the above inequality, we have

$$\frac{|(G \circ f)'(x)|^p}{(G \circ F)'(x)^{p-1}} \leq \frac{d}{dx}\Big\{G\Big(\int_a^x \frac{|f'(t)|^p}{F'(t)^{p-1}}\, dt\Big)\Big\} = \frac{|f'(x)|^p}{F'(x)^{p-1}} G'\Big(\int_a^x \frac{|f'(t)|^p}{F'(t)^{p-1}}\, dt\Big). \tag{A.64}$$

Integrating both sides of (A.64) on the interval $[a, b]$, we obtain the desired inequality (A.61) in view of (A.62). When $f(x)$ is of the form $F(\min(x, y))$, $y \in (a, b]$, it is obvious that equality holds since $F \geq 0$.

Remark A.6. As a special case of $G(x) \equiv x/(1-x)$, $F(x) \equiv x$, $p = 2$, $a = 0$ and $b = R = 1$ in Theorem A.4, (A.53) yields the prototype inequality. Note that $G'(x) = (1-x)^{-2}$ is geometrically convex on $(0, 1)$. Equality holds if and only if $f(x) = \min(x, y)$, $y \in [0, 1)$, which is seen from Theorem A.5 in the next section.

Remark A.7. Let H be a real vector space of functions $f \in \mathrm{AC}[a, b]$ with $f(a) = 0$. Then H becomes a reproducing kernel Hilbert space if we assign $f \in H$ the norm

$$\sqrt{\int_a^b |f'(t)|^2 \rho(t)\, dt}, \tag{A.65}$$

where the weight ρ is a positive continuous function on $[a, b]$. The reproducing kernel k of H is given by

$$k(x, y) = F(\min(x, y)) \tag{A.66}$$

with $F(x) = \displaystyle\int_a^x \rho(t)^{-1} dt$. Since the kernel determines its reproducing kernel Hilbert space uniquely, denote this RKHS by $H_k(E)$ and the above norm by $\|f\|_k$. Then, for $p = 2$ we can rewrite formally the inequality (A.61) of Theorem A.4 as

$$\|G \circ f\|^2_{G(k)} \leq G(\|f\|_k^2) \quad \text{for any } f \in H_k[a, b]. \tag{A.67}$$

Thus, it may seem that the inequality (A.61) is merely an example of general norm inequalities for RKHSs. This, however, is not the case, since the inequality does not require the real analyticity of the function G, while we need to assume this in general norm inequalities for RKHSs.

A.2.2 Equality Condition

For most cases, equality in the Main Inequality is attained only for functions stated in Theorem A.4, i.e., the function 0 or $F(\min(x, y))$. We are able to see this by adding further assumptions on the function $G(x)$.

Theorem A.5. *Under the same hypothesis as in Theorem A.4, assume, moreover, that G' is strictly monotone increasing on $(0, R)$;*

$$G'(t) > G'(s) \quad 0 < s < t < R. \tag{A.68}$$

Then, if equality holds in inequality (A.61), *then there exist constants C and y* ($a < y \leq b$) *such that*

$$f(x) = C \cdot F(\min(x, y)). \tag{A.69}$$

If, in addition, we assume that

1. *for some* $x \in (0, F(y))$,

$$|G'(-x)| \neq G'(x) \tag{A.70}$$

   and that
2. *there exist no constants* $\alpha > 0$ *and* $\beta \geq 0$ *such that*

$$G'(x) = \alpha x^{\beta} \tag{A.71}$$

   on $(0, F(y))$,

then equality holds in inequality (A.61) *if and only if f is of the form* (A.69) *with* $C \in \{0, 1\}$.

Proof. First, we remark that equality occurs on the right-hand side inequality of (A.63) if and only if $x = \sqrt[p]{y}\sqrt[q]{z}$. For, since G' is strictly monotone increasing, we have

$$G'(x) \leq G'(\sqrt[p]{y}\sqrt[q]{z}) \leq \sqrt[p]{G'(y)}\sqrt[q]{G'(z)} = G'(x), \tag{A.72}$$

which implies $x = \sqrt[p]{y}\sqrt[q]{z}$.

When $f = 0$ it suffices to take $C = 0$. Hence, we may assume $f \neq 0$. Putting $y = \operatorname{ess\,sup}\{x{:}f'(x+0) \neq 0,\ a \leq x < b\}$, we have $a < y \leq b$. If equality holds in (A.61), then y must be a cluster point of the set $\{x;\ f'(x) \neq 0\}$. Hence, by continuity

$$G'(|f(y)|) = \sqrt[q]{G'(F(y))}\sqrt[p]{G'\Big(\int_a^y \frac{|f'(t)|^p}{F'(t)^{p-1}}\,dt\Big)}, \tag{A.73}$$

and from the remark above we obtain

$$|f(y)| = \sqrt[q]{F(y)}\sqrt[p]{\int_a^y \frac{|f'(t)|^p}{F'(t)^{p-1}}\,dt}. \tag{A.74}$$

From the equality condition of Hölder's inequality, there exists a constant $C \neq 0$ such that $f(x) = CF(x)$, $(a \leq x \leq y)$. Since $f'(x) = 0$($y \leq x \leq b$ almost every) by definition of y, we conclude that f is of the form

$$f(x) = C \cdot F(\min(x, y)). \tag{A.75}$$

Thus we obtain (A.69). Now we prove the latter half of the assertion of Theorem A.4. For all $x \in [a, y]$ we have

$$\int_a^x \frac{|f'(t)|^p}{F'(t)^{p-1}}\,dt = |C|^p F(x), \tag{A.76}$$

and hence, the following identities must hold simultaneously: for all x, $a \le x \le y$,

$$|C|F(x) = \sqrt[q]{F(x)}\,\sqrt[p]{|C|^p F(x)},$$
$$G'(|C|F(x)) = \sqrt[q]{G'(F(x))}\,\sqrt[p]{G'(|C|^p F(x))}.$$

If $|C| \neq 1$, then $F(x) \neq |C|^p F(x)$ for $x > a$. Since the equality condition for Jensen's inequality for two distinct points imply linearity of the function on the interval between these points, one verifies easily that $G'(x)$ is of the form αx^β, $(\alpha > 0,\ \beta \ge 0)$ on the interval $(0, F(y))$. But we can exclude this by our assumption (A.71). Finally, if $C = -1$, then we must have equality, which contradicts the condition (A.70).

Remark A.8. If $G(x) \equiv \alpha|x|^\beta$ $(\alpha > 0,\ \beta > 1)$, then equality holds in inequality (A.61) for every $f(x)$ of the form $C \cdot F(\min(x, y))$ $(C \in \mathbb{R}, a < y \le b)$.

A.2.3 Applications

The main inequality allows us immediately to derive Opial-type inequalities [38, 39, 75, 307]. For brevity we restrict ourselves to the case that the constant R in Theorem A.4 is infinity.

Theorem A.6. *Let $p > 1$ and $q, r > 0$ satisfy $1/p + 1/r = 1/q$. Let s and t be nonnegative, measurable functions on $[a, b]$ such that*

$$\int_a^b \frac{1}{\sqrt[p-1]{t(x)}}\,dx < +\infty \tag{A.77}$$

for some $p > 1$. Set

$$F(x) \equiv \int_a^x \frac{1}{\sqrt[p-1]{t(y)}}\,dy \quad x \in [a, b]$$

and assume that the functions G, F and f satisfy the same conditions as Theorem A.4 with $R = +\infty$. We assume

$$K \equiv \sqrt[r]{\int_a^b \sqrt[p]{(G \circ F)'(x)^{r(p-1)}}\,\sqrt[q]{s(x)^r}\,dx} < \infty. \tag{A.78}$$

Then we have

$$\sqrt[q]{\int_a^b |(G\circ f)'(x)|^q s(x)\,dx} \le K\cdot\sqrt[p]{G\Big(\int_a^b |f'(x)|^p t(x)\,dx\Big)}. \tag{A.79}$$

If, in addition, we assume the conditions (A.68), (A.70) *and* (A.71) *in Theorem A.5, then equality holds in the inequality* (A.79) *if and only if either* $f = 0$ *or there exist constants* $C \in [0,\infty)$ *and* $y \in (a,b]$ *such that* $f(x) \equiv F(\min(x,y))$ *and that* $s(x) \equiv C\cdot(G\circ F)'(x)^{1-q}$.

Proof. Rewrite the integrand on the left-hand side of (A.79) as

$$|(G\circ f)'|^q\cdot s = \frac{|(G\circ f)'|^q}{(G\circ F)'^{\alpha}}\cdot(G\circ F)'^{\alpha}s,\quad \alpha\equiv\frac{q(p-1)}{p}, \tag{A.80}$$

use Hölder's inequality with conjugate exponents p/q and r/q, and then apply Theorem A.4. We obtain equality condition immediately from Theorem A.5.

Remark A.9. The existence of the multiplicative constant K is a merit of the inequality (A.79). See [75, 188].

Remark A.10. Let $-\infty < a < b < \infty$. Let $G(x) \equiv |x|^{p/q}$, $p > q$, $k > 1$ and $k > q > 0$. If

$$\int_a^b \frac{1}{\sqrt[k-1]{t(x)}}dx < +\infty,$$

then Theorem A.6 yields the Opial-type inequality

$$\int_a^b |f(x)|^{p-q}|f'(x)|^q s(x)\,dx \le K\cdot\sqrt[k]{\Big\{\int_a^b |f'(x)|^k t(x)\,dx\Big\}^p}, \tag{A.81}$$

where the constant

$$K \equiv \sqrt[k]{\Big(\frac{q}{p}\Big)^q\Big\{\int_a^b \sqrt[k-q]{\frac{s(x)^k}{t(x)^q}}\Big(\int_a^x \frac{dy}{\sqrt[k-1]{t(y)}}\Big)^{\frac{(p-q)(k-1)}{k-q}}dx\Big\}^{k-q}} \tag{A.82}$$

is finite. Note that this is exactly the same inequality as the equality condition that can be obtained easily as above. Remark that Beesack and Das employed a fundamental technique to prove (A.82) in [39].

A.3 Explicit Integral Representations of Implicit Functions

Here, we will refer to the explicit integral representation of implicit functions derived from Implicit Function Theory that may be considered as one of the master pieces in mathematical analysis.

A.3.1 2-Dimensional Case

Auxiliary Material and Framework

We first state in a precise way the representations of inverses of nonlinear mappings for the 2-dimensional and n dimensional spaces cases.

Let $D \subset \mathbb{R}^2$ be a bounded domain with a finite number of piecewise C^1-class boundary components. Let f be one-to-one C^1-class mapping from $\overline{D}$ into $\mathbb{R}^2$ and we assume that its Jacobian $J(x)$ is positive on D. We represent f in the form

$$\begin{pmatrix} y_1 \\ y_2 \end{pmatrix} = \begin{pmatrix} f_1(x) \\ f_2(x) \end{pmatrix} = \begin{pmatrix} f_1(x_1, x_2) \\ f_2(x_1, x_2) \end{pmatrix} \tag{A.83}$$

and the inverse mapping f^{-1} of f as follows:

$$\begin{pmatrix} x_1 \\ x_2 \end{pmatrix} = \begin{pmatrix} (f^{-1})_1(y) \\ (f^{-1})_2(y) \end{pmatrix} = \begin{pmatrix} f_1^{-1}(y_1, y_2) \\ f_2^{-1}(y_1, y_2) \end{pmatrix}, \tag{A.84}$$

where $y = (y_1, y_2)$ and f_1^{-1} and f_2^{-1} abbreviate $(f^{-1})_1$ and $(f^{-1})_2$, respectively. Then, we would like to represent ${}^t((f^{-1})_1(y), (f^{-1})_2(y))$ in terms of the direct mapping (A.83).

Additionally, we are also interested in some numerical and practical solutions of the nonlinear simultaneous equations (A.83).

We recall Theorem 8.10. For the mappings (A.83) and (A.84), we obtain the representation, for any $y^* = (y_1^*, y_2^*) \in f(D)$,

$$\begin{pmatrix} f_1^{-1}(y^*) \\ f_2^{-1}(y^*) \end{pmatrix} = \frac{1}{2\pi} \oint_{\partial D} \begin{pmatrix} x_1 \\ x_2 \end{pmatrix} d\text{Arctan} \left(\frac{f_2(x_1, x_2) - y_2^*}{f_1(x_1, x_2) - y_1^*} \right) \tag{A.85}$$
$$- \frac{1}{2\pi} \iint_D \frac{\text{adj}J(x_1, x_2)}{|f(x_1, x_2) - y^*|^2} \begin{pmatrix} f_1(x_1, x_2) - y_1^* \\ f_2(x_1, x_2) - y_2^* \end{pmatrix} dx_1 dx_2.$$

We can consider the singular integral in (A.85) (as well as all the others in the sequel) in the sense of the Cauchy principal value.

In order to obtain a general version, we recall the following general representation formula of real-valued functions of class C^1.

Let us discuss the general case for the time being. We denote by $*$ the *Hodge star operator*, by G_n the *fundamental solution* of the *Laplacian* $\Delta_n = \sum_{i=1}^{n} \frac{\partial^2}{\partial x_i^2}$. We also denote by $(\cdot, \cdot)$ the *inner product* of the *vector space* $A^n(D)$ consisting of the n *order differential forms* over D with finite L^2-norms, that is

$$(\omega, \eta) = \int_D \omega \wedge *\eta = \int_D \eta \wedge *\omega \quad (\omega, \eta \in A^n(D)).$$

We write

$$f^{-1}(y) = (f_1^{-1}(y), f_2^{-1}(y), \ldots, f_n^{-1}(y)).$$

We recall Theorem 8.8: Let D be a bounded domain in $\mathbb{R}^n$ whose boundary ∂D is made up of a finite number of C^1-class boundary components. Let f be a C^1-class real-valued function on $\overline{D}$. For any $\hat{x} \in D$, we have the representation

$$f(\hat{x}) = -c_n\big(df(x), dG_n(x - \hat{x})\big) + c_n \int_{\partial D} f(x) * dG_n(x - \hat{x}), \tag{A.86}$$

where for $c_n = \max(1, n-2)$.

By using the pullback f^* for the integral representations of the inversions, we obtain the following result: Let us consider the situation of Theorem 8.8 and, furthermore, assume that f is a preserving C^1-class function on $\overline{D}$ in $\mathbb{R}^n$ with a single-valued inverse. Then, for $\hat{y} \in f(D)$, we have the representation

$$f_i^{-1}(\hat{y}) = -\int_D dx_i \wedge f^*[*dG_n(\cdot - \hat{y})](x) + \int_{\partial D} x_i f^*[*dG_n(\cdot - \hat{y})](x). \tag{A.87}$$

Here, f_i^{-1} denotes the i component of f^{-1}.

Meanwhile, for $\hat{y}$ in the outside of the image $f(D)$, the right-hand sides of the integrals are zero for all i.

For $n = 2$, we have (A.85) and we can represent (A.87) as follows. For any $\hat{y} \in f(D)$, we have

$$f_i^{-1}(\hat{y}) = \frac{1}{2\pi}\left(\oint_{\partial D} x_i d\theta_i - \int_D dx_i \wedge d\theta_i\right), \quad i = 1, 2.$$

Here,

$$\theta_1 = -\mathrm{Arctan}\left(\frac{f_1(x) - \hat{y}_1}{f_2(x) - \hat{y}_2}\right), \quad \theta_2 = \mathrm{Arctan}\left(\frac{f_2(x) - \hat{y}_2}{f_1(x) - \hat{y}_1}\right).$$

In particular, furthermore, when $f(D)$ is a convex domain, we have the representation

$$f_i^{-1}(\hat{y}) = \frac{\hat{x}_i^{min} + \hat{x}_i^{max}}{2} + \frac{1}{2\pi}\left(\oint_{\partial D} \theta_i dx_i - \int_D dx_i \wedge d\theta_i\right), \quad i = 1, 2.$$

Here, we can determine $\hat{x}_i^{min}$ and $\hat{x}_i^{max}$ by $\hat{y}$ as the two points of ∂D [491].

A.3.2 Representations of Implicit Functions

We will now pay attention to the Implicit Function Theorem, and explain the main goal of the present session. For simplifying the statement, we assume some global properties: On a smooth bounded domain $U \subset \mathbb{R}^{n+k}$ surrounded by finite number of C^1-class and simple closed surfaces, for k functions

$$f_i(x_1, x_2, \dots, x_n, x_{n+1}, x_{n+2}, \dots, x_{n+k}), \quad i = 1, 2, \dots, k, \tag{A.88}$$

we assume that for some point on U, it holds that

$$f_i(x_1, x_2, \dots, x_n, x_{n+1}, x_{n+2}, \dots, x_{n+k}) = 0 \tag{A.89}$$

and on U we have

$$\det \frac{\partial(f_1, f_2, \dots, f_k)}{\partial(x_{n+1}, x_{n+2}, \dots, x_{n+k})}(x) > 0. \tag{A.90}$$

Then, we assume globally that there exist k C^1-class functions $g_j(x_1, x_2, \dots, x_n)$ for $j = 1, 2, \dots, k$ on $U \cap \mathbb{R}^n$ satisfying the properties.

We followed [90] when we write this section.

$$f_i(x_1, x_2, \dots, x_n, g_1, g_2, \dots, g_k) = 0, \quad i = 1, 2, \dots, k, \tag{A.91}$$

and

$$x_{n+j} = g_j(x_1, x_2, \dots, x_n), \quad j = 1, 2, \dots, k. \tag{A.92}$$

The main purpose is to represent the explicit functions g_j explicitly, in terms of the implicit functions $\{f_i\}$.

We are now in the position to present the main result.

Theorem A.7. *Let $U \subset \mathbb{R}^{n+k}$ be a smooth bounded domain surrounded by a finite number of C^1-class and simple closed hypersurfaces. For k functions*

$$f_i(x_1, x_2, \ldots, x_n, x_{n+1}, x_{n+2}, \ldots, x_{n+k}), \quad (i = 1, 2, \ldots, k), \tag{A.93}$$

we assume that for some point on U it holds that

$$f_i(x_1, x_2, \ldots, x_n, x_{n+1}, x_{n+2}, \ldots, x_{n+k}) = 0, \quad i = 1, 2, \ldots, k, \tag{A.94}$$

and that, on U, we have

$$\det \left(\frac{\partial (f_1, f_2, \ldots, f_k)}{\partial (x_{n+1}, x_{n+2}, \ldots, x_{n+k})}(x) \right) > 0. \tag{A.95}$$

In this way we also assume that C^1-functions $g_j(x_1, x_2, \ldots, x_n)$ on V globally satisfies

$$f_i(x_1, x_2, \ldots, x_n, g_1, g_2, \ldots, g_k) = 0, \quad i = 1, 2, \ldots, k, \tag{A.96}$$

and

$$x_{n+j} = g_j(x_1, x_2, \ldots, x_n), \quad j = 1, 2, \ldots, k \tag{A.97}$$

for each $j = 1, 2, \ldots, k$, where

$$V \equiv \left\{ (x_1, x_2, \ldots, x_n) \ : \ {}^t\begin{pmatrix} x_1 \\ x_2 \\ \vdots \\ x_n \\ g_1(x_1, x_2, \ldots, x_n) \\ g_2(x_1, x_2, \ldots, x_n) \\ \cdots \\ g_k(x_1, x_2, \ldots, x_n) \end{pmatrix} \in U \right\}.$$

Then, for $j = 1, 2, \ldots, k$, it holds that

$$\begin{aligned} & g_j(x_1, x_2, \ldots, x_n) \\ & = \sum_{i=1}^{n+k} \frac{(-1)^{n+j+i+1}}{c_{n+k}\omega_{n+k}} \\ & \quad \times \int_U \frac{(\eta - \eta_0)_i}{|\eta - \eta_0|^{n+k}} \det \frac{\partial(\eta_1, \eta_2, \ldots, \eta_{i-1}, \eta_{i+1}, \ldots, \eta_{n+k})}{\partial(\xi_1, \xi_2, \ldots, \xi_{n+j-1}, \xi_{n+j+1}, \ldots, \xi_{n+k})}(\xi)\, d\xi \\ & \quad + \int_{\partial U} \xi_{n+j} F^*[*dG_{n+k}(\eta - \eta_0)], \end{aligned}$$

where $\omega_n \equiv 2\pi^{\frac{n}{2}}/\Gamma(\frac{n}{2})$ *is the surface measure of the n dimensional unit disk and F is a mapping from U into* $\mathbb{R}^{n+k}$ *such that*

$$F(x_1, x_2, \ldots, x_{n+k}) = \begin{pmatrix} x_1 \\ x_2 \\ \vdots \\ x_n \\ f_1(x_1, x_2, \ldots, x_{n+k}) \\ f_2(x_1, x_2, \ldots, x_{n+k}) \\ \vdots \\ f_k(x_1, x_2, \ldots, x_{n+k}) \end{pmatrix}. \tag{A.98}$$

Proof. In order to apply Theorem 8.8, first we fix the direct mapping in Theorem 8.8 for our situation.

We will consider the C^1-class mapping F from U into $\mathbb{R}^{n+k}$ introduced in (A.98). It is clear that the Jacobian of this mapping is not vanishing on U as in

$$\det F'(x_1, x_2, \ldots, x_{n+k}) = \det\left(\frac{\partial(f_1, f_2, \ldots, f_k)}{\partial(x_{n+1}, x_{n+2}, \ldots, x_{n+k})}(x)\right) > 0. \tag{A.99}$$

By assumption, since the mapping F is injective on U, we can consider its inversion on its image domain. In particular, for any $(x_1, x_2, \ldots, x_n, 0, \ldots, 0)$ of the image domain that is the restriction to the domain U, we have from the situation in Sect. A.3.1, on $U \cap \mathbb{R}^n$ that

$$\begin{aligned} &F|_U^{-1}(x_1, x_2, \ldots, x_n, 0, \ldots, 0) \\ &= F|_U^{-1}\Big(x_1, x_2, \ldots, x_n, f_1(\mathbf{x}, \mathbf{g}), f_2(\mathbf{x}, \mathbf{g}), \ldots, f_k(\mathbf{x}, \mathbf{g})\Big) \\ &= \begin{pmatrix} x_1 \\ x_2 \\ \vdots \\ x_n \\ g_1 \\ g_2 \\ \vdots \\ g_k \end{pmatrix}, \end{aligned} \tag{A.100}$$

where

$$(\mathbf{x}, \mathbf{g}) = (x_1, x_2, \ldots, x_n, g_1, g_2, \ldots, g_k).$$

Therefore, by the representation in Theorem 8.8, we obtain the identities for the explicit functions g_i,

$$g_i(x_1, x_2, \ldots, x_n) = \int_{\partial U} \xi_{n+i} F^*[*dG_{n+k}](\eta - \eta_0) - \int_U d\xi_{n+i} \wedge F^*[*dG_{n+k}](\eta - \eta_0), \tag{A.101}$$

for $(x_1, x_2, \ldots, x_n) \in U \cap \mathbb{R}^n$. Here,

$$\xi = (\xi_1, \xi_2, \ldots, \xi_n, \xi_{n+1}, \ldots, \xi_{n+i}, \ldots, \xi_{n+k}), \quad (i = 1, 2, \ldots, k), \tag{A.102}$$

and

$$\eta - \eta_0 = (\xi_1 - x_1, \xi_2 - x_2, \ldots, \xi_n - x_n, f_1(\xi_1, \xi_2 \ldots, \xi_{n+k}), \ldots, f_k(\xi_1, \xi_2 \ldots, \xi_{n+k})). \tag{A.103}$$

Recalling that ω_n is the surface measure of the n dimensional unit disk, we have

$$G_n(x) = \frac{1}{c_n \omega_n} \begin{cases} |x|, & n = 1 \\ \log |x|, & n = 2 \quad \text{(logarithmic kernel)} \\ -|x|^{-n+2}, & n \geq 3 \quad \text{(Newton kernel)}. \end{cases}$$

Hence, on $\mathbb{R}^n \setminus U_\varepsilon(0)$ we have

$$dG_n(x) = \frac{1}{c_n \omega_n |x|^n} \sum_{i=1}^n x_i dx_i.$$

Therefore, by definition,

$$* \, dG_n(x) = \frac{1}{c_n \omega_n |x|^n} \sum_{i=1}^n (-1)^{i-1} x_i dx_1 \wedge \ldots \wedge dx_{i-1} \wedge dx_{i+1} \wedge \ldots \wedge dx_n, \tag{A.104}$$

for $x = (x_1, x_2 \ldots, x_n)$.

As a consequence, it holds that

$$*dG_{n+k}(\eta - \eta_0) = \sum_{i=1}^{n+k} \frac{(\eta - \eta_0)_i}{c_{n+k} \omega_{n+k} |\eta - \eta_0|^{n+k}} * d\eta_i.$$

Then, we can compute the pull back $F^*[*dG_{n+k}(\eta - \eta_0)]$ needed in the representation of the explicit functions by the following general formula and the Jacobian:

$$\begin{aligned} & y^*(dy_1 \wedge \ldots \wedge dy_{j-1} \wedge dy_{j+1} \wedge \ldots \wedge dy_n) \\ & = \sum_{k=1}^n (-1)^{k+} \det \left(\frac{\partial(y_1, y_2, \ldots, y_{j-1}, y_{j+1}, \ldots, y_n)}{\partial(x_1, x_2, \ldots, x_{k-1}, x_{k+1}, \ldots, x_n)} (x) \right) * dx_k. \end{aligned}$$

Indeed,

$$F^* * dG_{n+k}(\eta - \eta_0)$$

$$= \frac{1}{c_{n+k}\omega_{n+k}|\eta - \eta_0|^{n+k}}$$

$$\sum_{i=1}^{n+k}\sum_{p=1}^{n+k}(-1)^{i+p}(\eta - \eta_0)_i \det\left(\frac{\partial(\eta_1, \eta_2, \ldots, \eta_{i-1}, \eta_{i+1}, \ldots, \eta_{n+k})}{\partial(\xi_1, \xi_2, \ldots, \xi_{p-1}, \xi_{p+1}, \ldots, \xi_{n+k})}(\xi)\right) * d\xi_p$$

$$= \frac{1}{c_{n+k}\omega_{n+k}|\eta - \eta_0|^{n+k}}$$

$$\sum_{p=1}^{n+k}\left(\sum_{i=1}^{n+k}(-1)^{i+p}(\eta - \eta_0)_i \cdot \det\frac{\partial(\eta_1, \eta_2, \ldots, \eta_{i-1}, \eta_{i+1}, \ldots, \eta_{n+k})}{\partial(\xi_1, \xi_2, \ldots, \xi_{p-1}, \xi_{p+1}, \ldots, \xi_{n+k})}(\xi)\right) * d\xi_p.$$

Therefore, we obtain the desired representation of g_j for $j = 1, 2, \ldots, k$.

A.3.3 The 2-Dimensional Case: $n = 1, k = 1$

We will state the concrete representation formula for the 2-dimensional case. Realizing that

$$F^* * dG_2(\eta - \eta_0) = \frac{1}{\pi c_2((\xi_1 - x_1)^2 + f_1(\xi_1, \xi_2)^2)}$$
$$\times\left\{(\xi_1 - x_1)\frac{\partial f_1}{\partial \xi_1}d\xi_1 + (\xi_1 - x_1)\frac{\partial f_1}{\partial \xi_2}d\xi_2 - f_1(\xi_1, \xi_2)d\xi_1\right\},$$

we obtain

Theorem A.8. *For a C^1-class function $f(x_1, x_2)$ on a domain U in $\mathbb{R}^2$, we assume that for a point $x^0 = (x_1^0, x_2^0)$,*

$$f(x_1^0, x_2^0) = 0,$$
$$\frac{\partial f}{\partial x_2}(x_1^0, x_2^0) \neq 0.$$

- *There exist a neighbourhood $U_1 \times U_2$ $(\subset U)$ around the point x^0 and an explicit function $g : U_1 \to U_2$ determined by the implicit function $f = 0$ as $f(x_1, g(x_1)) = 0$ and, furthermore, it is represented as follows:*

$$g(x_1^*) = \frac{1}{2\pi}\left\{\oint_{\partial(U_1\times U_2)} x_2 d\theta - \int_{U_1\times U_2} dx_2 \wedge d\left[\text{Arctan}\left(\frac{f(x_1,x_2)}{x_1 - x_1^*}\right)\right]\right\},$$

for any $x_1^* \in U_1$.
- *For any $x_1^* \in U_1$, it holds that*

$$x_1^* = \frac{1}{2\pi}\left\{\oint_{\partial(U_1\times U_2)} x_1 d\left[\text{Arctan}\left(\frac{f(x_1,x_2)}{x_1 - x_1^*}\right)\right]\right.$$
$$\left. - \int_{U_1\times U_2} dx_1 \wedge d\left[\text{Arctan}\left(\frac{f(x_1,x_2)}{x_1 - x_1^*}\right)\right]\right\}.$$

Corollary A.2 (Representations of exact differential equations). *Let $f(x,y)$ be the C^1-class solution of the differential equation $f_x dx + f_y dy = 0$ on some domain $I_1 \times I_2$ on $\mathbb{R}^2$ satisfying $\dfrac{\partial f(x,y)}{\partial y} \neq 0$ and $y(x_0) = y_0$, $(x_0, y_0) \in I_1 \times I_2$. Then, we obtain the representation of the explicit function $y = y(x)$ which is determined by the implicit function $f(x,y) - f(x_0,y_0) = 0$ for any $x^* \in I_1$,*

$$y(x^*) = \frac{1}{2\pi}\oint_{\partial(I_1\times I_2)} y d\left[\text{Arctan}\left(\frac{f(x,y)-f(x_0,y_0)}{x - x^*}\right)\right]$$
$$- \frac{1}{2\pi}\iint_{I_1\times I_2} dy \wedge d\left[\text{Arctan}\left(\frac{f(x,y)-f(x_0,y_0)}{x - x^*}\right)\right].$$

Corollary A.3 (Representations of the inverse functions). *On an open interval $[a,b]$, for a C^1-class function f satisfying $f' > 0$ on $[a,b]$, its inverse function f^{-1} on $[f(a), f(b)]$ can be represented as follows:*

$$f^{-1}(y^*) = \frac{1}{2\pi}\iint_{[a,b]\times[f(a),f(b)]} dx \wedge d\left[\text{Arctan}\left(\frac{y - f(x)}{y - y^*}\right)\right]$$
$$-\frac{1}{2\pi}\oint_{\partial([a,b]\times[f(a),f(b)])} x d\left[\text{Arctan}\left(\frac{y - f(x)}{y - y^*}\right)\right]$$

for any $y^* \in [f(a), f(b)]$.

A.3.4 The 3-Dimensional Cases: $n = 1, k = 2$ or $n = 2, k = 1$

Finally, we will compute the 3-dimensional cases. Suppose $n = 2$ and $k = 1$. From

$$F^*[*dG_3(\eta - \eta_0)] = \frac{1}{4\pi c_3((\xi_1 - x_1)^2 + (\xi_2 - x_2)^2 + f_1(\xi_1, \xi_2, \xi_3)^2)^{3/2}}$$

$$\cdot\Big((\xi_1 - x_1)\frac{\partial f_1}{\partial \xi_3} d\xi_2 \wedge d\xi_3 - (\xi_2 - x_2)\frac{\partial f_1}{\partial \xi_3} d\xi_1 \wedge d\xi_3$$

$$+\big(f_1(\xi_1, \xi_2, \xi_3)\big)d\xi_1 \wedge d\xi_2 - (\xi_1 - x_1)\frac{\partial f_1}{\partial \xi_1} - (\xi_2 - x_2)\frac{\partial f_1}{\partial \xi_2}\Big),$$

we have:

$$g_1(x_1, x_2) = \frac{1}{4\pi c_3}\int_U \frac{1}{((\xi_1 - x_1)^2 + (\xi_2 - x_2)^2 + f_1(\xi_1, \xi_2, \xi_3)^2)^{3/2}}$$

$$\cdot\left((\xi_1 - x_1)\frac{\partial f_1}{\partial \xi_1} + (\xi_2 - x_2)\frac{\partial f_1}{\partial \xi_2} - f_1(\xi_1, \xi_2, \xi_3)\right) d\xi_1 \wedge d\xi_2 \wedge d\xi_3$$

$$+ \int_{\partial U} \xi_3 F^* * dG_3(\eta - \eta_0).$$

Next, suppose $n = 1$ and $k = 2$. From

$$F^* * dG_3(\eta - \eta_0) = \frac{1}{c_3\omega_3((\xi_1 - x_1)^2 + f_1(\xi_1, \xi_2, \xi_3)^2 + f_2(\xi_1, \xi_2, \xi_3)^2)^{3/2}}$$

$$\cdot\Big[(\xi_1 - x_1)\det\frac{\partial(f_1, f_2)}{\partial(\xi_2, \xi_3)} d\xi_2 \wedge d\xi_3 + (\xi_1 - x_1)\det\frac{\partial(f_1, f_2)}{\partial(\xi_1, \xi_3)} d\xi_1 \wedge d\xi_3$$

$$+\Big((\xi_1 - x_1)\det\frac{\partial(f_1, f_2)}{\partial(\xi_1, \xi_2)} - f_1(\xi_1, \xi_2, \xi_3)\frac{\partial f_2}{\partial \xi_2} + f_2(\xi_1, \xi_2, \xi_3)\frac{\partial f_1}{\partial \xi_2}\Big)d\xi_1 \wedge d\xi_2\Big],$$

we have:

$$g_1(x_1) = \frac{1}{c_3\omega_3}\int_U \frac{(\xi_1 - x_1)}{((\xi_1 - x_1)^2 + f_1(\xi_1, \xi_2, \xi_3)^2 + f_2(\xi_1, \xi_2, \xi_3)^2)^{3/2}}$$

$$\cdot\det\frac{\partial(f_1, f_2)}{\partial(\xi_1, \xi_3)} d\xi_1 \wedge d\xi_2 \wedge d\xi_3 + \int_{\partial U} \xi_2 F^* * dG_3(\eta - \eta_0),$$

and

$$g_2(x_1) = \int_{\partial U} \xi_3 F^* * dG_3(\eta - \eta_0)$$

$$-\frac{1}{4\pi c_3} \int_U \frac{d\xi_1 \wedge d\xi_2 \wedge d\xi_3}{((\xi_1 - x_1)^2 + f_1(\xi_1, \xi_2, \xi_3)^2 + f_2(\xi_1, \xi_2, \xi_3)^2)^{3/2}}$$

$$\cdot \left((\xi_1 - x_1) \det \frac{\partial(f_1, f_2)}{\partial(\xi_1, \xi_2)} - f_1(\xi_1, \xi_2, \xi_3) \frac{\partial f_2}{\partial \xi_2} + f_2(\xi_1, \xi_2, \xi_3) \frac{\partial f_1}{\partial \xi_2} \right).$$

Furthermore, for the singular integrals appeared in the above formulas, simple regularizations and their error estimates were given in [420]. By their regularizations, the above singular integrals may be calculated, easily by computers. We can see also numerical experiments there.

A.4 Overview on the Theory of Reproducing Kernels

For the theory of reproducing kernels in one-variable complex analysis, see the fundamental books [46, 146, 202]. For the advanced and profound theory, see the books D. A. Hejhal [202] and J. D. Fay [146] in connection with the Riemann theta functions and the Klein prime form. For their applications, see A. Yamada [484]. Their theory now seems to be, however, too complicated and advanced to deal with for any mathematician. For the theory of reproducing kernels in complex analysis with general variables, see the classical books [167, 168]. For the old history of reproducing kernels, see [28, 382]. In the proceedings [392] of an international conference, we can find various results on the theory of reproducing kernels. See [16] for wider topics on reproducing kernels.

We find many results in **learning theory** where applications of the theory of reproducing kernels are important; for example, the estimation of covering numbers by the disks of reproducing kernel Hilbert spaces as subspaces of a family of continuous functions, some detailed smoothness relationships between reproducing kernels, and reproducing kernel Hilbert spaces, and approximations of functions by Sobolev spaces. See, for example, S. Smale and Zhou [117, 502]. Indeed, we have many references for learning theory, see [471].

For **Support Vector Machines** that are favourable for engineers, see [116] and its research center at MIT.

We would like to note the active research results of K. Fukimizu [119, 169, 170] on statistic theory and reproducing kernels.

The article [15] summarizes various applications of the theory in connection with operator theory on Hilbert spaces whose research center is in Israel with a great group; see for example, D. Alpay at Ben-Gurion University of the Negev.

For the connection with stochastic theory and reproducing kernels, see Berlinet [48, 49] who was very active in the 5th ISAAC Catania Congress with his colleagues and we see many results in this field.

For some more recent general discretization principles with many concrete applications, see [92, 93]. A new global theory combining the fundamental relations among eigenfunctions, initial value problems in general linear partial differential operators, and reproducing kernels, see [96, 97].

References

1. L.D. Abreu, The reproducing kernel structure arising from a combination of continuous and discrete orthogonal polynomials into Fourier systems. Constr. Approx. **28**(2), 219–235 (2008)
2. L.D. Abreu, A q-sampling theorem related to the q-Hankel transform. Proc. Am. Math. Soc. **133**, 1197–1203 (2005)
3. L.D. Abreu, Sampling theory associated with q-divergence equations of the Sturm-Liouville type. J. Phys. A: Math. Gen. **38**, 10311–10319 (2005)
4. L.D. Abreu, A. Bandeira, Landau's necessary density conditions for the Hankel transform. J. Funct. Anal. **162**, 1845–1866 (2012)
5. L.D. Abreu, O. Ciaurri, J.L. Varona, Bilinear biorthogonal expansions and the Dunkl kernel on the real line. Expo. Math. **30**, 32–48 (2012)
6. A.A. Adamov, On the asymptotic expansion of the polynomials $e^{ax^2/2}d^n[e^{(-ax^2/2)}]/dx^n$ for large values of n (in Russian). Ann. Polytech. Inst. St. Petersb. **5**, 127–143 (1906)
7. H. Aikawa, Infinite order sobolev spaces, analytic continuation and polynomial expansions. Complex Var. **18**, 253–266 (1992)
8. H. Aikawa, N. Hayashi, S. Saitoh, Isometrical identities for the Bergman and the Szegö spaces on a sector. J. Math. Soc. Jpn. **43**, 195–201 (1991)
9. J. Agler, Nevanlinna-Pick interpolation on Sobolev space. Proc. Am. Math. Soc. **108**, 341–351 (1990)
10. J. Agler, J.E. McCarthy, *Pick Interpolation and Hilbert Function Spaces*. Graduate Studies in Mathematics, vol. 44 (American Mathematical Society, Providence, 2002)
11. G.D. Allen, F.J. Narcowich, J.P. Williams, An operator version of a theorem of Kolmogorov. Pac. J. Math. **61**(2), 305–312 (1975)
12. D. Alpay, P. Bruinsma, A. Dijksma, H.d. Snoo, Interpolation problems, extensions of symmetric operators and reproducing kernel spaces I. Integr. Equ. Oper. Theory **14**, 465–500 (1991)
13. D. Alpay, P. Bruinsma, A. Dijksma, H.d. Snoo, Interpolation problems, extensions of symmetric operators and reproducing kernel spaces II. Integr. Equ. Oper. Theory **14**, 378–388 (1992)
14. D. Alpay, H. Dym, On applications of reproducing kernel spaces to the Schur algorithm and rational J unitary factorization. Oper. Theory Adv. Appl. **18**, 89–159 (1986)
15. D. Alpay, *The Schur Algorithm, Reproducing Kernel Spaces and System Theory* (Translation of book: *Algorithme de Schur, espaces noyau reproduisant et théorie des systémes*. Panoramas et Synthéses, vol. 6 (Société Mathématique de France, 1998)). SMF/AMS Texts and Monographs of the Société Mathématique de France, vol. 5 (American Mathematical Society, Providence, 2001)

S. Saitoh, Y. Sawano, *Theory of Reproducing Kernels and Applications*,
Developments in Mathematics 44, DOI 10.1007/978-981-10-0530-5

16. D. Alpay, *Reproducing Kernel Spaces and Applications*. Operator Theory, Advances and Applications, vol. 143 (Birkhäuser, Basel, 2003)
17. L.V. Ahlfors, *Complex Analysis* (McGraw-Hill Science/Engineering/Mathematics, New York, 1978)
18. I.A. Aizenberg, A.P. Yuzhakov, *Integral Representations and Residues in Multidimensional Complex Analysis*, translated from the Russian by H.H. McFaden, translation ed. by Lev J. Leifman. Translations of Mathematical Monographs, vol. 58 (American Mathematical Society, Providence, 1983)
19. K. Amano, S. Saitoh, M. Yamamoto, Error estimates of the real inversion formulas of the laplace transform. Integr. Trans. Spec. Funct. **10**(3–4), 165–178 (2000)
20. K. Amano, D. Okano, H. Ogata, M. Sugihara, Numerical conformal mappings onto the linear slit domains. Jpn. J. Ind. Appl. Math. **29**, 165–186 (2012)
21. M.H. Annaby, A.I. Zayed, On the use of Green's function in sampling theory. J. Integr. Equ. Appl. **10**(2), 117–139 (1998)
22. B.D.O. Anderson, T. Kailath, Fast algorithms for the integral equations of the inverse scattering problem. Integr. Equ. Oper. Theory **1**(1), 132–136 (1978)
23. G.E. Andrews, R. Askey, R. Roy, *Special Functions*. Encyclopedia of Mathematics and Its Applications, vol. 71 (Cambridge University Press, Cambridge, 2009)
24. D.D. Ang, D.N. Thanh, V.V. Thanh, Regularized solutions of a Cauchy problem for the Laplace equation in an irregular strip. J. Integr. Equ. Appl. **5**, 429–441 (1993)
25. M.H. Annaby, G. Freiling, Sampling expansions associated with Kamke problems. Math. Z. **234**(1), 163–189 (2000)
26. J. Arazy, S. Fisher, J. Peetre, Hankel operators on weighted Bergman spaces. Am. J. Math. **110**, 989–1054 (1988)
27. G.B. Arfken, H.J. Weber, *Mathematical Methods for Physicists*, 5th edn. (Harcourt/Academic Press, Burlington, 2001)
28. N. Aronszajn, Theory of reproducing kernels. Trans. Am. Math. Soc. **68**, 337–404 (1950)
29. N. Aronszajn, Green's functions and reproducing kernels, in *Proceedings of the Symposium on Spectral Theory and Differential Problems* (Oklahoma Agricultural and Mechanical College, Oklahoma) (1951), pp. 164–187
30. N. Aronszajn, K.T. Smith, Theory of Bessel potentials. Part I. Ann. Inst. Fourier Grenoble **11**, 385–475 (1961)
31. M. Asaduzzaman, T. Matsuura, S. Saitoh, *Constructions of Approximate Solutions for Linear Differential Equations by Reproducing Kernels and Inverse Problems*. Advances in Analysis, Proceedings of the 4th International ISAAC Congress (World Scientific, Singapore/Hackensack, 2005), pp. 335–343
32. K.T. Atkinson, The numerical solution of the eigenvalue problem for compact integral operators. Trans. Am. Math. Soc. **129**, 458–465 (1967)
33. S. Axler, Bergman spaces and their operators, in *Survey of Some Recent Results in Operator Theory, Volume I*, ed. by J.B. Conway, B.B. Morrel. Pitman Research Notes in Mathematics, vol. 171 (Longman Scientific & Technical/Wiley, New York, 1988), pp. 1–50
34. S. Azizi, D. Cochran, Reproducing kernel structure and sampling on time-warped spaces with application to warped wavelets. IEEE Trans. Inf. Theory **48**(3), 789–790 (2002)
35. J.A. Ball, V. Vinnikov, *Formal Reproducing Kernel Hilbert Spaces: The Commutative and Noncommutative Settings. Reproducing Kernel Spaces and Applications*. Operator Theory: Advances and Applications, vol. 143 (Birkhäuser, Basel, 2003), pp. 77–134
36. R.B. Bapat, *Graphs and Matrices*. Universitext (Springer/Hindustan Book Agency, London/New Delhi, 2010). x+171pp.
37. M.G. Beaty, M.M. Dodson, S.P. Eveson, J.R. Higgins, On the approximate form of Kluvanek's theorem. J. Approx. Theory **160**, 281–303 (2009)
38. P.R. Beesack, Elementary proofs of some Opial-type integral inequalities. J. Anal. Math. **36**(1980), 1–14 (1979)
39. P.R. Beesack, K.M. Das, Extensions of Opial's inequality. Pac. J. Math. **26**, 215–232 (1968)

40. H. Begehr, R.P. Gilbert, *Transformations, Transmutations, and Kernel Functions*, vol. 1. Pitman Monographs and Surveys in Pure and Applied Mathematics, vol. 58 (Longman Scientific and Technical/Copublished in the United States with John Wiley and Sons, Harlow/New York, 1992)
41. H. Begehr, R.P. Gilbert, *Transformations, Transmutations, and Kernel Functions*, vol. 2. Pitman Monographs and Surveys in Pure and Applied Mathematics, vol. 59 (Longman Scientific and Technical/Copublished in the United States with John Wiley and Sons, Harlow/New York, 1993)
42. Ju.M. Berezanskii, *Expansions in Eigenfunctions of Selfadjoint Operators*. Translations of Mathematical Monographs, vol. 17 (American Mathematical Society, Providence, 1968)
43. F.A. Berezin, Quantization. Math. USSR IZv. **8**, 1109–1165 (1974)
44. F.A. Berezin, General concept of quantization. Commun. Math. Phys. **40**, 153–174 (1975)
45. S. Bergman, Über die Entwicklung der harmischen Funktionen der Ebene und des Rümes nach Orthogonal Functionen. Math. Ann. **86**, 238–271 (1922)
46. S. Bergman, *The Kernel Function and Conformal Mapping* (American Mathematical Society, Providence, 1950/1970)
47. B. Bergman, M.M. Schiffer, *Kernel Functions and Differential Equations* (Academic, New York, 1953)
48. A. Berlinet, C. Thomas-Agnan, *Reproducing Kernel Hilbert Spaces in Probability and Statistics* (Kluwer Akademic Publishers, Boston/Dordrecht/London, 2004)
49. A. Berlinet, Reproducing kernels in probability and statistics, in *More Progress In Analysis, Proceedings of the 5th International ISAAC Congress*, Catania (2010), pp. 153–162
50. F. Berntsson, L. Elden, Numerical solution of a Cauchy problem for the Laplace equation. Special issue to celebrarate Pierre Sabatier's 65th birthday (Montpellier, 2000). Inverse Probl. **17**, 10–87 (2005)
51. M. Bertero, Linear inverse and ill-posed problems. Adv. Electron. Electron Phys. **75**, 2–120 (1989)
52. R. Bhatia, *Matrix Analysis*. Graduate Texts in Mathematics, vol. 169 (Springer, New York, 1997)
53. G. Bogveradze, L.P. Castro, Toeplitz plus Hankel operators with infinite index. Integr. Equ. Oper. Theory **62**, 43–63 (2008)
54. B.E. Boser, I.M. Guyon, V.N. Vapnik, A training algorithm for optimal margin classifiers, in *Fifth Annual ACM Workshop on Computational Learning Theory*, ed. by D. Haussler (ACM Press, Pittsburgh, 1992), pp. 144–152
55. A. Boumenir, T.K. Tuan, Sampling eigenvalues in Hardy spaces. SIAM J. Numer. Anal. **45**, 473–483 (2007)
56. A. Boumenir, T.K. Tuan, The computation of eigenvalues of singular Sturm-Liouville operators. Adv. Appl. Math. **39**, 222–236 (2007)
57. A. Boumenir, T.K. Tuan, Representation and sampling of Hardy functions. Math. Methods Appl. Sci. **33**, 485–492 (2010)
58. L. Bourgeois, A mixed formulation of quasi-reversibility to solve the Cauchy problem for Laplace's equation. Inverse Probl. **21**, 1087–1104 (2005)
59. L. de Branges, J. Rovnyak, *Square Summable Power Series* (Holt, Rinehart and Winston, New York, 1966)
60. L. de Branges, *Hilbert Spaces of Entire Functions* (Prentice-Hall, Englewood Cliffs, 1968)
61. L. de Branges, *Underlying Concepts in the Proof of the Bieberbach Conjecture* (Proceedings of International Congress of Mathematics, Berkeley, 1986), pp. 25–42
62. J. Buescu, A.C. Paixao, On differentiability and analyticity of positive definite functions. J. Math. Anal. Appl. **375**(1), 336–341 (2011)
63. J. Buescu, A.C. Paixao, Algebraic, differential, integral and spectral properties of Mercer-like-kernels, in *More Progress in Analysis. Proceedings of the 5th International ISAAC Congress*, Catania, ed. by H.G.W. Begehr et al., 25–30 July 2005 (World Scientific, Hackensack, 2009), pp. 175–188
64. J. Burbea, Total positivity of certain reproducing kernels. Pac. J. Math. **67**, 101–130 (1976)

65. J. Burbea, A Dirichlet norm inequality and some inequalities for reproducing kernel spaces. Proc. Am. Math. Soc. **83**(2), 279–285 (1981)
66. J. Burbea, Norm inequalities of exponential type for holomorphic functions. Kodai Math. J. **5**(2), 339–354 (1982)
67. J. Burbea, Inequalities for reproducing kernel spaces. Ill. J. Math. **27**(1), 130–137 (1983)
68. J. Burbea, Inequalities for holomorphic functions of several complex variables. Trans. Am. Math. Soc. **276**(1), 247–266 (1983)
69. J. Burbea, P. Masani, Banach and Hilbert spaces of vector-valued functions. in *Their General Theory and Applications to Holomorphy*. Research Notes in Mathematics, vol. 90 (Pitman (Advanced Publishing Program), Boston, 1984)
70. T. Byczkowski, RKHS for Gaussian measures on metric vector spaces. Bull. Polish Acad. Sci. Math. **35**(1–2), 93–103 (1987)
71. D.W. Byun, S. Saitoh, A real inversion formula for the Laplace transform. Z. Anal. Anw. **12**, 597–603 (1993)
72. D.W. Byun, S. Saitoh, Approximation by the solutions of the heat equation. J. Approx. Theory **78**, 226–238 (1994)
73. D.W. Byun, S. Saitoh, Best approximation in reproducing kernel Hilbert spaces, in *Proceedings of the 2th International Colloquium on Numerical Analysis* (VSP-Holland, 1994), pp. 55–61
74. P.L. Butzer, P.J.S.G. Ferreira, J.R. Higgins, S. Saitoh, G. Schmeisser, R.L. Stens, Interpolation and Sampling: E.T. Whittaker, K. Ogura and their followers. J. Fourier Anal. Appl. **17**, 320–354 (2011)
75. J. Calvert, Some generalizations of Opial's inequality. Proc. Am. Math. Soc. **18**, 72–75 (1967)
76. J.R. Cannon, P. DuChateau, Approximating the solution to the Cauchy problem for Laplace's equation. SAM J. Numer. Anal. **14**, 473–483 (1977)
77. L.P. Castro, R. Duduchava, F.O. Speck, Singular integral equations on piecewise smooth curves in spaces of smooth functions. Oper. Theory Adv. Appl. **135**, 107–144 (2002)
78. L.P. Castro, Q. Chen, S. Saitoh, Source inversion of heat conduction from a finite number of observation data. Appl. Anal. **89**, 801–813 (2010)
79. L.P. Castro, E.M. Rojas, Reduction of singular integral operators with flip and their Fredholm property. Lobachevskii J. Math. **29**(3), 119–129 (2008)
80. L.P. Castro, E.M. Rojas, Similarity transformation methods for singular integral operators with reflection on weighted Lebesgue spaces. Int. J. Mod. Math. **3**, 295–313 (2008)
81. L.P. Castro, E.M. Rojas, A collocation method for singular integral operators with reflection. RIMS Kokyuroku **1719**, 155–167 (2010–2011)
82. L.P. Castro, E.M. Rojas, S. Saitoh, Inversion from different kinds of information and real inversion formulas of the Laplace transform from a finite number of data. MESA Math. Eng. Sci. Aerosp. **1**, 181–190 (2010)
83. L.P. Castro, S. Saitoh, N.M. Tuan, Convolutions, integral transforms and integral equations by means of the theory of reproducing kernels. Opusc. Math. **32**(4), 633–646 (2012)
84. L.P. Castro, S. Saitoh, Natural outputs and global inputs of linear systems with a finite number of input data. Appl. Anal. **91**(2), 225–236 (2012)
85. L.P. Castro, S. Saitoh, Y. Sawano, A. Simões, General inhomogeneous discrete linear partial differential equations with constant coefficients on the whole spaces. Complex Anal. Oper. Theory **6**(1), 307–324 (2012)
86. L.P. Castro, S. Saitoh, Y. Sawano, A.S. Silva, Discrete linear differential equations. Analysis **32**(3), 181–191 (2012)
87. L.P. Castro, S. Saitoh, New convolutions and norm inequalities. Math. Inequal. Appl. **15**(3), 707–716 (2012)
88. L.P. Castro, S. Saitoh, Fractional functions and their representations. Complex Anal. Oper. Theory **7**(4), 1049–1063 (2013)
89. L.P. Castro, H. Itou, S. Saitoh, Numerical solutions of linear singular integral equations by means of Tikhonov regularization and reproducing kernels. Houst. J. Math. **38**(4), 1261–1276 (2012)

90. L.P. Castro, K. Murata, S. Saitoh, M. Yamada, Explicit integral representations of implicit functions. Carpath. J. Math. **29**(2), 141–148 (2013)
91. L.P. Castro, H. Fujiwara, M.M. Rodrigues, S. Saitoh, A new discretization method by means of reproducing kernels, in *Interactions Between Real and Complex Analysis*, ed. by L.H. Son, W. Tutscheke (Science and Technics Publication House, Ha Noi, 2012), pp. 185–223
92. L.P. Castro, H. Fujiwara, M.M. Rodrigues, S. Saitoh, V.K. Tuan, Aveiro discretization method in mathematics: a new discretization principle, in *Mathematics Without Boundaries: Surveys in Pure Mathematics*, ed. by P. Pardalos, T.M. Rassias (Springer, New York, 2014), pp. 37–92
93. L.P. Castro, H. Fujiwara, T. Qian and S. Saitoh, How to catch smoothing properties and analyticity of functions by computers? in *Mathematics Without Boundaries: Surveys in Interdisciplinary Research* (Springer, New York, 2014), pp. 101–116
94. L.P. Castro, S. Saitoh, Moore-Penrose generalized solutions of convolution integral equations. Numer. Funct. Anal. Optim. (in press)
95. L.P. Castro, A. Silva, S. Saitoh, A reproducing kernel Hilbert space constructive approximation for integral equations with Toeplitz and Hankel kernels. Libertus Math. (New Ser.) **34**(1), 1–22 (2014)
96. L.P. Castro, M.M. Rodorigues, S. Saitoh, Initial value problems in linear integral operators equations, in *Topics in Mathematical Analysis and Applications*, vol. 94, ed. by L. Toth, T.M. Rassias (Springer, Cham, 2014), pp. 175–188
97. L.P. Castro, M.M. Rodorigues,S. Saitoh, A fundamental theorem on initial value problems by using the theory of reproducing kernels. Complex Anal. Oper. Theory **9**(1), 87–98 (2014)
98. A.L. Cauchy, Cours d'Analyse de l'École Royale Polytechnique, Irére partie: Analyse Algébrique, Paris, Debure frères, 1821, Also in Oeuvres Complètes d'Augustin Cauchy, Séric II, T. E (Gauthier-Villars, Paris, 1987)
99. K. Chadan, P.C. Sabatier, *Inverse Problems in Quantum Scattering Theory*, 2nd edn. Texts and Monographs in Physics (Springer, New York, 1989)
100. B.L. Chalmers, Subspace kernels and minimum problems in Hilbert spaces with kernel function. Pac. J. Math. **31**, 619–628 (1969)
101. Y.-S. Chan, G.H. Paulino, A.C. Fannjiang, The crack problem for nonhomogeneous materials under antiplane shear loading: a displacement based formulation. Int. J. Solids Struct. **38**, 2989–3005 (2001)
102. Y.-S. Chan, A.C. Fannjiang, G.H. Paulino, Integral equations with hypersingular kernels–theory and applications to fracture mechanics. Int. J. Eng. Sci. **41**, 683–720 (2003)
103. Z. Chena, Y. Zhoub, An efficient algorithm for solving Hilbert type singular integral equations of the second kind. Comput. Math. Appl. **58**, 632–640 (2009)
104. J. Cheng, Y.C. Hon, T. Wei, M. Yamamoto, Numerical computation of a Cauchy problem for Laplace's equation. Z. Angew. Math. Mech. **81**, 665–674 (2001)
105. V.G. Cherednichenko, *Inverse Logarithmic Potential Problem* (VSP, Utrecht, 1996)
106. T.S. Chihara, *An Introduction to Orthogonal Polynomials* (Gordon and Breach, New York, 1978)
107. D. Colton, R. Kress, *Inverse Acoustic and Electromagnetic Scattering Theory* (Springer, Berlin, 1992)
108. A.M. Cohen, *Numerical Methods for Laplace Transform Inversion* (Springer, New York, 2007)
109. D. Constales, R.S. Krausshar, Szegö and polymonogenic Bergman kernels for half-space and strip domains, and single-periodic functions in Clifford analysis. Complex Var. Theory Appl. **47**(4), 349–360 (2002)
110. D. Constales, R.S. Krausshar, Bergman kernels for rectangular domains and multiperiodic functions in Clifford analysis. Math. Meth. Appl. Sci. **25**(4), 1509–1526 (2002)
111. T. Constantinescu, A. Gheondea, Representations of Hermitian kernels by means of Krein spaces. II. Invariant kernels. Commun. Math. Phys. **216**(2), 409–430 (2001)
112. T. Constantinescu, A. Gheondea, Invariant Hermitian kernels and their Kolmogorov decompositions (English. Abridged French version). C. R. Acad. Sci. Paris Ser. I Math. **331**(10), 797–802 (2000)

113. T. Constantinescu, A. Gheondea, Representations of Hermitian kernels by means of Krein spaces. Publ. RIMS Kyoto Univ. **33**, 917–951 (1997)
114. C. Corduneanu, *Integral Equations and Applications* (Cambridge University Press, Cambridge, 1991)
115. B. Cotterell, J.R. Rice, Slightly curved or kinked cracks. Int. J. Fract. **16**, 155–169 (1980)
116. N. Cristianini, J. Shawe-Taylor, *An Introduction to Support Vector Machines and Other Kernel-Based Learning Methods* (Cambridge University Press, Cambridge, 2001)
117. F. Cucker, S. Smale, On the mathematical foundations of learning. Bull. Am. Math. Soc. **39**, 1–49 (2002)
118. M. Cui, Y. Lin, *Nonlinear Numerical Analysis in the Reproducing Kernel Space* (Nova Science Publisher, New York, 2009)
119. M. Cuturi, K. Fukumizu, J.P. Vert, Semigroup kernels on measures. J. Mach. Learn. Res. **6**, 1169–1198 (2005)
120. C.J. Dalzell, J.O. Ramsay, Computing reproducing kernels with arbitrary boundary constraints. SIAM J. Sci. Comput. **14**(3), 511–518 (1993)
121. I. Daubechies, *Ten Lectures on Wavelets and Applications* (Society for Industrial and Applied Mathematics, Philadelphia, 1992)
122. P.J. Davis, *Interpolation and Approximation* (Blaisdell, New York, 1963)
123. A. Devinatz, Integral representations of positive definite functions. Trans. Am. Math. Soc. **74**, 56–76 (1953)
124. A. Devinatz, On the extensions of positive definite functions. Acta Math. **102**, 109–134 (1959)
125. N.H. Dinh, M.H. Pham, Stability results for the Cauchy problems for the Laplace equation in a strip. Inverse Probl. **19**, 833–844 (2003)
126. G. Doetsch, *Handbuch der Laplace Transformation*, vol. 1 (Birkhäuser Verlag, Basel, 1950)
127. M.F. Driscoll, The reproducing kernel Hilbert space structure of the sample paths of a Gaussian process. Z. Wahrscheinlichkeitstheorie und Verw. Gebiete **26**, 309–316 (1973)
128. J. Du, M. Cui, Solving the forced duffing equation with integral boundary conditions in the reproducing kernel space. Int. J. Comput. Math. **87**, 2088–2100 (2010)
129. D.T. Duc, N.D.V. Nhan, On some convolution norm inequalities in weighted $L_p(\mathbb{R}^n, \rho)$ spaces and their applications. Math. Inequal. Appl. **11**(3), 495–505 (2008)
130. D.T. Duc, N.D.V. Nhan, Some applications of convolution inequalities in weighted L_p spaces. Integr. Transf. Spec. Funct. **19**(7–8), 471–480 (2008)
131. D.T. Duc, N.D.V. Nhan, On some reverse weighted $L_p(\mathbb{R}^n)$-norm inequalities in convolutions and their applications. Math. Inequal. Appl. **12**(1), 67–80 (2009)
132. D.T. Duc, N.D.V. Nhan, Generalize some norm inequalities of Saitoh. Kodai Math. J. **34**(2), 191–201 (2011)
133. D.T. Duc, N.D.V. Nhan, Norm inequalities for new convolutions and their applications. Appl. Anal. Discret. Math. **9**, 168–179 (2015)
134. J. Duoandikoetxea, *Fourier Analysis*. Translated and revised from the 1995 Spanish original by David Cruz-Uribe. Graduate Studies in Mathematics, vol. 29 (American Mathematical Society, Providence, 2001)
135. D.L. Duttweiler, T. Kailath, RKHS approach to detection and estimation problems. V. Parameter estimation. IEEE Trans. Inf. Theory **IT-19**(1), 29–36 (1973)
136. D.L. Duttweiler, T. Kailath, RKHS approach to detection and estimation problems. IV. Non-Gaussian detection. IEEE Trans. Inf. Theory **IT-19**(1), 19–28 (1973)
137. D.L. Duttweiler, T. Kailath, An RKHS approach to detection and estimation problems III. Generalized innovations representations and a likelihood-ratio formula. IEEE Trans. Inf. Theory **IT-18**, 730–745 (1972)
138. H. Dym, *Hermitian Block Toeplitz Matrices, Orthogonal Polynomials, Reproducing Kernel Pontryagin Spaces, Interpolation And Extension. Orthogonal Matrix-Valued Polynomials and Applications (Tel Aviv, 1987–1988)*. Operator Theory: Advances and Applications, vol. 34 (Birkhäuser, Basel, 1988), pp. 79–135
139. H. Dym, *J Contractive Matrix Functions, Reproducing Kernel Hilbert Spaces and Interpolation*. CBMS Regional Conference Series in Mathematics, vol. 71 (Published for the

Conference Board of the Mathematical Sciences/American Mathematical Society, Washington, DC/Providence, 1989)
140. I. Feldman, I. Gohberg, N. Krupnik, Convolution equations on finite intervals and factorization of matrix functions. Integr. Equ. Oper. Theory **36**, 201–211 (2000)
141. H.W. Engl, M. Hanke, A. Neubauer, *Regularization of Inverse Problems*. Mathematics and Its Applications, vol. 376 (Kluwer Academic Publishers, Dordrecht/Boston, 2000)
142. R. Estrada, R.P. Kanwal, *Singular Integral Equations* (Birkhäuser, Boston, 2000)
143. W.N. Everitt, G. Nasri–Roudsari, Sturm–Liouville problems with coupled boundary conditions and Lagrange interpolation series. J. Comput. Anal. Appl. **1**, 319–347 (1999)
144. A.C. Fannjiang, Y.-S. Chan, G.H. Paulino, Strain gradient elasticity for antiplane shear cracks: a hypersingular integrodifferential equation approach. SIAM J. Appl. Math. **62**, 1066–1091 (2002)
145. H.M. Farkas, I. Kra, *Riemann Surfaces*, 2nd edn. (Springer, New York, 1992)
146. J.D. Fay, *Theta Functions on Riemann Surfaces*. Lecture Notes in Mathematics, vol. 352 (Springer, Berlin/New York, 1973)
147. R.J.P. de Figueirredo, G. Chen, *Nonlinear Feedback Control Systems: An Operator Theory Approach* (Academic Press, Boston/San Diego/New York, 1993)
148. H.G. Feichtinger, Modulation spaces on locally compact Abelian groups. Technical report, University of Vienna (1983)
149. H.G. Feichtinger, Atomic characterization of modulation spaces through Gabor-type representation, in *Proceedings of the Conference Constructive Function Theory*, Edmonton, pp. 113–126 (1989)
150. H.G. Feichtinger, Gewichtsfunktionen auf lokalkompakten Gruppen. Österreich. Akad. Wiss. Math.-Natur. Kl. Sitzungsber. II **188**, 451–471 (1979)
151. H.G. Feichtinger, K. Gröchenig, *A Unified Approach to Atomic Decompositions via Integrable Group Representations. Function Spaces and Applications*, Lund, 1986. Lecture Notes in Mathematics, vol. 1302 (Springer, Berlin, 1988), pp. 52–73
152. H.G. Feichtinger, K. Gröchenig, Banach spaces related to integrable group representations and their atomic decompositions, I. J. Funct. Anal. **21**, 307–340 (1989)
153. H.G. Feichtinger, K. Gröchenig, Banach spaces related to integrable group representations and their atomic decompositions, II. Monatsh. Math. **108**, 129–148 (1989)
154. H.G. Feichtinger, K. Gröchenig, Gabor wavelets and the Heisenberg group: Gabor expansions and short time fourier transform from the group theoretical point of view, in *Wavelets A Tutorial in Theory and Applications*, ed. by C.K. Chui. Wavelet Analysis and Its Applications, vol. 2 (Academic Press, Boston, 1992), pp. 359–397
155. H.G. Feichtinger, K. Gröchenig, Gabor frames and time-frequency analysis of distributions. J. Funct. Anal. **146**(2), 464–495 (1997)
156. M. Frazier, B. Jawerth, *φ-Transform and Applications to Distribution Spaces, Function Spaces and Applications*, (Lund 1986). Lecture Notes in Mathematics, vol. 1302 (Springer, Berlin, 1988), pp. 223–246
157. M. Frazier, B. Jawerth, A discrete transform and decompositions of distribution spaces. J. Funct. Anal. **93**(1), 34–170 (1990)
158. H. Fujiwara, High-accurate numerical method for integral equation of the first kind under multiple-precision arithmetic. Theor. Appl. Mech. Jpn. **52**, 192–203 (2003)
159. H. Fujiwara, exflib – multiple-precision arithmetic library (2005). http://www-an.acs.i.kyoto-u.ac.jp/~fujiwara
160. H. Fujiwara, T. Matsuura, S. Saitoh, Y. Sawano, The real inversion of the Laplace transform by numerical singular value decomposition. J. Anal. Appl. **6**, 55–68 (2008)
161. H. Fujiwara, Numerical real inversions of the Laplace transform by multiple-precision arithmetic. RIMS Kokyuroku **1566**, 181–195 (2007)
162. H. Fujiwara, Applications of reproducing kernel spaces to real inversions of the Laplace transform. RIMS Kokyuroku **1618**, 188–209 (2008)

163. H. Fujiwara, T. Matsuura, S. Saitoh, Y. Sawano, *Numerical Real Inversion of the Laplace Transform by Using a High-Accuracy Numerical Method*. Further Progress in Analysis (World Scientific Publication, Hackensack, 2009), pp. 574–583
164. H. Fujiwara, Numerical real inversion of the Laplace transform by reproducing kernel and multiple-precision arithmetic, in *Progress in Analysis and Its Applications, Proceedings of the 7th International ISAAC Congress* (World Scientific, London, 2010), pp. 289–295
165. H. Fujiwara, N. Higashimori, Numerical inversion of the laplace transform by using multiple-precision arithmetic. Libertas Math. (new Ser.) **34**(2), 5–21 (2014)
166. H. Fujiwara, S. Saitoh, The general sampling theory by using reproducing kernels, in *Contributions in Mathematics and Engineering. Honor of Constantin Caratheodory*, ed. by T.M. Rassias, P. Pardalos (Springer, 2016)
167. B.A. Fuks, *Introduction to the Theory of Analytic Functions of Several Complex Variables*. Translations of Mathematical Monographs, vol. 8 (American Mathematical Society, Providence, 1963)
168. B.A. Fuks, *Special Chapters in the Theory of Analytic Functions of Several Complex Variables*. Translations of Mathematical Monographs, vol. 14 (American Mathematical Society, Providence, 1965)
169. K. Fukumizu, F.R. Bach, M.I. Jordan, Dimensionality reduction for supervised learning with reproducing kernel Hilbert spaces. J. Mach. Learn. Res. **5**, 73–99 (2004)
170. K. Fukumizu, F.R. Bach, A. Gretton, Statistical consistency of kernel canonical correlation analysis. J. Mach. Learn. Res. **8**, 361–383 (2007)
171. P.R. Garabedian, *Partial Differential Equations* (Wiley, New York, 1964)
172. A.G. Garcia, M.A. Hernandez-Medina, The discrete Kramer sampling theorem and indeterminate moment problems. J. Comput. Appl. Math. **134**, 13–22 (2001)
173. A.G. Garcia, A brief walk through sampling theory. Adv. Imaging Electron Phys. **124**, 63–137 (2002)
174. A.G. Garcia, F.H. Szafraniec, A converse of the Kramer sampling theorem. Sampl. Theorem Image Proc. **1**, 53–61 (2002)
175. J.B. Garnett, *Bounded Analytic Functions*. Pure and Applied Mathematics, vol. 96 (Academic Press Inc., New York, 1981)
176. A. Gasmi, F. Soltani, Fock spaces for the Bessel-Stuve kernel. J. Anal. Appl. **3**, 91–106 (2005)
177. F. Geng, M. Cui, Solving singular nonlinear second-order periodic boundary value problems in the reproducing kernel space. Appl. Math. Comput. **192**, 389–398 (2007)
178. F. Geng, M. Cui, A reproducing kernel method for solving nonlocal fractional boundary value problems. Appl. Math. Lett. **25**, 818–823 (2012)
179. F. Geng, M. Cui, Homotopy perturbation-reproducing kernel method for nonlinear systems of second order boundary value problems. J. Comput. Appl. Math. **235**, 2405–2411 (2011)
180. F. Geng, M. Cui, A novel method for nonlinear two-point boundary value problems: combination of ADM and RKM. Appl. Math. Comput. **217**, 4676–4681 (2011)
181. F. Geng, M. Cui, Solving singular nonlinear boundary value problems by combining the homotopy perturbation method and reproducing kernel Hilbert space method. Int. J. Comput. Math. **87**, 2024–2031 (2010)
182. B.T. Giang, N.V. Mau, N.M. Tuan, Operational properties of two integral transforms of Fourier type and their convolutions. Integr. Equ. Oper. Theory **65**(3), 363–386 (2009)
183. B.T. Giang, N.V. Mau, N.M. Tuan, Convolutions for the Fourier transforms with geometric variables and applications. Math. Nachr. **283**, 1758–1770 (2010)
184. B.T. Giang, N.M. Tuan, Generalized convolutions for the Fourier integral transforms and applications. J. Sib. Federal Univ. **1**(4), 371–379 (2008)
185. R.P. Gilbert, G.N. Hile, Hilbert function modules with reproducing kernels. Nonlinear Anal. **1**(2), 135–150 (1977)
186. R.P. Gilbert, R.J. Weinacht, Reproducing kernels for elliptic systems. J. Approx. Theory **15**, 243–255 (1975)
187. C.D. Godsil, *Algebraic Combinatorics* (Chapman and Hall Mathematics, New York, 1993)

188. E.K. Godunova, V.I. Levin, An inequality of Maroni (in Russian). Mat. Zametki **2**, 221–224 (1967)
189. I.S. Gradshlein, L.M. Ryzhik, *Table of Integrals, Series, and Products* (Elsevier/Academic Press, Amsterdam, 2007)
190. B.L. Granovsky, H.G. Müller, On the optimality of a class of polynomial kernel functions. Stat. Decis. **7**(4), 301–312 (1989)
191. A. Gretton, K.M. Borgwardt, M. Rasch, B. Schölkopf, A. Smola, A kernel method for the two-sample-problem, in *Advances in Neural Information Processing Systems*, vol. 19, ed. by B. Schölkopf, J. Platt, T. Hoffman (MIT Press, Cambridge, 2007)
192. C.W. Groetsch, *Inverse Problems in the Mathematical Sciences* (Vieweg Sohn Verlagsgesellschaft mbH, Braunschweig/Wiesbaden, 1993)
193. D. Han, M.Z. Nashed, Q. Sun, Sampling expansions in reproducing kernel Hilbert and Banach spaces. Numer. Funct. Anal. Optim. **30**(9–10), 971–987 (2009)
194. P.C. Hansen, Analysis of discrete ill-posed problems by means of the L-curve. SIAM Rev. **34**, 561–580 (1992)
195. D.N. Hao, N. Dinh, D. Lesnic, The Cauchy problem for Laplace's equation via the conjugate gradient method. IMA J. Appl. Math. **65**, 199–217 (2000)
196. N. Hayashi, Global existence of small analytic solutions to nonlinear Schrödinger equations. Duke Math. J. **60**, 717–727 (1990)
197. N. Hayashi, Solutions of the (generalized) Korteweg-de Vries equation in the Bergman and the Szegö spaces on a sector. Duke Math. J. **62**, 575–591 (1991)
198. N. Hayashi, K. Kato, Regularity of solutions in time to nonlinear Schrödinger equations. J. Funct. Anal. **128**, 255–277 (1995)
199. N. Hayashi, S. Saitoh, Analyticity and smoothing effect for the Schrödinger equation. Ann. Inst. Henri Poincaré **52**, 163–173 (1990)
200. N. Hayashi, S. Saitoh, Analyticity and global existence of small solutions to some nonlinear Schrödinger equation. Commun. Math. Phys. **139**, 27–41 (1990)
201. H. Hedenmalm, B. Korenblum, K. Zhu, *Theory of Bergman Spaces*. Graduate Texts in Mathematics, vol. 199 (Springer, New York, 2000)
202. D.A. Hejhal, *Theta Functions, Kernel Functions and Abel Integrals*, vol. 129 (Memoirs American Mathematical Society, Providence, 1972)
203. T. Hida, N. Ikeda, Analysis on Hilbert space with reproducing kernel arising from multiple Wiener integral, in *1967 Proceedings of the Fifth Berkeley Symposium on Mathematical Statistics and Probability*, Berkeley. Contributions to Probability Theory, Part 1, vol. II (University California Press, Berkeley, 1965/1966), pp. 117–143
204. J.R. Higgins, *Sampling Theory in Fourier and Signal Analysis: Foundations* (Clarendon Press, Oxford, 1996)
205. J.R. Higgins, R.L. Stens, *Sampling Theory in Fourier and Signal Analysis: Advanced Topics* (Clarendon Press, Oxford, 1999)
206. J.R. Higgins, A sampling principle associated with Saitoh's fundamental theory of linear transformations, in *Analytic Extension Formulas and Their Applications*. International Society for Analysis, Applications, and Computation, vol. 9 (Kluwer Academic Publishers, Dordrecht/Boston, 2001), pp. 73–86
207. J.R. Higgins, Converse sampling theorems in a reproducing kernel theory setting. Sampl. Theorem Signal Image Proc. **14**(2), 145–152 (2015)
208. J.W. Hilgers, On the equivalence of regularization and certain reproducing kernel Hilbert space approaches for solving first kind problems. SIAM J. Numer. Anal. **13**, 172–184 (1976)
209. E. Hille, Introduction to general theory of reproducing kernels. Rocky Mt. J. Math. **2**(3), 321–368 (1972)
210. D.A. Hills, D. Nowell, A. Sackfield, *Mechanics of Elastic Contacts* (Butterworth–Heinemann, Oxford, 1993)
211. D.A. Hills, P.A. Kelly, D.N. Dai, A.M. Korsunsky, *Solution of Crack Problems: The Distributed Dislocation Technique* (Kluwer Academic Publishers, Dordrecht, 1996)

212. I.I. Hirschman, D.V. Widder, *The Convolution Transform* (Princeton University Press, Princeton, 1955)
213. Y. Hishikawa, Fractional calculus on parabolic Bergman spaces. Hiroshima Math. J. **38**, 471–488 (2008)
214. Y. Hishikawa, The reproducing formula with fractional orders on the parabolic Bloch space. J. Math. Soc. Jpn. **62**, 1219–1255 (2010)
215. Y. Hishikawa, M. Nishio, M. Yamada, A conjugate system and tangential derivative norms on parabolic Bergman spaces. Hokkaido Math. J. **39**, 85–114 (2010)
216. Y. Hishikawa, Representing sequences on parabolic Bergman spaces. J. Korean Math. Soc. **48**, 1017–1041 (2011)
217. H. Hochstadt, *Integral Equations* (Wiley, New York, 1973)
218. Y.C. Hon, T. Wei, Backus-Gilbert algorithm for the cauchy problem of the laplace equation. Inverse Probl. **17**(2), 261–271 (2001)
219. L. Hörmander, *An Introduction to Complex Analysis in Several Variables* (D. Van Nostrand Company, Inc. Princeton, 1967)
220. L. Hörmander, L^2 estimates and existence theorems for the $\overline{\partial}$-operator. Acta Math. **113**, 89–152 (1965)
221. http://library.wolfram.com/inforcenter/MathSource/4738/
222. http://www.columbia.edu/ww2040/abate.html
223. L.K. Hua, *Harmonic Analysis of Functions of Several Complex Variables in the Classical Domains*. Translations of Mathematical Monographs, vol. 6 (American Mathematical Society, Providence, 1963)
224. T.J.R. Hughes, *The Finite Element Method* (Prentice Hall, Englewood Cliffs, 1987)
225. H. Imai, T. Takeuchi, M. Kushida, On numerical simulation of partial differential equations in infinite precision. Adv. Math. Sci. Appl. **9**, 1007–1016 (1999)
226. H. Imai, Multipy-precision arithmetic (in Japanese). Sūgaku Math. Soc. Jpn. Iwanami-Shoten Co. **55**, 316–325 (2003)
227. S.W. Indratno, A.G. Ramm, Inversion of the laplace transform from the real axis using an adaptive iterative method. Int. J. Math. Math. Sci. **2009**, 38. Article ID 898195
228. V. Isakov, *Inverse Problems for Partial Differential Equation* (Springer, New York, 1998)
229. P.D. Iseger, Numerical transform inversion using Gaussian quadrature. Probab. Eng. Inf. Sci. **20**, 1–44 (2006)
230. H. Itou, S. Saitoh, Analytical and numerical solutions of linear singular integral equations. Int. J. Appl. Math. Stat. **12**, 76–89 (2007)
231. K. Iwamura, T. Matsuura, S. Saitoh, A numerical construction of a natural inverse of any matrix by using the theory of reproducing kernels with the Tikhonov regularization. Far East J. Math. Edu. **4**, 141–149 (2010)
232. W. Jiang, M. Cui, Y. Lin, Anti-periodic solutions for Rayleigh-type equations via the reproducing kernel Hilbert space method. Commun. Nonlinear Sci. Numer. Simul. **15**, 1754–1758 (2010)
233. W. Jiang, M Cui, Solving nonlinear singular pseudoparabolic equations with nonlocal mixed conditions in the reproducing kernel space. Int. J. Comput. Math. **87**, 3430–3442 (2010)
234. A.J. Jerri, The Shannon sampling theorem – its various extensions and applications – a tutorial review. Proc. IEEE **65**, 1565–1596 (1977)
235. X.Q. Jin, L.M. Keer, Q. Wang, A practical method for singular integral equations of the second kind. Eng. Fract. Mech. **75**, 1005–1014 (2008)
236. P.E.T. Jorgensen, Integral representations for locally defined positive definite functions on Lie group. Int. J. Math. **2**, 257–286 (1991)
237. T. Kailath, Some integral equations with nonrational kernels. IEEE Trans. Inf. Theory **IT-12**, 442–447 (1966)
238. T. Kailath, D. Duttweiler, An RKHS approach to detection and estimation problems, III. Generalized innovations representations and a likelihood-ratio formula. IEEE Trans. Inf. Theory **IT-18**, 730–745 (1972)

239. T. Kailath, R. Geesey, H. Weinert, Some relations among RKHS norms, Fredholm equations, and innovations representations. IEEE Trans. Inf. Theory **IT-18**, 341–348 (1972)
240. T. Kailath, RKHS approach to detection and estimation problems, I. Deterministic signals in Gaussian noise. IEEE Trans. Inf. Theory **IT-17**, 530–549 (1971)
241. J.P. Kaipio, E. Somersalo, *Statistical and Computational Inverse Problems* (Springer, New York, 2004)
242. V.A. Kakichev, On the convolution for integral transforms. Izv. ANBSSR Ser. Fiz. Mat. **2**(2), 48–57 (1967) (in Russian)
243. G. Kallianpur, Abstract Wiener processes and their reproducing kernel Hilbert spaces. Z. Wahrscheinlichkeitstheorie und Verw. Gebiete **17**, 113–123 (1971)
244. G. Kallianpur, *The Role of Reproducing Kernel Hilbert Spaces in the Study of Gaussian Processes*. Advances in Probability and Related Topics, vol. 2 (Dekker, New York, 1970), pp. 49–53
245. Y. Kametaka, K. Watanabe, A. Nagai, The best constant of Sobolev inequality in an n dimensional Euclidean space. Proc. Jpn. Acad. Ser. A **81**, 57–60 (2005)
246. Y. Kametaka, K. Watanabe, A. Nagai, S. Pyatkov, The best constant of Sobolev inequality in an n dimensional Euclidean space. Sci. Math. Jpn. **e-2004**, 295–303 (2004); Sci. Math. Jpn. **61**(1), 15–23 (2005)
247. Y. Kametaka, Y. Oshime, K. Watanabe, H. Yamagishi, A. Nagai, K. Takemura, The best constant of L^p Sobolev inequality corresponding to the periodic boundary value problem for $(-1)^M(d/dx)^{2M}$. Sci. Math. Jpn. **e-2007**, 269–281 (2007)
248. Y. Kametaka, K. Watanabe, A. Nagai, H. Yamagishi, K. Takemura, The best constant of Sobolev inequality which corresponds to a bending problem of a string with periodic boundary condition. Sci. Math. Jpn. **66**(2), 151–168 (2007)
249. Y. Kametaka, H. Yamagishi, K. Watanabe, A. Nagai, K. Takemura, The best constant of Sobolev inequality corresponding to Dirichlet boundary value problem for $(-1)^M(d/dx)^{2M}$. Sci. Math. Jpn. **e-2008**, 439–451 (2008)
250. Y. Kametaka, H. Yamagishi, K. Watanabe, A. Nagai, K. Takemura, M. Arai, The best constant of Sobolev inequality which corresponds to Schrödinger operator with Dirac delta potential. Sci. Math. Jpn. **69**(2), 211–225 (2009)
251. Y. Kametaka, H. Yamagishi, K. Watanabe, A. Nagai, K. Takemura, M. Arai, H. Yama-gishi, The best constant of Sobolev inequality corresponding to Dirichlet-Neumann boundary value problem for $(-1)^M(d/dx)^{2M}$. Hiroshima Math. J. **39**, 421–442 (2009)
252. A.A. Karelin, Applications of operator equalities to singular integral operators and to Riemann boundary value problems. Math. Nachr. **280**, 1108–1117 (2007)
253. K. Kataoka, Quasi-positivity for pseudodiŁerential operators and microlocal energy methods, in *Proceedings of Taniguchi Symposium HERT*, Katata (1984), pp. 125–141
254. K. Kataoka, Microlocal energy methods and pseudo-differential operators. Invent. Math. **81**, 305–340 (1985)
255. K. Kataoka, Some applications of microlocal energy methods to analytic hypoellipticity. Prospect Algebr. Anal. **1**, 287–303 (1988)
256. G. Kimeldorf, G. Wahba, Some results on Tchebycheffian spline functions. J. Math. Anal. Appl. **33**, 82–95 (1971)
257. M.V. Klibanov, F. Santosa, A computational quasi-reversibility method for Cauchy problems for Laplace's equation. SIAM J. Appl. Math. **51**(6), 1653–1675 (1991)
258. D. Klusch, The sampling theorem, Dirichlet series and Hankel transforms. J. Comput. Appl. Math. **44**, 261–273 (1992)
259. M. Kobayashi, Y. Sawano, Molecular decomposition of the modulation spaces $M^{p,q}$ and its application to the pseudo-differential operators. Osaka J. Math. **47**(4), 1029–1053 (2010)
260. A.N. Kolmogoroff, Stationary sequences in Hilbert's space (in Russian). Bull. Math. Univ. Mosc. **2**, 40 (1941)
261. M.G. Kreĭn, Hermitian positive kernels on homogeneous spaces. I. Am. Math. Soc. Transl. (2) **34**, 69–108 (1963)

262. H. Körezlioğlu, Reproducing kernels in separable Hilbert spaces. Pac. J. Math. **25**, 305–314 (1968)
263. H. Kramer, A generalized sampling theorem. J. Math. Phys. **38**, 68–72 (1959)
264. S. Krantz, *Function Theory of Several Complex Variables*. Pure and Applied Mathematics. A Wiley-Interscience Publication (Wiley, New York, 1982)
265. S. Krantz, *Function Theory of Several Complex Variables*. Reprint of the 1992 edition (AMS Chelsea Publishing, Providence, 2001)
266. V.G. Kravchenko, G.S. Litvinchuk, *Introduction to the Theory of Singular Integral Operators with Shift* (Kluwer Academic Publishers, Dordrecht, 1994)
267. R. Kress, *Linear Integral Equations* (Springer, New York, 1998)
268. V.V. Kryzhniy, Regularized inversion of integral transformations of Mellin convolution type. Inverse Probl. **19**, 573–583 (2003)
269. V.V. Kryzhniy, Numerical inversion of the Laplace transform: analysis via regularized analytic continuation. Inverse Probl. **22**, 579–597 (2006)
270. A.B. Kuijlaas, M. Vanlessen, Universality for eigenvalue correlations from the modified Jacobi unitary ensemble. Int. Math. Res. Not. **30**, 1575–1600 (2002)
271. J.L. Lagrange, (Nouv. Mém. Acad., Berlin, 1773); Also in Oeuvres, T. 3 (Gauthier-Villars, Paris, 1869), p. 662f
272. T.L. Lai, Reproducing kernel Hilbert spaces and the law of the iterated logarithm for Gaussian processes. Z. Wahrscheinlichkeitstheorie und Verw. Gebiete **29**, 7–19 (1974)
273. I.E. Lagaris, A. Likas, D.I. Fotiadis, Artificial neural networks for solving ordinary and partial differential equations. IEEE Trans. Neural Netw. **9**, 987–1000 (1998)
274. I.E. Lagaris, A. Likas, D.I. Fotiadis, Neural-network methods for boundary value problems with irregular boundaries. IEEE Trans. Neural Netw. **11**, 1041–1049 (2000)
275. F.M. Larkin, Optimal approximation in Hilbert spaces with reproducing kernel functions. Math. Comput. **24**, 911–921 (1970)
276. C.L. Lawson, R.J. Hanson, *Solving Least Squares Problems* (Prentice-Hall, Englewood Cliffs, 1974)
277. J.Y. Li, S.W. Luo, Y.J Qi, Y.P. Huang, Numerical solution of elliptic partial differential equation using radial basis function neural networks. Neural Netw. **16**, 729–734 (2003)
278. A. Van Der Linde, Interpolation of regression functions in reproducing kernel Hilbert spaces. Statistics **16**, 351–361 (1985)
279. M.A. Lytrides, N.S. Stylianopoulos, Error analysis of the Bergman kernel method with singular based functions. Comput. Methods Funct. Theory **11**(2), 487–529 (2011)
280. R. Le Page, Subgroups of paths and reproducing kernels. Ann. Probab. **1**, 345–347 (1973)
281. N.A. Lebedev, I.M. Milin, An inequality. Vestnik Leningrad. Univ. **20**(19), 157–158 (1965) (Russian)
282. J.Y. Lee, J.R. Yoon, A numerical method for the Cauchy problem using singular decompositon, in *Second Japan-Korea Joint Seminar on Inverse Problems and Related Topics*, Seoul, 2001. Commun. Korean Math. Soc. **16**, 487–508 (2001)
283. Y. Lin, J. Niu, M. Cui, A numerical solution to nonlinear second order three-point boundary value problems in the reproducing kernel space. Appl. Math. Comput. **218**, 7362–7368 (2012)
284. Y. Lin, M. Cui, A numerical solution to nonlinear multi-point boundary value problems in the reproducing kernel space. Math. Methods Appl. Sci. **34**, 44–47 (2011)
285. G.S. Litvinchuk, *Solvability Theory of Boundary Value Problems and Singular Integral Equations with Shift* (Kluwer Academic, Dordrecht, 2000)
286. D. Lubinsky, A new approach to universality limits involving orthogonal polynomials. Ann. Math. **170**(2), 915–939 (2009)
287. D. Lubinsky, Universality for arbitrary measures on compact sets. www.math.gatech.edu/news/conferences/banff/abstracts.html-19k[O]
288. W. Magnus, F. Oberhettinger, R.P. Soni, *Formulas and Theorems for the Special Functions of Mathematical Physics*, Third enlarged edition (Springer, Berlin, 1966)
289. B.N. Mandal, G.H. Bera, Approximate solution of a class of singular integral equations of second kind. J. Comput. Appl. Math. **206**, 189–195 (2007)

290. J. Matkowski, L^p-like paranorms, in *Selected Topics in Functional Equations and Iteration Theory*, Graz, 1991. Grazer Math. Ber. **316**, 103–138 (1992)
291. J.T. Marti, *Introduction to the Theory of Bases* (Springer, New York/Berlin/Heidelberg, 1969)
292. T. Matsuura, S. Saitoh, Analytical and numerical solutions of the inhomogeneous wave equation. Aust. J. Math. Anal. Appl. **1**(1), Article 7, 18 (2004)
293. T. Matsuura, S. Saitoh, D.D. Trong, Numerical solutions of the Poisson equation. Appl. Anal. **83**, 1037–1051 (2004)
294. T. Matsuura, S. Saitoh, Analytical and numerical solutions of linear ordinary differential equations with constant coefficients. J. Anal. Appl. **3**, 1–17 (2005)
295. T. Matsuura, S. Saitoh, Numerical inversion formulas in the wave equation. J. Comput. Math. Optim. **1**, 1–19 (2005)
296. T. Matsuura, S. Saitoh, D.D. Trong, Approximate and analytical inversion formulas in heat conduction on multidimensional spaces. J. Inverse Ill-Posed Probl. **13**, 479–493 (2005)
297. T. Matsuura, S. Saitoh, Dirichlet's principle using computers. Appl. Anal. **84**, 989–1003 (2005)
298. T. Matsuura, S. Saitoh, Analytical and numerical inversion formulas in the Gaussian convolution by using the Paley-Wiener spaces. Appl. Anal. **85**, 901–915 (2006)
299. T. Matsuura, A. Al-Shuaibi, H. Fujiwara, S. Saitoh, Numerical real inversion formulas of the Laplace transform by using a Fredholm integral equation of the second kind. J. Anal. Appl. **5**, 123–136 (2007)
300. T. Matsuura, A. Al-Shuaibi, H. Fuijiwara, S. Saitoh, M. Sugihara, Numerical real inversion formulas of the Laplace transform by a sinc method. Far East J. Math. Sci. **27**, 1–14 (2007)
301. R. Meidan, Reproducing-kernel Hilbert spaces of distributions and generalized stochastic processes. SIAM J. Math. Anal. **10**(1), 62–70 (1979)
302. H. Meschkowski, *Hilbertsche Räume mit Kernfunktion*, (German) Die Grundlehren der mathematischen Wissenschaften, vol. 113 (Springer, Berlin/Göttingen/Heidelberg, 1962), viii+256pp
303. M. Mboup, *On the Structure of Self-Similar Systems: A Hilbert Space Approach*. Reproducing Kernel Spaces and Applications, Operator Theory: Advances and Applications, vol. 143 (Birkhäuser, Basel, 2003), pp. 273–302
304. R. Megginson, *A Introduction to Banach Space Theory*. Graduate Texts in Mathematics, vol. 183 (Springer, New York, 1998)
305. S.G. Mikhlin, K.L. Smolitskiy, *Approximate Methods for Solution of Differential and Integral Equations* (American Elsevier, New York, 1967)
306. S.G. Mikhlin, S. Prössdorf, *Singular Integral Operators* (Springer, Berlin, 1986)
307. D.S. Mitrinović, *Analytic Inequalities*. Die Grundlehren der mathematischen Wissenschaften, vol. 165 (Springer, New York/Berlin, 1970), xii+400pp
308. Y. Mo, T. Qian, Support vector machine adapted Tikhonov regularization method to solve Dirichlet problem. Appl. Math. Comput. **245**, 509–519 (2014)
309. B. Mohar, The Laplacian spectrum of graphs, in *Graph Theory, Combinatorics, and Applications*, ed. by Y. Alavi, G. Chartrand, O.R. Oellermann, A.J. Schwenk, vol. 2 (Wiley, 1991), pp. 871–898
310. E.H. Moore, *General Analysis*. Memoirs of the American Philosophical Society Part I; Memoirs of the American Philosophical Society Part II (The American Philosophical Society, Philadelphia, 1935–1939)
311. P.M. Morse, H. Feshbach, *Methods of Theoretical Physics* (McGraw-Hill Book Company Inc., New York, 1953)
312. N.I. Muskhelishvili, *Singular Integral Equations* (Noordhoff, Groningen, 1972)
313. A. Nagai, K. Takemura, Y. Kametaka, K. Watanabe, H. Yamagishi, Green function for boundary value problem of $2M$-th order linear ordinary equations with free boundary condition. Far East J. Appl. Math. **26**, 393–406 (2007)
314. K. Nakade, T. Ohwada, K.S. Saito, Kolmogorov's factorization theorem for von Neumann algebras. J. Math. Anal. Appl. **401**(1), 289–292 (2013)

315. M. Nakai, L. Sario, Square integrable harmonic functions on plane regons. Ann. Acad. Scxi. Fenn. **4**, 193–201 (1978/1979)
316. M.Z. Nashed, G. Wahba, Convergence rates of approximate least squares solutions of linear integral and operator equations of the first kind. Math. Comput. **28**, 69–80 (1974)
317. M.Z. Nashed, G. Wahba, Regularization and approximation of linear operator equations and reproducing kernel spaces. Bull. Am. Math. Soc. **80**, 1213–1218 (1974)
318. M.Z. Nashed, G. Wahba, Generalized inverses in reproducing kernel spaces: an approach to regularization of linear operator equations. SIAM J. Math. Anal. **5**, 974–987 (1974)
319. M.Z. Nashed, G.G. Walter, General sampling theorems for functions in reproducing kernel Hilbert spaces. Math. Control Signals Syst. **4**, 363–390 (1991)
320. M.Z. Nashed, Q. Sun, Sampling and reconstruction of signals in a reproducing kernel subspace of $L^p(\mathbb{R}^d)$. J. Funct. Anal. **258**(7), 2422–2452 (2010)
321. F. Natterer, *The Mathematics of Computerized Tomography*. Classics in Applied Mathematics, vol. 32 (Society for Industrial and Applied Mathematics (SIAM), Philadelphia, 2001)
322. Z. Nehari, *Conformal Mapping* (McGraw-Hill Book Company Inc., New York, 1952)
323. N.D.V. Nhan, D.T. Duc, V.K. Tuan, Weighted L_p-norm inequalities for various convolution type transformations and their applications. Armen. J. Math. **1**(4), 1–18 (2008)
324. N.D.V. Nhan, D.T. Duc, Fundamental inequalities for the iterated Laplace convolution in weighted L_p spaces and their applications. Integr. Trans. Spec. Funct. **19**(9–10), 655–664 (2008)
325. N.D.V. Nhan, D.T. Duc, Reverse weighted L_p-norm inequalities and their applications. J. Math. Inequal. **2**(1), 57–73 (2008)
326. N.D.V. Nhan, D.T. Duc, Weighted L_p-norm inequalities in convolutions and their applications. J. Math. Inequal. **2**(1), 45–55 (2008)
327. N.D.V. Nhan, D.T. Duc, V.K. Tuan, Reverse weighted L_p-norm inequalities for convolution type integrals. Armen. J. Math. **2**(3), 77–93 (2009)
328. N.D.V. Nhan, D.T. Duc, Fundamental iterated convolution inequalities in weighted L_p spaces and their applications. Math. Inequal. Appl. **12**(3), 487–498 (2009)
329. N.D.V. Nhan, D.T. Duc, V.K. Tuan, Weighted norm inequalities for a nonlinear transform. Comput. Math. Appl. **61**, 832–839 (2011)
330. N.D.V. Nhan, D.T. Duc, Convolution inequalities in l_p weighted spaces. Canad. Math. Bull. **55**(2), 355–367 (2012)
331. N.D.V. Nhan, D.T. Duc, Various inequalities in reproducing kernel Hilbert spaces. Taiwan. J. Math. **17**(1), 221–237 (2012)
332. M. Nishio, N. Suzuki, M. Yamada, Schatten class Toeplitz operators on the parabolic Bergman space. II. Kodai Math. J. **35**, 52–77 (2012)
333. M. Nishio, N. Suzuki, M. Yamada, Interpolating sequences of parabolic Bergman spaces. Potential Anal. **28**, 357–378 (2008)
334. M. Nishio, N. Suzuki, M. Yamada, Carleson inequalities on parabolic Bergman spaces. Tohoku Math. J. **62**(2), 269–286 (2010)
335. M. Nishio, N. Suzuki, M. Yamada, Weighted Berezin transformations with application to Toeplitz operators of Schatten class on parabolic Bergman spaces. Kodai Math. J. **32**, 501–520 (2009)
336. M. Nishio, N. Suzuki, M. Yamada, Compact Toeplitz operators on parabolic Bergman spaces. Hiroshima Math. J. **38**, 177–192 (2008)
337. M. Nishio, K. Shimomura, N. Suzuki, α-parabolic Bergman spaces. Osaka J. Math. **42**, 133–162 (2005)
338. H. Ogawa, N.-E. Berrached, Theory of extended pseudo-biorthogonal bases. IEICE Trans. Inf. Syst. **E76-D**(8), 890–897 (1993)
339. H. Ogawa, N.-E. Berrached, A theory of extended pseudo-biorthogonal bases and its application to generalized sampling theorem, in *Mathematical Analysis, Wavelets, and Signal Processing. An International Conference on Mathematical Analysis and Signal Processing*, Cairo University, Cairo, ed. by M.E.H. Ismail et al., Jan 1994. Contemporary Mathematics, vol. 190 (American Mathematical Society, Providence, 1995), pp. 305–321

340. H. Ogawa, Theory of pseudo-biorthogonal bases and its application. The theory of reproducing kernels and their applications (Japanese) (Kyoto, 1998). RIMS Kokyuroku No. **1067**, 24–38 (1998)
341. H. Ogawa and A. Hirabayashi, Sampling theorem with optimum noise suppression. Sampl. Theory Signal Image Process. **6**(2), 167–184 (2007)
342. H. Ogawa, *Functional Analysis for Engineers (in Japanese*) (Morikita Co., 2009), 292pp
343. K. Oguiso, *Daisu kyokusen ron (in Japanese*) (Asakura, 2002)
344. T. Ohe, K. Ohnaka, Uniqueness and convergence of numerical solution of the Cauchy problem for the Laplace equation by a charge simulation method. Jpn. J. Indust. Appl. Math. **21**, 339–559 (2004)
345. T. Ohsawa, T. Takegoshi, On the extension of L^2 holomorphic functions. Math. Zeit. **195**, 197–204 (1987)
346. T. Ohsawa, A remark on Kazhdan's theorem on sequences of Bergman metrics. Kyushu J. Math. **63**, 133–137 (2009)
347. B. Okutmustur, A. Gheondea, *Reproducing Kernel Hilbert Spaces* (Lambert Academic Publishing, Saarbücken, 2010)
348. K. Onishi, Q. Wang, *Numerical Solutions of the Cauchy Problem in Potential and Elastostatics*. Inverse problems and related topics (Koebe, 1998). Chapman & Hall/CRC Research Notes in Mathematics, vol. 419 (Chapman & Hall/CRC, Boca Raton, 2000), pp. 115–132
349. Z. Opial, Sur une inégalité. Ann. Polon. Math. **8**, 29–32 (1960)
350. Y. Oshime, Y. Kametaka, H. Yamagishi, The best constant of L^p Sobolev inequality corresponding to Dirichlet boundary value problem for $(d/dx)^{4m}$. Sci. Math. Jpn. **e-2008**, 461–469 (2008)
351. Y. Oshime, K. Watanabe, The best constant of L^p Sobolev inequality corresponding to Dirichlet boundary value problem II. Tokyo J. Math. **34**, 115–133 (2011)
352. Y. Oshime, H. Yamagishi, K.Watanabe, The best constant of L^p Sobolev inequality corresponding to Neumann boundary value problem for $(-1)^M(d/dx)^{2M}$. Hiroshima Math. J. **42**, 1–7 (2012)
353. K.R. Parthasarathy, K. Schmidt, *Positive Definite Kernels, Continuous Tensor Products, and Central Limit Theorems of Probability Theory*, vol. 107 (Springer Lecture Notes in Mathematics, Berlin, 1972)
354. E. Parzen, *Probability Density Functionals and Reproducing Kernel Hilbert Spaces*. 1963 Proceedings of Symposium on Time Series Analysis (Wiley, New York, Brown University, 1962), pp. 155–169
355. E. Parzen, Extraction and detection problems and reproducing Kernel Hilbert spaces. J. SIAM control Ser. A **1**, 35–62 (1962)
356. E. Parzen, Statical inference on time series by RKHS methods, in *Proceedings of 12th Biennial Seminar of the Canadian Mathematical Congress* (American Mathematical Society, Providence, 1971), pp. 1–37
357. J. Peng, S.K. Chung, Laplace transforms and generators of semigroups of operators. Proc. Am. Math. Soc. **126**(8), 2407–2416 (1998)
358. A.D. Polyanin, *Handbook of Linear Partial Differential Equations for Engineers and Scientists* (Chapman & Hall/CRC, Boca Raton, 2002)
359. A.D. Polyanin, V.F. Zaitsev, *Handbook of Exact Solutions for Ordinary Differential Equations* (CRC Press, Boca Raton, 2003)
360. A.D. Polyanin, A.V. Manzhirov, *Handbook of Integral Equations* (CRC Press, Boca Raton, 2008)
361. E.L. Post, Generalized differentiation. Trans. Am. Math. Soc. **32**(4), 723–781 (1930)
362. A.P. Prudnikov, Yu.A. Brychkov, O.L. Marichev, *More Special Functions*. Integrals and Series, vol. 3 (Gordon and Breach Publisher, New York, 1990)
363. T. Qian, L. Tan, Backward shift invariant subspaces with applications to band preserving and phase retrival problems. Math. Math. Appl. Sci. 7/2015. doi:10.1002/mma3591
364. A.G. Ramm, *Random Fields Estimation Theory*. Pitman Monographs and Surveys in Pure and Applied Mathematics, vol. 48 (Longman Scientific and Technical, Harlow, 1990)

365. A.G. Ramm, *Multidimensional Inverse Scattering Problems*. Monographs and Surveys in Pure and Applied Mathematics, vol. 51 (Longman Scientific and Technical, Harlow, 1992)
366. T. Ransford, *Potential Theory in the Complex Plane*. London Mathematical Society, Student Texts, vol. 28 (Press Syndicate of the University of Cambridge, Cambridge/New York, 1995)
367. Th.M. Rassias, S. Saitoh, The Pythagorean theorem and linear mappings. PanAmerican Math. J. **12**, 1–10 (2002)
368. H. Rauhut, T. Ullrich, Generalized coorbit space theory and inhomogeneous function spaces of Besov-Lizorkin-Triebel type. J. Funct. Anal. **260**(11), 3299–3362 (2011)
369. M. Reimer, *Constructive Theory of Multivariate Functions with an Application to Tomography* (Wissenschaftsverlag, Mannheim/Wien/Zürich, 1990)
370. H.J. Reinhardt, H. Han, N.H. Dinh, Stability and regularization of a discrete approximation to the Cauchy problem for the Laplace's equation. SIAM J. Numer. Anal. **36**(3), 890–905 (1999)
371. E.M. Rocha, A reproducing kernel Hilbert discretization method for linear PDEs with nonlinear right-hand side. Lib. Math. (New Ser.) **34**(2), 91–104 (2014)
372. W. Rudin, *Functional Analysis*, 2nd edn. (McGraw Hill, New York, 1991)
373. S. Saitoh, The Bergman norm and the Szegö norm. Trans. Am. Math. Soc. **249**(2), 261–279 (1979)
374. S. Saitoh, The Dirichlet norm and the norm of Szegö type. Trans. Am. Math. Soc. **254**, 355–364 (1979)
375. S. Saitoh, Some inequalities for analytic functions with a finite Dirichlet integral on the unit disc. Math. Ann. **246**(1), 69–77 (1979)
376. S. Saitoh, Some inequalities for entire functions. Proc. Am. Math. Soc. **80**(2), 254–258 (1980)
377. S. Saitoh, Reproducing kernels of the direct product of two Hilbert spaces. Riazi J. Karachi Math. Assoc. **4**, 1–20 (1982)
378. S. Saitoh, Hilbert spaces induced by Hilbert space valued functions. Proc. Am. Math. Soc. **89**, 74–78 (1983)
379. S. Saitoh, The Weierstrass transform and an isometry in the heat equation. Appl. Anal. **16**, 1–6 (1983)
380. S. Saitoh, A fundamental inequality in the convolution of L_2 functions on the half line. Proc. Am. Math. Soc. **91**(2), 285–286 (1984)
381. S. Saitoh, Hilbert spaces admitting reproducing kernels on the real line and related fundamental inequalities. Riazi J. Karachi Math. Assoc. **6**, 25–31 (1984)
382. S. Saitoh, *Theory of Reproducing Kernels and Its Applications*. Pitman Research Notes in Mathematics Series, vol. 189 (Longman Scientific & Technical, Harlow, 1988)
383. S. Saitoh, Interpolation problems of Pick-Nevanlinna type. Pitman Res. Notes Math. Ser. **212**, 253–262 (1989)
384. S. Saitoh, Representations of the norms in Bergman-Selberg spaces on strips and half planes. Complex Variab. Theory Appl. **19**(4), 231–241 (1992)
385. S. Saitoh, One approach to some general integral transforms and its applications. Integr. Trans. Spec. Funct. **3**, 49–84 (1995)
386. S. Saitoh, Natural norm inequalities in nonlinear transforms, in *General Inequalities, 7*, Oberwolfach, 1995. International Series of Numerical Mathematics, vol. 123 (Birkhäuser, Basel, 1997), pp. 39–52
387. S. Saitoh, Representations of inverse functions. Proc. Am. Math. Soc. **125**, 3633–3639 (1997)
388. S. Saitoh, *Integral Transforms, Reproducing Kernels and Their Applications*. Pitman Research Notes in Mathematics Series, vol. 369 (Addison Wesley Longman Ltd, 1998). CRC Press/Taylor & Francis Group, Boca Raton/London/New York (in hard cover)
389. S. Saitoh, *Nonlinear Transforms and Analyticity of Functions*. Nonlinear Mathematical Analysis and Applications (Hadronic Press, Palm Harbor, 1998), pp. 223–234
390. S. Saitoh, Various operators in Hilbert space introduced by transforms. Int. J. Appl. Math. **1**, 111–126 (1999)
391. S. Saitoh, Applications of the general theory of reproducing kernels, *Reproducing Kernels and Their Applications* (Kluwer Academic Publishers, 1999), pp. 165–188

392. S. Saitoh, D. Alpay, J.A. Ball, T. Ohsawa (eds), *Reproducing Kernels and Their Applications* (Kluwer Academic Publishers, 1999)
393. S. Saitoh, M. Yamamoto, Integral transforms involving smooth functions, in *Reproducing Kernels and Their Applications* (Kluwer Academic Publishers, 1999), pp. 149–164
394. S. Saitoh, Linear integro-differential equations and the theory of reproducing kernels, in *Volterra Equations and Applications*, ed. by C. Corduneanu, I.W. Sandberg (Gordon and Breach Science Publishers, Amsterdam, 2000)
395. S. Saitoh, Weighted L_p-norm inequalities in convolutions, in *Survey on Classical Inequalities*. Mathematics and Its Applications, vol. 517 (Kluwer Academic Publisher, Dordrecht, 2000), pp. 225–234
396. S. Saitoh, V.K. Tuan, M. Yamamoto, Reverse weighted L_p-norm inequalities in convolutions. J. Inequal. Pure Appl. Math. **1**(1), Article 7, 7 (2000)
397. S. Saitoh, Analytic extension formulas, integral transforms and reproducing kernels, in *Analytic Extension Formulas and Their Applications* (Kluwer Academic Publishers, Dordrecht/Boston, 2001), pp. 207–232
398. S. Saitoh, Applications of the reproducing kernel theory to inverse problems. Comm. Korean Math. Soc. **16**, 371–383 (2001)
399. S. Saitoh, Principle of telethoscope,in *Functional-Analytic and Complex Methods, Their Interaction and Applications to Partial Differential Equations – Proceedings of the International Graz Workshop*, Graz, 12–16 Feb 2001 (World Scientific, 2001), pp. 101–117
400. S. Saitoh, N. Hayashi, M. Yamamoto (eds.), *Analytic Extension Formulas and Their Applications* (Kluwer Academic Publishers, Dordrecht/Boston, 2001)
401. S. Saitoh, V.K. Tuan, M. Yamamoto, Conditional stability of a real inverse formula for the laplace transform. Z. Anal. Anw. **20**, 193–202 (2001)
402. S. Saitoh, V.K. Tuan, M. Yamamoto, Reverse convolution inequalities and applications to inverse heat source problems. J. Inequal. Pure Appl. Math. **3**(5), Article 80, 11 (2002)
403. S. Saitoh, *Introduction to the Theory of Reproducing Kernels* (in Japanese) (Makino-Shoten Co., 2002)
404. S. Saitoh, T. Matsuura, M. Asaduzzaman, Operator equations and best approximation problems in reproducing kernel Hilbert spaces. J. Anal. Appl. **1**, 131–142 (2003)
405. S. Saitoh, Constructions by reproducing kernels of approximate solutions for linear differential equations with L_2 integrable coefficients. Int. J. Math. Sci. **2**, 261–273 (2003)
406. S. Saitoh, Approximate real inversion formulas of the Laplace transform. Far East J. Math. Sci. **11**, 53–64 (2003)
407. S. Saitoh, Reproducing kernels and a family of bounded linear operators. Oper. Theory Adv. Appl. **143**, 303–312 (2003)
408. S. Saitoh, Generalizations of the triangle inequality. J. Inequal. Pure Appl. Math. **4**(3), Article 62 (2003)
409. S. Saitoh, Approximate real inversion formulas of the Gaussian convolution. Appl. Anal. **83**, 727–733 (2004)
410. S. Saitoh, Best approximation, Tikhonov regularization and reproducing kernels. Kodai. Math. J. **28**, 359–367 (2005)
411. S. Saitoh, Tikhonov regularization and the theory of reproducing kernels, in *Finite or Infinite Dimensional Complex Analysis and Applications (Proceedings of the 12th ICFIDCAA)*, (Kyushu University Press, Fukuoka-shi, 2005), pp. 291–298
412. S. Saitoh, Applications of reproducing kernels to best approximations, Tikhonov regularizations and inverse problems, in *Advances in Analysis, Proceedings of the 4th International ISAAC Congress* (World Scientific, Singapore/Hackensack, 2005), pp. 439–445
413. S. Saitoh, T. Matsuura, M. Asaduzzaman, *Operator Equations and Best Approximation Problems in Reproducing Kernel Hilbert Spaces with Tikhonov Regularization* (Advances in Analysis World Scientific Publication, Hackensack, 2005), pp. 99–107
414. S. Saitoh, M. Yamada, Inversion formulas for a linear system determined by input and response relations, by using suitable function spaces. Hokkaido Univ. Tech. Report Ser. Math. **118**, 18–21 (2007)

415. S. Saitoh, Theory of reproducing kernels; applications to approximate solutions of bounded linear operator equations on Hilbert spaces. Am. Math. Soc. Transl. **230**, 107–134 (2010)
416. S. Saitoh, Generalized inversions of Hadamard and tensor products for matrices. Adv. Linear Algebra Matrix Theory **4**(2), 87–95 (2014)
417. S. Saitoh, Y. Sawano, *Generalized Delta Functions as Generalized Reproducing Kernels* (manuscript)
418. S. Saitoh, Y. Sawano, *General Initial Value Problems Using Eigenfunctions and Reproducing Kernels* (manuscript)
419. Y. Sawano, H. Fujiwara, S. Saitoh, Real inversion formulas of the Laplace transform on weighted function spaces. Complex Anal. Oper. Theory **2**, 511–521 (2008)
420. Y. Sawano, M. Yamada, S. Saitoh, Singular integrals and natural regularizations. Math. Inequal. Appl. **13**, 289–303 (2010)
421. Y. Sawano, Pasting reproducing kernel Hilbert spaces. Jaen J. Approx. **3**(1), 135–141 (2011)
422. Y. Sawano, H. Tanaka, Decompositions of Besov-Morrey spaces and Triebel-Lizorkin-Morrey spaces. Math. Z. **257**(4), 871–905 (2007)
423. Y. Sawano, A quick introduction to vector analysis, Kyoritsu (2014, in Japanese)
424. S. Scheinberg, Uniform approximation by meromorphic functions having prescribed poles. Math. Ann. **243**(1), 83–93 (1979)
425. M. Schiffer, D.C. Spencer, Functionals of finite Riemann surfaces (Princeton University Press, Princeton, 1954)
426. E. Schmidt, Über die Charlier–Jordansche Entwicklung einer willkürlichen Funktion nach der Poissonschen Funktion und ihren Ableitungen. Z. Angew. Math. Mech. **13**, 139–142 (1933)
427. B. Schölkopf, A.J. Smola, *Learning with Kernels* (MIT Press, Cambridge, 2002)
428. L. Schwartz, Sous-espaces hilbertiens d'espaces vectoriels topologiques et noyaux associés (noyaux reproduisants). J. Anal. Math. **13**, 115–256 (1964)
429. C.E. Shannon, A mathematical theory of communication. Bell Syst. Tech. J. **27**, 379–423, 623–656 (1948)
430. T. Shirai, Fermion measure and related topics (in Japanese). www.math.kyoto-u.ac.jp/probability/sympo/shirai.pdf
431. H.S. Shapiro. A.L. Shields, On the zeros of functions with finite Dirichlet integral and some related function spaces. Math. Z. **80**, 217–229 (1962)
432. B. Simon, Weak convergence of CD kernels and applications. Duke Math. J. **146**, 305–330 (2009)
433. B. Simon, *The Christoffel-Darboux Kernel. Perspectives in Partial Differential Equations, Harmonic Analysis and Applications*. Proceedings of Symposia in Pure Mathematics, vol. 79 (American Mathematical Society, Providence, 2008), pp. 295–335
434. G.D. Smith, *Numerical Solution of Partial Differential Equations: Finite Difference Methods* (Clarendon Press, Oxford, 1978)
435. F. Soltani, Inversion formulas in the Dunkl-type heat conduction on $\mathbb{R}^d$. Appl. Anal. **84**, 541–553 (2005)
436. F. Soltani, Practical inversion formulas in a quantum mechanical system. Appl. Anal. **84**, 759–767 (2005)
437. F. Soltani, Multiplication and translation operators on the Fock spaces for the q-modified Bessel function. Adv. Pure Math. (APM) **1**, 221–227 (2011)
438. F. Soltani, Toeplitz and translation operators on the q-Fock spaces. Adv. Pure Math. (APM) **1**, 325–333 (2011)
439. F. Soltani, Best approximation formulas for the Dunkl L^2-multiplier operators on $\mathbb{R}^d$. Rocky Mt. J. Math. (RMJM) **42**, 305–328 (2012)
440. F. Soltani, Fock spaces for the q-Dunkl kernel. Adv. Pure Math. (APM) **2**, 169–176 (2012)
441. F. Soltani, Multiplier operators and extremal functions related to the dual Dunkl-Sonine operator. Acta Math. Sci. **33**(2), 430–442 (2013)
442. F. Soltani, Extremal functions on Sturm-Liouville hypergroups. Complex Anal. Oper. Theory **8**(1), 311–325 (2014)

443. E.M. Stein, *Singular Integral and Differential Property of Functions* (Princeton University Press, Princeton, 1970)
444. E.M. Stein, *Harmonic Analysis: Real-Variable Methods, Orthogonality, and Oscillatory Integrals* (Princeton University Press, Princeton, 1993)
445. I. Steinwart, A. Christmann, *Support Vector Machine* (Springer, New York, 2008)
446. F. Stenger, *Numerical Methods Based on Sinc and Analytic Functions*. Springer Series in Computational Mathematics, vol. 20 (Springer, New York, 1993)
447. J. Stewart, Positive definite functions and generalizations, an historical survey. Rocky Mt. J. Math. **6**(3), 409–434 (1976)
448. K. Stroethoff, The Berezin transform and operators on spaces of analytic functions. Linear Oper. Banach Center Publ. **38**, 361–380 (1997). Institute of Mathematics Polish Academy of Sciences, Warsaw
449. S. Szegö, Über orthogonale Polynome, die zu einer gegebenen Kurve der Komplexen Ebene gehören. Math. Z. **9**, 218–270 (1921)
450. F.H. Szafraniec, Multipliers in the reproducing kernel Hilbert space, subnormality and noncommutative complex analysis, in *Reproducing Kernel Spaces and Applications*. Operator Theory: Advances and Applications, vol. 143 (Birkhäuser, Basel, 2003), pp. 313–331
451. G. Szegö, Orthogonal polynomials (American Mathematical Society, Providence, 1959)
452. H. Takahasi, M. Mori, Double exponential formulas for numerical integration. Publ. RIMS. Kyoto Univ. **10**, 721–741 (1974)
453. M. Seto, S. Suda, T. Taniguchi, Gram matrics of reproducing kernel Hilbert spaces over graphs. Linear Algebra Appl. **445**, 56–68 (2014)
454. S.-E. Takahasi, J.M. Rassias, S. Saitoh, Y. Takahashi, Refined generalizations of the triangle inequality on Banach spaces. Math. Inequal. Appl. **13**(4), 733–741 (2010)
455. K. Takemura, H. Yamagishi, Y. Kametaka, K. Watanabe, A. Nagai, The best constant of Sobolev inequality correspondence to a bending problem of a beam on an interval. Tsukuba J. Math. **33**(2), 253–280 (2009)
456. K. Takemura, The best constant of Sobolev inequality corresponding to clamped-free boundary value problem for $(-1)^M(d/dx)^{2M}$. Proc. Jpn. Acad. Ser. A Math. Sci. **85**(8), 112–117 (2009)
457. K.Takemura, Y. Kametaka, K. Watanabe, A. Nagai, H. Yamagishi, The best constant of Sobolev inequality corresponding to a bending problem of a beam on a half line. Far East J. Appl. Math. **51**(1), 45–71 (2011)
458. H. Tanabe, *Functional Analysis* (in Japanese) (Jikkyo, Tokyo, 2003)
459. C. Thomas-Agnan, Computing a family of reproducing kernels for statistical applications. Numer. Algorithms **13**(1–2), 21–32 (1996)
460. A.N. Tikhonov, V.Y. Arsenin, *Solutions of Ill-Posed Problems*, Translated from the Russian. Preface by translation editor Fritz John. Scripta Series in Mathematics (V.H. Winston & Sons/Wiley, Washington, DC/New York/Toronto/London, 1977)
461. V. Totik, Asymptotics for Christoffel functions for general measures on the real line. J. Anal. Math. **81**, 282–303 (2000)
462. H. Triebel, *The Structure of Functions* (Birkhäuser, Basel, 2000)
463. J.N. Tsitsiklis, B.C. Levy, Integral equations and resolvents of Toeplitz plus Hankel kernels. Technical report LIDS-P-1170, Laboratory for Information and Decision Systems, M.I.T., Silver Edition (1981)
464. N.M. Tuan, N.T.T. Huyen, The solvability and explicit solutions of two integral equations via generalized convolutions. J. Math. Anal. Appl. **369**, 712–718 (2010)
465. N.M. Tuan, N.T.T. Huyen, The Hermite functions related to infinite series of generalized convolutions and applications. Complex Anal. Oper. Theory **6**, 219–236 (2012)
466. N.M. Tuan, P.D. Tuan, Generalized convolutions relative to the Hartley transforms with applications. Sci. Math. Jpn. **70**, 77–89 (2009)
467. V.K. Tuan, D.T. Duc, A new real inversion formula of the Laplace transform and its convergence rate. Dedicated to the 60th anniversary of Prof. Francesco Mainardi. Fract. Calc. Appl. Anal. **5**(4), 387–394 (2002)

468. V.K. Tuan, N.T. Hong, Interpolation in the Hardy space. Integr. Transf. Spec. Funct. **24**(8), 664–671 (2013)
469. W. Ulmer, W. Kaissl, The inverse problem of a Gaussian convolution and its application to the finite size of measurement chambers/detectors in photon and proton dosimetry. Phys. Med. Biol. **48**, 707–727 (2003)
470. W. Ulmer, Inverse problem of linear combinations of Gaussian convolution kernels (deconvolution) and some applications to proton/photon dosimetry and image processing. Inverse Probl. **26**(8), Article ID 085002 (2010)
471. V. Vapnik, *Statistical Learning Theory* (Wiley, New York, 1998)
472. G. Wahba, On the approximate solution of Fredholm integral equations of the first kind. Technical summit report 990, Mathematics Research Center, University of Wisconsin-Madison (1969)
473. G. Wahba, Convergence rates of certain approximate solutions to Fredholm integral equations of the first kind. J. Approx. Theory **7**, 167–185 (1973)
474. G. Wahba, A class of approximate solutions to linear operator equations. J. Approx. Theory **9**, 61–77 (1973)
475. G. Wahba, Practical approximate solutions to linear operators equations when the data are noisy. SIAM J. Numer. Anal. **14**, 651–667 (1977)
476. G. Wahba, *Spline Models for Observational Data*. CBMS-NSF Regional Conference Series in Applied Mathematics, vol. 59 (SIAM, Philadelphia, 1990)
477. W. Wang, B. Han, M. Yamamoto, Inverse heat problem of determining time-dependent source parameter in reproducing kernel space. Nonlinear Anal. Real World Appl. **14**(1), 875–887 (2013)
478. K. Watanabe, Y. Kametaka, A. Nagai, K. Takemura, H. Yamagishi, The best constant of Sobolev inequality on a bounded interval. J. Math. Anal. Appl. **340**, 699–706 (2008)
479. K. Watanabe, Y. Kametaka, H. Yamagishi, A. Nagai, K. Takemura, The best constant of Sobolev inequality corresponding to clamped boundary value problem. Bound. Value Probl. **2011**(Article ID 875057), 17
480. K. Watanabe, Y. Kametaka, A. Nagai, K. Takemura, H. Yamagishi, The best constants of Sobolev and Kolmogorov type inequalities on a half line. Far East J. Appl. Math. **52**(2), 101–129 (2011)
481. T. Wei, Y.C. Hon, J. Cheng, Computation for multidimensional Cauchy problem. SIAM J. Control Optim. **42**(2), 381–396 (2003) (electronic)
482. D.V. Widder, *The Laplace Transform* (Princeton University Press, Princeton, 1972)
483. K. Yajima, *Lebesgue Integral and Functional Analysis* (in Japanese) (Asakura Shoten, 2002)
484. A. Yamada, Fay's trisecant formula and Hardy H^2 reproducing kernels, in *Reproducing Kernels and Their Applications* (Kluwer Academic Publishers, 1999), pp. 165–188
485. A. Yamada, Equality conditions for general norm inequalities in reproducing kernel Hilbert spaces, in *Advances in Analysis* (World Scientific, 2005), pp. 447–455
486. A. Yamada, Saitoh's inequality and Opial's inequality. Math. Inequal. Appl. **14**, 523–528 (2011)
487. A. Yamada, Oppenheim's inequality and RKHS. Math. Inequal. Appl. **15**(2), 449–456 (2012)
488. A. Yamada, Inequalities for Gram matrices and their applications to reproducing kernel Hilbert spaces. Taiwan. J. Math. **17**(2), 427–430 (2013)
489. M. Yamada, T. Matsuura, S. Saitoh, Representations of inverse functions by the integral transform with the sign kernel. Fract. Calc. Appl. Anal. **10**, 161–168 (2007)
490. M. Yamada, S. Saitoh, Identification of non-linear systems. J. Comput. Math. Optim. **4**, 47–60 (2008)
491. M. Yamada, S. Saitoh, Numerical solutions of two non-linear simultaneous equations. Appl. Anal. **88**, 151–160 (2009)
492. M. Yamada, S. Saitoh, Explicit and direct representations of the solutions of nonlinear simultaneous equations, in *Progress in Analysis and Its Applications Proceedings of the 7th International ISAAC Congress* (World Scientific, Singapore/Hackensack/London, 2010), pp. 372–378

493. K. Yao, Application of reproducing kernel Hilbert spaces – bandlimited signal models. Inf. Control **11**, 429–444 (1967)
494. L. Yingzhen, Z. Yongfang, Solving nonlinear pseudoparabolic equations with nonlocal boundary conditions in reproducing kernel space. Numer. Algor. **52**, 173–186 (2009)
495. N. Young, *An Introduction to Hilbert Space*. Cambridge Mathematical Textbooks (Cambridge University Press, Cambridge/New York, 1988)
496. K. Yoshida, Functional Analysis (Springer, Berlin, 1974)
497. S. Zaremba, *L'equation biharminique et une class remarquable de fonctions foundamentals harmoniques*. Bull. Int. de l'Academie des Sci. de Cracovie **39**, 147–196 (1907)
498. A.I. Zayed, On Kramer's sampling theorem associated with general Sturm–Liouville problems and Lagrange interpolation. SIAM J. Appl. Math. **51**, 575–604 (1991)
499. A.I. Zayed, *Advances in Shannon's Sampling Theory* (CRC Press, Boca Raton, 1993)
500. M. Zedek, On the Lebedev-Milin inequality. Proc. Am. Math. Soc. **33**, 395–397 (1972)
501. X.D. Zhang, Vertex degrees and doubly stochastic graph matrices. J. Graph Theory **66**(2), 104–114 (2011)
502. D.X. Zhou, Capacity of reproducing kernel spaces in learning theory. IEEE Trans. Inform. Theory **49**, 1743–1752 (2003)

Index

Symbols
L^*, 134
$\int_X^{\oplus} H_x \, d\mu(x)$, 174, 175
$\otimes^n H_K(E)$, 113
$\overline{\partial}$-equation, 134
$\overline{\partial}$-operator, 285
$\wedge^n H_K(E)$, 113
$\ker(L)$, 135
$\ker(L)^{\perp}$, 135
$\mathbb{C}^E$, 390
$\mathscr{B}(X)$, xvii
$\mathscr{F}(E)$, 1, 65
$\mathscr{F} = \{\psi_x\}_{x \in X}$, 161, 162, 165
$\mathscr{F}_a(\mathbb{R})$, 170
$\mathscr{H}_1[+]\mathscr{H}_2$, 101
$\mathscr{H}\mathscr{S}(H)$, 186
AC$[a, b]$, xviii
Erfc, 279
ev, 65
$\mathrm{ev}_p(f) = f(p)$, 66
$\mathrm{sinc}(x)$, 3
$\mathbf{f}_1[+]\mathbf{f}_2$, 100
$H^s(\mathbb{R}^n)$, 255
$L^{\dagger}\mathbf{d} = f_{\mathbf{d}}$, 169
$W^{1,2}[a, b]$, 80
$A \lesssim B$, xvii
$A^{\dagger} \equiv VSW^*$, 165
$A^n(D)$, 414
$f_{\mathbf{d},\alpha(\delta)}$, 184
H-bounded homomorphism, 393
H-hull, 393
$H^2_{\mathrm{S}}(\mathbb{R})$, 217
$H^s(\mathbb{R})$, 219
$H^s(\mathbb{R}^n)$, 245
$H_{K_h}(\mathbb{R})$, 4, 6
$H_{K_q}(R^+)$, 146
$H_{K_q}(S_r)$, 145
$H_{\mathrm{S,e}}(\mathbb{R})$, 274
H_K, 2, 68
$H_K(E)$, 2, 3, 66, 68, 70, 137
H_S, 254
$H_S(\mathbb{R})$, 16, 254, 274
$K_1 \ll K_2$, 96
L^*, 61
$L^*\mathbf{d} = L^*L\tilde{f}$, 167
L^*L, 208
$L^{\dagger}\mathbf{d}$, 169
R-dense, 393
R-maximal, 394
$R_1 \cdot R_2$, 392
$W\left(\frac{\pi}{h}\right)$, 7
$W_2^{\ell}(G)$, 293
$W^{1,2}(a, b)$, 9
$W^{1,2}(I)$, 9
$W^{1,2}(\mathbb{R})$, 9
$W^{1,2}[a, b]$, 11, 80, 342
$W^{2,1}(\mathbb{R})$, 190, 191
$W^{2,2}(\mathbb{R})$, 21, 192
$W^{m,2}(\mathbb{R}^n)$, 22
W_h, 7
$W_h(\mathbb{R})$, 295

A
adjoint operator, 61, 134
ADMM, 133
algebra of meromorphic functions, 400
algebra-dense, 389, 393
algorithm by the iteration method, 247
analytic extension, 146

S. Saitoh, Y. Sawano, *Theory of Reproducing Kernels and Applications*,
Developments in Mathematics 44, DOI 10.1007/978-981-10-0530-5

analyticity, 197
ANNs, 252
applications to ordinary differential equations, 217
applications to partial differential equations, 231
approximate fractional function, 195
approximate Hilbert transform, 301
approximation error estimate, 153
approximation of functions, 170
approximations using the first order Sobolev Hilbert space, 190
approximations using the second order Sobolev Hilbert space, 192
Aronszajn, N., v, 322
artificial neural networks, 252
Aveiro discretization method, 131, 378
Aveiro Discretization Method in Mathematics, 372

B

balloon of reproducing kernel Hilbert spaces, 82
Banach Alaoglu theorem, 94
band preserving, 365
basic properties of RKHS, 65
Berezin transform, 36
Bergman kernel, v, 24, 40, 146, 404
Bergman kernels on the unit disk, 35
Bergman norm, 321
Bergman Selberg reproducing kernel, 37
Bergman space, 36, 39, 321
Bergman, S., v, 24, 40, 119
Bergman–Selberg space, 145
Bessel differential equations, 373
Bessel potential, 24
best approximation, 161
best approximation problem, 166, 169
best approximation with point data, 193
Birkhoff interpolation, 24
Branges, L. de, 7

C

Carleman equation, 295
Cauchy Binet formula, 115
Cauchy principal value, 413
Cauchy problem, 249
Cauchy's principal value, 295
Chebyshev polynomials, 47
Christoffel Darboux formula, 43
classical sampling theorem, 350
Clifford analysis, 25
compactness of the modified Laplace transform, 200
complete orthonormal system, 186
computerized tomography, 150
congruent, 152
CONS, 115, 116
constructing solutions satisfying boundary conditions, 243
construction of approximate solutions by iteration, 338
construction of approximate solutions by using a CONS expansion, 303
construction of RKHS, 115
continuity, 73
continuous frame, 161, 162
convergence properties of the approximate solutions, 287
convolution inequality, 315, 317
convolution integral equation, 313
convolution norm inequalities, 330
cross-section, 175

D

data analysis, 151
Daubechies, I., 140
DE formula, 210
definition for graphs, 53
determination of linear systems, 159
difference method, 133
Dini convergence theorem, 123
Dini's theorem, 99
Dirac delta, 211
Dirac's delta function, 60
direct representations of the sources by using a finite data, 240
direct sum of Hilbert spaces, 175
Dirichlet integral, 341
Dirichlet norm, 388
Dirichlet principle, 245
Dirichlet space, 388
discrete inverse source problem, 229
discrete ordinary linear differential equation, 224
discrete singular integral equation, 308
distribution delta, 210
distribution theory, 60
Dixon's equation, 376
DM, 133
double helix structure of gene, 381
dual basis, 354
dual frame, 164
Duc, D. T., 325, 333, 334, 407
Dunkel transform, 289

E
eigenfunction, 366
eigenvalue of Hermitian matrices, 335
elementary properties of RKHS, 65
embedding norm, 97
empirical risk minimization, 253
equality problem, 322, 330, 333, 387
ERM, 253
error and convergence rate estimate, 357
error estimate, 183, 252, 305
errorness data, 253
Euler differential equations, 374
evaluation operator, 65
exact frame, 349
Examples of orthonormal systems, 46
exceptional case, 390
extremal, 389

F
factorization theory, 308
FDM, 252
FEM, 133, 252
filtering and estimation theory in signal and image processing, 151
finite difference method, 252
finite element method, 133, 252
finite interval case, 221
Fischer space, 170
frame, 161, 348, 349
frame operator, 162
Fredholm integral equation, 159, 210, 365
Fujiwara's infinite precision algorithm, 365
Fujiwara's infinite precision method, 253
Fujiwara, H., vi, 129, 131, 210–215, 253, 365
fully general triangle inequalities, 103
fundamental properties of RKHS, 65
fundamental solution, 24, 61
fundamental solution of the Laplacian, 343

G
Gauss transform, 253
Gaussian convolution (the Weierstrass transform), 255
Gaussian convolution formula, 255
general discrete PDE, 289
general fractional function, 193
general fundamental theorem on inversions, 154
general inhomogeneous PDEs, 284
general Kramer–type Lemma, 354
general nonlinear transforms, 325
general properties of orthonormal systems, 41
general reproducing kernel, 381
general sampling principle, 348
general sampling theorem, 129
general singular integral equations, 306
generalizations of Opial's inequality, 406
generalized convolution inequality, 332
generalized delta function, 381
generalized fractional function, 195
generalized isoperimetric inequality, 321
generalized reproducing kernel, 378, 381
generalized triangle inequality, 103
geometrically convex, 407, 408
global inputs of linear systems, 269
Gram determinant inequality, 334
Gram matrix, 58, 334
Gram Schmidt orthogonalization, 40, 46
Gram Schmidt orthogonalization procedure, 119
graph, 53
Green function, 24
Green's function, 24, 60, 160
Green's function $G(p, q)$, 61

H
Hankel kernel, 316
Hankel operator, 36
Hardy H^2-norm, 388
Hartog's theorem, 77
heat equation, 253
Hejhal, D. A., 322
Helmholtz equation, 289
Hermite polynomials, 47, 48
Hermitian positive definite, 188
Higgins, J. R., 7, 347
Hilbert scale, 215
Hilbert Schmidt operator, 187, 189
Hodge star operator, 343, 414
how to obtain eigenfunctions, 373
hypercircle inequality, 350, 353

I
identification of the image of a linear mapping, 134
identification of the images of linear transforms, 134
identifications of non-linear systems, 346
ill-posed problem, 160
Image identification by reproducing kernels, 141
implicit function, 413
implicit function theorem, 346, 415
impulse response, 61

inequalities for Gram matrices, 334
infinite precision, 210
infinite precision arithmetic algorithm, 129
information-loss error in sampling theory, 350
initial value problem, 366
integral equation, 295
integral equations with the mixed Toeplitz Hankel kernel, 315
integral representations of implicit functions, 413
integral representations of the inversions, 414
integro-differential equation, 147
inverse formulas using CONS, 153
inverse problems by use of a finite number of data, 282
inverse problems in the 1-dimensional wave equation, 273
inverse source problems by a finite number of data, 311
inversion for any matrix, 337
inversion formula, 155
inversion formula in the Weierstrass transform, 156
inversion mapping from many types of information data, 148

J
Jacobian $J(x)$, 345
Jensen's inequality, 411

K
Kaczmarz's method, 150
Kataoka, K., 153
KDV equation, 329
kernel method, 152
kinked crack problem, 296
Klein prime form, 422
known real inversion formulas of the Laplace transform, 198
Kolmogorov factorization theorem, 140, 141
Kolmogorov theorem on measures, 140
Kramer sampling lemma, 348
Kramer's formula for regular matrices, 343
Krawtchouk polynomials, 52
Krein, M. G., 78

L
Laguerre polynomials, 48
Lalesco-Picard integral equation, 375
Laplace's equation, 245
Laplacian, 414
learning theory, 422
least squares method, 151
Lebedev and Milin inequality, 387
Lebedev Milin inequality, 389
Legendre polynomials, 47
Loeve theorem, 152

M
Mathematica, 210
Matsuura, T., 255
maximal, 394
Meixner-type polynomials, 51
membership problem for RKHS, 363
Mercer expansion, 120, 121
Mercer kernel, 361
meromorphically dense, 400
method of L-curve, 185
Meyer wavelet, 140
Meyer Y., 140
microlocal enery method, 153
modified Bessel functions, 373
modified Laplace tranform, 200
modulation spaces, 138
Moore Penrose generalized inverse, 161, 169
Moore Penrose generalized inverses and Tikhonov regularization, 161
Moore Penrose inverse, 165
Moore-Penrose generalized inverse, 259
multiple-precision arithmetic, 210

N
natural output, 269
Nhan, N. D. V., 325, 333, 334, 407
noise reduction, 215
non-linear simultaneous equations, 413
nonharmonic integral transform, 158
nonharmonic transform, 158
nonlinear simultaneous equations, 343
norm inequality, 321
normalized basis, 348
numerical Dirichlet problem, 245
numerical real inversion of the Laplace transform, 198

O
Ogawa, H., 53, 348
one point boundary condition case, 223
operator equations with a parameter, 174
Opial's inequality, 406
Opial-type inequalities, 411
optimal real inversion formula, 212

ordinary differential equations, 217
ordinary linear differential equation, 217
orthonormal basis, 74, 349
overview on the theory of reproducing kernels, 422

P
Paley Wiener reproducing kernel, 3, 6
Paley Wiener space, 3, 239, 307
Paley Wiener space $H_{K_h}(\mathbb{R})$, 240
parabolic Bergman and Bloch spaces, 141
pasting of reproducing kernel Hilbert spaces, 103
PDEs and inverse problems, 293
phase retrieval, 365
Plemeli formula, 295
Poisson Carlier Polynomials, 52
Poisson's equation, 231
polynomial reproducing kernel Hilbert spaces, 49
polynomial ring, 399
polynomially dense, 399
positive definite function, 3, 67
positive definite Hermite matrix, 2
positive definite Hermitian matrix, 130
positive definite matrix, 67, 107
positive definite quadratic form function, 3, 67
positive matrix, 67
positive type function, 67
pseudodifferential operator, 153
pullback of a reproducing kernel, 81
pullback of reproducing kernels, 92
pullback of RKHS, 81
Pythagorean theorem, 149

R
radiation pattern, 290, 291
random fields estimation, 150
real inversion formulas of the Laplace transform, 197
real inversion of the Laplace transform, 208, 209
realization by discrete points of RKHS, 125
recurrence relation, 42
regular, 390, 400
representation of inverse functions, 340
representations of exact differential equations, 420
representations of implicit functions, 415
representations of the inverse functions, 420
representations of the solutions of the Tikhonov functional equation, 186
reproducing kernel, 1, 136
reproducing kernel Hilbert space, 1, 65, 137
reproducing kernel Hilbert spaces of Sobolev type, 8
reproducing property, 1
restriction of a reproducing kernel, 78
restriction of RKHS, 78
Riemann boundary value problem, 308
Riemann Hilbert method, 51
Riemann mapping function, 40
Riemann Roch theorem, 401
Riemann theta functions, 422
Riesz basis, 349
RKHS, 1
RKHS and complex analysis, 24
RKHS and linear mappings, 133
RKHS for trigonometric polynomials, 52
RKHS in the spaces of polynomials, 41
RKHS of fractional order, 23
RKHS of order 1 associated to the wave equation, 19
RKHS of second-order, 21
RKHS of Sobolev type, 8
RKHS of Sobolev type of order m, 22
RKHS of the first order on the closed interval, 11
RKHS on $\mathbb{C}$, 29
RKHS on the half line, 12
RKHS on the half line with boundary condition, 13

S
sampling theorem, 53
sampling theory, 5–7, 347
scaling limit of Christoffel Darboux kernels, 50
Scheinberg, S., 404
Schwartz, L., v, 60
self-adjoint operator, 61, 208
sensibility to the system, 270
sensible interpolation of the input function, 271
Sinc function, 264
singular integral equation, 295
singular system, 209
singular value decomposition, 215
singular values decomposition, 208
size estimates of the algorithm, 340
smoothness of functions and kernels of RKHS, 73
Sobolev embedding theorem, 293
Sobolev space, 214
Sobolev spaces on an interval, 9
Sommerfeld radiation condition, 290

source inversions in the Poisson equation, 238
source inversions of the heat conduction, 262
spectral analysis, 179
spectral decomposition, 358
squeezing of RKHS, 87
stable sampling, 349
statistical learning theory, 153, 357
Stenger, F. , 7
sum of arbitrary abstract Hilbert spaces, 100
sum of reproducing kernels, 93
support vector machine, 151, 152, 253, 422
support vector machine method, 252
SVM, 253
Szegö kernel, v, 24
Szego kernels, 404
Szego norm, 321
Szego space, 38, 146, 322
Szegö, S., v

T
Tchebycheffian spline function, 24
tensor product, 389
tensor product of Hilbert spaces, 105
tight, 161
Tikhonov extremal function, 184
Tikhonov functional, 182, 195, 254, 303
Tikhonov functional equation, 186
Tikhonov regularization, 179, 184, 209, 214, 239
Tikhonov Regularization Method, 372
Toeplitz kernel, 316
transform of RKHS, 89
transforms of RKHS by operators, 90
triangle inequality, 101
Tricomi equations, 377
truncation error, 350
Tuan, V. K., 325, 407

U
ultimate realization of RKHS, 127
ultimate sampling theorem, 356
unconditional basis, 348
use of a finite number of data, 224
use of a finite number of data for the Cauchy problem, 250
use of Paley-Wiener spaces (sinc method), 257

V
variable coefficients case, 220
visible component, 135, 158, 290
Volterra integral equations, 376
Volterra series, 346

W
$W^{1,2}(I)$, 9
$W^{2,2}(\mathbb{R})$, 21
wave equation, 18, 273
wavelet analysis, 140
weak-derivatives and Sobolev space, 8
weakly regular, 390, 395
wedge product, 113
wedge product of RKHS, 113
Weierstrass approximation theorem, 121
Weierstrass integral transform, 157
Weierstrass transform, 253
weighted Bergman space, 36, 403
weighted Dirichlet space, 403
weighted Szego space, 403
Weyl's monotonicity principle, 335
Wiener, N., 151
windowed Fourier transform, 138

Y
Yamada, A., vii, 322, 325, 330, 333, 334, 387, 389, 407

Z
Zaremba, S., v
Zero points and RKHS, 49

MIX
Papier aus verantwortungsvollen Quellen
Paper from responsible sources
FSC® C105338

If you have any concerns about our products,
you can contact us on
ProductSafety@springernature.com

In case Publisher is established outside the EU,
the EU authorized representative is:
Springer Nature Customer Service Center GmbH
Europaplatz 3, 69115 Heidelberg, Germany

Printed by Libri Plureos GmbH
in Hamburg, Germany